Reliability Modelling and Analysis in Discrete Time

Reliability Modelling and Analysis in Discrete Time

N. Unnikrishnan Nair

Department of Statistics, Cochin University of Science and Technology, Cochin, India

P.G. Sankaran

Department of Statistics, Cochin University of Science and Technology, Cochin, India

N. Balakrishnan

Department of Mathematics & Statistics, McMaster University, Hamilton, ON, Canada

ELSEVIER

ACADEMIC PRESS

An imprint of Elsevier

Academic Press is an imprint of Elsevier
125 London Wall, London EC2Y 5AS, United Kingdom
525 B Street, Suite 1800, San Diego, CA 92101-4495, United States
50 Hampshire Street, 5th Floor, Cambridge, MA 02139, United States
The Boulevard, Langford Lane, Kidlington, Oxford OX5 1GB, United Kingdom

Notices

Knowledge and best practice in this field are constantly changing. As new research and experience broaden our understanding, changes in research methods, professional practices, or medical treatment may become necessary.

Practitioners and researchers must always rely on their own experience and knowledge in evaluating and using any information, methods, compounds, or experiments described herein. In using such information or methods they should be mindful of their own safety and the safety of others, including parties for whom they have a professional responsibility.

To the fullest extent of the law, neither the Publisher nor the authors, contributors, or editors, assume any liability for any injury and/or damage to persons or property as a matter of products liability, negligence or otherwise, or from any use or operation of any methods, products, instructions, or ideas contained in the material herein.

Library of Congress Cataloging-in-Publication Data
A catalog record for this book is available from the Library of Congress

British Library Cataloguing-in-Publication Data
A catalogue record for this book is available from the British Library

ISBN: 978-0-12-801913-9

Publisher: Candice Janco
Acquisition Editor: Graham Nisbet
Editorial Project Manager: Susan Ikeda
Production Project Manager: Paul Prasad Chandramohan
Designer: Miles Hitchen

Typeset by VTeX

Dedication

To Jamuna, Anoop, Arun
N. Unnikrishnan Nair

To Sandhya, Revathi, Nandana
P.G. Sankaran

To my mother, Mrs. N. Lakshmi
N. Balakrishnan

Contents

Authors Biographies

N. Unnikrishnan Nair – Dr. N. Unnikrishnan Nair was Professor of Statistics and Vice Chancellor of the Cochin University of Science and Technology, India. He was also President of the Indian Society for Probability and Statistics. He received Honorary Doctor of Human Letters from Juniata College, USA. His research interest are reliability analysis, distribution theory and statistical inference and has published over hundred and fifty research papers in international journals.

P.G. Sankaran – Dr. P.G. Sankaran is Professor of Statistics of the Cochin University of Science and Technology, India. He is an elected member of International Statistical Institute. He received young researcher award of International Indian Statistical Association in 2010. His research interest are reliability theory, survival analysis, distribution theory and statistical inference. He has published more than 130 research papers in international journals.

N. Balakrishnan – Dr. N. Balakrishnan is a Distinguished University Professor at McMaster University, Hamilton, Ontario, Canada. He is a Fellow of the American Statistical Association, the Institute of Mathematical Statistics and the Turkish Statistical Association. He is also an Elected Member of the Greek Statistical Institute and recently received an Honorary Doctorate from the National University of Athens, Greece. He is serving as the Editor-in-Chief of the 3 series of Communications in Statistics in addition to being in the editorial board of several reputed statistical journals.

Preface

Reliability theory has developed into a major discipline during the last six decades, thanks to the efforts of researchers from various subjects like mathematics,statistics, engineering, operations research, material science etc. At the same time, the ideas and concepts in reliability analysis have contributed to open new areas of study in several disciplines such as economics, actuarial science, medicine, biology, survival analysis, to mention a few. Technology has made tremendous progress in devising new products, devices and systems to the service of mankind for which the tools of reliability are indispensable to ensure their failure-free operation. During the early phase of the development of reliability theory and methods, discussions were mainly confined to continuous lifetimes with only occasional reference to discrete models. Though the trend on emphasis on continuous models still continues, the role of discrete models in lifetime studies is receiving much recognition in the last two decades. The fact that there are natural cases in which lifetimes emerge as a counting variable and also that there are situations in which discrete lifetimes are more appropriate than clock time when the latter is available, have made this shift appealing. Inspite of the growing interest in this field, a reference book exclusively for study and research in discrete reliability analysis does not appear to be available. The present monograph is an attempt to provide the material needed for a theoretical study of the concepts and results in modelling and analysis of reliability data in the discrete domain.

While writing this book we noticed that different kinds of definitions were used even in the basic notions concerning discrete lifetimes. We have presented the results in this connection under a uniform framework for easier understanding and application. Also there were no discrete analogues in literature for several important concepts and theorems in the continuous case: We have tried to include them to provide a continuity in the flow of discussions without gaps.

The book is organized into nine chapters. Chapter 1 deals with some basic tools in the form of definition and results that are required to develop the mathematical discussions in the subsequent chapters. This is followed by Chapter 2 in which the basic reliability functions and their properties are studied.

Various distributions that could be used as discrete lifetime models are presented in Chapter 3. The ageing concepts, the implications among them and the feature of classes of distributions based on them are discussed in Chapter 4. In Chapter 5 we consider non-monotonic ageing behaviour of the hazard rate and mean residual life along with models exhibiting such behaviour and their characteristics. Multivariate reliability functions form the material in Chapter 6. Various reliability functions in the univariate case in Chapter 2 are extended to the multivariate case in Chapter 7. Then we present in Chapter 8 some basic multivariate distributions which possess simple functional forms for the reliability functions. The concepts and methodology in reliability theory have contributed to open new areas of study in other disciplines. In Chapter 9, we review such contributions in survival analysis, economics, risk analysis, information theory, mathematics and statistics.

Many of the works reported in the book are analogues of the ideas and results in the continuous case proposed by various researchers. We could not make references to all such contributors due to limitations of space. Our sincere apologies to them as well as to those whose work do not find a place in this monograph. We whole-heartedly welcome all suggestions for improvements of the contents of this volume. Finally, we wish to thank Sanjay Varma and the Colleagues in the Department of Statistics of the Cochin University of Science and Technology for their invaluable contributions to our cause.

N. Unnikrishnan Nair
Cochin, India

P.G. Sankaran
Cochin, India

N. Balakrishnan
Hamilton, ON, Canada

About the Book

This book is the first of its kind devoted exclusively to the study of concepts and methods to analyze discrete lifetime data. Different chapters in the work consist of Basic results in mathematics and statistics needed to follow the rest of the contents.

- Various reliability concepts in discrete time
- Discrete lifetime models
- Ageing criteria and their properties
- Bathtub distributions
- Multivariate reliability concepts
- Multivariate ageing concepts
- Multivariate life distributions
- Applications of reliability theory

The book provides a continuous flow of materials by including several new results that are not available in the current literature. The book can be used as a text book for graduate course in reliability and as a reference for higher level research work.

Chapter 1

Reliability Theory

1.1 RELIABILITY THEORY

In ordinary usage, an object is meant to be reliable if it works upto expectation in times of need. However, for the development of a proper theory out of this intuitive meaning, one needs to have a more precise definition that would be capable of utilizing our scientific knowledge to characterize, quantify and measure reliability. A commonly used definition of reliability is that it is the probability that an equipment or device will perform the functions intended of it under conditions specified for its operation for a given period of time. When the device does not carry out its intended function, we may say that it has failed. The notion of failure can be more elaborately described, of course. Failure can be any incident or condition that causes an industrial plant, manufactured product, process, material or service to degrade or become unsuitable for performing satisfactorily, safely, reliably and cost-effectively (Witherell, 1994). An important objective of reliability theory is to understand the pattern in which failures occur for different mechanisms under different operating conditions. The uncertainty in the times at which failures occur demands a probabilistic framework for dealing with reliability of a device or system.

Rapid strides in technology during the last few decades have given rise to new products, devices and services that aim to make human life more comfortable. These include a wide range of products, from consumer goods for daily use to sophisticated devices for inter-planetary missions. From the producer to the customer, everyone is interested in ensuring that the product undergoes failure-free operation at least for a specified period of time. Reliability theory is a body of concepts, practices and methodology that attempts to meet this objective. Blischke and Murthy (2000) have identified five main areas through which the tasks in reliability can be accomplished, and these are *reliability modelling, reliability analysis, reliability engineering, reliability science* and *reliability management*. Of these, reliability modelling deals with the identification of mechanism that generates observations on the failure times in a specific study. Various concepts that describe the failure mechanism and their properties can shed some light into the patterns in the observations thus enabling the determination of the law that governs the lifetimes of the device.

Reliability Modelling and Analysis in Discrete Time. http://dx.doi.org/10.1016/B978-0-12-801913-9.00001-4

Many results in probability theory, statistics and stochastic processes become the main inputs in this connection. A suitable mathematical model, along with the real data, can then be used to draw inferences about the reliability. This is the main topic in reliability analysis. Reliability engineering aims at producing more reliable products through testing, designing and construction of products. Some key properties of materials that may cause failure and the effects of the manufacturing process are studied in reliability science. Emphasis in reliability management is on the consequences of unreliable products such as cost, loss of goodwill and repair, and to ensure quality of the product through management of design, operation and manufacture. Of these different areas, this book focuses only on reliability modelling and analysis. The relevant mathematical theory is described in detail with numerous examples in Chapters 2 through 8.

In the discussions held so far, emphasis has been made on the reliability of products and the role of reliability theory in assessing it. The time interval of failure free operation of a device is often termed as the *lifetime of the device*. However, lifetime can be viewed within a more general framework. If the duration spent by an element or object in a given state is taken to be the lifetime, then the theory of reliability can readily find applications in life tables, nuptiality studies, period of service before leaving a particular job in manpower planning, and so on. Reliability concepts have also found many useful applications in the construction of income distributions, analysis of human settlements, under-reported incomes, and bibliometry. A survey of some important applications is presented in Chapter 9. Thus, a discussion of the basic concepts and methodology in reliability theory assume importance in a plethora of areas of scientific activity.

In most of the discussions on reliability theory, the lifetime is treated as continuous. The modelling and analysis aspects use distributions and processes that commonly represent a non-negative continuous random variable. Comparatively, much less literature is available when the lifetime is discrete. However, there are some compelling reasons to consider failure times as discrete random variables taking on non-negative integer values. When a piece of equipment operates in cycles and the observation is the number of cycles completed before failure, the lifetime is clearly discrete. So also is the case when the device is monitored only in completed units of time, like how many failures have occurred at the completion of one hour, two hours, and so on. The lack of accuracy of measuring devices may also generate discrete lives. There are occasions to prefer counts over clock time even when the latter is available. In weapons reliability, the number of rounds fired is more important than the age at failure. The same is the case with lifetimes of car tires wherein the number of kilometres run before it becomes out of use is preferred to the number of days before failure. These types of problems provide a strong impetus for studying reliability in discrete time. The concepts in continuous and discrete times are the same, but the definitions and interpretations may differ between the two. To derive reliability properties analogous to the continuous case, occasionally continuous distributions are discretized. But, it is not necessary in such cases that the distributional properties are the same nor the discretization process will always result in meaningful discrete models. Consequently, there are conceptual and mathematical problems in developing discrete reliability theory. These points will be deliberated at appropriate places in the ensuing chapters.

The build-up of reliability theory for discrete lifetimes requires inputs from other disciplines, predominantly tools from mathematics, statistics and probability theory. These are described briefly in the remaining sections of this chapter.

1.2 DISCRETE LIFE DISTRIBUTIONS

Let X be a discrete random variable, assuming values $0, 1, \ldots, b$, where b can be finite or infinite, representing the lifetime of a device. Reliability, being defined in terms of probability, involves the basic function of interest in describing the probabilities of X, which is the *distribution function*.

$$F(x) = P(X \leq x), \quad x = 0, 1, \ldots, b,$$

which is a non-decreasing step function satisfying $P(X < 0) = 0$ and $F(b) = 1$. Of equal importance is the *survival function* defined by

$$S(x) = P(X \geq x)$$

representing the probability that the device has a lifetime of at least x. In view of this, $S(x)$ is also called the *reliability function*. Note that

$$S(x + 1) = 1 - F(x).$$

The *probability mass function* of X is

$$f(x) = P(X = x)$$

which satisfies the conditions $f(x) \geq 0$ and $\sum_{x=0}^{b} f(x) = 1$. Conversely, if $\langle f(x) \rangle$ is a collection of non-negative real numbers satisfying $\sum f(x) = 1$, then $f(x)$ is the probability mass function of some discrete random variable X. In the sequel, we will often make use of the following relationships:

$$F(x) = \sum_{t=0}^{x} f(t), \quad S(x) = \sum_{t=x}^{b} f(t),$$
$$f(x) = F(x) - F(x - 1) \text{ and } f(x) = S(x) - S(x + 1).$$

In modelling and other practical applications, it is important to describe the main features of a distribution by means of a few simple parameters or characteristics. One way to accomplish this is through the moments of a distribution. The expected value of a function $\Psi(X)$ of X is

$$E[\Psi(X)] = \sum_{x=0}^{b} \Psi(x) f(x),$$

provided the series on the right side is absolutely convergent. When $\Psi(X) = X^r$, we get the *rth moment*

$$\mu_r' = E(X^r) = \sum_{x=0}^{b} x^r f(x), \quad r = 1, 2, \ldots, \tag{1.1}$$

which is said to exist when the series in (1.1) it is absolutely convergent. The mean of the distribution, conceived as a measure of *location* or central point of the distribution, is the first moment

$$\mu_1' = \mu = E(X) \equiv \sum_{x=1}^{\infty} S(t). \tag{1.2}$$

To define other measures, we need the *central moments* defined by

$$\mu_r = E(X - \mu)^r = \sum_{x=0}^{b} (x - \mu)^r f(x).$$

For measuring the *dispersion*, or spread of the values from the mean, the variance is defined as

$$\sigma^2 = \mu_2 = E(X^2) - E^2(X). \tag{1.3}$$

Similarly, for capturing the shape characteristics of the distribution, measures of skewness and kurtosis are defined as

$$\beta_1 = \frac{\mu_3}{\mu_2^2} \text{ and } \beta_2 = \frac{\mu_4}{\mu_2^2}. \tag{1.4}$$

These measures are generally used though there are some other measures introduced in the literature for this purpose. For discrete models, sometimes the *descending factorial moments* defined by

$$\mu_{(r)} = E(X_{(r)}), \tag{1.5}$$

where

$$X_{(r)} = X(X-1)\cdots(X-r+1)$$

is the descending factorial expression, may be more convenient to derive and use. Another similar type of moments are the *ascending factorial moments* expressed as

$$\mu^{(r)} = E(X^{(r)}) = E[X(X+1)\cdots(X+r-1)]. \tag{1.6}$$

There are some functions that generate the above-mentioned moments. Of these, the important ones are the characteristic function $\phi(t)$, the probability generating function $P(t)$, and the factorial moment generating function $W(t)$. The characteristic function is defined, for all real t, as

$$\phi(t) = E(e^{itX}) = \sum_{x=0}^{b} e^{itx} f(x). \tag{1.7}$$

If the moment of order r exists, we have for $0 \le p \le r$,

$$\frac{d^p\phi}{dt^p} = i^p \sum_{x=0}^{b} x^p e^{itx} f(x),$$

or

$$i^p \mu'_p = \left[\frac{d^p\phi}{dt^p}\right]_{t=0}.$$

Moreover, $f(x)$ can be recovered from $\phi(t)$ through the inversion formula (Johnson et al., 1992)

$$f(x) = \frac{1}{2\pi}\int_{-\pi}^{\pi} e^{-itx}\phi(t)dt.$$

The probability generating function is defined as

$$P(t) = E(t^X) = \sum_{x=0}^{b} t^x f(x). \tag{1.8}$$

We can obtain the distribution of X from $P(t)$ using

$$f(x) = \left[\frac{1}{x!}\frac{d^x P(t)}{dt^x}\right]_{t=0}, \quad x = 0, 1, \ldots, b, \tag{1.9}$$

and also the moments from

$$\mu'_r = \left[\frac{d^r P(e^t)}{dt^r}\right]_{t=0}, \quad r = 1, 2, \ldots .$$

Similarly, the factorial moment generating function is defined as

$$W(t) = E[(1+t)^X] = \sum_{x=0}^{b} (1+t)^x f(x) = P(1+t). \tag{1.10}$$

From (1.10), the rth descending factorial moment is calculated from the formula

$$\mu_{(r)} = \left[\frac{d^r P(1+t)}{dt^r}\right]_{t=0} = \left[\frac{d^r W(t)}{dt^r}\right]_{t=0}. \tag{1.11}$$

Given the factorial moments, the survival function can be expressed as

$$S(x) = \sum_{t \geq x} (-1)^{x+t} \binom{t-1}{x-1} \frac{\mu_{(r)}}{r!}.$$

1.3 MIXTURE DISTRIBUTIONS

A *mixture* of distributions is a convex combination of distributions of different functional forms or of different parameters in specified proportions. Let $\langle F_i(x) \rangle$, $i = 1, 2, \ldots, n$, be a collection of distribution functions and $0 \leq \alpha_i \leq 1$, be a collection of numbers such that $\sum_{i=1}^{n} \alpha_i = 1$. Then,

$$F(x) = \alpha_1 F_1(x) + \alpha_2 F_2(x) + \cdots + \alpha_n F_n(x) \tag{1.12}$$

is called a *finite mixture* if n is finite or an *infinite mixture* if n is infinite. The simplest form of mixture is a two-component mixture of the form

$$F(x) = \alpha F_1(x) + (1-\alpha) F_2(x), \quad 0 \leq \alpha \leq 1, \tag{1.13}$$

where F_1 and F_2 are called the components of the mixture and α is said to be the mixing constant.

When the distribution function is indexed by a parameter of the form $F(x|\theta_i)$ and θ_i is the realization of a random variable Θ with probability mass function $p_i = P(\Theta = \theta_i)$, $i = 0, 1, \ldots$, the mixture is formed as

$$G(x) = \sum_{i=0}^{\infty} p_i F(x|\theta_i). \tag{1.14}$$

One can easily recognize that (1.12) and (1.13) are special cases of (1.14). On the other hand, if Θ is continuous with distribution function $C(\theta)$, the resultant mixture has its distribution function as

$$G(x) = \int F(x|\theta) dC(\theta), \tag{1.15}$$

or equivalently, has its probability mass function as $g(x) = \int f(x|\theta) dC(\theta)$.

In either case,

$$G(x) = E_\Theta[F(x|\theta)],$$

where the expectation is taken with respect to the distribution of Θ. Barlow and Proschan (1965) have provided some examples in which mixture distributions arise naturally in reliability studies. A manufacturer produces 60% of units in one assembly line and 40% in another. The units from both assembly lines flow into a common shipping room, making a random mixture with life-length distributions $F_1(x)$ and $F_2(x)$. A unit selected at random from the shipping room has life distribution $F(x) = 0.6F_1(x) + 0.4F_2(x)$. As a second example, an important quality characteristic of a product depends on the amount θ of impurity present in the raw material. The distribution of the quality characteristic is $F(x|\theta)$ with θ as a random variable possessing distribution function $C(\theta)$. Then, the resultant quality characteristic is governed by the probability law in (1.15). It may be noted from (1.15) that the probability mass function $g(x)$ of $G(x)$ satisfies the equation

$$g(x) = \int f(x|\theta)dC(\theta).$$

Some of the standard distributions of importance in reliability analysis can also arise as mixtures. For example, let X be geometric with

$$f(x|p) = p(1-p)^x, \quad x = 0, 1, \ldots, \quad 0 < p < 1,$$

where p has a beta distribution with probability density function,

$$c(p) = \frac{p^{c-1}(1-p)^{b-1}}{B(c,b)}, \quad 0 < p < 1, \ b, c > 0,$$

with $B(c, b)$ denoting the complete beta function. Then, from (1.15), we get

$$\begin{aligned} g(x) &= \int_0^1 p(1-p)^x \frac{p^{c-1}(1-p)^{b-1}}{B(c,b)} dp \\ &= \frac{B(c+1, x+b)}{B(c,b)} = \frac{c \cdot b(b+1)\cdots(b+x-1)}{(c+b)\cdots(c+b+x)} \\ &= \frac{(a-b)(b)_x}{(a)_{x+1}}, \end{aligned} \tag{1.16}$$

where $a = c + b$ and $(b)_x = b(b+1)\cdots(b+x-1)$. Eq. (1.16) is the so-called *Waring distribution*, whose properties are studied in the subsequent chapters.

As a second example, let X be binomial with

$$f(x|n,p) = \binom{n}{x} p^x (1-p)^{n-x}, \quad x = 0, 1, \ldots, n, \ 0 < p < 1.$$

Assuming beta distribution $g(p)$ with parameters a and k for p, we get

$$\begin{aligned}
g(x) &= \int_0^1 \binom{n}{x} p^x (1-p)^{n-x} \frac{p^{a-1}(1-p)^{k-1}}{B(a,k)} dp \\
&= \binom{n}{x} \frac{B(x+a, n-x+k)}{B(a,k)} \\
&= \binom{n}{x} \frac{a(a+1)\cdots(a+x-1)\cdot k(k+1)\cdots(k+n-x-1)}{(a+k)(a+k+1)\cdots(a+k+n-1)} \\
&= \binom{n}{x} \frac{(-1)^x(-a)(-a-1)\cdots(-a-x+1)(-1)^{n-x}(-k)(-k-1)\cdots(-k-n+x+1)}{(-1)^n(-a-k)(-a-k-1)\cdots(-a-k-n+1)} \\
&= \binom{n}{x} x! \binom{-a}{x} (n-x)! \binom{-k}{n-x} \cdot \frac{\binom{-a-k}{n}}{n!} \\
&= \frac{\binom{-a}{x}\binom{-k}{n-x}}{\binom{-a-k}{n}}, \quad x = 0, 1, \ldots, n. \qquad (1.17)
\end{aligned}$$

Eq. (1.17) is the so-called *negative hyper geometric distribution*. When $a = 1$, the distribution with

$$g(x) = \frac{\binom{-1}{x}\binom{-k}{n-x}}{\binom{-1-k}{n}}, \quad x = 0, 1, \ldots, n, \qquad (1.18)$$

is of special interest in the sequel.

The reliability functions of mixture models will be discussed in Chapter 2 and the role of mixtures in generating models with non-monotonic hazard rates will be discussed in Chapter 5.

1.4 WEIGHTED DISTRIBUTIONS

The concept of weighted distributions was introduced first by Fisher (1934) and was later popularized by Rao (1965); see also Patil (2002). When observations are recorded, they will not have the assumed model unless every observation has the same probability of being included in the set of observations. When there is a probability of recording an observation x of a discrete random variable X with probability mass function $f(x)$, then the distribution of the recorded observation X_w has its mass function as

$$f_w(x) = Cw(x)f(x), \qquad (1.19)$$

where C is such that $\sum f_w(x) = 1$. When $w(x)$ is an arbitrary non-negative weight function which may exceed unity, (1.19) is called the weighed distribution of X with weight function $w(x)$. Patil et al. (1986) provided a structural perspective of discrete weighed model and have then presented several examples to show how such models arise in various real-life situations. They have also provided some general properties of weighted distributions and weight functions that are of specific interest. Gupta and Kirmani (1990) have pointed out several instances from survival analysis, wherein weighted distributions are of natural interest, while Pakes et al. (2003) have established many characterization results using weighted distributions, and Kokonendji et al. (2009) have developed semi-parametric estimation method for count data through weighted distributions. In this and the rest of this work, we designate X as the original random variable and its distribution specified by either $f(x)$, $F(x)$ or $S(x)$ as the original distribution. The distribution in (1.19) can also be presented as

$$f_w(x) = \frac{w(x)f(x)}{Ew(X)} \tag{1.20}$$

provided $E(w(X)) < \infty$, or by the survival function

$$S_w(x) = \frac{\sum_{t=x+1}^{\infty} w(x)f(t)}{Ew(X)} = \frac{E[w(X)|X \geq x]S(x)}{Ew(X)} . \tag{1.21}$$

Apart from the general forms in (1.20) or (1.21), there are two special cases that are of interest in reliability modelling. The first one is the *length-biased model* defined as

$$f_L(x) = \frac{xf(x)}{\mu} , \tag{1.22}$$

when $\mu < \infty$, which arises when the sampling mechanism selects units with probability proportional to the lengths of lifetimes of the units. Blumenthal (1967) and Scheaffer (1972) have discussed the role of such models in life-length studies. The data on chronic diseases reveal that patients with longer pre-clinical disease period have a greater probability of being identified, which illustrates a practical situation resulting in length-biased sampling. Some general results that hold for model (1.20) and its specialized form in (1.22) wherein $w(x) = x$, are the following (Patil and Rao, 1971 and Patil et al., 1986):

(a) Truncated distributions may be considered as weighted versions with $w(x) = I_{(x_0,\infty)}(x)$, where $I_{(x_0,\infty)}(x) = 1$ if $X > x_0$, and zero if $X \leq x_0$;

(b) $E(X_w) \geq (\leq) E(X)$ according as $w(x)$ is monotone increasing or decreasing in x. Note that monotonicity condition on $w(x)$ is sufficient, but not necessary. The mean of the length-biased version X_L exceeds $E(X)$ by the variance to mean ratio of X and further, the harmonic mean of X_L is $E(X)$;

(c) If $\phi(t)$ and $\phi_L(t)$ denote the characteristic functions of X and X_L, respectively, then

$$\phi_L(t) = (i\mu)^{-1}\phi'(t), \quad \mu = E(X)$$

(Sen and Khattree, 1996);

(d) $E(X_w) > E(X)$ if $\mathrm{Cov}(X, w(X)) > 0$ and $E(X_w) < E(X)$ if $\mathrm{Cov}(X, w(X)) < 0$ (Patil and Rao, 1978);

(e) When $w(x) = x^n$ in (1.20), we say that X_w is the *length-biased version of order* n of X. A distribution of X_w is form-invariant if it has the same form of distribution of X. Let X be a random variable with probability mass function $f(x|\theta)$ which is form-invariant under length-bias of order n, with $E(X|\theta) = \lambda\theta^{\beta}$. Similarly, Θ has distribution $c(\theta)$ which is also form invariant under length-bias of order n. Then, the mixture distribution in (1.15) is form-invariant under length-bias of order n.

One particular family of weighted distributions that has received considerable attention in the literature is the weighted Poisson distributions; see, for example, Castillo and Pérez-Casany (1998, 2005) and Kokonendji et al. (2008). These authors have all established several connections of the Poisson weight function $w(x)$ to overdispersion and underdispersion corresponding to the cases when the population variance is larger than the mean and smaller than the mean, respectively. One of the results that Castillo and Pérez-Casany (2005) proved is that if the weight function is of the form $w(x) = e^{rt(x)}$, $x \in \mathbb{N}$, where $r \in \mathbb{R}$ and $t(x)$ is a convex function, the then corresponding weighted Poisson distribution is overdispersed, equidispersed and underdispersed if $r > 0$, $r = 0$ and $r < 0$, respectively. Observe that the length-biased Poisson distribution with $w(x) = x$, $x \in \mathbb{N}$, is not covered by the above result.

To further examine the properties of the weighted Poisson distribution, Kokonendji et al. (2008) started with its general form of the probability mass function given by

$$f_w(x; \theta, \phi) = \varphi(x; \phi)e^{\theta x - K(\theta,\phi)}, \quad x \in \mathbb{N},$$

where $\theta \in \Theta \subset \mathbb{R}$ is the natural parameter of the Poisson (λ) distribution with $\theta = \ln\lambda$ and $\phi > 0$, corresponding to a weighted Poisson distribution with weight function $w(x; \phi) = x!\varphi(x; \phi)$, $x \in \mathbb{N}$, and established the mean-variance relationship

$$\mathrm{Var}(X_w) = E(X_w) + \lambda^2 \frac{d^2}{d\lambda^2} \ln(E(w(X; \phi))).$$

Then, by utilizing this relationship, these authors proved that the mean weight function $E(w(X; \phi))$ is log-convex or log-concave if and only if the weighted

Poisson random variable X_w is overdispersed or underdispersed, respectively. They also presented a weaker version of this result, that is easier for verification, that the weighted Poisson random variable X_w is overdispersed or underdispersed if the Poisson weight function $w(x)$ is log-convex or log-concave, respectively.

A useful and interesting example of the Weighted Poisson distribution is the *COM-Poisson family*, originally introduced by Conway and Maxwell (1962) with probability mass function

$$f(x) = \frac{1}{\sum_{j=0}^{\infty} \frac{\lambda^j}{(j!)^{\phi}}} \cdot \frac{\lambda^x}{(x!)^{\phi}}, \qquad x \in \mathbb{N},\ \lambda > 0,\ \phi \geq 0.$$

Evidently, this is a weighted Poisson distribution with weight function $w(x;\phi) = (x!)^{1-\phi}$, $x \in \mathbb{N}$, $\phi \geq 0$, and is overdispersed, equidispersed and underdispersed when $0 \leq \phi < 1$, $\phi = 1$ (Poisson case) and $\phi > 1$, respectively. As demonstrated by Shmueli et al. (2005), this is a very flexible family of distributions as it includes geometric (when $\phi = 0$ and $\lambda < 1$), Poisson (when $\phi = 1$) and Bernoulli (when $\phi \to \infty$) distributions as special cases, and is therefore quite useful for fitting count data which may be overdispersed or underdispersed. For example, it has been used in developing flexible cure rate models in survival analysis; see Rodrigues et al. (2009), and Balakrishnan and Pal (2013, 2016).

A further extension has been provided by Minkova and Balakrishnan (2013) in terms of *compound weighted Poisson distribution* by considering the compound random variables of the form $N_w = \sum_{i=1}^{X_w} Y_i$, where X_w is a weighted Poisson random variable and Y_i's are i.i.d. discrete random variables. They have then discussed the overdispersion and underdispersion of these distributions as well as properties of several special cases of these compound weighted Poisson distributions.

As mentioned earlier, a second special case of X_w that is of extensive application in lifetime analysis is the equilibrium distribution. A random variable X_E has equilibrium distribution corresponding to X if it has its probability mass function of the form

$$f_E(x) = \frac{S(x+1)}{\mu}, \qquad x = 0, 1, \ldots . \tag{1.23}$$

Applications of (1.23) are found in many diverse topics that include characterizations of probability distributions, criteria for ageing, maintenance policies, income analysis, and risk theory. There are different interpretations for (1.23). As a weighted distribution, (1.23) has a slightly different meaning than in the continuous case, where the weight function is the reciprocal of the hazard rate of X. However, the discrete counterpart has weight function

$$w(x) = \frac{S(x+1)}{f(x)} = \frac{1-h(x)}{h(x)}$$

where $h(x)$ is the discrete hazard rate defined in Section 2.2. Noticing that $E\frac{S(x+1)}{f(x)} = \mu$, we have (1.23) from the basic definition in (1.20). A second interpretation is through the bivariate distribution of a random vector (Y_1, Y_2) specified by

$$P(Y_1 = x_1, Y_2 = x_2) = \mu^{-1} f(x_2), \quad x_1 = 0, 1, \ldots, x_2 - 1, \quad x_2 = x_1 + 1, \ldots, \tag{1.24}$$

and $f(\cdot)$ is the probability mass function of X defined on $(0, 1, 2\ldots)$. The marginal distributions of (1.24) are seen to be

$$P(Y_1 = x_1) = \sum_{x_2} \mu^{-1} f(t) = f_E(x_1), \quad x = 0, 1, \ldots,$$

and

$$P(Y_2 = x_2) = \sum_{x_1} \mu^{-1} f(t) = \mu^{-1} x_1 f_L(x_2), \quad x_2 = 1, 2, \ldots.$$

Notice that

$$P(Y_1 = x_1 | Y_2 = x_2) = \frac{1}{x_2}, \quad x_1 = 0, 1, \ldots, x_2 - 1,$$

the discrete uniform distribution. In other words, if the sampling is done proportional to the lifelength x and given that sample points fall in an interval of length x, it is equally likely to fall anywhere in the interval. Thence, the conditional probability of X_E is uniform over $(0, x]$. Thus, the unconditional probability of X_E is given by

$$P[X_E = x_E] = \sum_{x > x_E} \frac{1}{x} f_L(x) = f_E(x_E)$$

(see Gupta, 1979). For the definition and properties of the bivariate concepts used here, see Section 1.9.

Some important properties of equilibrium distributions are as follows:

(i) The characteristic functions of X_E and X are related by

$$\phi_E(t) = (i\mu t)^{-1}(\phi(t) - 1),$$

and further

$$\phi_E(t) = \frac{1}{t}\int_0^t \phi_L(y)dy;$$

(ii) X_E and X_L are identically distributed if and only if X is degenerate at zero;

(iii) If X belongs to the class of modified power series distributions given by

$$P(X = x) = \frac{a(x)(g(\theta))^x}{b(\theta)}, \; x \in T \quad \text{(see Section 1.3)},$$

where T is a subset of non-negative integers, $a(x) > 0$, $g(\theta)$ and $b(\theta)$ are positive, finite and differentiable, then the geometric distribution is the only one having $E(X) = E(X_E)$. The results in (i) and (ii) are due to Sen and Khattree (1996) while (iii) is from Gupta (1979);

(iv) If X is geometric with parameter p (Waring with parameter (a, b), negative hypergeometric with parameter (k, n)), then X_E has the same geometric (Waring $(a, b+1)$, negative hypergeometric $(k+1, n-1)$) distribution.

The *zero-modified version* of the random variable X is specified by

$$P(X = 0) = \alpha + (1-\alpha) f(0),$$
$$P(X = x) = (1-\alpha) f(x).$$

Now, consider the zero modified geometric law. It its has mean as

$$\mu = \sum_{x=0}^{\infty} x f(x) = \sum_{x=1}^{\infty} (1-\alpha) f(x) = (1-\alpha)\frac{q}{p},$$

and so

$$f_E(x) = \frac{S(x+1)}{\mu} = \sum_{t=x+1}^{\infty} \frac{(1-\alpha)q^x p}{(1-\alpha)q/p} = q^x p.$$

Thus, the converse of the statement in (iv) need not be true, as the zero-modified geometric also has the same $f_E(x)$. However, note that the zero-modified version includes the original distribution when $\alpha = 0$.

Since $f_E(x)$ defines a proper distribution, we can talk about its equilibrium model which gives a second-order equilibrium distribution of X. By successive application, the equilibrium distributions of order n of X are defined recursively, for $n = 1, 2, 3, \ldots$, by the probability mass function

$$f_n(x) = (\mu_{n-1})^{-1} S_{n-1}(x) \tag{1.25}$$

with $\mu_{n-1} = \sum_{x=0}^{\infty} S_{n-1}(x+1) < \infty$, $S_0(x) = S_E(x)$ and $\mu_0 = \mu = E(X)$. In the sequel, we denote by Y_n the random variable with distribution (1.25), so that $Y_0 = X_E$. The survival functions of X and Y_n are connected by (Willmot et al., 2005, and Nair et al., 2012b)

$$S_n(x+1) = [\mu_{(n)}]^{-1} E[(X-x-1)^{(n)} | X > x+n] S(x+n+1), \tag{1.26}$$

where $a_{(r)}$ is the descending factorial

$$a_{(r)} = a(a-1)\cdots(a-r+1)$$

and $\mu_{(n)}$ is the nth factorial moment as defined in (1.5). The usual moment of Y_n is then

$$\mu_n = \frac{\mu_{(n+1)}}{(n+1)\mu_{(n)}}.$$

The factorial moment generating function defined in (1.10), of Y_n is given by

$$W_n(t) = \left[\frac{t^n \mu_{(n)}}{n!}\right]^{-1} \left[W(t) - \sum_{r=0}^{n-1} \frac{t^r \mu_{(r)}}{r!}\right], \tag{1.27}$$

where $W(t)$ is the factorial moment generating function of X. An alternative expression for (1.27) is

$$W_n(t) = 1 + \sum_{r=1}^{\infty} t^r \mu_n \cdots \mu_{n+r-1}.$$

More discussions on the reliability aspects of the models presented in this section can be found in Chapter 4.

Example 1.1. Let X have a uniform distribution over $(1, 2, \ldots, b)$ with

$$f(x) = \frac{1}{b}.$$

Then, $\mu = E(X) = \frac{b+1}{2}$ so that the length-biased version of X is distributed as

$$f_L(x) = \frac{2x}{b(b+1)},$$

which will be referred to as the *arithmetic distribution*, as the probabilities are in arithmetic progression.

Example 1.2. Assume X to be geometric with

$$f(x) = q^x p, \quad q = 1-p, \quad x = 0, 1, 2, \ldots, \quad 0 < p < 1.$$

Then, as observed earlier,

$$f_E(x) = q^x p = f(x).$$

Consequently, equilibrium distributions of all orders of the geometric are the same as the original distribution of X.

Li (2011) observed that when X represents the claim size, the first-order equilibrium distribution can be interpreted as the distribution of the amount of first drop below the initial reserve, given that there is such a drop. He also gave some properties of the equilibrium distribution useful in connection with ruin theory. The rth factorial moment of X_E is

$$\mu_{1(r)} = \frac{\mu_{(r+1)}}{(r+1)\mu}$$

arising from the probability generating function of X_E given by

$$P_E(t) = \frac{1 - P(t)}{\mu(1-t)}.$$

In general, Y_n has its generating function as

$$P_n(t) = \frac{(-1)^{n-1} n![1 - P(t)]}{\mu_{(n)}(1-t)^n} + \frac{n!}{\mu_{(n)}} \sum_{k=1}^{n-r} \frac{(-1)^{n-k-1}}{k!} \frac{\mu_{(k)}}{(1-t)^{n-k}}$$

from which $P_E(t)$ is obtained when $n = 1$.

1.5 CONVOLUTION

Let X_1 and X_2 be two non-negative independent discrete random variables with probability mass functions $f_1(x)$ and $f_2(x)$, respectively. The distribution of $Z = X_1 + X_2$ is called the *convolution* of the distributions of X_1 and X_2. We note that the random variable $Z = z$ is the union of mutually exclusive events $(X_1 = 0, X_2 = z), (X_1 = 1, X_2 = z - 1), \ldots, (X_1 = z, X_2 = 0)$ so that

$$\begin{aligned} f_Z(z) = P(Z = z) &= f_1(0) f_2(z) + f_1(1) f_2(z-1) + \cdots + f_1(z) f_2(0) \\ &= \sum_{t=0}^{z} f_1(t) f_2(z-t), \end{aligned} \tag{1.28}$$

and we write $f_Z(x) = f_1(x) * f_2(x)$, or simply $f_Z = f_1 * f_2$. Alternatively, if $P_1(t)$ and $P_2(t)$ are probability generating functions of X_1 and X_2, respectively, Z has its generating function as

$$P_Z(t) = P_1(t) P_2(t). \tag{1.29}$$

The above definition has a natural extension to n random variables. If $X_1, \ldots, X_n$ are mutually independent random variables with generating function $P_i(t)$ for X_i, $i = 1, 2, \ldots, n$, then $Z = \sum_{i=1}^{n} X_i$ has its generating function as

$$P_Z(t) = \prod_{i=1}^{n} P_i(t). \tag{1.30}$$

If the X_i' is have the same distribution with probability mass function $f(x)$ and generating function $P(t)$, then evidently

$$P_Z(t) = [P(t)]^n, \tag{1.31}$$

and we write this as $f_Z = f^{n*}$ and call f_Z as the *n-fold convolution* of f.

Example 1.3. Let X_1 and X_2 have Poisson distributions (Section 1.3)

$$f_1(x_1) = \frac{e^{-\lambda_1}\lambda_1^{x_1}}{x_1!} \text{ and } f_2(x_2) = e^{-\lambda_2}\frac{\lambda^{x_2}}{x_2!}, \quad x_1, x_2 = 0, 1, 2, \ldots.$$

Then,

$$\begin{aligned} f_Z(z) &= \sum_{k=0}^{z} e^{-\lambda_1}\frac{\lambda_1^k}{k!} e^{-\lambda_2}\frac{\lambda_2^{z-k}}{(z-k)!} \\ &= \frac{e^{-\lambda_1-\lambda_2}}{z!}\sum_{k=0}^{z}\frac{z!}{k!(z-k)!}\lambda_1^k\lambda_2^{z-k} \\ &= e^{-(\lambda_1+\lambda_2)}\frac{(\lambda_1+\lambda_2)^z}{z!}, \end{aligned}$$

which is just a Poisson distribution with parameter $(\lambda_1 + \lambda_2)$.

Example 1.4. Take $Z = X_1 + X_2 + \ldots + X_n$, where X_i's are independent with the same geometric law

$$f(x) = q^x p.$$

Then, the generating function of each X_i is

$$P(t) = \sum t^x q^x p = \frac{p}{1-qt}.$$

Accordingly,

$$P_Z(t) = \left(\frac{P}{1-qt}\right)^n,$$

and so

$$f(z) = \frac{1}{z!} \frac{d^z P_Z(t)}{dt^z}\bigg|_{t=0}$$
$$= \binom{-n}{x} p^n (-q)^x, \quad x = 0, 1, 2, \ldots ,$$

which is the negative binomial distribution (see Section 1.3). In reliability analysis, convolution of lifetimes has important application when a spare part is used in maintaining the working of a system, when the original part fails. When a failed component is replaced by a spare, the total life accumulated is obtained by the addition of the lifetimes of the original and spare parts. Such an operation is an essential activity in formulating maintenance policy. It is also customary to check whether a particular ageing property is preserved under such convolutions. However, except for a few distributions, analytical derivation of convolutions of distributions is a difficult task. In applied problems, methods have been devised to obtain convolutions of standard forms recursively through algorithms. A common procedure to obtain the n-fold convolution of independent and identically distributed lifelengths is to use the recurrence

$$f_Z(z) = f_x^{n*}(z),$$

where

$$f_X^{(k+1)*}(z) = \sum_{x=0}^{z} f_X(x) f_x^{k*}(y - x).$$

For particular families and in the general case, recurrence relations have been given in Panjer (1981), Pril (1984), Woodward and Palmer (1997), and Evans and Leemis (2004).

A more general type of model will result if we consider a random number of variables to constitute Z; that is, when

$$Z_N = X_1 + X_2 + \cdots + X_N,$$

where N is a positive integer-valued random variable, $X_i^{\prime}s$ are independent and identically distributed and are independent of N. Then, the distribution of Z_N is given by

$$g(x) = \sum_{n=1}^{\infty} f^{n*}(x)\, P(N = n),\ x = 0, 1, 2, \ldots \tag{1.32}$$

Models of the form (1.32) are called compound distributions. Willmot and Panjer (1987) and Hesselager (1994) have discussed procedures for the evaluation of compound distributions. If the $X_i^{\prime}$s have generating function $P(t)$ and N has generating function $P_N(t)$, then the generating function of Z_n is evidently given by

$$P_{Z_N}(t) = P_N(P(t)).$$

A classic example of compound distributions is the so-called Pólya-Aeppli distribution, formed as the distribution of $N = \sum_{i=1}^{Y} X_i$, where Y, independent of X_i's, is a Poisson(λ) random variable and X_i's are i.i.d. Geometric($1-\rho$) random variables with probability mass function

$$P(X = x) = (1-\rho)p^{x-1}, \quad x = 1, 2, \ldots .$$

Then, the probability generating function of N can be shown to be

$$P_N(s) = E(s^N) = \exp\left[-\lambda\left\{1 - \frac{(1-\rho)s}{1-\rho s}\right\}\right];$$

see, for example, Johnson et al. (1992) and Minkova (2002). Using the above probability generating function, it can be readily shown that

$$E(N) = \frac{\lambda}{1-\rho} \quad \text{and} \quad \mathrm{Var}(N) = \frac{\lambda(1+\rho)}{(1-\rho)^2}$$

from which it can be seen that $\mathrm{Var}(X) > E(X)$, meaning that the distribution is overdispersed with respect to the Poisson distribution. It should be mentioned that if X_i's above instead had been taken as Geometric($1-\rho$) random variables taking values $0, 1, \ldots$, then the distribution of N will be still Pólya-Aeppli with parameters $\lambda\rho$ and ρ (instead of λ and ρ, as above).

1.6 SHOCK MODELS

We now look at the reliability aspects through models that represent the state of a system as it changes over time, by assuming that the system is subject to shocks that occur randomly in time. The essential elements in such models are the time between two consecutive shocks, the extent of damage caused by a shock, the failure of the system and the interrelationship between these elements. Let $\{N(t),\ t \geq 0\}$ be the number of shocks received by a component in the time interval $[0, t]$ and $\overline{P}_K$ be the probability that the component survives k shocks, $k = 0, 1, 2, \ldots$. Then, the survival function of the device has the form

$$S(x) = \sum_{k=0}^{\infty} \overline{P}_K\, P[N(x) = k],\ x \geq 0.$$

According to the form of times between consecutive shocks, these models assume $N(t)$ to be (i) a homogeneous Poisson process (Esary et al., 1973), (ii) non-homogeneous Poisson process (A-Hameed and Proschan, 1973), (iii) nonstationary pure birth process (A-Hameed and Proschan, 1975), and (iv) a renewal process (Skoulakis, 2000).

There are further improvements in the basic models, by incorporating additional concepts like recovery from shocks, ageing, and so on; see, for example, Finkelstein and Zarudnij (2001) and Fan et al. (2000). The magnitude of the shocks is modelled by a random variable which gives rise to (i) extreme shock model, wherein the system fails as soon as the magnitude of an individual shock exceeds a given level, (ii) cumulative shock model in which the system fails when the cumulative magnitude exceeds some threshold, and (iii) run shock model wherein the system works until k consecutive shocks with critical magnitudes occur. In the models cited above, the distribution of the times between successive shocks are continuous. Comparatively, there are very few models in the discrete case, and some papers that deal with discrete shock models are Aven and Gaarder (1987), Gut (2001), Cha and Finkelstein (2012), and Eryilmaz (2012, 2013).

Eryilmaz (2012) has considered a system subjected to periodic random shocks which occur with probability p in each period, $n = 1, 2, \ldots$. The magnitude of the shock at period j is a random variable X_j and the system fails if the sum of magnitudes exceed a level $k > 0$. Let us define $Y_j = X_j$, if a shock occurs at period j and $Y_j = 0$ otherwise. Assuming $\langle X_j \rangle$, $j \geq 1$, to be a sequence of strictly positive random variables which are independent and identically distributed with probability mass function $f(x)$, the failure time under the cumulative damage model is defined by the random variable

$$Y_k = \min\left(n \Big| \sum_{j=1}^{n} Y_j > k\right).$$

Accordingly, Y_k has its distribution as

$$P(Y_k > n) = p \sum_{x=1}^{\min(k,b)} P(Y_{k-x} > n-1) f(x) + (1-p)\, P(Y_k > n-1)$$

and $P(Y_k > 0) = 1$, where b is the right hand limit of the support of f, with mean

$$E(Y_k) = \frac{1}{p} \sum_{x=1}^{\min(k,b)} E(Y_{k-x}) f(x).$$

Eryilmaz (2012) has also discussed the mixed shock model, in which the system fails if either a single shock with large magnitude or if the cumulative shock exceeds a critical level. Another shock model wherein the shocks occur according to interarrival times between successive shocks following a geometric law with mean $\frac{1}{p}$ has been discussed by Eryilmaz (2013). The system fails when the time between two consecutive shocks is less than a prespecified level δ and the system lifetime is the time up to the occurrence of this event, with

$$T_\delta = \sum_{i=0}^{M} X_i, \quad P(X_i = x) = p(1-p)^{x-1}, \quad x = 1, 2, \ldots$$
$$M = m \quad \text{iff } [X_1 > \delta, \ldots, X_{m-1} > \delta, X_m \le \delta].$$

The distribution of T_δ, its mean and variance have been derived by Eryilmaz (2013). Further, a Markov chain is assumed for the occurrences of shocks to relax the assumption of independence of such events assumed above and the distribution of T_δ has been obtained in such a general situation.

Continuing along these line, Parvardeh and Balakrishnan (2015) considered a general setup with $\{(X_k, Z_k),\ k = 1, 2, \ldots\}$ being a sequence of i.i.d. bivariate random vectors wherein X_k represents the time between the $(k-1)$th and kth shocks and Z_k represents the magnitude of the kth shock. Suppose (X, Z) has a joint distribution H and marginal distributions F and G, respectively. Then, as their Model I, Parvardeh and Balakrishnan (2015) considered the case of system failing when the time between two successive shocks is less than a given threshold δ, or the magnitude of single shock is larger than a given threshold γ. In this case, the lifetime of the system is given by $T = \sum_{k=1}^{N}$, where

$$(N = n) = (X_1 > \delta, Z_1 \le \gamma, \ldots, X_{n-1} > \delta, Z_{n-1} \le \gamma, X_n \le \delta)$$
$$\bigcup (X_1 > \delta, Z_1 \le \gamma, \ldots, X_{n-1} > \delta, Z_{n-1} \le \gamma, X_n > \delta, Z_n > \gamma)$$

and, in fact, N has the geometric distribution with parameter $F(\delta) + \bar{H}(\delta, \gamma)$. These authors have then presented an explicit formula for the survival function of the system lifetime T and the mean lifetime $E(T)$, and also their simplified expressions in the special case when X's and Z's are independent with distributions F and G, respectively.

Next as their Model II, Parvardeh and Balakrishnan (2015) further considered the case of system failing when the time between two successive shocks is less than a given threshold δ, or the cumulative magnitude of shocks is larger than a given threshold γ. In this case, the lifetime of the system is given by $t = \sum_{k=1}^{N} X_k$, where

$$(N = n) = (X_1 > \delta, Y_1 \le \gamma, \ldots, X_{n-1} > \delta, Y_{n-1} \le \gamma, X_n \le \delta)$$
$$\bigcup (X_1 > \delta, Y_1 \le \gamma, \ldots, X_{n-1} > \delta, Y_{n-1} \le \gamma, X_n > \delta, Y_n > \gamma)$$

with $Y_n = \sum_{k=1}^{n} Z_k$, $n = 1, 2, \ldots$, being the cumulative magnitude of first n shocks. These authors have presented an explicit formula for the survival function of the system lifetime T as well as a simplified expression in the special case when X's and Z's are independent with distributions F and G, respectively.

1.7 CONVEXITY AND RELATED CONCEPTS

Consider a sequence $\langle g(x)\rangle$, $x = 1, 2, \ldots$. Then, we say that $\langle g(x)\rangle$ is a *convex sequence* if for any

$$x_1 < x_2 < x_3 \Rightarrow \frac{g(x_2) - g(x_1)}{x_2 - x_1} \leq \frac{g(x_3) - g(x_1)}{x_3 - x_1} \leq \frac{g(x_3) - g(x_2)}{x_3 - x_2}$$

wherein x_1, x_2, x_3 are equi-spaced points with $x_2 = x_1 + 1$, $x_3 = x_1 + 2$. The above definition is equivalent to

$$g(x_1 + 2) - 2g(x_1 + 1) + g(x_1) \geq 0, \quad x_1 \geq 1. \tag{1.33}$$

In the sequel, we shall use the term increasing (decreasing) to mean non-decreasing (non-increasing). Accordingly, the function $g(x_1)$ is increasing if $\Delta g(x_1) \geq 0$, where Δ is the difference operator $\Delta g(x_1) = g(x_1 + 1) - g(x_1)$. Also, (1.33) means that $\Delta^2 g(x_1) \geq 0$ or $\langle \Delta(g(x_1))\rangle$ is an increasing sequence. Further $\langle g(x)\rangle$ is concave if the inequality in (1.33) is reversed or $\Delta^2 g(x_1) \leq 0$ meaning $\langle \Delta g(x_1)\rangle$ is a decreasing sequence. An associated concept is that of log-convexity (log-concavity). We say that $\langle g(x_1)\rangle$ is log-convex (log-concave) iff $\langle \log g(x_1)\rangle$ is convex (concave). Simple calculation reveals that the condition for log-convexity (log-concavity) becomes

$$[g(x_1 + 1)]^2 \leq (\geq) g(x_1) g(x_1 + 2), \; x \geq 1 \tag{1.34}$$

The geometric concepts in (1.33) and (1.34) are important in reliability theory as they facilitate expressions of some reliability concepts in terms of them. Further, their properties become handy in interpreting reliability concepts by means of geometric terms. In non-parametric inferential problems, convexity or log-convexity provides additional inputs about the underlying distribution.

Another geometric concept of interest is *star-shapedness* (*anti-star-shapedness*). The sequence $\langle g(x)\rangle$ is star-shaped (anti-star-shaped) if, for $x_1, x_2 \geq 1$,

$$x_1 \leq x_2 \Rightarrow g(x_1) \leq (\geq) \frac{x_1}{x_2} g(x_2) \tag{1.35}$$

One can interpret (1.35) by saying that $\frac{g(x)}{x}$ is an increasing (decreasing) function for it to be star (anti-star) shaped. A convex sequence $\langle g(x)\rangle$, such that $g(x)$ passes through the origin, is star-shaped.

Finally, we say that $\langle g(x)\rangle$ is super-additive (sub-additive) iff

$$g(x + y) \geq (\leq) g(x) + g(y), \quad x, y \geq 1. \tag{1.36}$$

Some properties resulting from the above definitions are as follows:

(i) If $g(\cdot)$ is convex and $a(\cdot)$ is increasing and convex, then $a(g(\cdot))$ is convex;
(ii) If $g(\cdot)$ is concave and $a(\cdot)$ is increasing and concave, then $a(g(\cdot))$ is concave;
(iii) If $g(\cdot)$ is convex and $a(\cdot)$ is decreasing and concave, then $a(g(\cdot))$ is concave;
(iv) If $g(\cdot)$ is concave and $a(\cdot)$ is decreasing and convex, then $a(g(\cdot))$ is convex;
(v) The sum of convex (concave) functions is convex (concave);
(vi) The product of log-convex (log-concave) functions is log-convex (log-concave);
(vii) If $g(\cdot)$ is a positive function and $g(\cdot)$ is log-convex, then $g(\cdot)$ is convex, while if $g(\cdot)$ is a positive concave function, then $g(\cdot)$ is log-concave;
(viii) If $g(\cdot)$ is non-negative and $g(0) \le 0$, then $g(\cdot)$ is convex implies $g(\cdot)$ is star-shaped;
(ix) If $g(\cdot)$ is star-shaped, it is super-additive.

A real-valued function $a(x)$ is PF_2 (*Polya frequency function of order* 2) if

(a) $a(x) \ge 0$,
(b) $\begin{vmatrix} a(x_1 - y_1) & a(x_1 - y_2) \\ a(x_2 - y_1) & a(x_2 - y_2) \end{vmatrix} \ge 0$ for all $x_1 < x_2$ and $y_1 < y_2$.

Two other equivalent conditions for $a(x)$ to be PF_2 are

(c) $\log a(x)$ is concave, and
(d) for any fixed $t > 0$, $\frac{a(x+t)}{a(x)}$ is decreasing within a range $[a, b]$, where $a = \inf_{a(y)>0} y$ and $b = \sup_{a(y)>0} y$. Notice that a PF_2 function $a(x)$ need not be a probability mass function.

Let $a(x, y)$ be a function defined on $A \times B$, where A and B are subsets of the real line. Then, $b(x, y)$ is TP_2 (*Totally positive of order* 2) if, for all $x_1 < x_2$ and $y_1 < y_2$.

$$\begin{vmatrix} b(x_1, y_1) & b(x_2, y_1) \\ b(x_1, y_1) & b(x_2, y_2) \end{vmatrix} \ge 0.$$

It is easy to see from the definition that if $b(x, y) = a(x - y)$ is TP_2, then the function $a(\cdot)$ is PF_2.

1.8 MULTIVARIATE DISTRIBUTIONS

The concept of discrete distribution and various characteristics associated with a discrete random variable X can be extended to the case of several random variables. In reliability theory, they arise with system (device) consisting of several components and each component has a random lifetime. If there are p components that make up the system and X_i is the lifetime of the ith component, the lifetime of the system is a vector $\mathbf{X} = (X_1, \ldots, X_p)$. Since we are considering only discrete lifetime, each of X_i takes an at most countable number of values $(0, 1, 2, \ldots, b_i)$, $b_i \leq \infty$, for $i = 1, 2, \ldots, p$. The realization of $(X_1, \ldots, X_p)$ is the ages x_i, $i = 1, 2, \ldots, p$, at which the ith component fails, and is denoted by the vector $\mathbf{x} = (x_1, \ldots, x_p)$. The joint probability mass function of $\mathbf{X}$ is defined as

$$f(\mathbf{x}) = P(\mathbf{X} = \mathbf{x}) = P(X_1 = x_1, \ldots, X_p = x_p). \tag{1.37}$$

In the sequel, equality or inequality among vectors means componentwise equality or inequality. Thus, $\mathbf{X} \geq \mathbf{x} = (X_1 \geq x_1, \ldots, X_p \geq x_p)$, for example. The function $f(\mathbf{x})$ must be such that

$$f(\mathbf{x}) \geq 0 \text{ and } \sum f(\mathbf{x}) = 1,$$

where the summation extends over the range of values of $x_1, \ldots, x_p$. The distribution function and survival function of $\mathbf{X}$ are, respectively,

$$F(\mathbf{x}) = P(\mathbf{X} \leq \mathbf{x}) = P(X_1 \leq x_1, \ldots, X_p \leq x_p)$$

and

$$S(\mathbf{x}) = P(\mathbf{X} \geq x) = P(X_1 \geq x_1, \ldots, X_p \geq x_p).$$

Notice that

$$S(\mathbf{x}) = \sum_{t_1=x_1}^{b_1} \cdots \sum_{t_p=x_p}^{b_p} f(t_1, \ldots, t_p) \tag{1.38}$$

and $S(\mathbf{x}) = 0$ if at least one of the x_i's is equal to ∞ and $S(\mathbf{x}) = 1$ if $\mathbf{x} = \mathbf{0}$, where $\mathbf{0} = (0, 0, 0, \ldots, 0)$. The distribution of any subset $\mathbf{X}_m = (X_{i_1}, \cdots, X_{i_m})$ where $X_{i_1}, X_{i_2} \ldots, X_{i_m}$ are some selected components among $X_1, \ldots, X_p$, $m < p$, is called the m-dimensional marginal distributions of $\mathbf{X}$ for $m = 1, 2, \ldots$. They are obtained either as

$$\begin{aligned} f_{i_1,i_2,\ldots,i_m}(x_{i_1} x_{i_2} \ldots x_{i_m}) &= P(X_{i_1} = x_{i_1} \ldots X_{i_m} = x_{i_m}) \\ &= \sum P(X_1 = x_1, \ldots, X_p = x_p,\ x_j \geq 0,\ j \neq i_1, \ldots i_m) \\ &= \sum_{x_j} f(x_1, \ldots x_p), \quad j \neq i_1, i_2 \ldots, i_m, \end{aligned} \tag{1.39}$$

or from the survival function

$$S_{i_1,\ldots,i_m}(x_{i_1},\ldots,x_{i_m}) = P(X_{i_1} \geq x_{i_1},\ldots,X_{i_m} \geq x_{i_m},\ X_j \geq 0 \quad \text{for } j \neq i_1,\ldots,i_m). \tag{1.40}$$

The one-dimensional marginals, which are of considerable interest, is thus obtained by summation of $f(\mathbf{x})$ over all other x_j's. A second set of distributions considered in connection with multivariate distributions are conditional distributions. The conditional probability mass function of $X_{i_1}, X_{i_2}, \ldots, X_{i_m}$, given that $X_{j_1} = x_{j_1}, \ldots, X_{j_{p-m}} = x_{j_{p-m}}$, $j \neq i$, is defined by

$$P(X_{i_1} = x_{i_1},\ldots,X_{i_m} = x_{i_m} | X_{j_1} = x_{j_1},\ldots,X_{j_{p-m}} = x_{j_{p-m}}) = \frac{f(\mathbf{x})}{f_{j_1,\ldots,j_{p-m}}(x_{j_1},\ldots,x_{j_{p-m}})}. \tag{1.41}$$

This will be denoted by $f_{i_1,\ldots,i_m|j_1,\ldots,j_{p-m}}(x_{i_1},\ldots,x_{i_m})$. Both the functions $f_{i_1,\ldots,i_m}$ and $f_{i_1,\ldots,i_m|j_1,\ldots,j_{p-m}}$ are non-negative and sum to unity.

There are many practical situations in which the constituent random variables in $\mathbf{X}$ are independent. In this case, we have

$$f(\mathbf{x}) = f_1(x_1)f_2(x_2)\ldots f_p(x_p)$$

where $f_i(x)$, $i = 1, 2, \ldots, p$, is the one-dimensional marginal probability mass function of X_i. Also,

$$S(\mathbf{x}) = S_1(x_1)S_2(x_2)\ldots S_p(x_p)$$

with $S_i(x_i)$ being the survival function of X_i.

The expected value of a function $g(\mathbf{X})$ is defined in an analogous way to the univariate case in (1.1), as

$$Eg(\mathbf{X}) = \sum g(\mathbf{x})\, f(\mathbf{x}), \tag{1.42}$$

where the summation is over for all values of $x_1, \ldots, x_p$. In (1.42), the expectation exists only when the series on the right hand side is absolutely convergent. In particular,

$$E(\mathbf{X}) = E(X_1, \ldots, X_p),$$

where

$$\begin{aligned}\mu_i = E(X_i) &= \sum x_i\, f(x_1, x_2, \ldots, x_p), \quad i = 1, \ldots p.\\ &= \sum x_i\, f_i(x_i).\end{aligned}$$

Further, if $g_j(\mathbf{x})$ are functions for which expectations exist, for $j = 1, 2, \ldots, p$, we have

$$E\left[\sum_{j=1}^{p} k_j g_j(\mathbf{X})\right] = k_j \sum_{j=1}^{p} E g_j(\mathbf{X}),$$

and if $X_1, X_2, \ldots$ are independent, then

$$E\left[\prod_{j=1}^{p} g_j(x_j)\right] = \prod_{j=1}^{p} E[g_j(X_j)].$$

The expected value

$$\mu'_{r_1,\ldots,r_p} = E\left[\prod_{j=1}^{p} X_j^{r_j}\right] \tag{1.43}$$

is called *product-moment about the origin*, and

$$\mu_{r_1,\ldots,r_p} = E\left[\prod_{j=1}^{p}[X_j - E(X_j)\right]^{r_j} \tag{1.44}$$

is called *central product moment*. In particular,

$$\mu_{jk} = E(X_j - E(X_j))(X_k - E(X_k)) \tag{1.45}$$

is called the *covariance* between X_j and X_k, and is also written in short as $\text{Cov}(X_j, X_k)$. The coefficient of correlation between X_j and X_k is defined as

$$r_{j,k} = \frac{\text{Cov}(X_j, X_k)}{[V(X_j)V(X_k)]^{\frac{1}{2}}}, \tag{1.46}$$

where $V(X)$ denotes for the variance of X. We say that X_j and X_k are uncorrelated if $\text{Cov}(X_j, X_k) = 0$.

The expectations with respect to conditional distributions are also defined similarly. For example,

$$E(X_i|X_j = x_j,\ j \neq i) = \sum x_i\, f(x_i|x_j) \tag{1.47}$$

Of the conditional expectations, that of X conditioned on $(X_1, \ldots, X_m)$

$$E(X|X_1, \ldots, X_m)$$

is called the *regression function* of X on $(X_1, \ldots, X_m)$, and it is an important tool in prediction. Distributions with linear regression functions are of special interest in distribution theory.

The probability generating functions of $\mathbf{X}$ is

$$P_{\mathbf{X}}(\mathbf{t}) = E\left(\prod_{j=1}^{p} t_j^{X_j}\right)$$
$$= \sum t_1^{x_1} t_2^{x_2} \cdots t_p^{x_p} f(x_1, \ldots, x_p)$$

where $\sum$ stands for $\sum_{x_1=0}^{b_1} \sum_{x_2=0}^{b_2} \cdots \sum_{x_p=0}^{b_p}$ and $\mathbf{t} = (t_1, \ldots, t_p)$, with $|t_j| < 1$.

The probability generating function of the marginal distributions are

$$P_{X_1}(t_1) = P(t_1, 1, \ldots 1), \ldots, P_{X_p}(t_p) = P(1, 1, \ldots t_p)$$

and the variables $X_1, \ldots, X_p$ are independent if and only if

$$P_{\mathbf{X}}(\mathbf{t}) = P_{X_1}(t_1) P_{X_2}(t_2) \ldots P_{X_p}(t_p)$$
$$= P(t_1, 1, \ldots, 1)\, P(1, t_2, 1 \ldots 1) \ldots P(1, 1, \ldots, t_p).$$

If $\langle X_n \rangle$ is a sequence of independent and identically distributed random variables with common generating function $P(t)$ and $S_N = X_1 + \ldots + X_N$ and N is a random variable independent of the X_i's with probability mass function $g(n) = P(N = n)$, then the probability generating function of S_N is given by

$$P_{S_N}(t) = G_N(P(t)),$$

where $G_N(t) = \sum_0^\infty t^n g(n)$, $|t| < 1$, is the probability generating function of N. Let $r = \sum_{j=1}^{p} r_i$. Then, the factorial moments of $\mathbf{X}$ are obtained from

$$\mu_{(r_1, \ldots, r_p)} = \left[\frac{\partial^r P_{\mathbf{X}}(\mathbf{t})}{\partial t_1^{r_1} \cdots \partial t_p^{r_p}}\right]_{t_1 = \cdots = t_p = 1}$$

Among multivariate models, more discussions on reliability aspects are made usually on bivariate distributions. Therefore, we specialize some of the above results to the bivariate case for ease of reference. The joint probability mass function of (X_1, X_2) is specified by

$$f(x_1, x_2) = P(X_1 = x_1, X_2 = x_2) \tag{1.48}$$

with survival function

$$S(x_1, x_2) = P(X_1 \geq x_1, X_2 \geq x_2). \tag{1.49}$$

The two are related by the equations

$$S(X_1, X_2) = \sum_{t_1=x_1}^{b_1} \sum_{t_2=x_2}^{b_2} f(t_1, t_2)$$

and

$$f(x_1, x_2) = S(x_1, x_2) - S(x_1 + 1, x_2) - S(x_1, x_2 + 1) + S(x_1 + 1, x_2 + 1).$$

Moreover, the distribution function is

$$F(x_1, x_2) = P(X_1 \leq x_1, X_2 \leq x_2). \tag{1.50}$$

Notice that $S(x_1, x_2)$ is decreasing in x_1 and x_2, $S(0, 0) = 1$, $S(x_1, 0) = P(X_1 \geq x_1)$ and $S(0, x_2) = P(X_2 \geq x_2)$ are the marginal survival functions of X_1 and X_2 denoted by $S_1(x_1)$ and $S_2(x_2)$, and $S(b_1+1, x_2) = S(x_1, b_2+1) = S(b_1+1, b_2+1) = 0$. The marginal probability mass functions are

$$f_1(x_1) = \sum_{t=0}^{b_2} f(x_1, t) \text{ and } f_2(x_2) = \sum_{t=0}^{b_1} f(t, x_2). \tag{1.51}$$

We define the conditional distributions of X_1, given $X_2 = x_2$, and of X_2, given $X_1 = x_1$, as

$$f_{1|2}(x_1|x_2) = \frac{P(X_1 = x_1, X_2 = x_2)}{P(X_2 = x_2)} \text{ when } P(X_2 = x_2) > 0, \tag{1.52}$$

and

$$f_{2|1}(x_2|x_1) = \frac{P(X_1 = x_1, X_2 = X_2)}{P(X_1 = x_1)} \text{ when } P(X_1 = x_1) > 0, \tag{1.53}$$

so that

$$f(x_1, x_2) = f_1(x_1) f_{2|1}(x_2|x_1) = f_2(x_2) f_{1|2}(x_1|x_2).$$

Earlier in Section 1.5, as an example of the univariate compound distributions, we presented the Pólya-Aeppli distribution as the distribution of $N = \sum_{i=1}^{Y} X_i$, where X_i's are i.i.d. Geometric$(1-\rho)$ random variables, having support $0, 1, \ldots,$ and Y is an independent Poisson random variable. By adopting a similar compounding method, Minkova and Balakrishnan (2014a) have constructed a natural bivariate analogue of the Pólya-Aeppli distribution by considering $(N_1, N_2) = (U + W, V + R)$, where $U = \sum_{i=1}^{Z_1} V_i$, $W = \sum_{i=1}^{Z_3} W_i$ and $R = \sum_{i=1}^{R_3} R_i$, with U_i's, V_i's and R_i's being i.i.d. Geometric$(1-\rho)$ random variables having support $0, 1, \ldots,$ and $(Y_1, Y_2) = (Z_1, +Z_3, Z_2 + Z_3)$ being a bivariate Poisson random vector with Y_i's being independent Poisson(λ_i) random variables $i = 1, 2, 3$. The distribution so constructed, referred to as *Type I bivariate Pólya-Aeppli distribution*, has its joint probability generating function as

$$\begin{aligned} P_{N_1,N_2}(s_1, s_2) &= E\left(s_1^{N_1} s_2^{N_2}\right) \\ &= e^{-(\lambda_1+\lambda_2+\lambda_3)} \\ &\quad \cdot \exp\left\{\lambda_1 \frac{(1-\rho)s_1}{1-\rho s_1} + \lambda_2 \frac{(1-\rho)s_2}{1-\rho s_2} + \lambda_3 \frac{(1-\rho)^2 s_1 s_2}{(1-\rho s_1)(1-\rho s_2)}\right\}. \end{aligned}$$

As shown by Minkova and Balakrishnan (2014a), this bivariate distribution is overdispersed with respect to bivariate Poisson distribution. These authors (Minkova and Balakrishnan, 2014b) have also constructed another bivariate analogue of the Pólya-Aeppli distribution by considering $(N_1, N_2) = \left(\sum_{i=1}^{Y} X_{1i}, \sum_{i=1}^{Y} X_{2i}\right)$, where (X_{1i}, X_{2i}) are i.i.d bivariate variables with joint probability generating function $\theta/(1-\theta_1 s_1 - \theta_2 s_2)$, with $0 < \theta_1, \theta_2 < 1$ and $\theta = 1 - \theta_1 - \theta_2 > 0$, and Y is an independent Poisson(λ) random variable. The distribution so constructed, referred to as *Type II bivariate Pólya-Aeppli distribution*, has its joint probability generating function as

$$P_{N_1,N_2}(s_1, s_2) = E\left(s_1^{N_1} s_2^{N_2}\right) = \exp\left\{-\lambda\left(1 - \frac{\theta}{1-\theta_1 s_1 - \theta_2 s_2}\right)\right\}.$$

Interested readers may also refer to the works of Qin et al. (2017) and Balakrishnan et al. (2017) for likelihood inference for these two bivariate Polya-Aeppli distributions.

1.9 MULTIVARIATE WEIGHTED DISTRIBUTIONS

The notion of weighted distributions presented earlier in Section 1.4 has a natural extension to higher dimensions. For a given weight function $w(\mathbf{x})$ which is positive with a non-zero finite expectation, the multivariate weighted probability function corresponding to $f(\mathbf{x})$ is defined as

$$f_w(\mathbf{x}) = \frac{w(\mathbf{x}) f(\mathbf{x})}{E w(\mathbf{X})}. \tag{1.54}$$

The random variable that corresponds to f_w is denoted by $\mathbf{X}_w$. When $w(\mathbf{x}) = x_1 \cdots x_p$, (1.54) is called the *multivariate length-biased distribution* with mass function.

$$f_L(\mathbf{x}) = \frac{x_1 \cdots x_p f(x_1, \ldots, x_p)}{E(X_1 \cdots X_p)}.$$

The marginal and conditional distributions are

$$f_{w,i_1,i_2\ldots,i_m}(x_{i_1},\ldots,x_{i_m})=\frac{Ew(\mathbf{X}|x_{i_1},\ldots,x_{i_m})}{Ew(\mathbf{X})}$$
$$f_{i_1 i_2\ldots i_m}(x_{i_1},\ldots,x_{i_m}),\quad m<p,$$

and

$$f_{w,i_1,\ldots,i_m|j_1,\ldots,j_{p-m}}(x_{i_1},\ldots,x_{i_m}|x_{j_1},\ldots,x_{j_{p-m}})$$
$$=\frac{w(\mathbf{x})f_{i_1,\ldots,i_m|j_1,\ldots,j_{p-m}}(x_{i_1},\ldots,x_{i_m}|x_{j_1},\ldots,x_{j_{p-m}})}{Ew(\mathbf{X}|X=x_{j_1},\ldots,X=x_{j_{p-m}})}.$$

Accordingly the one-dimensional marginal and conditional versions given by

$$f_i^w(x_i)=\frac{Ew(\mathbf{X}|X_i=x_i)}{Ew(\mathbf{X})}f_i(x_i),\ i=1,\ldots,p, \tag{1.55}$$

and

$$f_{w,i|(i)}(x_i|(\mathbf{x}_{(i)}))=\frac{w(x_i)}{Ew(\mathbf{X}|\mathbf{X}_{(i)}=\mathbf{x}_{(i)})}f_{i|(i)}(x_i|\mathbf{x}_{(i)}), \tag{1.56}$$

where

$$\mathbf{x}_{(i)}=(x_1,\ldots,x_{i-1},x_{i+1},\ldots,x_p)\text{ and }(i)=(1,\ldots,i-1,i+1,\ldots,p). \tag{1.57}$$

Earlier in Section 1.4, as an example of the univariate weighted Poisson distribution, we presented the COM-Poisson distribution with probability mass function

$$f(x)=\frac{1}{Z[\lambda,\phi]}\cdot\frac{\lambda^x}{(x!)^\phi},\qquad x=0,1,\ldots,\ \lambda>0,\ \phi\geq 0,$$

where $Z[\lambda,\phi]=\sum_{j=0}^{\infty}\frac{\lambda^j}{(j!)^\phi}$, which includes the well-known geometric, Poisson and Bernoulli distributions as special cases. Recently, Sellers et al. (2016) constructed a natural bivariate generalization of the COM-Poisson distribution as one with joint probability generating function of the form

$$\begin{aligned}P_{X_1,X_2}(t_1,t_2)&\\ &=E\left(t_1^{X_1}t_2^{X_2}\right)\\ &=\frac{Z[\lambda\{1+p_{1+}(t_1-1)+p_{+1}(t_2-1)+p_{11}(t_1-1)(t_2-1)\},\phi]}{Z[\lambda,\phi]},\end{aligned}$$

where $p_{1+}=\frac{\lambda_1+\lambda_3}{\lambda}$, $p_{+1}=\frac{\lambda_2+\lambda_3}{\lambda}$ and $p_{11}=\frac{\lambda_3}{\lambda}$, with $\lambda>\lambda_1+\lambda_2+\lambda_3$. This construct does, in fact, include the three special cases of the bivariate COM-

Poisson distribution just as the univariate case does. For example, when $\phi = 1$, the above bivariate probability generating function $P_{X_1,X_2}(t_1, t_2)$ reduces to

$$\exp\{\lambda p_{1+}(t_1 - 1) + \lambda p_{+1}(t_2 - 1) + \lambda p_{11}(t_1 - 1)(t_2 - 1)\},$$

which is precisely the probability generating function of the bivariate Poisson distribution. Similarly, when $\phi \to \infty$, the above bivariate probability generating function $P_{X_1,X_2}(t_1, t_2)$ reduces to

$$p^*_{00} + p^*_{10}t_1 + p^*_{01}t_2 + p^*_{11}t_1t_2,$$

where $p^*_{00} = \frac{1+\lambda p_{00}}{1+\lambda}$, $p^*_{10} = \frac{1+\lambda p_{10}}{1+\lambda}$, $p^*_{01} = \frac{1+\lambda p_{01}}{1+\lambda}$, and $p^*_{11} = \frac{1+\lambda p_{11}}{1+\lambda}$, which is the probability generating function of the bivariate Bernoulli distribution. Finally, for the case when $\phi = 0$, the above bivariate probability generating function $P_{X_1,X_2}(t_1, t_2)$ becomes

$$\left[1 - \frac{\lambda}{1-\lambda}\{p_{1+}(t_1 - 1) + p_{+1}(t_2 - 1) + p_{11}(t_1 - 1)(t_2 - 1)\}\right]^{-1}$$

for $\lambda < 1$ and $\lambda\{p_{1+}(t_1 - 1) + p_{+1}(t_2 - 1) + p_{11}(t_1 - 1)(t_2 - 1)\} < 1$, which is indeed the probability generating function of the bivariate geometric distribution. Sellers et al. (2016) have then discussed dispersion and inferential issues of the bivariate COM-Poisson distribution, and have demonstrated its flexibility and suitability in analyzing bivariate count data.

In the univariate case, the equilibrium corresponding to X with survival function $S(x)$ was found to be (1.23)

$$f_E(x) = \frac{S(x+1)}{\mu}, \qquad x = 0, 1, 2, \ldots,$$

which is a special case of the univariate weighted distribution in (1.20). When this definition is extended to the multivariate case, there is no unique representation and also the multivariate extensions do not enjoy some of the properties in the univariate case listed in Section 1.4. Discrete equilibrium distributions in higher dimensions do not appear to have received much attention in the literature. The following definitions in the discrete case are, therefore, based on the discussions in the continuous case that appeared in Gupta and Sankaran (1998), Chatterjee and Mukherjee (2001), Nair and Preeth (2008), and Navarro and Sarabia (2010).

Definition 1.1. The *bivariate equilibrium distribution* of a random vector (X_1, X_2) with survival function $S(x_1, x_2)$ is defined as

$$f_E(x_1, x_2) = \frac{S(x_1 + 1, x_2 + 1)}{E(X_1X_2)}, \qquad x_1, x_2 = 0, 1, 2, \ldots, \tag{1.58}$$

provided $E(X_1X_2) < \infty$.

Definition 1.2. The bivariate equilibrium distribution of (X_1, X_2) is defined as the distribution of a random vector (Y_1, Y_2) satisfying

$$g_{1|2}(y_1|Y_2 > y_2) = \frac{P(X_1 > y_1|X_2 > y_2)}{E(X_1|X_2 > y_2)} \tag{1.59}$$

and

$$g_{2|1}(y_2|Y_1 > y_1) = \frac{P(X_2 > y_2|X_1 > y_1)}{E(X_2|X_1 > y_1)} \tag{1.60}$$

where $g_{1|2}$ and $g_{2|1}$ are the conditional probability mass functions of $(Y_1|Y_2 > y_2)$ and $(Y_2|Y_1 > y_1)$, and provided that (1.59) and (1.60) satisfy the compatibility conditions for leading to a bivariate distribution.

Definition 1.3. The equilibrium distribution of (X_1, X_2) is the distribution of a random vector (Z_1, Z_2) such that the conditional probability mass functions

$$g_{1|2}(z_1|Z_2 = z_2) = \frac{P(X_1 > z_1|X_2 = z_2)}{E(X_1|X_2 = z_2)} \tag{1.61}$$

and

$$g_{2|1}(z_2|Z_1 = z_1) = \frac{P(X_2 > z_2|X_1 = z_1)}{E(X_2|X_1 = z_1)} \tag{1.62}$$

provided $g_{1|2}$ and $g_{2|1}$ satisfy the compatibility conditions for them to produce a bivariate distribution.

Definition 1.2 easily extends to multi-dimensional case.

Definition 1.4. The *multivariate equilibrium distribution* of $\mathbf{X}$ with survival function $S(\mathbf{x})$ is defined as the joint distribution of $\mathbf{Y} = (Y_1, \ldots, Y_p)$ determined by the conditional survival functions.

$$R_{1,i}(\mathbf{x}) = P(Y_{1,i} > x_i|\mathbf{Y}_{1,(i)} > \mathbf{x}_{(i)})$$

where $P(X_i > x_i|\mathbf{X}_{(i)} > \mathbf{x}_{(i)})$ are subject to the compatibility condition

$$\frac{P(X_i > x_i|\mathbf{X}_{(i)} > \mathbf{x}_{(i)}))}{P(X_j > x_j|\mathbf{X}_{(j)} > \mathbf{x}_{(j)}} = \frac{A_j(\mathbf{x}_{(j)})}{A_i(\mathbf{x}_{(i)})} \tag{1.63}$$

with A_j and A_i being survival functions.

Considerable literature is available on properties of weighted distributions in the context of reliability analysis in the continuous case. For an in-depth analysis in both cases, the works of those cited earlier along with those of Jain and Nanda (1995), Sunoj and Nair (1998a, 1998b, 1999), and Navarro et al. (2001, 2006) will be most helpful.

1.10 DEPENDENCE MEASURES AND CONCEPTS

While modelling data using multivariate data, an important aspect to be considered is the dependence relations between the constituent variables. A suitable model has to be chosen such that it nearly matches the dependence relations in the observed data. This will safeguard the proximity between the empirical evidence and the data generating mechanism revealed through the model. In general, there are three approaches to assess the dependence or association between the random variables. In the first approach, certain scalar measures or indices are used to indicate the extent of association. These are global measures that give an overall picture of the association and can be useful for comparison purposes. A more involved method is through dependence concepts that are stochastic in nature, being defined in terms of survival or distribution functions or expected values. A third approach is to use time-dependent measures. In various studies on bivariate survival times, measures of association indexed by the ages of the organisms or systems provide means of assessing the influence of certain important factors on their lives; see Anderson et al. (1992) for examples and detailed discussion. Measures that are functions of x and y associated with the lifetime random vector (X, Y) are called time-dependent measures in our future discussions. A redeeming feature of time-dependent measures in the context of reliability and survival analysis is that sometimes they may be related to ageing properties.

1.10.1 Measures of Dependence

The most popular measure of dependence is the coefficient of correlation between two random variables X_1 and X_2 given by

$$r = \frac{\mathrm{Cov}\,(X_1, X_2)}{[V(X_1)V(X_2)]^{\frac{1}{2}}} \tag{1.64}$$

defined earlier in (1.46). The range of values of r is $[-1, 1]$, where $r = \pm 1$ occurs only when X_1 and X_2 are perfectly linearly related. When X and Y are independent $r = 0$, but the converse is not true. Correlation coefficient is invariant under a change of origin in the variables or multiplication by constants of the same sign. We say that X_1 and X_2 are positively associated if $0 < r \leq 1$, negatively associated if $-1 \leq r < 0$, and uncorrelated if $r = 0$. However, caution has to be exercised in assessing the degree of association based on the values of r alone. For a detailed discussion on the demerits of correlation coefficient, see Barnett (1979) and Elffers (1980). The coefficient of correlation is estimated from a sample of n bivariate observations, $(x_{11}, x_{21}), \ldots, (x_{1n}, x_{2n})$, by the formula

$$\hat{r} = \frac{\sum_{i=1}^{n}(x_{1i} - \overline{x}_1)(x_{2i} - \overline{x}_2)}{[\sum_{i=1}^{n}(x_{1i} - \overline{x}_1)^2 \sum_{i=1}^{n}(x_{2i} - \overline{x}_2)^2]^{\frac{1}{2}}},$$

where $\overline{x}_1$ and $\overline{x}_2$ are the sample means of $(x_{11}, \ldots, x_{1n})$ and $(x_{21}, \ldots, x_{2n})$, respectively.

Kendall's Tau

Let (x_1, y_1) and (x_2, y_2) be two observations from (X_1, X_2). Then, we say that the observations are concordant if $(x_1 - x_2)(y_1 - y_2) > 0$ and discordant if $(x_1 - x_2)(y_1 - y_2) < 0$. In the population version, let (X_1, Y_1) and (X_2, Y_2) be independent and identically distributed continuous random variables with distribution function $F(x, y)$. In the continuous case, Kendall's tau τ is defined as the difference between probability of concordance and probability of discordance; that is,

$$\begin{aligned}\tau &= P[(X_1 - X_2)(Y_1 - Y_2) > 0] - P[(X_1 - X_2)(Y_1 - Y_2) < 0] \\ &= 2P[(X_1 - X_2)(Y_1 - Y_2) > 0] - 1.\end{aligned}$$

In the discrete case, the above definition does not suffice as $P(X_1 = X_2$ or $Y_1 = Y_2)$ may have positive value. Since

$$P(\text{concordance}) + P(\text{discordance}) + P(\text{ties}) = 1,$$

the definition in the continuous case is modified as (Bouezmarni et al., 2009)

$$\begin{aligned}\tau &= 2P(\text{concordance}) - 1 + P(\text{ties}) \\ &= 4P(X_2 < X_1, Y_2 < Y_1) - 1 + P(X_1 = X_2 \text{ or } Y_1 = Y_2).\end{aligned} \tag{1.65}$$

If X_1 and X_2 are independent, $\tau = 0$, but the converse need not be true. Also, τ lies between -1 and 1 with $\tau = \pm 1$ when there is perfect concordance or discordance.

Spearman's Rho

This measure is also based on concordance and discordance, but with independent random vectors (X_1, Y_1), (X_2, Y_2) and (X_3, Y_3) which have a common distribution function F and marginals F_1, F_2 and F_3. The Spearman's rho is proportional to the probability of concordances minus the probability of discordance for the pairs (X_1, Y_1) and (X_2, Y_3) or (X_3, Y_3), i.e.,

$$\rho = 3[P(X_1 - X_2)(Y_1 - Y_2) > 0 - P((X_1 - X_3)(Y_1 - Y_3) < 0)]. \tag{1.66}$$

For discrete marginals, the possible non-zero values of ties create problems in the application of the above formula for the continuous case. Neslehova (2007) suggests a rescaling of the expression for ρ to arrive at

$$\rho_1 = \frac{\rho(X_1, X_2)}{[\{1 - \sum_{x_1} f_1(x_1)^3\}\{1 - \sum_{x_2} f_2(x_2)^3\}]^{\frac{1}{2}}},$$

where $f_1(x_1) = P(X_1 = x_1)$ and $f_2(x_2) = P(X_2 = x_2)$. A distinguishing property of ρ_1 is that its sample analogue is the correlation coefficient r of the mid-ranks $\frac{1}{n}\sum_{k=1}^{n} p_k$, where $p_1, \ldots, p_n$ are ranks (if there were no ties). If r_i and s_i are the ranks of x_i and y_i, respectively, in a random sample (x_i, y_i), $i = 1, \ldots, n$, a commonly used formula for the sample Spearman's measure is

$$\hat{\rho} = 1 - \sum_{i=1}^{n} \frac{6(r_i - s_i)^2}{n(n^2 - 1)}.$$

In the case of ties, the mid-ranks are used. As is the case with τ, ρ is defined for every pair of random variables, taking values in $[-1, 1]$ and $\rho = 0$ if X and Y are independent.

Blomqvist's β

The Blomqvist measure is also based on concordance, but with the difference that instead of two random vectors here we choose a random vector and a fixed point (x_0, y_0) with x_0 as the median of X_1 and y_0 as the median of X_2. Thus, the measure in the case of a continuous random vector (X_1, X_2) becomes

$$\begin{aligned}\beta &= P[(X_1 - x_0)(X_2 - y_0) > 0] - P[(X_1 - x_0)(X_2 - y_0) < 0] \\ &= 2\{P[X_1 < x_0, X_2 < y_0] + P[X_1 > x_0, X_2 > y_0]\} - 1.\end{aligned}$$

In the discrete case, this modifies to

$$\beta = 4P[X_1 < x_0, X_2 < y_0] - 1 + P[X_1 = x_0 \text{ or } X_2 = y_0]. \tag{1.67}$$

The medians are chosen to be the nearest integer values.

1.10.2 Dependence Concepts

The concept of dependence is usually seen to belong to three categories. We say that X_1 and X_2 are positively dependent when large values of X_1 correspond to large values of X_2. Similarly, negative dependence implies that large values of X_1 correspond to small values of X_2. The third is lack of dependence or independence when $F(x, y) = F_1(x)F_2(y)$ or its other equivalent forms. We mention only the positive dependence concepts here as the definition of negative dependence will be obvious from the former.

(a) *Positive correlation.* As mentioned earlier, X_1 and X_2 are positively associated if $\text{Cov}(X_1, X_2) \geq 0$;

(b) *Positively Quadrant Dependent.* We say that two discrete random variables X_1 and X_2 are positively quadrant dependent $(PQD) \iff$ for all $x_1, x_2 = 0, 1, 2, \ldots,$

$$P(X_1 \leq x_1, X_2 \leq x_2) \geq P(X_1 \leq x_1)P(X_2 \leq x_2), \tag{1.68}$$

or equivalently

$$P(X_1 > x_1, X_2 > x_2) \geq P(X_1 > x_1)P(X_2 > x_2).$$

When the inequality is reversed, we have negatively quadrant dependence (NQD). The sense conveyed by PQD is that the probability that X_1 and X_2 are simultaneously small or large is at least as large as it would be if X_1 and X_2 were independent;

(c) *Positive association.* The random variables X_1 and X_2 are positively associated if for every pair of increasing functions p and q, we have

$$\text{Cov}(p(X_1, X_2), q(X_1, X_2)) \geq 0;$$

(d) *Tail monotonicity.*

(i) X_2 is right-tail increasing in $X_1(RTI(X_2|X_1))$ iff $P(X_2 > x_2|X_1 > x_1)$ is a non-decreasing function of x_1 for all x_2;

(ii) X_1 is right-tail increasing in $X_2(RTI(X_1|X_2))$ iff $P(X_1 > x_1|X_2 > x_2)$ is a non-decreasing function in x_2 for all x_1;

(iii) X_2 is left-tail decreasing in $X_1(LTD(X_2|X_1))$ iff $P(X_2 \leq x_2|X_1 \leq x_1)$ is a non-increasing function in x_1 for all x_2;

(iv) X_1 is left-tail decreasing in $X_2(LTD(X_1|X_2))$ iff $P(X_1 \leq x_1|X_1 \leq x_2)$ is a non-increasing function in x_2 for all x_1.

Intuitively, (i) indicates that the probability that X_2 has a small value decreases as the value of x_1 is non-decreasing. Just as in (i)–(iv), we can have left-tail increasing (LTI) and right-tail decreasing (RTD) concepts by replacing non-decreasing with non-increasing and vice-versa;

(e) *Stochastic monotonicity.* Replacing the conditioning events $X_i > x_i$ by $X_i = x_i$ in (d), we have various notions of stochastic monotonicity:

(i) X_2 is stochastic increasing in X_1 $(SI(X_2|X_1))$ if $P[X_2 > x_2|X_1 = x_1]$ is non-decreasing in x_1 for all x_2, or if $P[X_2 \leq x_2|X_1 = x_1]$ is non-increasing in x_1 for all x_2;

(ii) X_1 is stochastic increasing in X_2 $(SI(X_1|X_2))$ if $P[X_1 > x_1|X_2 = x_2]$ is non-decreasing in x_2 for all x_1, or if $P[X_1 \leq x_1|X_2 = x_2]$ is non-increasing in x_2 for all x_1.

We can define stochastic decreasing (SD($X_i|X_{3-i}$)), $i = 1, 2$, similarly by saying that $P[X_i > x_i|X_{3-i} = x_{3-i})$ is non-increasing in x_{3-i} for all x_i. In the literature, the above property is also referred to as *positive regression dependent* (PRD). Thus, for example, $SD(X_2|X_1) \iff PRD(X_2|X_1)$, and so on.

(f) *Total positivity*. Recall the definition of total positivity of order 2 (TP-2) given in Section 1.7. The joint probability mass function $f(x_1, x_2)$ of a bivariate random vector is totally positive of order 2 if, for all $x_1 < x_2$ and $y_1 < y_2$,

$$f(x_1, y_1)f(x_2, y_2) \geq f(x_1, y_2)f(x_2, y_1). \tag{1.69}$$

When f is TP-2, the distribution functions and survival functions are also TP-2. Hence,

$$F(x_1, y_1)F(x_2, y_2) \geq F(x_1, y_2)F(x_2, y_1)$$

and

$$S(x_1, y_1)S(x_2, y_2) \geq S(x_1, y_2)S(x_2, y_1).$$

Sometimes, an alternative name positively likelihood ratio dependent (PLRD) is also used for TP-2. The negative dependence concept corresponding to TP-2 is called *reverse regular of order* 2, abbreviated as RR-2. We define RR-2 by reversing the inequality in (1.69) as

$$f(x_1, y_1)f(x_2, y_2) \leq f(x_1, y_2)f(x_2, y_1) \tag{1.70}$$

(g) *Corner set monotonicity.* The definition of corner set monotonicities bear analogy with tail monotonicities with the change that the joint probabilities are considered here:

(i) The pair (X_1, X_2) is left-corner set decreasing, LCSD (X_1, X_2) (increasing, LCSI(X_1, X_2)), if

$$P[X_1 \leq x_1, X_2 \leq x_2|X_1 \leq x_1', X_2 \leq x_2']$$

is non-increasing (non-decreasing) in x_1' and in x_2' for all x_1 and x_2;

(ii) The pair (X_1, X_2) is right-corner set increasing, RCSI (X_1, X_2), if

$$P[X_1 > x_1, X_2 > x_2|X_1 > x_1', X_2 > x_2']$$

is non-decreasing in x_1' and in x_2' for all x_1 and x_2. The dual concept right corner set decreasing, RCSD (X_1, X_2), is obtained by changing non-decreasing in RCSI to non-increasing;

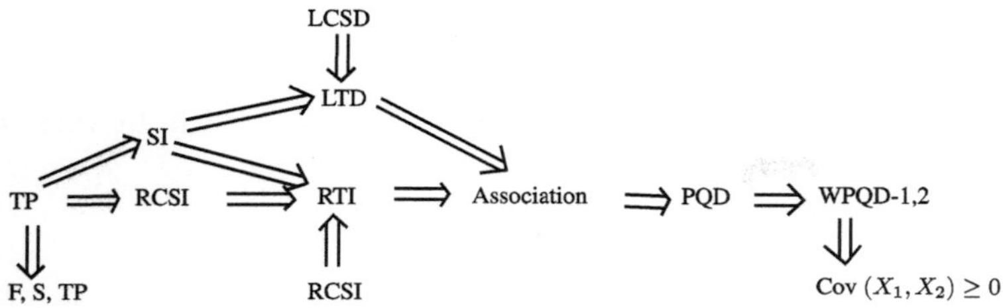

FIGURE 1.1 Implications among different measures of association.

Obviously, LCSD is equivalent to

$$F(x_1, x_2)F(x_1', x_2') \geq F(x_1, x_2')F(x_1', x_2),$$

while RCSI is equivalent to

$$S(x_1, x_2)S(x_1', x_2') \geq S(x_1, x_2')S(x_1', x_2).$$

(h) *Weak quadrant dependence.* A weaker concept than PQD is obtained by considering an integrated version of the PQD definition. Accordingly, (X_1, X_2) is weakly positive quadrant dependent of type I (WPQD-1) if

$$\sum_{t_1=x_1}^{\infty} \sum_{t_2=x_2}^{\infty} [P(X_1 > t_1, X_2 > t_2) - P(X_1 > t_1)P(X_2 > t_2)] \geq 0$$

for all (x_1, x_2), and weakly quadrant dependent of type II (WPQD-2)

$$\sum_{t_1=-\infty}^{x_1} \sum_{t_1=-\infty}^{x_2} [P(X_1 > t_1, X_2, t_2) - P(X_1 > t_1)P(X_2 > t_2)] \geq 0.$$

A detailed discussion of the above concepts, their properties and interrelationships can be found in Lehmann (1966), Esary et al. (1967), Harris (1970) and Alzaid (1990) where they all deal with the continuous case. All these concepts are interrelated as shown in Fig. 1.1.

Notice that the same implications apply when the random variables involved are discrete or continuous.

1.10.3 Multivariate Dependence

Many of the dependence concepts in the multivariate case are straightforward extensions of the bivariate case. In this section, we take $\mathbf{X} = (X_1, \ldots, X_p)$ and $\mathbf{Y} = (Y_1, \ldots, Y_p)$ to denote two p-dimensional vectors.

1. The vector $\mathbf{X}$ has conditionally increasing sequence (CIS) if $P(X_i > x_i | X_1 = x_1, \ldots, X_{i-1} = x_{i-1})$ is increasing in $x_1, \ldots, x_{i-1}$, for $i = 1, \ldots, p$. Also, $\mathbf{X}$ is conditionally increasing (CI) if $X_{i_1} \cdots X_{i_p}$ is CIS for every permutation $(i_1, \ldots, i_p)$ of $(1, \ldots, p)$;
2. $\mathbf{X}$ is left-tail decreasing in sequence (LTDS) if

$$P(X_i > x_i | X_1 \leq x_1, \ldots, X_{i-1} \leq x_{i-1})$$

is decreasing in $x_1, \ldots, x_{i-1}$ for all $x_i, i = 1, \ldots, p$, and right-tail increasing in sequence (RTIS) if

$$P(X_i > x_i | X_1 > x_1, \ldots, X_{i-1} > x_{i-1})$$

is increasing in $x_1, \ldots, x_{i-1}$ for all $x_i, i = 1, \ldots, p$. We say $\mathbf{X}$ is multivariate left-tail decreasing (MLTD) if $(X_{i_1} \cdots X_{i_p})$ is LTDS for all permutations $(i_1, \ldots, i_p)$ of $(1, \ldots p)$. Similarly, MRTI is defined using RTIS;
3. We say that $\mathbf{X}$ is multivariate LCSD (MLCSD) if $P(\mathbf{X} \leq \mathbf{x} | \mathbf{X} < \mathbf{x}')$ is decreasing in $\mathbf{x}'$ for all $\mathbf{x}$. Also, $\mathbf{X}$ is multivariate RCSI (MRCSI) if $P(\mathbf{X} > \mathbf{x} | \mathbf{X} > x')$ is increasing in $\mathbf{x}'$ for all $\mathbf{x}$;
4. $\mathbf{X}$ is positively lower orthant dependent (PLOD), if for all $\mathbf{x}$,

$$P(\mathbf{X} \leq \mathbf{x}) \geq \prod_{i=1}^{p} P(X_i \leq x_i),$$

and $\mathbf{X}$ is positively upper orthant dependent (PUOD), if for all $\mathbf{x}$,

$$P(\mathbf{X} > \mathbf{x}) \geq \prod_{i=1}^{p} P(X_i > x_i).$$

If $\mathbf{X}$ is both PLOD and PUOD, we say that $\mathbf{X}$ is positively orthant dependent or POD. Analogously, negative dependence NLOD, NUOD and NOD are defined by reversing the inequalities in the relationships involving $P(\mathbf{X} \leq \mathbf{x})$ and $P(\mathbf{X} > \mathbf{x})$ with $P(X_i \leq x_i)$ and $P(X_i > x_i)$, respectively;
5. The random vector $\mathbf{X}$ is
   - **(i)** LTD $(\mathbf{X}_A | \mathbf{X}_{A^c})$ if $P(\mathbf{X}_A \leq \mathbf{x}_A | \mathbf{X}_{A^c} \leq \mathbf{x}_{A^c})$ is non-increasing in $\mathbf{x}_{A^c}$ for all $\mathbf{x}_A$, where A is a subset of $(1, \ldots, p)$ and A^C is the complement of A;
   - **(ii)** RTI $(\mathbf{X}_A | \mathbf{X}_{A^c})$ if $P(\mathbf{X}_A > \mathbf{x}_A | \mathbf{X}_{A^c} > \mathbf{x}_{A^c})$ is non-decreasing in $\mathbf{x}_{A^c}$ for all $\mathbf{x}_A$;
   - **(iii)** SI $(\mathbf{X}_A | \mathbf{X}_{A^c})$ if $P(\mathbf{X}_A > \mathbf{x}_A | \mathbf{X}_{A^c} = \mathbf{x}_{A^c})$ is non-decreasing in $\mathbf{x}_{A^c}$ for all $\mathbf{x}_A$;

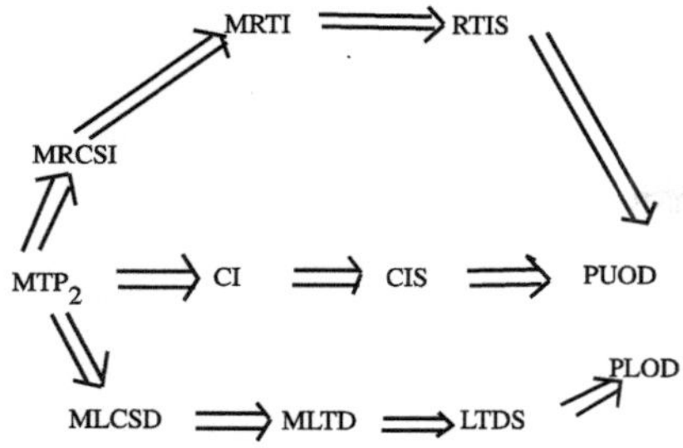

FIGURE 1.2 Interdependence between various dependence concepts.

(iv) Positive dependent through stochastic ordering (PDS) when $\mathrm{SI}(\mathbf{X}_A|\mathbf{X}_{A^c})$ holds for all $A^c = (i)$, $i = 1, \ldots, p$. In the special case when $p = 2$, both POD and CIS in (i) above are equivalent to $\mathrm{SI}(X_2|X_1)$ and $\mathrm{SI}(X_1|X_2)$, respectively;

(v) A function $h(.)$ is said to be multivariate totally positive of order $2(\mathrm{MTP}_2)$ if

$$h(\mathbf{x})h(\mathbf{y}) \le h(\mathbf{x} \vee \mathbf{y})h(\mathbf{x} \wedge \mathbf{y}), \qquad \text{for all } \mathbf{x}, \mathbf{y},$$

where

$$\mathbf{x} \vee \mathbf{y} = (\max(x_1, y_1), \ldots, \max(x_p, y_p))$$

and

$$\mathbf{x} \wedge \mathbf{y} = (\min(x_1, y_1), \ldots, \min(x_p, y_p)).$$

Using this definition, the joint probability mass function f is MTP_2 (positively likelihood ratio dependent) if the above inequality holds good for f.

The interdependence between the various concepts are shown in Fig. 1.2. For a detailed discussion on the above concepts and their properties, we refer to Colangelo et al. (2005) and Joe (1997).

1.10.4 Time-Dependent Measures

A. Clayton (1978) proposed a time-dependent measure whose discrete version is taken as

$$\begin{aligned}\theta(x_1, x_2) &= \frac{P(X_1 = x_1, X_2 = x_2)\, P(X_1 \ge x_1, X_2 \ge x_2)}{P(X_1 = x_1, X_2 \ge x_2)\, P(X_1 \ge x_1, X_2 = x_2)} \\ &= \frac{f(x_1, x_2)S(x_1, x_2)}{[S(x_1, x_2) - S(x_1 + 1, x_2)][S(x_1, x_2) - S(x_1, x_2 + 1)]}.\end{aligned} \tag{1.71}$$

When X_1 and X_2 are positively (negatively) associated, $\theta(x_1, x_2) > (<)1$ and $\theta = 1$ implies independence of X_1 and X_2. The measure in (1.71) is related to the positive dependence concept stated as $\theta(x_1, x_2) > 1$ if and only if (X_1, X_2) is RCSI;

B. Anderson et al. (1992) suggested two measures one of which is given by

$$\phi(x_1, x_2) = \frac{E(X_1 - x_1 | X_1 > x_1, X_2 > x_2)}{E(X_1 - x_1 | X_1 > x_1)}. \tag{1.72}$$

Values of ϕ very different from unity indicate strong association between X_1 and X_2. If X_1 and X_2 are positively associated, $\phi(x_1, x_2)$ should increase for increasing values of x_2. Also, $\phi(x_1, x_2) = 1$ if and only if X_1 and X_2 are independent and geometrically distributed. The second measure is a ratio of survival functions

$$\Psi(x_1, x_2) = \frac{P(X_1 \geq x_1 | X_2 \geq x_2)}{P(X_1 \geq x_1)} = \frac{S(x_1, x_2)}{S_1(x_1)S_2(x_2)}. \tag{1.73}$$

When X_1 and X_2 are independent, $\Psi = 1$ and large values of Ψ indicate positive association;

C. Bjerve and Doksum (1993) defined a local non-parametric dependence function which measures the strength of association between X_1 and X_2 as a function of x_1. It is defined by

$$\rho(x_1) = \frac{\sigma_1 \mu_1'(x_1)}{[(\sigma_1 \mu_1'(x_1))^2 + \sigma^2(x_1)]^{\frac{1}{2}}} \tag{1.74}$$

where $\mu_1^1(x_1) = E[X_2 | X_1 = x_1], \sigma^2(x_1) = V(X_2 | X_1 = x_1)$ with V standing for the variance, and $\sigma_1^2 = V(x_1)$. The measure $\rho(x_1)$ provides the strength of relationship between X_1 and X_2 and is useful in the case of non-normal models. It reduces to the usual correlation coefficient in the normal case. We note that $-1 \leq \rho(x) \leq 1$ and $\rho(x) = 0$ for all x when X_1 and X_2 are independent. The measure in (1.74) was intended for use in the continuous case as it involves the derivative $\mu_1'(x_1)$ of the regression function $\mu_1(x_1)$. However, in the discrete case, when $X_1, X_2 = 0, 1, \ldots,$ the approximation $\mu_1'(x_1) = \mu_1(x_1 + 1) - \mu_1(x_1)$ can be employed. The same authors have proposed another dependence function, called the conditional correlation curve, as

$$\zeta(x_1) = \frac{\sigma_1 \mu_1'(x_1)}{\sigma(x_1)}, \tag{1.75}$$

which can also be used in the discrete case upon replacing $\mu_1'(x_1)$ by $\mu_1(x + 1) - \mu_1(x_1)$ in (1.75);

D. Replacing $E(X_1)$ and $E(X_2)$ by the conditional expectations $E(X_1|X_2 = x_2)$ and $E(X_2|X_1 = x_1)$, respectively, in the Pearson correlation coefficient, Bairamov et al. (2003) considered the function

$$H(x_1, x_2) = \frac{E[(X_1 - E(X_1|X_2 = x_2))(X_2 - E(X_2|X_1 = x_1))]}{[E(X_1 - E(X_1|X_2 = x_2))^2 E(X_2 - E(X_2|X_1 = x_1))^2]^{\frac{1}{2}}}, \tag{1.76}$$

which was interpreted as a local dependence function. Alternatively,

$$H(x_1, x_2) = \frac{r + a_{X_1}(x_2)a_{X_2}(x_1)}{[1 + a^2_{X_1}(x_2)]^{\frac{1}{2}}[1 + a^2_{X_2}(x_1)]^{\frac{1}{2}}},$$

where r is the Pearson correlation coefficient and

$$a_{X_i}(x_{3-i}) = \frac{E(X_i - E(X_i|x_{3-i} = x_{3-i})}{\sigma_{X_i}}, \quad i = 1, 2,$$

and $\sigma^2_{X_i}$ is the variance of X_i. When X_1 and X_2 are independent, $H(x_1, x_2) = 0$, $-1 \leq H(x_1, x_2) \leq 1$ for all (x_1, x_2), and if $\rho = \pm 1$, $H(x_1, x_2) = \pm 1$.

E. Another measure of association in terms of conditional expectations proposed by Nair and Sankaran (2010a), which is valid in the discrete case also, is

$$\alpha(x_1, x_2) = \frac{E[(X_1 - x_1)(X_2 - x_2)|X_1 > x_1, X_2 > x_2]}{E(X_1 - x_1|X_1 > x_1 X_2 > x_2)E(X_2 - x_2|X_1 > x_1, X_2 > x_2)}. \tag{1.77}$$

The vector (X_1, X_2) is positively (negatively) associated if $\alpha(x_1, x_2) > (<) 1$, while there is no association if $\alpha(x_1, x_2) = 1$.

1.11 SCHUR CONVEXITY AND CONCAVITY

Let $g(X_1, X_2)$ be a real-valued function defined on N^2, $N = 0, 1, \ldots$. Then, $g(x_1, x_2)$ is said to be a *Schur convex function* if

$$(x_1 - x_2)[\Delta_1 g(x_1, x_2) - \Delta_2 g(x_1, x_2)] \geq 0,$$

Δ_i being the difference with respect to x_i, or

$$(x_1 - x_2)[g(x_1 + 1, x_2) - g(x_1, x_2 + 1)] \leq 0.$$

When the above inequality is reversed, we say that $g(x_1, x_2)$ is Schur-concave; that is, if g is Schur-convex, $-g$ is Schur-concave. A Schur-constant function is one which is both Schur-concave and Schur-convex. For example,

$$g(x_1, x_2) = x_1^2 + x_2^2$$

is Schur-convex, while

$$\sum_{i=1}^{2} p_i \log \frac{1}{p_i}$$

is Schur-concave. It is easy to see that $g(x_1, x_2) = k(x_1 + x_2)$ is a Schur-constant function.

Suppose $\mathbf{a} = (a_1, \ldots, a_p)$ and $\mathbf{b} = (b_1, \ldots, b_p)$ are two vectors where the components are arranged in descending order. Then we say that **a** weakly dominates or majorizes **b** if $\sum_{i=1}^{q} a_i \leq \sum_{i=1}^{q} b_i$ for $q = 1, \ldots, p$, and write $\mathbf{a} >_w \mathbf{b}$. Further, if $\mathbf{a} >_w \mathbf{b}$ and $\sum_{i=1}^{p} a_i = \sum_{i=1}^{p} b_i$, the vector **a** is said to be majorized by **b**, denoted by $\mathbf{a} > \mathbf{b}$. For example,

$$(6, 0, 0) > (3, 3, 0).$$

The majorization order given above is related to Schur-convexity. Our function $g(x_1, x_2)$ is such that when (x_1, x_2) is majorized by (y_1, y_2), $g(x_1, x_2) \leq g(y_1, y_2)$ and g is Schur-convex. The interest in majorization concept in statistics is that if $\mathbf{a} > \mathbf{b}$, then **a** is more spread out than **b**.

Chapter 2

Basic Reliability Concepts

2.1 INTRODUCTION

As mentioned in the last chapter, the reliability of a device is the probability that the device performs its intended function for a given period of time under conditions specified for its operation. When the device does not perform its function satisfactorily, we say that it has failed. When the random variable X represents the lifetime of a device, the observation on X is realized as the time of failure. The primary concern in reliability theory is then to understand the pattern in which failures occur for different devices and under varying operating environments. This is often done by analyzing the observed failure times or ages at failure with the help of a model that satisfactorily represents the predominant features of the data. One direct method is to find a probability distribution that provides a reasonable fit to the observations. Sometimes, there may exist more than one distribution that can pass an appropriate goodness-of-fit test. In any case, it is more desirable to find a probability model that manifests certain physical properties of the failure mechanism. In reliability theory, some basic concepts that help in the study of failure patterns have been developed. The objective of this chapter is to define such concepts and discuss their properties and inter-relationships. Two important aspects that necessitate the study of these concepts are (a) various functions considered in this context determine the life distribution uniquely, so that the knowledge of their functional form is equivalent to that of the distribution itself, and (b) it should be easier to deal with these functions than the distribution function or probability density function of the corresponding distributions.

2.2 HAZARD RATE FUNCTION

Let X be a discrete random variable assuming values in $N = (0, 1, \ldots)$ with probability mass function $f(x)$ and survival function $S(x) = P(X \geq x)$. We will think of X as the random lifetime of a device that can fail only at times (ages) in N. The *hazard rate function* of X is defined as

$$h(x) = \frac{f(x)}{S(x)} \tag{2.1}$$

Reliability Modelling and Analysis in Discrete Time. http://dx.doi.org/10.1016/B978-0-12-801913-9.00002-6

at points x for which $S(x) > 0$. Treated as a function of x, the hazard rate is also called failure rate, instantaneous death rate, force of mortality and intensity function in other disciplines such as survival analysis, actuarial science, demography, extreme value theory and bio-sciences. Although in the continuous case, the concept of hazard rate dates back to historical studies in human mortality, its discrete version came up much later in the works of Barlow and Proschan (1965), Cox (1972) and Kalbfleisch and Prentice (2002), to mention a few.

When X has a finite support $(0, 1, \ldots, n)$, $n < \infty$, then $h(n) = 1$. As a convention we take $h(x) = 1$ for $x > n$. The hazard function $h(x)$ is interpreted as the conditional probability of the failure of the device at age x, given that it did not fail before age x. Thus, $0 \leq h(x) \leq 1$. The interpretation and boundedness of the discrete hazard rate is thus different from that of the continuous case.

We see from (2.1) that $h(x)$ is determined from $f(x)$ or $S(x)$. The converse that the hazard rate function determines the distribution of X uniquely is also true. To see this, we note that

$$h(x) = \frac{S(x) - S(x+1)}{S(x)},$$

or

$$\frac{S(x+1)}{S(x)} = 1 - h(x).$$

So,

$$S(x) = \begin{cases} \prod_{t=0}^{x-1} (1 - h(t)), & x \geq 1, \\ 1, & x = 0. \end{cases} \tag{2.2}$$

Eq. (2.2) reveals also that $h(x)$ can be used as a tool to model the life distribution. When a functional form for $h(x)$ is assumed as model for a given data set, one has to ensure that the assumed form conforms to the hazard rate function of a distribution. The following theorem is useful in this regard.

Theorem 2.1. *A necessary and sufficient condition that $h : N \to [0, 1]$ is the hazard rate function of a distribution with support N is that $h(x) \in [0, 1]$ for $x \in N$ and $\sum_{x=0}^{\infty} h(t) = \infty$.*

If the probability mass function is required from (2.1) and (2.2), we see that

$$f(x) = h(x) \prod_{t=0}^{x-1} (1 - h(t)). \tag{2.3}$$

The above results have appeared repeatedly in several papers; see, for example, Gupta (1979), Shaked et al. (1995) and Kemp (2004).

Example 2.1. Let X follow the geometric distribution with parameter p $(G(p))$ specified by

$$f(x) = q^x p, \quad x \in N,\ 0 < p < 1,\ q = 1 - p. \tag{2.4}$$

Then,

$$S(x) = q^x$$

so that $h(x) = p$, a constant for all x.

Example 2.2. Consider the Waring distribution with parameters a and b $(W(a, b))$ having probability mass function

$$f(x) = \frac{(a-b)(b)_x}{(a)_{x+1}}, \quad x \in N,\ a > b, \tag{2.5}$$

developed by Irwin (1968, 1975a, 1975b, 1975c), where

$$(b)_x = b(b-1)\cdots(b-x+1) = \frac{\Gamma(b+1)}{\Gamma(b-x+1)}$$

is the Pochammer's symbol. In studying the properties of this distribution, we need the Waring expansion

$$\frac{1}{x-a} = \frac{1}{x} + \frac{a}{x(x+1)} + \frac{a(a+1)}{x(x+1)(x+2)} + \cdots \tag{2.6}$$

which converges for $x > a$. Then,

$$\begin{aligned} S(x) &= (a-b)\sum_{t=x}^{\infty} \frac{(b)_t}{(a)_{t+1}} \\ &= (a-b)\frac{(b)_x}{(a)_x}\left[\frac{1}{a+x} + \frac{b+x}{(a+x)(a+x+1)} + \cdots\right] \\ &= \frac{(b)_x}{(a)_x}, \quad \text{by (2.6).} \end{aligned} \tag{2.7}$$

Hence, we get in this case

$$h(x) = \frac{a-b}{a+x}. \tag{2.8}$$

Example 2.3. A special case of the negative hyper geometric law with parameters n and k is defined by the probability mass function

$$f(x) = \frac{\binom{-1}{x}\binom{-k}{n-x}}{\binom{-1-k}{n}}, \quad x = 0, 1, \ldots, n,\ k > 0. \tag{2.9}$$

Then,

$$S(x) = \sum_{t=x}^{n} f(t) = \sum_{t=x}^{n} \binom{k+n-t-1}{n-t} \Big/ \binom{k+n}{n} \tag{2.10}$$

on using the identity

$$\binom{-p}{k} = (-1)^k \binom{p+k-1}{k}.$$

So,

$$S(x) = \sum_{t=0}^{n-x} \binom{k+t-1}{t} \Big/ \binom{k+n}{n}.$$

Now to find the sum on the right hand side, the combinatorial expression (Riordan, 1968)

$$\sum_{x=0}^{n} \binom{a+n-x-1}{n-x} = \binom{a+n}{n}$$

is employed in order to obtain

$$S(x) = \binom{k+n-x}{n-x} \Big/ \binom{k+n}{n}. \tag{2.11}$$

We then find

$$h(x) = \frac{\binom{k+n-x-1}{n-x}}{\binom{k+n-x}{n-x}} = \frac{k}{k+n-x}.$$

The distribution in (2.11) will be denoted by NH (n, k). For more functional forms of $h(x)$ that characterize various distributions, see Table 3.2.

Remark 2.1. The geometric, Waring and negative hyper-geometric models form a set of models possessing some attractive properties for their reliability characteristics, in as much the same way as the exponential, Pareto II and rescaled beta distributions in the continuous case. The results in the above examples show that the models (2.4), (2.5) and (2.8) have hazard rates of the form

$$h(x) = (lx + m)^{-1} \tag{2.12}$$

with $l = 0$ for geometric, $l = \frac{a}{a-b} > 0$ for Waring, and $l = -\frac{1}{k} < 0$ for negative hyper-geometric distributions.

Remark 2.2. Using (2.3), it can be seen that a reciprocal linear hazard rate function in (2.12) characterizes the above three distributions. A direct proof of this fact is available in Xekalaki (1983).

Remark 2.3. When Remark 2.1 is employed in a practical problem, it should be borne in mind that the support of X is N. Thus, the geometric distribution has a constant hazard rate means that a device with such a lifetime distribution does not age. However, when the support of X is $n < \infty$, $h(x) = p$ leads to

$$f(x) = \begin{cases} q^x p, & x = 0, 1, 2, \ldots, n-1, \\ q^n, & x = n. \end{cases}$$

If X has bounded support $(0, 1, 2, \ldots, n)$, (2.12) is satisfied with $f(0) > 0$, then $l > 1$ and $m < 0$ and $n = \frac{1-m}{l}$. This follows from the facts that at $x = 0$, $\frac{1}{l} < 1$ and at $x = n$, $ln + m = 1$.

The sum of the hazard rates from 0 through $x - 1$ is of interest in reliability theory and is called the cumulative hazard rate, defined by

$$H(x) = \sum_{t=0}^{x-1} h(t). \tag{2.13}$$

Also define $H(0) = 0$. Graphically, the cumulative hazard rate represents the area under the step function representing $h(x)$. Definition (2.13) does not satisfy properties analogous to the continuous case in which the cumulative hazard rate satisfies the identity

$$H(x) = -\log \bar{F}(x), \quad \bar{F}(x) = P(X > x).$$

Therefore, Cox and Oakes (1984) proposed an alternative definition of cumulative hazard rate in the form

$$H_1(x) = -\log S(x). \tag{2.14}$$

This means that

$$H_1(x) = -\log \sum_{t \geq x} f(t) = -\log \prod_{t=0}^{x-1} (1 - h(t)). \tag{2.15}$$

If

$$H_1(x) = \sum_{t=0}^{x-1} h_1(t), \quad x = 1, 2, \ldots \tag{2.16}$$

then $H_1(x)$ is a cumulative hazard rate corresponding to an alternative hazard rate function defined by

$$h_1(x) = \log \frac{S(x)}{S(x+1)}, \quad x = 0, 1, 2, \ldots. \tag{2.17}$$

Xie et al. (2002a) advocated the use of (2.17) as the hazard rate function instead of (2.1) by citing the following arguments. In the continuous case, the hazard rate is not a probability, but (2.1) is a conditional probability which is bounded. Consequently, (2.1) cannot increase too fast either linearly or exponentially to provide models of lifetimes of components in the wear-out phase. When cumulative hazard rate is defined as the negative logarithm of the survival function, $-\log S(x) \neq \sum_{t=0}^{x-1} h(t)$. This causes problems in defining discrete ageing concepts that are analogues of their continuous counterparts, such as increasing hazard rate average (see Chapter 4). Thus, discrete ageing concepts based on $h(x)$ may not convey the same meaning as those in the continuous case. Similar problems persist with the construction of proportional hazards models and with series systems. Further details about these are provided in Sections 2.10 and 2.11. It may also be noted that unlike $h(x)$, the definition of $h_1(x)$ does not have any interpretation. Xie et al. (2002a) and Kemp (2004) have obtained the following interrelationships among the two hazard rate functions and the other reliability functions discussed so far:

$$
\begin{aligned}
H_1(x) &= -\sum_{t=0}^{x-1} \log(1 - H(t) + H(t-1)), \\
H(x) &= \sum_{t=0}^{x} (1 - \exp[H_1(x) - H_1(x+1)]), \\
S(x) &= \exp[-H_1(x)] = \prod_{t=0}^{x-1} (1 - H(t) + H(t-1)), \\
f(x) &= \exp[-H_1(x)] - \exp[-H_1(x+1)] \\
&= (H(x) - H(x-1)) \times \prod_{t=0}^{x-1} (1 - H(t) + H(t-1)), \\
h(x) &= 1 - \exp[H_1(x) - H_1(x+1)] = H(x) - H(x-1),
\end{aligned}
$$

and

$$h(x) = 1 - e^{-h_1(x)}. \tag{2.18}$$

Thus, the function $h(x)(H(x))$ determines $h_1(x)(H_1(x))$ uniquely and hence is useful in characterizing life distributions. We give some examples that compare the expressions of $h(x)$ and $h_1(x)$.

Example 2.4. The discrete Pareto distribution

$$S(x) = \left(\frac{\alpha}{x + \alpha - 1}\right)^{\beta}, \quad x = 1, 2, \ldots,$$

provides

$$h_1(x) = \beta \log \left[\frac{x+\alpha}{x+\alpha-1} \right]$$

whereas

$$h(x) = 1 - \left(\frac{x+\alpha-1}{x+\alpha} \right)^{\beta}.$$

Example 2.5. The Waring distribution with

$$S(x) = \frac{(b)_x}{(a)_x}$$

in Example 2.2 gives

$$h_1(x) = -\log \left(\frac{b+x}{a+x} \right)$$

compared to $h(x) = \frac{a-b}{a+x}$ from (2.8).

2.3 MEAN RESIDUAL LIFE

The analysis of the lifetime of a device after it has attained age x is of special relevance in reliability and survival analysis. Thus, if X is the original lifetime with survival function $S(x) = P(X \geq x)$, the corresponding residual lifetime after age x is the random variable $X_x = (X - x|X > x)$. From the definition of conditional probability, one can arrive at the distribution of X_x as

$$S_x(t) = \frac{S(x+t+1)}{S(x+1)}, \quad t = 0, 1, \ldots. \tag{2.19}$$

The mean, variance, partial moments, coefficient of variation and percentiles of the distribution in (2.19) have been discussed extensively in the literature in the continuous case.

The mean residual life of X is defined as

$$\begin{aligned} m(x) &= E(X - x|X > x) \\ &= \frac{1}{S(x+1)} \sum_{t=x+1}^{\infty} (t-x) f(t) \\ &= \frac{1}{S(x+1)} \sum_{t=x+1}^{\infty} S(t), \quad x = -1, 0, 1, \ldots \end{aligned} \tag{2.20}$$

It is easy to see that $m(x)$ is the mean of the distribution in (2.19). When $x = -1$,

$$m(-1) = E(X+1) = \sum_{t=0}^{\infty} S(t) = 1 + \mu.$$

Characterizations of the distribution of X in terms of $m(x)$ and the hazard rate have been studied by Nair and Hitha (1989). From (2.20),

$$S(x+1)m(x) = \sum_{t=x+1}^{\infty} S(t)$$

which leads to

$$S(x)m(x-1) - S(x+1)m(x) = S(x)$$

and to the identity

$$h(x) = \frac{1 + m(x) - m(x-1)}{m(x)}, \quad x = 0, 1, 2, \ldots. \tag{2.21}$$

Now, from (2.2), we have

$$S(x) = \begin{cases} \prod_{t=0}^{x-1} \frac{m(t-1)-1}{m(t)}, & x \geq 1, \\ 1, & x = 0. \end{cases} \tag{2.22}$$

Thus, the three functions $h(x)$, $m(x)$ and $S(x)$ determine each other uniquely. Though $h(x)$ can be determined from $m(x)$ and vice-versa, both have unique features that ensure their necessity in reliability theory. The mean residual life function may exist when the hazard function does not exist and vice-versa, as will be seen in some of the examples that are considered later. While $h(x)$ is a local measure of failure patterns for any x, the mean residual life function depends upon the life history of devices at all ages beyond x and hence the latter is more informative. At the same time, $m(x)$ as a summary measure is highly sensitive to a single long-term survivor in a data set, which is not desirable. There are other interesting properties that distinguish the two concepts as will be seen in the subsequent chapters.

Since $0 \leq h(x) \leq 1$, we can note from (2.21) that

$$1 + m(x) \geq m(x-1) \tag{2.23}$$

and $m(x) \geq 1$. For many discrete distributions, the expression for $m(x)$ is not of a simple form. In the following examples, we have simple forms for $m(x)$ that have many desirable properties. For more examples of the $m(x)$ function, see Table 2.4.

Example 2.6. For the geometric distribution $G(p)$ in Example 2.1,

$$m(x) = \frac{1}{q^{x+1}} \sum_{t=x+1}^{\infty} q^{t+1} = \frac{1}{p}$$

is the reciprocal of the hazard rate function so that $h(x)m(x) = 1$.

Example 2.7. The Waring distribution $W(a, b)$ in Example 2.2 yields

$$\begin{aligned} m(x) &= \frac{(a)_{x+1}}{(b)_{x+1}} \sum_{t=x+1}^{\infty} \frac{(a)_t}{(b)_t} \\ &= \frac{(a)_{x+1}}{(a)_x} \left[\frac{1}{a+x} + \frac{b+x+1}{(a+x)(a+x+1)} + \cdots \right] \\ &= \frac{a+x}{a-b-1}, \quad \text{by Waring expansion in (2.6).} \end{aligned}$$

Notice that $m(x)$ is a linearly increasing function and satisfies $h(x)m(x) = \frac{a-b}{a-b-1} > 1$.

Example 2.8. In the case of the negative hypergeometric distribution $\text{NH}(n, k)$ in Example 2.3, we have

$$\begin{aligned} m(x) &= \frac{\binom{k+n}{n}}{\binom{k+n-x-1}{n-x-1}} \sum_{t=x+1}^{\infty} \frac{\binom{k+n-t}{n-t}}{\binom{k+n}{n}} \\ &= \frac{\binom{k+n-x}{n-x-1}}{\binom{k+n-x-1}{n-x-1}} = \frac{k+n-x}{k+1}. \end{aligned}$$

We can see that $m(x)$ is linear and decreasing in x and $h(x)m(x) = \frac{k}{k+1} < 1$.

Nair and Hitha (1989) and Hitha and Nair (1989) have established the following characterizations of the above three models.

Theorem 2.2. *A necessary and sufficient that a non-negative discrete random variable has a mean residual life function of the form*

$$m(x) = lx + m, \quad m > 0,$$

is that X is geometric ($l = 0$), or Waring ($l > 0$), or negative hyper-geometric ($l < 0$).

Theorem 2.3. *The relationship*

$$h(x)m(x) = C,$$

where C is a constant, is satisfied by a non-negative discrete random variable X if and only if X is distributed as geometric for $C = 1$, Waring for $C > 1$, and negative hyper-geometric for $C < 1$.

Identification of lifetime distributions that are uniquely determined by simple relationships between various reliability concepts has attracted several studies in the past and continues to be a fertile area of research. Many results pertaining to individual distributions and families of distributions are available in this context. Theorem 2.3 belongs to this category concerning three specific distributions and its applications will be discussed later in Chapter 4. A more general result in this connection is the following.

Let $k(x)$ and $c(x)$ be real-valued functions defined on N such that $E(k(X)) = \mu_k$ and $V(k(X)) = \sigma_k^2$. We define

$$m_k^*(x) = E[k(X)|X > x].$$

Further, assume that $Ec^2(X) < \infty$ and $E[\Delta c(X)g(X)] < \infty$, where Δ is the usual difference operator, $\Delta a(x) = a(x+1) - a(x)$, and $g(x)$ is some real-valued function.

Theorem 2.4 (Nair and Sudheesh, 2008). *Let X be discrete random variable taking values in N and $k(x)$, $c(x)$ and $g(x)$ be real-valued functions satisfying the above conditions. Then, for every $c(x)$ and some $g(x)$ and $k(x)$, the following statements are equivalent:*

$$\textit{(i)} \quad \frac{f(x+1)}{f(x)} = \frac{\sigma_k g_k(x)}{\sigma_k g_k(x+1) - \mu_k + k(x+1)} \tag{2.24}$$

with $g_k(0) = \frac{\mu_k - k(0)}{\sigma_k}$ and $f(0)$ is evaluated from $\sum_{x=0}^{\infty} f(x) = 1$;

$$\textit{(ii)} \quad m_k^*(x) = \mu_k + \sigma_k \frac{h(x)g(x)}{1-h(x)}; \tag{2.25}$$

$$\textit{(iii)} \quad V(c(X)) \geq E^2(g(X)\Delta c(X)), \tag{2.26}$$

where $h(x)$ in (2.25) is the hazard rate function.

When $k(x) = x$, we write

$$m^*(x) = E(X|X > x) \tag{2.27}$$

and call it the *vitality function*, which is in fact a conditional mean life function. It satisfies

$$m^*(-1) = \mu \text{ and } m^*(x) = m(x) + x.$$

Eq. (2.25) becomes a relationship between the mean residual life function and hazard rate function of the form

$$m(x) = x + \mu + \sigma \frac{h(x)g(x)}{1 - h(x)}$$

that characterizes the class of discrete distributions satisfying

$$\frac{f(x+1)}{f(x)} = \frac{\sigma g(x)}{\sigma g(x+1) - \mu + x + 1}, \tag{2.28}$$

where $\sigma^2 = V(x)$. It is further observed in Nair and Sudheesh (2008) that

(1) the equality in (2.26) holds if and only if $c(x)$ is linear in $k(x)$;
(2) $Eg(X) = \sigma_k^{-1}\mathrm{Cov}\,(X, k(X))$;
(3) the expression for $g(x)$ is unique for a particular choice of $k(x)$, but one can have different choices for $g(x)$ for the same distribution when $k(x)$ is different;
(4) for a given $k(x)$, the corresponding $g(x)$ characterizes the distribution of X.

Specialization of the identity in (2.25) for various distributions and their implications will be considered later in Chapter 3 wherein we discuss discrete lifetime models. The variance inequality is also useful in the unbiased estimation of functions of X.

Remark 2.4. Since $h(x) = 1 - e^{-h_1(x)}$, all the characteristic properties mentioned above can be translated in terms of $m(x)$ and $h_1(x)$.

Though closely related to the mean residual life function, the vitality function introduced in (2.28) has some importance in its own right in lifetime studies. By definition,

$$m^*(x) = \frac{1}{S(x+1)} \sum_{t=x+1}^{\infty} tf(t), \quad x = -1, 0, 1, \ldots .$$

Some properties of $m^*(x)$ that are of interest in the sequel are given below. From the definition

$$m^*(x-1)S(x) - m^*(x)S(x+1) = xf(x).$$

Dividing by $S(x)$ and simplifying, we get

$$m^*(x) - m^*(x-1) = h(x)[m^*(x) - x] = h(x)m(x). \tag{2.29}$$

Further,

$$m^*(x) > x \quad \text{for all } x \text{ in } N.$$

Analogous to the presentation on vitality function given in Kupka and Loo (1989), we note that in the integer interval $[0, x]$, $m([0, x]) = m^*(-1) - m^*(x - 1)$ represents the increment in the conditional mean life which is achieved by surviving from age 0 to x. Thus, low (high) values of $m^*([0, x])$ means that the device is ageing rapidly (slowly) during $[0, x]$ and this justifies the name vitality function of the lifetime X. However, it may be noticed that vitality function is always non-decreasing, unlike the mean residual life function which can be either decreasing or increasing.

Returning to the reliability functions based on real-valued functions $k(X)$ of X, more characterizations and useful relationships can be seen. Let $F(x)$ be the distribution function of X such that $F(A) = 1$ for some $A > 0$.

Theorem 2.5 (Gupta, 1975). *If* $E[k(X)] < \infty$, *then*

$$S(x+1) = \prod_{y=0}^{x-1} \frac{m_k(y)}{m_k(y+1) + k(y+2) - k(y+1)}, \qquad 0 \le x < A,$$

if and only if

$$m_k(y) = E(k(X)|X > y) - k(y+1), \quad 0 \le y < A.$$

In particular, X has geometric distribution $G(p)$ *if and only if*

$$E(X|X > y) = E(X) + y + 1.$$

Ruiz and Navarro (1995) considered discrete random variables taking values in $C = (x_a, x_{a+1}, \ldots, x_b)$, $x_i < x_{i+1}$, and defined for a monotonic $k(x)$, the doubly truncated mean function

$$\begin{aligned} M(x, y) &= E[k(X)|x \le X \le y] \\ &= \frac{1}{F(y) - F(x-)} \sum_{x \le x_i \le y} f(x_i)k(x_i) \end{aligned}$$

for $(x, y) \in D = \{(s, t)|F(t-) < F(s)\}$, $F(x) = P(X < x)$ and $f(x_i) = F(x_i) - F(x_i - 1)$. Then,

$$F(x_i) = \frac{P_1(x_i, x_i)}{P_1(x_i, x_i) + P_2(x_i, x_i) - 1},$$

where

$$P_1(x, y) = \prod_{x_i < x} \frac{M(x_{i+1}, y) - k(x_i)}{M(x_i, y) - k(x_i)}$$

and

$$P_2(x, y) = \prod_{y<x_i} \frac{k(x_i) - M(x, x_{i-1})}{k(x_i) - M(x, x_i)}.$$

They also obtained a necessary and sufficient condition for a given function in (x, y) to be a doubly truncated mean function. This result is of relevance in reliability modelling since $M(x, y)$ tends to the mean residual life as $x \to \infty$ and to the reversed mean residual life (Section 2.7) as $x \to 0$ when $k(X) = X$ and the range of X is N.

Example 2.9. For $k(X) = X$, the relationship

$$M(x, y) = \frac{q}{p} + \frac{xq^{x-1} - (y+1)q^y}{q^{x-1} - q^y}$$

characterizes the geometric distribution $G(p)$ with $S(x) = q^x$, $x = 0, 1, 2, \ldots$.

The conditional moments

$$m_r^*(x) = E(X^r | X > x) = \frac{1}{S(x+1)} \sum_{x+1}^{\infty} t^r f(t)$$

satisfy

$$h(x) = \frac{m_r^*(x-1) - m_r^*(x)}{x^r - m_r^*(x)} \tag{2.30}$$

so that the distribution of X is determined from $m_r^*(x)$ by virtue of (2.2). Further, since the left hand side of (2.30) is independent of r,

$$\frac{m_r^*(x-1) - m_r^*(x)}{x^r - m_r^*(x)} = \frac{m_{r-1}^*(x-1) - m_{r-1}^*(x)}{x^{r-1} - m_{r-1}^*(x)}$$

is a recurrence relation connecting two consecutive moments.

Glanzel et al. (1984) considered $k(x)$ to be a monotonic function such that $k(x) \neq$ a constant in $(n, n+1, \ldots) \cap N$ for any finite n. Then,

$$f(x) = \begin{cases} \left(\prod_{t=0}^{x-1} \frac{m_k(t)-k(t)}{m_k(t+1)-k(t)}\right) \frac{m_k(x+1)-m_k(x)}{m_k(x+1)-k(x)}, & \text{for } x \in N \text{ or } x < n \\ \prod_{t=0}^{n-1} \frac{m_k(t)-k(t)}{m_k(t+1)-k(t)}, & \text{for } x = n \end{cases}$$

Further, if $k(x)$ and $g(x)$ are two functions defined on N such that $l(x) = \frac{m_0(x)}{m_k(x)}$ is strictly monotonic, then the distribution of X has probability mass function $f(x)$ satisfying

$$A(x) = \frac{f(x)}{f(0)} \frac{l(x+1) - l(x)}{l(1) - l(0)} \prod_{t=1}^{x} \frac{l(t-1)k(t-1) - g(t-1)}{l(t+1)k(t) - g(t)}$$

and $f(0) = (1 + \sum_{x=1}^{\infty} A(x))^{-1}$. For a further refinement of the results on conditional expectations, see Su and Huang (2000). For characterizations of the geometric, Waring and negative hyper-geometric laws by doubly truncated mean residual life, one may see to Khorashadizadeh et al. (2012).

The alternative hazard rate $h_1(x)$ can also be related to $m(x)$, as shown in Xie et al. (2002a) when $x = 1, 2, 3, \ldots$. From the definition of $m(x)$ in (2.19) and Eq. (2.2), we have

$$\begin{aligned} m(x) &= \frac{1}{S(x+1)} \sum_{t=x+1}^{\infty} \prod_{u=0}^{t-1} (1 - h(u)) \\ &= \frac{1}{S(x+1)} \sum_{t=x+1}^{\infty} \exp\left[-\sum_{a=0}^{t-1} h_1(u) \right], \quad \text{using (2.18)} \\ &= \sum_{t=x+1}^{\infty} \exp[H_1(x+1) - H_1(t-1)], \quad \text{since } S(x) = e^{-H_1(x)}. \end{aligned}$$

Example 2.10. Let X be distributed as geometric $G(p)$. Then,

$$h_1(x) = \log \frac{S(x)}{S(x+1)} = -\log q$$

and

$$H_1(x) = -x \log q.$$

Thus,

$$m(x) = \sum_{t=x+1}^{\infty} \left[q^{-(x+1)} \big/ q^{-(t-1)} \right] = \frac{1}{p},$$

as has been observed earlier.

Some relationships involving $h(x)$ and $m(x)$ needed in the subsequent discussions are

$$E\left(\frac{1}{h(X)}\right) = 1 + \mu \tag{2.31}$$

and

$$E\left(\frac{1}{h(X)} | X > x\right) = \frac{1}{S(x+1)} \sum_{t=x+1}^{\infty} S(t) = m(x). \tag{2.32}$$

The expression $h(X)$ is referred to as the hazard rate at random time. We now prove a characteristic property of the mean of $h(X)$.

Theorem 2.6.

$$E(h(X)) \geq \frac{1}{1+\mu} \tag{2.33}$$

and the equality holds if and only if X is geometric.

Proof. Consider

$$E(h(X))E\left(\frac{1}{h(X)}\right) = \left[\sum_{x=0}^{\infty} h(x)f(x)\right]\left[\sum_{x=0}^{\infty} \frac{1}{h(x)}f(x)\right].$$

Applying the Cauchy-Schwarz inequality for real sequences $\langle a(x)\rangle$ and $\langle b(x)\rangle$, viz.,

$$\sum [a(x)]^2 \sum [b(x)]^2 \geq \left[\sum a(x)b(x)\right]^2 \tag{2.34}$$

with $a(x) = [h(x)f(x)]^{\frac{1}{2}}$ and $b(x) = [\frac{1}{h(x)}f(x)]^{\frac{1}{2}}$, we readily obtain

$$E[h(X)]E\left[\frac{1}{h(X)}\right] \geq 1.$$

From (2.31), the inequality in (2.33) follows. The inequality in (2.34) becomes equality if and only if $a(x) = kb(x)$, in which case $h(x) = k$, $0 < k < 1$, for all x. A constant hazard rate characterizes the geometric law. □

Theorem 2.7.

$$E\left(\frac{1}{m(X)}\right) \geq \frac{1}{E(m(X))} \tag{2.35}$$

and the equality holds if and only if X is geometric.

Proof.

$$E\left(\frac{1}{m(X)}\right) E(m(X)) = \left[\sum_{x=0}^{\infty} m(x)f(x)\right]\left[\sum_{x=0}^{\infty} \frac{1}{m(x)}f(x)\right].$$

Taking $a(x) = [m(x)f(x)]^{\frac{1}{2}}$ and $b(x) = [\frac{f(x)}{m(x)}]^{\frac{1}{2}}$ and applying the Cauchy-Schwarz inequality, (2.35) follows. For the equality sign to hold, $a(x) = kb(x)$ which implies $m(x) = k$, a characteristic property of the geometric distribution. □

TABLE 2.1 Model for the deaths of solders by horse-kicks

No. of deaths per year	Observed frequency	$\hat{m}(x)$	Expected frequency
0	129	1.4926	123.08
1	45	1.4002	53.79
2	22	1.1201	18.35
3	3	1.0000	3.78

2.3.1 Modelling Data

Although we have defined the hazard rate function and the mean residual life function in terms of the time to failure of a device, the definitions and properties can also be used to model data on the time to occurrence of an 'event' with appropriate modifications in interpretations that best suit the context. Accordingly, the results mentioned so far have been extensively applied in other disciplines as well. A detailed examination of this aspect will be taken up later in Chapter 9. In the present section, we consider some real data and illustrate the application of the characterizations established above in finding suitable models for them.

Example 2.11. In this example, we consider the famous Bortkiewicz data on the number of soldiers of the Prussian army who died of horse-kicks in a period of 20 consecutive years, with the modification of the data value suggested by Cohen (1960). The last observation in Cohen's data is omitted for the analysis; see Table 2.1 for details. From the data, the probability mass function of X, the time of death, is estimated as

$$\hat{f}(0) = \frac{129}{199} = 0.6482, \quad \hat{f}(1) = \frac{45}{199} = 0.2261,$$
$$\hat{f}(2) = \frac{22}{199} = 0.1106, \quad \hat{f}(3) = \frac{3}{199} = 0.0151.$$

These values give the estimate of the mean residual life as

$$\hat{m}(x) = \frac{1}{\hat{S}(x+1)} \sum_{t=x+1}^{3} \hat{S}(t), \quad x = -1, 0, 1, 2,$$

where $\hat{S}(x) = \sum_{t \geq x} \hat{f}(t)$. The values of $\hat{m}(x)$ can be seen in column 2 of Table 2.1. The next step is to seek a functional form for $\hat{m}(x)$. Accordingly, a straight line fit is considered for $m(x)$ by the method of least-squares. This yields

$$m(x) = 1.34112 - 0.17579x.$$

TABLE 2.2 Estimated values of $\hat{h}(x)$, $\hat{m}(x)$ and $\hat{g}(x)$ for the data in Table 2.3

x	$\hat{h}(x)$	$\hat{m}(x)$	$\hat{g}(x)$
0	0.029	3.9569	2.00
1	0.0795	3.2124	2.06
2	0.1632	2.6437	2.07
3	0.2672	2.2431	1.96
4	0.3695	1.9716	1.87
5	0.4493	1.7642	1.85
6	0.5462	1.6839	1.65
7	0.6615	1.7604	1.61
8	0.5518	1.5576	2.81
9	0.6306	1.5082	2.01
10	0.6223	1.3478	2.23
11	0.6522	1.000	2.12
12	1.000		

Being a linearly decreasing mean residual life, from Theorem 2.1, we propose the negative hyper-geometric distribution as a model for the data with $n = 3$. Comparing with the mean residual life function of the negative hyper-geometric law

$$m(x) = \frac{k+n-x}{k+1},$$

we obtain the estimate of k as $\hat{k} = 4.863$. Thus the proposed model has survival function

$$\hat{S}(x) = \binom{\hat{k}+n-x}{n-x} \Big/ \binom{\hat{k}+n}{n}$$

with $\hat{k} = 4.863$ and $n = 3$. The expected probability mass function is

$$\hat{f}(x) = \hat{S}(x) - \hat{S}(x+1).$$

Obviously, the chosen model rests on the assumption of linearity of $m(x)$. Hence the model is validated by calculating the expected frequencies and then verifying their closeness by the chi-square goodness-of-fit test. The chi-square value of 2.4949 does not reject the model at 5% level of significance.

Example 2.12. The results of the well-known Rutherford-Geiger experiment is produced in Table 2.2. It describes the number of α-particles remitted from

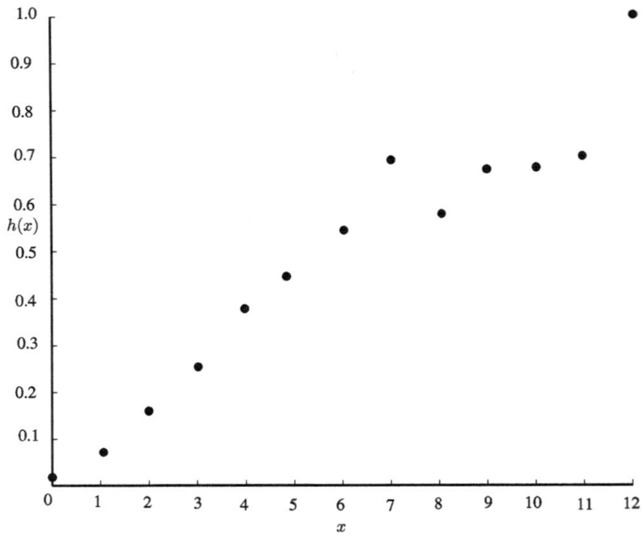

FIGURE 2.1 Empirical hazard function for Rutherford-Geiger experiment.

radioactive substances in 2608 fractions of 75 seconds. As in the previous example, the estimates $\hat{f}(x)$ and $\hat{S}(x)$ of the probability mass function and the survival function and there from $\hat{m}(x)$ are found out. In addition, we also require the estimated hazard rate function $\hat{h}(x) = \hat{f}(x)/\hat{S}(x)$. The observed values of $\hat{h}(x)$ and $\hat{m}(x)$ are shown in Table 2.2.

We also have

$$m(-1) = \sum_{t=0}^{\infty} \hat{S}(t) = 1 + \mu = 4.8702$$

so that an estimate of the mean becomes

$$\hat{\mu} = 3.8702.$$

Neither an examination of the values of $\hat{h}(x)$ and $\hat{m}(x)$ nor their plots in Figs 2.1 and 2.2 exhibit an obvious choice for the functional forms of $h(x)$ and $m(x)$. In such cases, Theorem 2.4 can be of some assistance in the determination of a suitable model. Since we have the empirical mean residual life from the data, choose $k(x) = x$ so that (2.24) becomes

$$m^*(x) = m(x) + x = \mu + \frac{\sigma h(x) g(x)}{1 - h(x)}.$$

Hence, we can calculate $\hat{g}(x)$ from

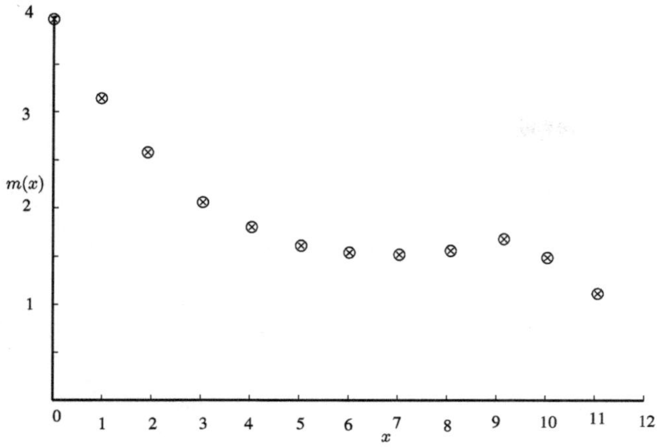

FIGURE 2.2 Empirical mean residual life function for Rutherford-Geiger experiment.

$$\hat{g}(x) = \left(\frac{\hat{m}(x) + x - \hat{\mu}}{\hat{\sigma}}\right)\left(\frac{1}{\hat{h}(x)} - 1\right). \tag{2.36}$$

In (2.36), the estimate of the standard deviation σ is taken as the sample standard deviation. The values of $\hat{g}(x)$ so arrived at are shown in Table 2.2. Notice that the value at $x = 8$ is significantly larger than the rest. Barring this value, the rest appears to be clustering around the average $\bar{g}(x) = 1.94$. This is the same as assuming $g(x) = c$, a constant for all x, since the least-square estimate of c is $\bar{g}(x)$. Recall from Theorem 2.4 that the value of $g(x)$ uniquely determines the distribution of X and further, $g(x) = c$ if and only if the distribution is Poisson with mean c^2. Thus, the data follows a Poisson distribution

$$f(x) = e^{-\lambda}\frac{\lambda^x}{x!}, \quad x = 0, 1, 2, \ldots$$

with $\lambda = c^2 = 3.8$. To validate the assumption made on $g(x)$, we compare the observed and expected frequencies, when $\lambda = 3.8$, presented in Table 2.3.

The chi-square value obtained above does not reject the hypothesis that the data follow the Poisson distribution. The following facts can be further observed:

(i) The characterization theorems used in Examples 2.11 and 2.12 provide some quick estimates of the parameters of the hypothesized models. In more sophisticated models that require iterative procedures to find estimates, the above estimates are good choice may serve as initial values;

(ii) The contribution to the chi-square value or the discrepancy between the observed and expected frequencies at $x = 8$ is considerably larger compared to the rest. The plot of the hazard function shows that the function is

TABLE 2.3 Goodness-of-fit for Rutherford-Geiger data

x	Observed frequency	Expected frequency	Contribution to χ^2-value
0	57	58.4	0.0036
1	203	221.7	1.5773
2	383	421.2	3.4644
3	525	533.6	0.1386
4	532	507.0	1.2327
5	408	385.2	1.3495
6	273	244.1	3.4215
7	139	132.5	0.3189
8	45	62.9	5.0934
9	27	26.6	0.0060
10	10	10.27	0.0039
11 12	4 2 } 6	4.4	0.5819
Total	2608	2608	16.2216

decreasing markedly at this point which is uncharacteristic of the Poisson model where the hazard rate is increasing. If we use the maximum likelihood estimate $\hat{\lambda} = 3.87$ for the Poisson distribution, the difference is still higher with expected frequency 67. Thus, the $\hat{g}(x)$ value at $x = 8$ is taken as a discordant one and omitted in the calculation of $\bar{g}(x)$.

2.4 VARIANCE RESIDUAL LIFE FUNCTION

The variance of the residual life $(X - x|X > x) = X_x$ is studied in reliability theory in various contexts. Primarily, its role is to define ageing concepts that are weaker than some ageing criteria based on the hazard rate and the mean residual life. Chapter 4 provides details in this direction. Secondly, variance of residual life has the same role as the usual variance when estimators of mean residual life are discussed. It is also required in the study of coefficient of variation of residual life. Assuming that $E(X^2) < \infty$, we define the variance residual life function as

$$\sigma^2(x) = E[(X - x)^2|X > x] - m^2(x), \tag{2.37}$$

where $m(x)$ is the mean residual life function defined in (2.20). Alternatively,

$$\sigma^2(x) = E(X^2|X > x) - E^2(X|X > x). \tag{2.38}$$

The second factorial moment of the residual life X_x, given by

$$\mu_{(2)}(x) = E[(X-x)(X-x-1)|X>x],$$

will frequently appear in the sequel. We have the following expression for $\mu_{(2)}(x)$.

Theorem 2.8.

$$\mu_{(2)}(x) = \frac{2}{S(x+1)} \sum_{t=x+1}^{\infty} \sum_{u=t+1}^{\infty} S(u). \tag{2.39}$$

Proof.

$$\begin{aligned} E(X^2|X>x) &= \frac{1}{S(x+1)} \sum_{t=x+1}^{\infty} t^2[S(t)-S(t+1)] \\ &= x^2 + \frac{1}{S(x+1)} \sum_{t=x+1}^{\infty} (2t-1)S(t). \end{aligned} \tag{2.40}$$

Hence,

$$\begin{aligned} E[(X-x)^2|X>x] &= \frac{1}{S(x+1)} \sum_{t=x+1}^{\infty} (2t-2x-1)S(t), \\ &= \frac{1}{S(x+1)} \left[2 \sum_{t=x+1}^{\infty} \sum_{u=t+1}^{\infty} S(u) + \sum_{t=x+1}^{\infty} S(t) \right] \qquad (2.41) \\ &= \frac{2}{S(x+1)} \sum_{t=x+1}^{\infty} \sum_{u=t+1}^{\infty} S(u) + E(X-x|X>x), \end{aligned}$$

which is equivalent to (2.39). Thus, the variance residual life function can be written in the form

$$\sigma^2(x) = \frac{2}{S(x+1)} \sum_{t=x+1}^{\infty} \sum_{u=t+1}^{\infty} S(u) - m(x)(m(x)-1). \tag{2.42}$$

An equivalent expression for (2.40), by defining the survival function as $P(X>x)$, has been obtained in Khorashadizadeh et al. (2010). □

Example 2.13. For the geometric law considered in Example 2.6, we have

$$S(x) = q^x, \quad m(x) = \frac{1}{p},$$

$$\sum_{u=t+1}^{\infty} S(t) = \frac{q^{t+1}}{p} \text{ and } \sum_{t=x+1}^{\infty} \frac{q^{t+1}}{p} = \frac{q}{p^2}$$

so that from (2.42), we have

$$\sigma^2(x) = \frac{2q}{p^2} - \frac{1}{p}\left(\frac{1}{p} - 1\right) = \frac{q}{p^2},$$

which is the same as the variance of X.

There exist some inter-relationships between the reliability functions discussed so far.

Theorem 2.9.

$$\sigma^2(x+1) - \sigma^2(x) = h(x+1)[\sigma^2(x+1) - m(x+1)(m(x)-1)]. \quad (2.43)$$

Proof. From (2.42), we have

$$\begin{aligned}&[\sigma^2(x) + m(x)(m(x)-1)]S(x+1)\\&\quad - [\sigma^2(x+1) + m(x+1)(m(x+1)-1)]S(x+2)\\&\quad = 2S(x+2)m(x+1).\end{aligned}$$

Dividing by $S(x+2)$ and using (2.3), we get

$$\sigma^2(x) + m(x)(m(x)-1) = (1-h(x+1))[\sigma^2(x+1) + m^2(x+1) + m(x+1)]. \quad (2.44)$$

Since

$$h(x+1) = \frac{1 + m(x+1) - m(x)}{m(x+1)},$$

further simplification of (2.44) yields (2.43). □

Remark 2.5. The identity in (2.43) can be expressed in terms of $\sigma^2(x)$ and $m(x)$ as

$$\frac{\sigma^2(x)}{m(x)-1} = \frac{\sigma^2(x+1)}{m(x+1)} + m(x+1) - m(x) + 1.$$

Theorem 2.10.

$$\sigma^2(x) = E[m(X)(m(X-1)-1)|X > x], \quad x = 0, 1, 2, \ldots \quad (2.45)$$

Proof. Starting from (2.43), we have

$$\sigma^2(x+1) - \sigma^2(x) = \frac{f(x+1)}{S(x+1)}[\sigma^2(x+1) - m(x+1)(m(x)-1)],$$

or

$$S(x+1)\sigma^2(x) - S(x+2)\sigma^2(x+1) = m(x+1)(m(x)-1)f(x+1).$$

Adding the above identity for $x+1, x+2, \ldots$, we get

$$S(x+1)\sigma^2(x) = \sum_{t=x+1}^{\infty} m(t)(m(t-1)-1)f(t)$$

which is the same as (2.45). □

Remark 2.6. Theorem 2.10 can be employed to find quick estimates of the variance residual life when the estimated mean residual life is known.

Unlike the cases of hazard rate and mean residual life functions, there is no inversion formula that expresses the survival function in terms of the variance residual life. Also, there are only a few standard distributions for which $\sigma^2(x)$ has simple tractable forms, as could be seen from Table 2.3. Therefore, characterizations of life distributions involving $\sigma^2(x)$ take the form of its relationship with other concepts. Hitha and Nair (1989) have shown that

$$\frac{\sigma^2(x)}{m(x)(m(x)-1)} = C, \text{ a constant,}$$

if and only if X is geometric for $C = 1$, negative hyper-geometric for $C > 1$, and Waring for $C < 1$. We also have a more general result satisfied by a class of distribution due to Sudheesh and Nair (2010). We retain the notations used for Theorem 2.4.

Theorem 2.11. *Let $k(X)$ be a real-valued function of X that is non-decreasing and satisfying the following conditions:*

(i) *$E[k^2(X)]$ and $E[|\Delta k(X)|g_k(X)]$ are finite;*
(ii) *$f(0) > 0$ and the support of X is an integer interval;*
(iii) *$\Delta k(x) \neq 0$ for all x.*

Then, X has distribution specified by

$$\frac{f(x+1)}{f(x)} = \frac{\sigma_k g_k(x)}{\sigma_k g_k(x+1) - \mu_k + k(x+1)}$$

for some real-valued function $g_k(x)$ if and only if for all x

$$V(k(X)|X > x) = \sigma_k E(\Delta k(X) g_k(X)|X > x) + (\mu_k - m_k(x))(m_k(x) - k(x+1)),$$

where $V(\cdot)$ stands for variance.

Remark 2.7. Setting $k(X) = X$, we have

$$\sigma^2(x) = \sigma E[g(X)|X > x] + (\mu - m^*(x))(m^*(x) - x - 1) \tag{2.46}$$

for all distributions in which the probability mass function satisfies

$$\frac{f(x+1)}{f(x)} = \frac{\sigma g(x)}{\sigma g(x+1) - \mu + x + 1}.$$

Remark 2.8. From Eqs (2.39) and (2.42), it is evident that the property

$$\frac{\mu_{(2)}(x)}{m(x)(m(x)-1)} = k,$$

a positive constant, characterize the geometric, Waring and negative hypergeometric models.

2.5 UPPER PARTIAL MOMENTS

A concept that is closely related to moments of residual life is that of partial moments, which can also be interpreted as moments of a different kind of residual life. The *rth upper partial moment* of X about a point x is defined as

$$\alpha_r(x) = E((X - x)_+^r), \quad r = 0, 1, \ldots; \; x = 0, 1, 2, \ldots, \tag{2.47}$$

where $(X - x)_+ = \max(X - x, 0)$. In the case of discrete models, it is sometimes more convenient to work with factorial partial moments defined as

$$\alpha_{(r)}(x) = E((X - x)_{+(r)}), \quad X > x + r - 1 \tag{2.48}$$

where

$$t_{(r)} = t(t-1)\ldots(t-r+1) \tag{2.49}$$

is the descending factorial expression. By virtue of the relationship

$$(t - x)_+^r = \sum_{k=0}^{r} (X - x)_{+(r)} S(r, k),$$

where $S(r,k)$ is the Stirling number of the second kind, α_r can be computed in terms of $\alpha_{(r)}$. Nair et al. (2000) have studied several properties of $\alpha_{(r)}$ and their implications to reliability modelling. First, we note that

$$\alpha_1(x) = \alpha_{(1)}(x) = \sum_{t=x+1}^{\infty} S(t),$$

and hence

$$S(x) = \alpha_1(x-1) - \alpha_1(x), \quad x = 0, 1, \ldots,$$

and

$$f(x) = \alpha_1(x-1) - 2\alpha_1(x) + \alpha_1(x+1).$$

The upper partial moments satisfy the recurrence formula

$$\alpha_{(r)}(x) - \alpha_{(r)}(x+1) = r\alpha_{(r-1)}(x+1), \quad r = 1, 2, \ldots .$$

Thus, any one partial moment sequence, particularly $\langle \alpha_1(x) \rangle$, $x = 0, 1, 2, \ldots,$ determines all other partial moments. Since only the first two partial moments are of importance in reliability analysis we concentrate on their properties. Notice that

$$\alpha_1(x) - \alpha_1(x+1) = S(x+1) \tag{2.50}$$

so that

$$m(x) = \frac{\alpha_1(x)}{\alpha_1(x) - \alpha_1(x+1)} \tag{2.51}$$

and

$$1 - h(x+1) = \frac{S(x+2)}{S(x+1)} = \frac{\alpha_1(x+1) - \alpha_1(x+2)}{\alpha_1(x) - \alpha_1(x+1)}. \tag{2.52}$$

From (2.51), the ratio of the upper partial means at consecutive values gives

$$\frac{\alpha_1(x+1)}{\alpha_1(x)} = \frac{m(x)-1}{m(x)}.$$

Thus, from Theorem 2.2, we deduce the following result.

Theorem 2.12. *If $E(X) < \infty$, then*

$$\frac{\alpha_1(x+1)}{\alpha_1(x)} = \frac{(l-1)+ax}{l+ax}, \quad l > 1,$$

if and only if X has geometric, Waring, negative hyper-geometric distributions when $a = 0$, $a > 0$, and $a < 0$, respectively.

The second factorial partial moment is

$$\alpha_{(2)}(x) = E[(X-x)(X-x-1)]^{+} = S(x+1)\mu_{(2)}(x)$$
$$= 2\sum_{t=x+1}^{\infty}\sum_{u=t+1}^{\infty} S(u),$$

while

$$E[(X-x)^2|X>x] = \frac{\alpha_2(x)}{\alpha_1(x)-\alpha_1(x+1)}.$$

Thus, the variance residual life is calculated as

$$\sigma^2(x) = \frac{\alpha_2(x)(\alpha_1(x)-\alpha_1(x+1)) - \alpha_1^2(x)}{[\alpha_1(x)-\alpha_1(x+1)]^2}.$$

Example 2.14. Let

$$f(x) = (a-b)\frac{(b)_x}{(a)_x}, \quad x = 0, 1, 2, \ldots, \ a > b.$$

Then, from Example 2.2, we have

$$S(x) = \frac{(b)_x}{(a)_x},$$
$$\alpha_1(x) = \sum_{x+1}^{\infty}\frac{(b)_t}{(a)_t} = \frac{(b)_{x+1}}{(a)_{x-1}}\left[\frac{1}{a+x} + \frac{b+x+1}{(a+x)(a+x+1)} + \cdots\right]$$
$$= \frac{(b)_{x+1}}{(a-b-1)(a)_x},$$
$$\alpha_1(x) = 2\sum_{t=x+1}^{\infty}\sum_{u=t+1}^{\infty} S(u)$$
$$= 2\sum_{t=x+1}^{\infty}\frac{(b)_{t+1}}{(a-b-1)(a)_t} = \frac{2(b)_{x+2}}{(a)_x(a-b-1)(a-b-2)}.$$

The ratios of partial moments admit simple forms in this case. For example,

$$\frac{\alpha_1(x+1)}{\alpha_1(x)} = \frac{b+x+1}{a+x}$$

and

$$\frac{\alpha_2(x+1)}{\alpha_2(x)} = \frac{b+x+2}{a+x},$$

both being homographic functions. A similar discussion of the properties is possible when the ascending factorial expression

$$x^{(r)} = x(x+1)\dots(x+r-1)$$

is used instead of $x_{(r)}$ in (2.48); for details see Priya et al. (2000).

So far, we have considered reliability functions specified by the survival function $S(x)$. A parallel theory that builds up when the event $X \leq x$, defined by the distribution function, also has been of interest in reliability literature. Referred to by the name reliability functions in reversed time, they are found to be useful in modelling and analysis of lifetime data and also in other fields of study. We discuss some important functions in this category in the next few sections.

2.6 REVERSED HAZARD RATE

The reversed hazard rate of X is defined as

$$\lambda(x) = P(X = x | X \leq x) = \frac{f(x)}{F(x)}. \tag{2.53}$$

Thus, $\lambda(x)$ in the discrete case is interpreted as the conditional probability that a device fails at age x, given that its lifetime is at most x. Being a conditional probability, $0 \leq \lambda(x) \leq 1$. Keilson and Sumita (1982), who first defined the reversed hazard rate in continuous time, called it the *dual failure function* by the property that X has reversed hazard rate $\lambda(x)$, $-\infty \leq a \leq x \leq b < \infty$ if and only if the random variable $-X$ has a hazard rate $\lambda(-x)$ on $(-b, -a)$.

Finkelstein (2002) observed that, in reliability, one often works with non-negative random variables and therefore the above duality is not applicable. Further, the upper point of support is generally infinite. Thus, the properties of reversed hazard rate of non-negative random variables with infinite support cannot be formally obtained from those of the hazard rates. This makes a study of $\lambda(x)$ becomes necessary in its own right. Since

$$\lambda(x+1) = \frac{f(x+1)}{F(x+1)} = \frac{F(x+1) - F(x)}{F(x)},$$

it follows that

$$F(x) = \prod_{t=x+1}^{\infty} (1 - \lambda(t)). \tag{2.54}$$

Notice further that $\lambda(0) = 1$ and

$$f(x) = \lambda(x) \prod_{t=x+1}^{\infty} (1 - \lambda(t)). \tag{2.55}$$

Thus, $\lambda(x)$ determines the life distribution uniquely; see Nair and Asha (2004) for details and examples.

Example 2.15. We say that X follows arithmetic distribution if it has a probability mass function of the form

$$f(x) = \frac{2x}{b(b+1)}, \quad x = 1, 2, \ldots, b.$$

The corresponding distribution function is

$$F(x) = \frac{x(x+1)}{b(b+1)},$$

and hence

$$\lambda(x) = \frac{2}{x+1},$$

a reciprocal linear function.

In the continuous case, Block et al. (1998) have shown that for an absolutely continuous random variable X with interval, of support (a, b), $-\infty \leq a < b \leq \infty$, if the reversed hazard rate is a constant $c > 0$ for all $a < x < b$, then $a = -\infty$, $b < \infty$ and

$$F(x) = \exp[c(x-b)], \quad x < b,$$

and conversely. Thus, there does not exist an absolutely continuous distribution with constant reversed hazard rate on the positive real axis. We will now show that in the discrete case, reversed hazard rate can be constant when a subset of the set of nonnegative integers is as the support of X.

Example 2.16. Let X be distributed with probability mass function

$$f(x) = \begin{cases} (1+c)^{-b}, & x = 0, \\ c(1+c)^{x-b-1}, & x = 1, 2, \ldots, b, \ c > 0, \end{cases} \tag{2.56}$$

where b is a finite positive integer greater than unity. Then,

$$F(x) = \begin{cases} (1+c)^{x-b}, & x = 0, 1, 2, \ldots, b \\ 1, & x \geq b. \end{cases} \tag{2.57}$$

The reversed hazard rate in this case is

$$\lambda(x+1) = \frac{c(1+c)^{x-b-1}}{(1+c)^{x-b}} = \frac{c}{1+c}, \quad \text{for all } x = 0, 1, 2, \ldots,$$

and $\lambda(0) = 1$.

Remark 2.9. The terms in (2.56), for $x = 1, 2, \ldots$, are in geometric progression with a common ratio $(1 + c)$ and therefore, the successive terms are increasing, as opposed to the usual geometric distribution where the monotonicity is in the opposite direction. We will term (2.56) as the *reversed geometric distribution* with parameter c.

This distribution has an important role in the sequel. Since it does not form part of the standard distributions discussed in the next chapter, some properties of the model (2.56) are presented here. First,

$$\frac{f(x+1)}{f(x)} = \begin{cases} c, & x = 0, \\ c+1, & x = 1, 2, \ldots \end{cases}.$$

Thus, the distribution can be identified in practice for data for which the ratio of the successive frequencies, except the first, are nearly constant. The mean and variance are given by

$$\begin{aligned}\mu = E(X) &= (1+c)^{-b}\left[1 + c\sum_{t=1}^{b} t(1+c)^{t-1}\right]\\ &= \frac{bc-1}{c} + \frac{1}{c(1+c)^{b-1}}\end{aligned}$$

and

$$\begin{aligned}\sigma^2 &= (1+c)^{-b}\left[1 + c\sum_{t=1}^{b} t^2(1+c)^{t-1}\right] - \mu^2\\ &= \frac{1}{c} + \frac{2}{c^2} + b^2 - \frac{2b}{c} - (\frac{1}{c} + \frac{1}{c^2})(1+c)^{-b} - \mu^2.\end{aligned}$$

For the geometric distribution, the hazard rate is constant which is equivalent to the lack of memory property. An analogous property in a reverse sense is satisfied by the reversed geometric law, as the next theorem shows.

Theorem 2.13. *Let X be a discrete random variable with a finite set* $(0, 1, \ldots, b)$ *as support. Then,*

$$P(X \le t | X \le t + s) = P(X \le 0 | X \le s) \tag{2.58}$$

for all t, s in the support of X if and only if X has reversed geometric distribution.

Proof. Eq. (2.58) can be restated as

$$F(t+s)F(0) = F(t)F(s). \tag{2.59}$$

Substituting (2.57) in (2.59), we readily have the 'if' part. Conversely, (2.59) is equivalent to the functional equation

$$a(t+s)=a(t)a(s),$$

where $a(t)=\frac{F(t)}{F(0)}$. To solve for $a(t)$, we set $s=1$, so that

$$a(t+1)=a(t)a(1).$$

Iterating for t,

$$a(t)=[a(1)]^t \text{ for all } t=1,2,\ldots,b,$$

and so

$$F(x+1)=\frac{[F(1)]^{x+1}}{[f(0)]^x}. \tag{2.60}$$

Thus,

$$f(x+1)=\frac{p^{x+1}}{[f(0)]^x}-\frac{p^x}{[f(0)]^{x-1}}, \quad p=F(1).$$

Summing for $x=0,1,2,\ldots$ and using the fact that $\sum_{x=0}^{b} f(x)=1$, we get $f(0)=p^{\frac{b}{b-1}}$. Inserting the values of $f(0)$ in (2.60), we get

$$F(x)=p^{-(x-b)}.$$

Since $F(1)<1$, we should have $p<1$ and setting $p=(1+c)^{-1}$, the distribution in (2.57) is recovered, as needed. □

Remark 2.10. The reversed lack of memory property implies that, given the lifetime of device is upto age x, the probability that the device fails at any age in $[0,x]$ is the same. Further, the distribution of such a lifetime is governed by the reversed geometric law.

Remark 2.11. From the above discussions, we observe that the following are equivalent:

(i) X has constant reversed hazard rate;
(ii) X follows reversed geometric law;
(iii) X has reversed lack of memory property.

Further discussions and application of these results can be seen in Nair and Sankaran (2013).

The definition in (2.53), when applied to the continuous case, has the form $\lambda(x)=\frac{d\log F(x)}{dx}$. This property is not shared in the discrete case. For reasons similar to those explained in Section 2.2 regarding the alternative hazard rate, a second definition for reversed hazard rate can be put forward as

$$\lambda_1(x) = \log \frac{F(x)}{F(x-1)}, \quad x = 1, 2, \ldots . \tag{2.61}$$

In this case,

$$F(x-1) = F(x)e^{-\lambda_1(x)},$$

or

$$F(x) = \exp\left[-\sum_{t=x+1}^{\infty} \lambda_1(t)\right].$$

Also,

$$f(x) = \exp\left[-\sum_{t=x+1}^{\infty} \lambda_1(t)\right][1 - e^{-\lambda_1(x)}], \quad x = 1, 2, \ldots,$$

and $f(0)$ is determined from

$$f(0) = 1 - \sum_{x=1}^{\infty} f(x).$$

Example 2.17. The reversed geometric distribution specified by

$$F(x) = (1+c)^{x-b}, \quad x = 0, 1, \ldots, b,$$

has

$$\lambda_1(x) = \log \frac{(1+c)^{x-b+1}}{(1+c)^{x-b}} = \log(1+c),$$

which is a constant for all x.

From the definitions in (2.61) and (2.53), the relationship between $\lambda(x)$ and $\lambda_1(x)$ is found to be

$$\lambda(x) = 1 - e^{-\lambda_1(x)}. \tag{2.62}$$

It may be noticed that $\lambda_1(0) = \infty$, is consistent with the value $\lambda(0) = 1$.

2.7 REVERSED MEAN RESIDUAL LIFE

A second measure of interest in reversed time is the reversed mean residual life. Suppose a device has failed before attaining age t. Then, the random variable ${}_tX = t - X|(X < t)$ is the time elapsed since the device has failed, conditioned on the fact that its lifetime is less than t, and this is referred to as the *reversed residual life* or *inactivity time* of X. It is easy to see that ${}_tX$ has the distribution function

$$ {}_tF_x = \frac{F(t-1) - F(t-x-1)}{F(t-1)}, \qquad x = 0, 1, 2, \ldots. \tag{2.63}$$

The mean of this distribution is called the *reversed mean residual life* or *mean inactivity time*, and is denoted by $r(x)$. One can also define $r(x)$ as

$$ r(x) = E(x - X|X < x) = \frac{1}{F(x-1)} \sum_{t=1}^{x-1} tf(t), \quad x = 1, 2, \ldots. \tag{2.64}$$

We define $r(0) = 0$. Note also that $r(1) = 1$. Goliforushani and Asadi (2008) have established the following properties of $r(x)$:

(i) $x - r(x)$ is an increasing function with $0 \leq r(x) \leq x$;

(ii) $\mu = x - F(x-1)r(x) + S(x)m(x-1)$;

(iii) $r(x)$ cannot be a decreasing function for all x;

(iv)

$$ \lambda(x) = \frac{1 + r(x) - r(x+1)}{r(x)}, \quad x = 1, 2, \ldots; $$

(v) If X has a finite support as $(0, 1, 2, \ldots, b)$, then

$$ F(x) = \frac{\prod_{t=1}^{x} \left(\frac{r(t)}{r(t+1)-1} \right)}{F(0)}, \quad x = 1, 2, \ldots, b, \tag{2.65}$$

and

$$ F(0) = \left[\prod_{t=1}^{b} \frac{r(t)}{r(t+1) - 1} \right]^{-1}; \tag{2.66}$$

(vi) If $\lambda(x)$ is decreasing function of x, then $r(x)$ is an increasing function of x;

(vii)

$$ r(x) = \frac{\sum_{t=1}^{x} F(t-1)}{F(x-1)}. $$

It is of interest to mention here that the notions of residual life and inactivity time have been extended to general coherent systems and their properties, mixture representations and stochastic orderings of various forms have been discussed by numerous authors; see, for example, Navarro et al. (2008, 2013), Zhang (2010), Goliforushani and Asadi (2011), Goliforushani et al. (2012), and Parvardeh and Balakrishnan (2013, 2014).

Example 2.18. For the geometric distribution $G(p)$, we have

$$F(x) = 1 - q^{x+1},$$

and hence

$$\begin{aligned} r(x) &= \frac{1}{F(x-1)}[F(0) + \cdots + F(x-1)] \\ &= \frac{xp - q(1-q^x)}{p(1-q^x)}. \end{aligned}$$

Example 2.19. Consider the discrete uniform distribution

$$f(x) = \frac{1}{b}, \quad x = 1, \ldots, b.$$

In this case,

$$F(x) = \frac{x}{b}$$

and

$$r(x) = \frac{b}{x-1}\left(\frac{1}{b} + \frac{2}{b} + \cdots + \frac{x-1}{b}\right) = \frac{x}{2}.$$

Notice that in this case $\lambda(x) = \frac{1}{x}$ and that the product x of $r(x)$ and $\lambda(x)$ is $\frac{1}{2}$ for all x.

In the case of geometric distribution, the hazard rate $h(x)$ is constant and so is the mean residual life function $m(x)$, the two being related by $m(x)h(x) = 1$. We will show that a different scenario exists in the relationship between $\lambda(x)$ and $r(x)$. For the reversed geometric law, $\lambda(x) = \frac{c}{1+c}$, whereas

$$r(x) = \frac{1+c}{c}\left[1 - \frac{1}{(1+c)^x}\right]$$

which is a strictly increasing function of x. However, we can recover a distribution with constant reversed mean residual life by modifying the probability at $x = 0$.

Theorem 2.14. *A random variable with the support* $(0, 1, 2, \ldots, b)$ *has reversed mean residual life*

$$r(x) = \frac{c+1}{c}, \quad c > 0, x = 2, 3, \ldots, b, \tag{2.67}$$

for all x if and only if its distribution is given by

$$F(x) = \begin{cases} c^{-1}(1+c)^{1-b}, & x = 0, \\ (1+c)^{x-b}, & x = 1, 2, \ldots, b. \end{cases} \tag{2.68}$$

Proof. For the distribution in (2.68), we have

$$\begin{aligned} r(x) &= (1+c)^{b-x+1}\left[\frac{1}{c(1+c)^{b-1}} + (1+c)^{1-b} + \cdots + (1+c)^{x-1-b}\right] \\ &= (1+c)^{b-x+1}\left[\frac{1}{c(1+c)^{b-1}} + \frac{(1+c)}{c}\frac{((1+c)^{x-1}-1)}{(1+c)^{b}}\right] \\ &= \frac{1+c}{c}. \end{aligned}$$

The converse follows upon noting from (2.67) that

$$\frac{r(x)}{r(x+1)-1} = (1+c)$$

and then applying (2.65) and the fact that $\sum_0^b f(x) = 1$. □

Remark 2.12. The probability mass function corresponding to (2.68) is

$$f(x) = \begin{cases} c^{-1}(1+c)^{1-b}, & x = 0, \\ (c-1)c^{-1}(1+c)^{1-b}, & x = 1, \\ c(1+c)^{x-b-1}, & x = 2, 3, \ldots, b. \end{cases}$$

Accordingly, we have

$$\lambda(x) = \begin{cases} 1, & x = 0, \\ (c-1)c^{-1}, & x = 1, \\ c(c+1)^{-1}, & x = 2, 3, \ldots, b. \end{cases}$$

Remark 2.13. In comparison with the reversed geometric model, there is a difference in the value of $\lambda(1)$ for the model in (2.68). It may also be noted that

$$\lambda(x)r(x) = 1, \quad x = 2, 3, \ldots, b,$$

for the distribution in (2.68). This is a characteristic property as evidenced by the identity

$$\lambda(x)r(x) = 1 + r(x) - r(x+1)$$

which gives

$$r(x+1) - r(x) = 0$$

for all x, or $r(x)$ is a constant.

The reversed mean residual life can also be related to the alternative reversed hazard rate. From (2.62) and the identity in (iv) above, we have

$$\lambda_1(x) = \log \frac{r(x)}{r(x+1)-1}, \quad x = 0, 1, 2, \dots .$$

In certain problems, it is more convenient to deal with the conditional mean

$$r^*(x) = E(X|X \le x) = x - r(x+1).$$

The corresponding results for $r(x)$ is easily derived from those of $r^*(x)$ by using the above identity.

Relationship between $\lambda(x)$ and $r(x)$ of a different nature, than those indicated above, that characterizes families of discrete distributions has been proposed in literature. The main results reviewed here are from Gupta et al. (2006) and Nair and Sudheesh (2008). We retain the same notation as in Section 2.3. Let $k(x)$ be real-valued function such that $E k^2(X) < \infty$. Then, the probability mass function of X will be of the form

$$\frac{f(x+1)}{f(x)} = \frac{\sigma_k g_k(x)}{\sigma_k g_k(x+1) - \mu_k + k(x+1)}, \tag{2.69}$$

where μ_k and σ_k are the mean and standard deviation of $k(X)$ satisfying $\sigma_k g_k(0) = \mu_k - k(0)$ for some real-valued function $g(\cdot)$ if and only if

$$r_k^*(x) = \mu_k - \sigma_k g_k(x)\lambda(x) \tag{2.70}$$

with $r_k^*(x) = E(k(X)|X \le x)$. The g_k function appearing in the above relationships is unique for a particular distribution and often assumes simple forms. Special cases of the form in (2.69) that includes various families like the discrete Pearson and Katz families and several individual distributions are discussed in Chapter 3. In practice, the formulas in (2.69) and (2.70) will work if we replace $\sigma_k g_k(x)$ by function $a_k(x)$, which can be determined from the data, without actually using σ_k.

When any two of the functions $h(x)$, $\lambda(x)$ and $F(x)$ are known, the third can be determined from the identity

$$F(x) = h(x)[h(x) + \lambda(x) - \lambda(x)h(x)]^{-1}.$$

Similarly, we also have

$$S(x) = \lambda(x)[h(x) + \lambda(x) - \lambda(x)h(x)]^{-1}$$

and

$$f(x) = \lambda(x)h(x)[h(x) + \lambda(x) - \lambda(x)h(x)]^{-1}.$$

From the last three forms, we arrive at

$$\frac{f(x+1)}{f(x)} = \frac{h(x+1)(1-h(x))}{h(x)} = \frac{\lambda(x+1)}{\lambda[1-\lambda(x+1)]}, \tag{2.71}$$

which will be useful in later chapters.

2.8 REVERSED VARIANCE RESIDUAL LIFE

Just as the mean of the reversed residual life ${}_xX$, the variance of ${}_xX$ is also an important function reliability analysis, called the *reversed variance residual life* or *variance inactivity time*, and is denoted by $v(x)$. In algebraic manipulations, different expressions for $v(x)$ have been employed. These are

$$\begin{aligned} v(x) &= V(x - X|X < x) = V(X|X < x) \\ &= E(X^2|X < x) - E^2(X|X < x) \end{aligned} \tag{2.72}$$

and

$$v(x) = E[(x - X)^2|X < x] - r^2(x). \tag{2.73}$$

Further,

$$\begin{aligned} E((x-X)^2|X<x) &= \frac{1}{F(x-1)} \sum_{t=0}^{x-1} (x-t)^2 f(t) \\ &= \frac{1}{F(x-1)} \sum_{t=1}^{x-1} (x-t)^2 [F(t) - F(t-1)] \\ &= \frac{2}{F(x-1)} \sum_{t=1}^{x} (x-t) F(t) - r(x) \\ &= \frac{2}{F(x-1)} \sum_{t=1}^{x} \sum_{u=1}^{t} F(u-1) - r(x), \end{aligned}$$

or

$$v(x) = \frac{2}{F(x-1)} \sum_{t=1}^{x} \sum_{u=1}^{t} F(u-1) - r(x)(r(x)+1). \tag{2.74}$$

Example 2.20. Consider the geometric distribution $G(p)$ in Example 2.2, for which $F(x) = 1 - q^{x+1}$. Then, the conditional probability mass function of $X|(X < x)$ is given by

$$f_X(t) = \frac{pq^t}{1-q^x}, \quad t = 0, 1, \ldots, x-1. \tag{2.75}$$

We use (2.72) to calculate $v(x)$. For this, from (2.75), we have

$$\sum_{t=0}^{x-1} pq^t = 1 - q^x.$$

Differentiating with respect to q, we get

$$\sum tpq^{t-1} = \sum_{0}^{x-1} q^t - xq^{x-1}.$$

Simplifying so as to make the left hand side an expected value, we get

$$\sum_{t=0}^{x-1} \frac{tpq^t}{1-q^x} = E(X|X < x) = \frac{q}{p} - \frac{xq^x}{1-q^x}. \tag{2.76}$$

Differentiating (2.76) again with respect to q and rearranging the terms in the same manner, we get

$$E(X^2|X < x) = \frac{q(1+q) - q^x((1+q)(px+q) + x(x-1)p^2)}{p^2(1-q^x)}. \tag{2.77}$$

The variance is now found from (2.72).

Various reliability functions in reversed residual lifetime for some distributions are exhibited in Table 2.4 and the graph of these functions for arithmetic distribution is presented in Fig. 2.3.

In the continuous case, identities that connect the three functions $\lambda(x)$, $r(x)$ and $v(x)$ have been established. The corresponding results in the discrete case are

$$v(x+1) - v(x) = \lambda(x)[r(x(r(x+1) - 1) - v(x)] \tag{2.78}$$

and

$$\frac{v(x+1)}{r(x+1)-1} + r(x+1) - 1 = \frac{v(x)}{r(x)} + r(x). \tag{2.79}$$

TABLE 2.4 The reversed hazard function, mean residual life and variance residual life for some distributions

Distribution	$F(x)$	$\lambda(x)$	$r(x)$	$v(x)$
geometric	$1-q^{x+1}$	$\frac{q^x p}{1-q^{x+1}}$	$\frac{xp-q(1-q^x)}{p(1-q^x)}$	(2.76) and (2.77)
uniform	$\frac{x}{b}$, $x=1,2,\ldots,b$	$\frac{1}{x}$	$\frac{x}{2}$	$\frac{x(x-2)}{12}$
arithmetic	$\frac{x(x+1)}{b(b+1)}$, $x=1,2,\ldots,b$	$\frac{2}{x+1}$	$\frac{x+1}{3}$	$\frac{(x+1)(x-2)}{18}$
reversed geometric	$(1+c)^{x-b}$, $x=0,1,\ldots,b$	$\frac{c}{1+c}$	$\frac{1+c}{c}[1-\frac{1}{(1+c)^x}]$	$\frac{[(1+c)^x+1][(1+c)^{x-1}+1]-2(cx+1)}{c^2(1+c)^{2(x-1)}}$

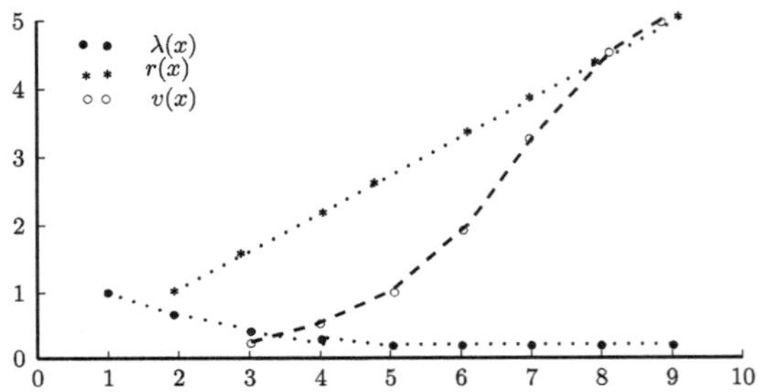

FIGURE 2.3 Reliability functions in reversed time for the arithmetic distribution.

Eqs (2.78) and (2.79) are employed in finding $v(x)$ when the others are known, especially in characterization problems and also in the discussions on the monotonicity of the reversed variance residual life function.

As in the case of the usual mean and variance residual lives, we have

$$r(x) = E\left[\frac{1}{\lambda(x)}|X < x\right]$$

and

$$v(x+1) = E[r(x)(r(x+1)-1)|X < x].$$

There are some special relationships between $\lambda(x)$, $r(x)$ and $v(x)$ that characterize certain families of distributions. Some important results in this connection are presented in the next two theorems.

Theorem 2.15. *For a random variable X with the support* $(1,2,\ldots,b)$*, the relationship*

$$\lambda(x)r(x) = c, \quad x = 2,3,\ldots,b, \quad 0 < c < 1, \tag{2.80}$$

holds if and only if the distribution of X is

$$F(x) = \frac{(b-1)!(\theta)_x}{(x-1)!(\theta)_b}, \quad x = 1, 2, \ldots, b, \tag{2.81}$$

with $\theta = \frac{c}{1-c}$, *a positive integer.*

Proof. Suppose (2.80) holds for X. From

$$\lambda(x) = \frac{1 + r(x) - r(x+1)}{r(x)},$$

we have the difference equation

$$r(x+1) - r(x) = 1 - c.$$

Iterating for x, with the boundary condition $r(2) = 1$, we get

$$r(x+1) = c + (1-c)x$$

and hence

$$\frac{r(x+1) - 1}{r(x)} = \frac{x-1}{\theta + x - 1}.$$

Substituting in (2.65), we obtain (2.81). To prove the converse, from (2.81), we have

$$\lambda(x) = \frac{\theta}{\theta + x - 1} = \frac{c}{c + (1-c)(x-1)}$$

so that $r(x)\lambda(x) = c$. □

Theorem 2.16. *The random variable X in Theorem 2.15 satisfies the property*

$$\frac{v(x)}{r(x)(r(x) - 1)} = k, \tag{2.82}$$

a constant in $(0, 1)$, *if and only if its distribution is* (2.81) *when* $k < 1$ *and the distribution is* (2.68) *when* $k = 1$.

Proof. First, we note that

$$r_2(x) = E[(x - X)^2 | X < x] = \frac{1}{F(x-1)} \sum_{t=1}^{x-1} (x-t)^2 f(t)$$

$$= \frac{2}{F(x-1)} \sum_{t=1}^{x-1} (x-t) F(t) - r(x),$$

or

$$[r_2(x) + r(x)]F(x-1) = 2\sum_{t=1}^{x-1} F(t). \tag{2.83}$$

Changing x to $x+1$ in (2.83) and then subtracting (2.83), we get

$$r_2(x+1)r(x) - r_2(x)(r(x+1)-1) - r(x)(r(x+1)-1) = r(x)r(x+1). \tag{2.84}$$

When (2.82) is satisfied with $k = 1$, from

$$v(x) + r^2(x) = r_2(x)$$

and (2.82), we obtain

$$r_2(x) = r(x)(r(x)-1) + r^2(x).$$

Substituting in (2.84) and simplifying, we get

$$r(x+1) - r(x) = 0$$

and hence $r(x)$ is a constant and the distribution is (2.68). On the other hand, when the distribution is (2.68), $r(x) = \frac{1+c}{c}$ and then (2.74) yields

$$\begin{aligned} v(x) &= \frac{2(c+1)}{cF(x-1)}\left[\sum_{t=2}^{x}(1+c)^{t-1-b} + \frac{(1+c)^{t-b}}{c}\right] - \frac{(1+c)^2}{c^2} - \frac{1+c}{c} \\ &= \frac{2(c+1)}{c(1+c)^{x-1}}\left[(1+c)((1+c)^{x-1}-1)\frac{1+c}{c}\right] - \frac{(1+c)^2}{c^2} - \frac{1+c}{c} \\ &= \frac{c+1}{c^2}. \end{aligned}$$

Also, $r(x)[r(x)-1] = \frac{c+1}{c}$, so that the theorem is proved when $k = 1$. In the more general case, when $k < 1$, assume that X has distribution (2.81). Then,

$$r(x) = c + (1-c)(x-1) = \frac{\theta + x - 1}{\theta + 1}. \tag{2.85}$$

Further, in formula (2.74), we have

$$\frac{2}{F(x-1)}\sum_{t=1}^{x}\sum_{u=1}^{t}F(u-1)=\frac{2}{F(x-1)}\sum_{t=2}^{x}r(t)F(t-1)$$
$$=\frac{2(b-1)!}{F(x-1)(\theta+1)(\theta)_b}\sum_{t=2}^{x}\frac{(\theta+t-1)(\theta)_{t-1}}{(t-2)!}$$
$$=\frac{2(b-1)!}{(\theta+1)(\theta)_b F(x-1)}\sum_{t=2}^{x}\frac{(\theta)_t}{(t-2)!}. \qquad (2.86)$$

Upon using the combinational identity

$$\sum_{j=0}^{r}\binom{j+a-1}{a-1}=\binom{r+a}{a},$$

we have

$$\sum_{t=x}^{x}\frac{(\theta)_t}{(t-2)!}=\frac{(x+\theta)!}{(\theta+2)!(x-2)!}.$$

Substituting for $F(x-1)$ from (2.81) and simplifying (2.86), we get

$$\frac{2}{F(x-1)}\sum_{t=1}^{x}\sum_{u=1}^{t}F(u-1)=\frac{2(\theta+x-1)(\theta+x)}{(\theta+1)(\theta+2)}.$$

Now from (2.85),

$$r(x)[r(x)+1]=\frac{(20+x)(\theta+x-1)}{(\theta+1)^2}$$

and so the formula (2.74) yields

$$v(x)=\frac{\theta(\theta+x-1)(x-2)}{(\theta+1)^2(\theta+2)}.$$

Thus,

$$\frac{v(x)}{r(x)(r(x)-1)}=\frac{\theta}{\theta+2}=k<1.$$

Conversely, under the hypothesis of the theorem, we have

$$v(x)=kr(x)(r(x)-1).$$

With the aid of (2.79), we write

$$k\frac{r(x+1)(r(x+1)-1)}{r(x+1)-1}+r(x+1)-1=\frac{kr(x)(r(x)-1)}{r(x)}+r(x).$$

The last equation leads to

$$r(x+1)-r(x)=\frac{1-k}{1+k}$$

which on solving using the boundary condition $r(2)=1$, gives

$$r(x)=1+\frac{1-k}{1+k}(x-2).$$

Setting $k=\frac{c}{2-c}$ so that $0<c<1$ as stipulated, we have

$$r(x)=c+(1-c)(x-1)$$

and hence the distribution is (2.81). □

Remark 2.14. We can write the distribution function in (2.81) also as

$$F(x)=\frac{\binom{\theta+x-1}{x-1}}{\binom{\theta+b-1}{b-1}}, \quad x=1,2,\ldots,b. \tag{2.87}$$

Remark 2.15. Eq. (2.84) represents a family of finite range distributions that contains the uniform distribution for $\theta=1$ and the arithmetic law for $\theta=2$.

Arising from characteristics of the finite range laws discussed above, we have some further characterizations. These results can be proved by invoking Cauchy-Schwarz inequality, as done in Theorems 2.6 and 2.7.

Theorem 2.17. *Let X be a discrete random variable with the support $(0,1,\ldots,b)$. Then:*

(i) $E(\lambda(X))\geq\frac{1}{E(\lambda(X))}$ *with the equality holding if and only if X has reversed geometric law;*
(ii) $E(r(X))\geq\frac{1}{Er(X)}$ *and the equality holds if and only if X follows distribution* (2.68).

Theorem 2.18. *Let X be a discrete random variable with the support $(1,2,\ldots,b)$. Then,*

$$E\left(\frac{\lambda(x)}{r(X)}\right)\geq\frac{1}{E(\frac{\lambda(X)}{r(X)})}$$

if and only if the distribution of X is (2.81).

2.9 ODDS FUNCTION

Along with the traditional reliability functions presented so far, there has been some interest in discovering the potential of odds function and log odds function in reliability analysis. The motivation for the consideration of these two functions are (i) they are easy to compute and interpret (ii) the estimation of these functions is relatively simpler, and (iii) the behaviour of other reliability functions can be ascertained through them.

The concept of odds ratio originated from gambling wherein the odds of an event A against another event B is defined as the ratio $P(A)/P(B)$. In reliability theory, we can take the event A as survival of age x and B as failure by age x. Then, the odds ratio of the events become a function of x. The odds ratio for surviving age x is defined as

$$\omega(x) = \frac{P(X > x)}{P(X \leq x)} = \frac{S(x+1)}{F(x)}, \quad x = -1, 0, 1, \ldots, \tag{2.88}$$

and it is called the odds function for survival. Similarly, the odds function for failure by age x is

$$\bar{\omega}(x) = \frac{F(x)}{S(x+1)}.$$

From the definitions, it follows that

(i) $\omega(-1) = \infty$, $\omega(\infty) = 0$ and $\omega(x)$ is decreasing,
(ii) $\bar{\omega}(-1) = 0$, $\bar{\omega}(\infty) = \infty$ and $\bar{\omega}(x)$ is increasing.

Odds functions ω and $\bar{\omega}$ are important tools in survival analysis and medical studies in developing models for survival data and in comparing a treatment group with a control group. We refer to Collett (1994) and Kirmani and Gupta (2001) for further details. There has not been much study about the role of odds functions in reliability analysis, especially in the discrete case. We note that

$$\omega(x) = \frac{1}{F(x)} - 1, \quad x = 0, 1, \ldots,$$

and

$$\bar{\omega}(x) = \frac{1}{S(x+1)} - 1, \quad x = 0, 1, \ldots, \ \bar{\omega}(-1) = 0.$$

Accordingly, both $\omega(x)$ and $\bar{\omega}(x)$ determine the distribution of X uniquely through the expressions

$$F(x) = \frac{1}{1+\omega(x)} = \frac{\bar{\omega}(x)}{1+\bar{\omega}(x)} \tag{2.89}$$

and

$$S(x+1) = \frac{\omega(x)}{1+\omega(x)} = \frac{1}{1+\bar{\omega}(x)}. \tag{2.90}$$

It is easy to see that the hazard and reversed hazard rates are

$$h(x) = \frac{\bar{\omega}(x) - \bar{\omega}(x-1)}{1+\bar{\omega}(x)}, \quad x = 0, 1, 2, \ldots, \tag{2.91}$$

and

$$\lambda(x) = \frac{\omega(x-1) - \omega(x)}{1+\omega(x)}. \tag{2.92}$$

We can express the monotonicities of $h(x)$ and $\lambda(x)$ in terms of $\omega(x)$ and $\bar{\omega}(x)$.

Theorem 2.19. *(i) $h(x)$ is decreasing $\Rightarrow \lambda(x)$ is decreasing $\Rightarrow \omega(x)$ is convex; (ii) $\lambda(x)$ is increasing $\Rightarrow h(x)$ is increasing $\Rightarrow \bar{\omega}(x)$ is convex.*

Proof. $h(x)$ is decreasing

$$\begin{aligned}
&\Rightarrow \frac{f(x+1)}{S(x+1)} - \frac{f(x)}{S(x)} \leq 0 \\
&\Rightarrow \frac{f(x+1)}{1-F(x+1)} - \frac{f(x)}{1-F(x-1)} \leq 0 \\
&\Rightarrow \lambda(x+1)\frac{F(x+1)}{1-F(x)} - \lambda(x)\frac{F(x)}{1-F(x-1)} \leq 0.
\end{aligned}$$

Since $F(x)$ is non-decreasing $\frac{F(x+1)}{1-F(x)} \geq \frac{F(x)}{1-F(x-1)}$ and so $\lambda(x+1) \leq \lambda(x)$, which proves that $\lambda(x)$ is decreasing. Next,

$$\begin{aligned}
\omega(x+1) - \omega(x) &= \frac{1-F(x+1)}{F(x+1)} - \frac{1-F(x)}{F(x)} = \frac{1}{F(x+1)} - \frac{1}{F(x)} \\
&= \Big(\prod_{x+2}^{\infty}(1-\lambda(t))\Big)^{-1} - \Big(\prod_{x+1}^{\infty}(1-\lambda(t))\Big)^{-1} \\
&= -\lambda(x+1)\prod_{x+1}^{\infty}(1-\lambda(t))^{-1}.
\end{aligned}$$

Similarly,

$$\omega(x) - \omega(x-1) = \frac{S(x+1)}{F(x)} - \frac{S(x)}{F(x-1)} = -\lambda(x)\left[\prod_{x}^{\infty}(1-\lambda(t))\right]^{-1}.$$

Thus,

$$(\omega(x+1)-\omega(x))-(\omega(x)-\omega(x-1))$$
$$=\left[\lambda(x)A(x+1)-(\lambda(x)-\lambda(x+1))\right]\left[\prod_{x+1}^{\infty}(1-\lambda(t)\right]^{-1}$$

which is non-negative whenever $\lambda(x)$ is decreasing, and so $\omega(x)$ is convex.

The proof of (ii) is similar. □

Example 2.21. Consider the uniform distribution in Example 2.2. In this case,

$$\omega(x)=\frac{b-x}{x};\quad \bar{\omega}(x)=\frac{x}{b-x}.$$

The hazard and reversed hazard rates are obtained by using (2.91) and (2.92) as

$$h(x)=\frac{1}{b-x+1}\text{ and }\lambda(x)=\frac{1}{x}.$$

Notice that $\lambda(x)$ is decreasing and $h(x)$ is increasing, so that the reverse implication $h(x)$ increasing means $\lambda(x)$ is increasing in (ii) is not true. However, it can be easily verified that $\bar{\omega}(x)$ is convex.

The concepts based on residual life and reversed residual life can also be expressed in terms of odds functions. Recall that the survival function of the residual life $(X-x|X>x)$ is

$$S_x(t)=\frac{S(x+t+1)}{S(x+1)},\qquad t=0,1,\ldots.$$

The residual odds functions are then

$$\omega_x(t)=\frac{1}{F_x(t)}-1=\frac{S(x+t+2)}{S(x+1)-S(x+t+2)}=\frac{1-F_x(t)}{F_x(t)},$$

or

$$F_x(t)=\frac{1}{1+\omega_x(t)}$$

and similarly

$$\bar{\omega}_x(t)=\frac{1}{S_x(t+1)}-1\text{ or }S_x(t+1)=\frac{1}{1+\bar{\omega}_x(t)}.$$

In terms of $\omega(x)$,

$$\bar{\omega}_x(t)=\frac{1+\bar{\omega}(x)}{\bar{\omega}(x+t+1)-\bar{\omega}(x)}$$

and

$$\omega_x(t) = \frac{F(x+t) - F(x)}{1 - F(x)} = \frac{[1+\omega(x)]\omega(x+t+1)}{\omega(x) - \omega(x+t+1)}.$$

Also, the mean residual life function is

$$m(x) = \frac{1}{S(x+1)} \sum_{x+1}^{\infty} S(t) = 1 + \bar{\omega}(x) \sum_{t=x+1}^{\infty} \frac{1}{1+\bar{\omega}(t-1)}. \tag{2.93}$$

Eq. (2.93) leads to the recurrence relation

$$m(x-1) = 1 + \frac{1+\bar{\omega}(x-1)}{1+\bar{\omega}(x)} m(x), \quad x = 0, 1, 2, \ldots \tag{2.94}$$

Theorem 2.20. *X has decreasing (increasing) mean residual life if and only if*

$$m(x) \leq (\geq) \frac{1+\bar{\omega}(x)}{\bar{\omega}(x) - \bar{\omega}(x-1)}.$$

Proof. By means of (2.91), we can write

$$\begin{aligned} X \text{ has decreasing mean residual life } &\Leftrightarrow m(x-1) - m(x) \geq 0 \\ &\Leftrightarrow 1 + \left(\frac{1+\bar{\omega}(x-1)}{1+\bar{\omega}(x)} - 1\right) m(x) \geq 0 \\ &\Leftrightarrow m(x) \leq \frac{1+\bar{\omega}}{\bar{\omega}(x) - \bar{\omega}(x-1)}. \end{aligned}$$

In terms of the odds functions, the reversed mean residual life is

$$r(x) = \frac{1}{F(x-1)} \sum_{t=1}^{x} F(t-1) = (1+\bar{\omega}(x-1)) \sum_{t=1}^{x} \frac{1}{1+\omega(t-1)}$$

and so

$$r(x+1) = 1 + \frac{1+\omega(x)}{1+\omega(x-1)} r(x), \quad x = 0, 1, \ldots \qquad \square$$

In the same manner as we have proved Theorem 2.3, we note the following.

Theorem 2.21. *X has increasing reversed mean residual life if and only if, for all x,*

$$r(x) \leq \frac{1+\omega(x-1)}{\omega(x-1) - \omega(x)}.$$

Instead of considering the odds functions, it is sometimes beneficial to use the odds rate defined as

$$\bar{\omega}_R(x) = \bar{\omega}(x) - \bar{\omega}(x-1). \tag{2.95}$$

The function $\bar{\omega}_R(x)$ is interpreted as the rate at which $\omega(x-1)$ changes at age x. It has the following properties:

(i) $\bar{\omega}_R(x) = \frac{1}{S(x+1)} - \frac{1}{S(x)} = \frac{h(x)}{S(x+1)}$;
(ii) The sequence of rates $\langle \bar{\omega}_R(x) \rangle$ determine the distribution of X uniquely as

$$S(x) = \left[1 + \sum_{t=0}^{x-1} \bar{\omega}_R(t)\right]^{-1}; \tag{2.96}$$

(iii) If $\bar{\omega}_R(x)$ is decreasing ($\bar{\omega}$ is concave), then $h(x)$ is decreasing;
(iv) If $h(x)$ is increasing, then $\bar{\omega}_R(x)$ is also increasing.

Example 2.22. Assume that $\bar{\omega}_R(x) = \frac{b}{(b-x)(b-x+1)}$, $x = 0, 1, 2, \ldots$. Writing

$$\bar{\omega}_R(x) = \frac{b}{b-x} - \frac{b}{b-x+1},$$
$$\sum_{t=1}^{x-1} \bar{\omega}_R(x) = \frac{x-1}{b-x+1}.$$

Employing formula (2.96), we have

$$S(x) = \frac{b}{b-x+1},$$

the survival function of the discrete uniform distribution.

Property (iii) of $\bar{\omega}_R(x)$ given above tells us that the class of distributions with decreasing $\bar{\omega}_R(x)$ is a subset of the decreasing failure rate class and also that a sufficient condition for failure rate to be decreasing is that $\bar{\omega}(x)$ is concave. Thus, we have a stronger condition that assists in characterizing and modelling failure time data. Further, when $h(x)$ is increasing, $\bar{\omega}_R(x)$ is also increasing, providing an alternative proof of (ii) in Theorem 2.3. For detailed discussion of the results in this section, we refer to Nair and Sankaran (2015b).

2.10 LOG-ODDS FUNCTIONS AND RATES

The log-odds functions and rates were introduced by Zimmer et al. (1998) as an alternative reliability measure and further propagated in the work of Wang et al. (2003, 2008). These functions were proposed as an alternative to the hazard rate in situations under which the device considered have high reliability or the corresponding hazard rate is non-monotone. When X is a discrete lifetime, Khorashadizadeh et al. (2013a) defined the log-odds function as

$$\bar{L}(x) = \log \frac{F(x)}{S(x+1)} = \log \bar{\omega}(x), \quad x = 0, 1, 2, \ldots, \tag{2.97}$$

and the corresponding log-odds rate as

$$\bar{L}_R(x) = \bar{L}(x) - \bar{L}(x-1).$$

They have shown the following:

(a) $\bar{L}_R(x) = \lambda_1(x) + h_1(x)$, where λ_1 and h_1 are the alternative reversed hazard and hazard rates of X discussed in Section 2.2;

(b) $F(x)$ is characterized by

$$F(x) = \frac{e^{a(x)+b}}{1+e^{a(x)+b}}, \quad x = 0, 1, \ldots,$$

where $a(x) = \sum_{t=1}^{x} \bar{L}(t)$ and $b = \log \frac{F(0)}{S(1)}$;

(c) In terms of $Y = \log X$, we have

$$\bar{L}_Y(x) = \bar{L}_X(e^x),$$

and hence

$$\bar{L}_{R,Y}(x) = \sum_{t=1}^{x} (\lambda_1(x) + h_1(x)) - \log \frac{F_X(xe^{-1})}{S_X((x+1)e^{-1})^{+b}},$$

where $b_1 = \log \frac{f_X(0)}{1-f_X(0)}$;

(d) X has increasing log-odds rate in terms of x $(y = \log x)$ if and only if $L(x)$ is convex with respect to x $(y = \log x)$.

From the above discussions, it is clear that $\bar{L}_R(x)$ is directly related to the alternative hazard and reversed hazard rates whereas $\bar{\omega}_R(x)$ is related to the usual hazard rate. There are other mathematical properties that make $\bar{\omega}(x)$ and $\omega(x)$ more desirable than their logarithms.

Theorem 2.22. *If X is a discrete random variable with odds function $\bar{\omega}(x)$, then there exists a random variable X^* with the same support as X with alternative hazard rate $h_1^*(x) = \bar{\omega}(x)$. The survival functions $S(x)$ of X and $S^*(x)$ of X^* are related by*

$$S(x) = (1 - \log S^*(x))^{-1}. \tag{2.98}$$

Proof. By virtue of the properties of $\bar{\omega}(x)$,

$$S^*(x) = \exp[-\bar{\omega}(x-1)], \quad x = 0, 1, 2, \ldots \tag{2.99}$$

is a survival function. Let X^* denote the corresponding random variable. From the representation in (2.14), the cumulative hazard rate of X^* is

$$H_1^*(x) = \sum_0^{x-1} h_1^*(t), \tag{2.100}$$

where $h_1^*(t) = \log \frac{S^*(x+1)}{S^*(x)}$ is the alternative hazard rate of X^*, and it satisfies

$$H_1^*(x) = -\log S^*(x) = \bar{\omega}(x-1) \tag{2.101}$$

Now, from (2.100) and (2.101), we have

$$h_1^*(x) = H_1^*(x+1) - H_1^*(x) = \bar{\omega}(x) - \bar{\omega}(x-1) = \bar{\omega}_R(x),$$

proving the first part. Using (2.99),

$$S^*(x) = \exp\left[-\frac{1}{S(x)} + 1\right]$$

so that (2.98) holds. Since there is a one-to-one correspondence between $S(x)$ and $S^*(x)$, they have the same support, and this completes the proof. □

Some observations from Theorem 2.22 are as follows:

(a) The converse of Theorem 2.22 is also true; that is, if $S^*(x)$ is the survival function of a random variable X^* satisfying (2.98), then $S(x)$ is the survival function of a random variable X for which $h_1^*(x) = \bar{\omega}_R(x)$;

(b) Since

$$\bar{\omega}_R(x) = \frac{h(x)}{S(x+1)}$$

we have

$$h^*(x) = \frac{h(x)}{S(x+1)},$$

and

$$h(x) = h^*(x) \exp\left[-\sum_0^x h_1^*(x)\right],$$

$$h_1^*(x) = \bar{\omega}(x) - \bar{\omega}(x-1);$$

(c) A parallel result that involves the odds function $\omega(x)$ and the alternative reversed hazard rate $\lambda_1^*(x)$ is also possible. To see this, it is easy to recognize that $F^*(x) = \exp[-\omega(x)]$ is a distribution function, with alternative reversed hazard rate

$$\lambda_1^*(x) = \log \frac{F^*(x)}{F^*(x-1)} = \omega(x-1) - \omega(x).$$

Since $\omega(x)$ is a decreasing function, $\omega(x-1) - \omega(x) = \omega_R(x)$ is the rate at which $\omega(x)$ is decreasing and is therefore an odds rate. Thus, there exists a distribution specified by $F^*(x)$ for which the odds rate $\omega_R(x)$ of $F(x)$ is the reversed hazard rate of $F^*(x)$ and the connection between the two is explained in the next theorem.

Theorem 2.23. *Corresponding to a discrete random variable X with distribution function $F(x)$ and odds rate $\omega(x)$, there exist another distribution function $F^*(x)$ with the same support as $F(x)$ and alternative reversed hazard rate $\lambda_1^*(x)$ that satisfies*

$$\lambda_1^*(x) = \omega_R(x). \tag{2.102}$$

Also, $F(x)$ and $F^*(x)$ are related through

$$F(x) = [1 - \log F^*(x)]^{-1}. \tag{2.103}$$

Conversely, if for two distribution functions $F(x)$ and $F^*(x)$ (2.103) *holds, then $\lambda_1^*(x) = \omega_R(x)$.*

Remark 2.16.

(1) The distribution function $F^*(x)$ represents the random variable X^* of Theorem 2.22;
(2) $\omega_R(x) = \frac{\lambda(x)}{F(x-1)} = \lambda^*(x)$;
(3) $\lambda(x) = \lambda^*(x)F(x-1)$ or $\lambda^*(x) = \lambda(x)/\prod_{t=x}^{\infty}(1-\lambda(t))$.

Example 2.23. The distributions of X and X^* can be mutually characterized. If X^* is geometric with

$$S^*(x) = q^x, \quad x = 0, 1, 2, \ldots, \quad 0 < q < 1,$$

the distribution of X is

$$S(x) = \frac{1}{1 - \alpha \log q}, \quad x = 0, 1, 2, \ldots.$$

In this case,

$$\bar{\omega}(x) = -(x+1)\log q$$

TABLE 2.5 Estimation of $\omega(x)$ and $\bar{\omega}(x)$

Age	Number dying	$\hat{S}(x)$	$\hat{F}(x)$	$\bar{\omega}(x)$	$\omega(x)$
0–4	34	1.0000	0.0556	0.0589	17.9856
5–14	9	0.9444	0.0704	0.0757	13.2100
15–24	23	0.9296	0.1080	0.1211	8.2576
25–34	19	0.8920	0.1319	0.1616	6.1811
35–44	16	0.8609	0.1653	0.1980	5.0505
45–54	59	0.8347	0.2619	0.3548	2.8185
55–64	101	0.7581	0.4272	0.7458	1.3408
65–75	185	0.5728	0.7300	2.7037	0.3699
76–83	97	0.2700	0.8870	7.8496	0.1274
83+	68	0.1112	1.0000	8.0917	0.1236

and

$$\omega_R(x) = -\log q = h_1^*(x).$$

Example 2.24. In this example, we present an analysis of real data on the number of deaths following surgery in a hospital in the US classified according to the age of the patients (Mosteller and Tukey, 1977). The survival function is estimated from the sample as

$$\hat{S}(x) = \frac{\text{number of observations } \geq x}{\text{total number of observations}},$$

while

$$\hat{F}(x) = \frac{\text{number of observations } \leq x}{\text{total number of observations}}.$$

$$\hat{\bar{\omega}}(x) = \frac{1}{\hat{S}(x+1)} - 1 \text{ and } \hat{\omega}(x) = \frac{1}{\hat{F}(x)} - 1.$$

Table 2.5 presents the necessary calculations and Fig. 2.4 shows the shapes of the $\omega(w)$ and $\bar{\omega}(x)$ curves.

2.11 MIXTURE DISTRIBUTIONS

The role of mixture distributions in reliability studies was mentioned earlier in Section 2.1. When the mixture is of the form (1.11), we have

$$f(x) = \alpha f_1(x) + (1-\alpha) f_1(x)$$

and

$$S(x) = \alpha S_1(x) + (1-\alpha) S_2(x), \tag{2.104}$$

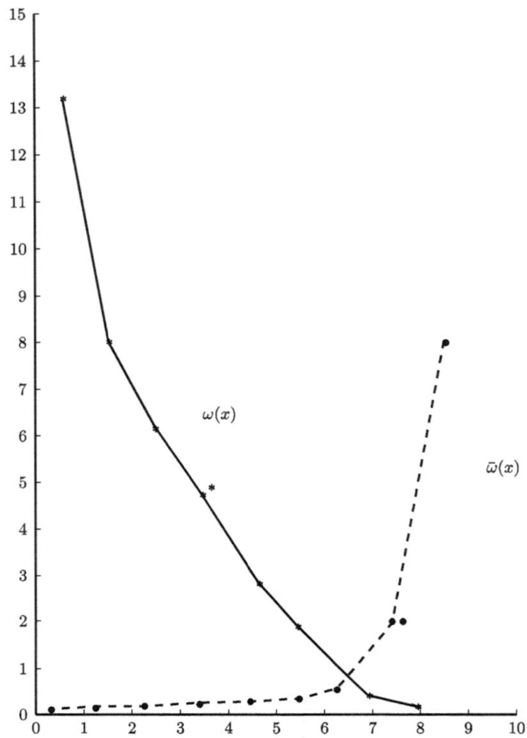

FIGURE 2.4 Plots of $\omega(x)$ and $\bar{\omega}(x)$.

where $f_1(f_2)$ is the probability mass function and S_1 (S_2) is the survival function corresponding to F_1 (F_2). If $h(x)$, $k_1(x)$ and $k_2(x)$, respectively, denote the hazard rates of f, f_1 and f_2, it is easy to see that

$$h(x) = p(x)k_1(x) + (1 - p(x))k_2(x), \tag{2.105}$$

where

$$p(x) = \frac{\alpha S_1(x)}{\alpha S_1(x) + (1-\alpha)S_2(x)}.$$

Likewise, for the reversed hazard rates λ, $\bar{\lambda}_1$ and $\bar{\lambda}_2$ of f, f_1 and f_2, respectively, we have

$$\lambda(x) = p_1(x)\bar{\lambda}_1(x) + (1 - p_1(x))\bar{\lambda}_2(x), \tag{2.106}$$

with

$$p_1(x) = \frac{\alpha F_1(x)}{\alpha F_1(x) + (1-\alpha)F_2(x)}.$$

Finite mixtures of two components are commonly used in heterogeneous populations in which the elements are classified into two categories. It follows from (2.105) that

$$\min(k_1(x), k_2(x)) \le h(x) \le \max(k_1(x), k_2(x)),$$

and also that

$$k_1(x) \le h(x) \le k_2(x)$$

whenever $k_1(x) \le k_2(x)$ for all x.

Theorem 2.24. *If $k_1(x) \le k_2(x)$ for all x, $p(x)$ is an increasing function.*

Proof. We have

$$\begin{aligned} p(x+1) - p(x) &= \frac{\alpha S_1(x+1)}{\alpha S_1(x+1) + (1-\alpha)S_2(x+1)} \\ &\quad - \frac{\alpha S_1(x)}{\alpha S_1(x) + (1-\alpha)S_2(x)}. \end{aligned}$$

The sign of the previous equation depends on

$$\begin{aligned} t(x) &= \alpha S_1(x+1)[\alpha S_1(x) + (1-\alpha)S_2(x)] - \alpha S_1(x)[\alpha S_1(x+1) \\ &\quad - (1-\alpha)S_2(x+1)] \\ &= (1-\alpha)\alpha[S_1(x+1)S_2(x) - S_1(x)S_2(x+1)] \\ &= (1-\alpha)\alpha\left[\frac{S_1(x+1)}{S_1(x)} - \frac{S_2(x+1)}{S_2(x)}\right] S_1(x)S_2(x) \\ &= \alpha(1-\alpha)S_1(x)S_2(x)[k_2(x) - k_1(x)] \ge 0, \end{aligned}$$

since $k_1(x) \le k_2(x)$. Thus, $p(x)$ is increasing. This result has the interpretation that when the lifetime distribution is a mixture, the weakest items will die out first.

When the distribution of X is indexed by a parameter θ, where θ is the value of a random variable Θ defined on $[0, \infty)$ with distribution function $C(\theta)$, from (1.3), the survival function and the probability mass function of X are given by

$$\bar{G}(x) = \int S(x|\theta)dC(\theta)$$

and

$$g(x) = \int f(x|\theta)dC(\theta)$$

so that the mixture has its hazard rate as

$$h^*(x) = \frac{\int_\Theta f(x|\theta)dC(\theta)}{\int_\Theta S(x|\theta)dC(\theta)}$$
$$= \int_\Theta h(x|\theta)dC(\theta|x),$$

where

$$h(x|\theta) = \frac{f(x|\theta)}{F(x|\theta)}$$

is the conditional hazard rate of X, given $\Theta = \theta$, and

$$dC(\theta|x) = \frac{S(x|\theta)dC(\theta)}{\int_\Theta S(x|\theta)dC(\theta)}$$

is the conditional density function of θ, given $X \geq x$. When $c(\theta) = \frac{dC(\theta)}{d\theta}$, the density function of Θ, $c(\theta|x)$, defines the conditional probability density function of θ with the same support as that of Θ. In a Bayesian context, $c(\theta)$ can be viewed as a prior distribution of θ, and $C(\theta|x)$ as the posterior distribution of θ after observing the data on X. Models in which θ is regarded as random are called *frailty models* which are extensively discussed in survival analysis. In particular, if a representation of the form

$$S_1(x) = S_1(x|\theta) = [S(x)]^\theta$$

holds, then from (2.17), we have the alternative hazard rates of S_1, say $\bar{h}_1(x)$, and S satisfy the relationship

$$\bar{h}_1(x) = \theta h_1(x).$$

We say that a random variable X_1 with survival function $S_1(x)$ is the proportional hazard rates model corresponding to X with survival function $S(x)$. □

Some similar results exist for reversed hazard rates as well.

Theorem 2.25. *In the case of reversed rate functions, if $\bar{\lambda}_1(x) > \bar{\lambda}_2(x)$ for all x, then $p_1(x)$ is an increasing function.*

Proof. We have

$$p_1(x+1) - p_1(x)$$
$$= \frac{\alpha F_1(x+1)}{\alpha F_1(x+1) + (1-\alpha)F_2(x+1)} - \frac{\alpha F_1(x)}{\alpha F_1(x) + (1-\alpha)F_2(x)}.$$

The sign of the left hand side depends on

$$\alpha(1-\alpha)F_1(x+1)F_2(x+1)\left[\frac{F_2(x)}{F_2(x+1)} - \frac{F_1(x)}{F_1(x+1)}\right]$$
$$= \alpha(1-\alpha)F_1(x+1)F_2(x+1)[\lambda_1(x+1) - \lambda_2(x+1)]$$

which is non-negative since $\bar{\lambda}_1(x) > \bar{\lambda}_2(x)$ for all x. Hence, $p_1(x)$ is an increasing function.

Assuming θ to be random with distribution function $C(\theta)$, the mixture has reversed hazard rate

$$\lambda^*(x) = \frac{\int_\Theta f(x|\theta)dC(\theta)}{\int_\Theta F(x|\theta)dC(\theta)}$$
$$= \int_\Theta \lambda(x|\theta)dC^*(\theta|x),$$

where

$$\lambda(x|\theta) = \frac{f(x|\theta)}{F(x|\theta)}$$

is the conditional reversed hazard rate of X, given $\Theta = \theta$, and

$$C^*(\theta|x) = \frac{F(x|\theta)dC(\theta)}{\int_\Theta F(x|\theta)dC(\theta)}$$

is the conditional distribution function of θ, given $X \le x$. As before, if a representation of the form

$$F_1(x) = F_1(x|\theta) = [F(x)]^\theta$$

holds, then the reversed hazard rates λ of F_1 and F satisfy the relationship

$$\lambda_{F_1}(x) = \theta\lambda_F(x).$$

In this case, we say that F_1 is the reversed proportional hazard rates model of F. □

Nelson (1982) has mentioned that units manufactured in different production periods may have different life distributions due to difference in design, raw materials, handling, etc., and it may therefore be necessary to identify production period, customer environment, etc. that result in poor units, for remedial action on that part of the population. Cox (1959) analyzed data on failure times using mixture models by classifying the cause of failure as identified or not; see also Mendenhall and Hader (1958) and Cheng et al. (1985). Identification of the life distribution is crucial in such cases. One way in which such identification is possible is to use characterization theorems that involve various reliability functions or relationships between them. Nair et al. (1999) proposed certain relationships between the hazard rate and the mean residual life that characterize some mixture distributions for the purpose.

Theorem 2.26. ***(a)*** *A discrete random variable X taking values in $(0, 1, \dots)$ follows a mixture of geometric laws with probability mass function*

$$f(x) = \alpha p_1 q_1^x + (1-\alpha) p_2 q_2^x, \qquad 0 \le \alpha \le 1, 0 < p_i < 1, q_i = 1 - p_i, i = 1, 2, \tag{2.107}$$

if and only if, for all x,

$$r(x) = \left(\frac{1}{p_1} + \frac{1}{p_2}\right) - \frac{1}{p_1 p_2} h(x+1);$$

(b) *X is a mixture of Waring distributions with probability mass function*

$$\begin{aligned} f(x) &= \alpha(a-b_1)\frac{(b_1)_x}{(a)_{x+1}} + (1-\alpha)(a-b_2)\frac{(b_2)_x}{(a)_{x+1}}, \\ &\quad a > b_1,\ b_1 \ge b_2,\ b_2 > 0, \end{aligned}$$

if and only if for all x

$$\begin{aligned} m(x) &= \frac{(2a - b_1 - b_2 - 1)}{(a - b_1 - 1)(a - b_2 - 1)}(a+x) \\ &\quad \times \frac{(a+x)(a+x+1)}{(a - b_1 - 1)(a - b_2 - 1)} h(x+1). \end{aligned}$$

Corollary 2.27. *When $p_1 = p_2 = p$, the geometric law is characterized by the property*

$$m(x) = 2p^{-1} - p^{-2} h(x+1),$$

and when $b_1 = b_2 = b$, the Waring distribution

$$f(x) = (a-b)\frac{(b)_x}{(a)_{x+1}}, \qquad x = 0, 1, 2, \dots,$$

is characterized by the property

$$r(x) = \frac{(2a - 2b - 1)}{(a-b-1)^2}(a+x) - \frac{(a+x)(a+x+1)}{(a-b-1)^2} h(x+1).$$

In Theorem 2.26, we have used the mean residual life of the mixture of the form in (2.104). According to Definition 2.20, it is calculated as

$$r(x) = \frac{1}{S(x+1)} \sum_{t=x+1}^{\infty} S(t) = \sum_{t=x+1}^{\infty} \frac{\alpha S_1(t) + (1-\alpha)S_2(t)}{\alpha S_1(x+1) + (1-\alpha)S_2(x+1)}, \tag{2.108}$$

where S_1 and S_2 are the survival functions of the component distributions. Eq. (2.108) is expressible as

$$m(x) = q(x)m_1(x) + (1 - q(x))m_2(x),$$

with

$$q(x) = \frac{\alpha S_1(x+1)}{\alpha S_1(x+1) + (1-\alpha)S_2(x+1)} = p(x+1)$$

and $m_1(x)$ and $m_2(x)$ are the mean residual life functions of the component distributions. When $S(x)$ is indexed by the values of a non-negative continuous random variable Θ, the residual life distribution becomes

$$S_x(t) = \frac{\int_\Theta S(x+t+1|\theta)dC(\theta)}{\int_\Theta S(x+1|\theta)dC(\theta)} \tag{2.109}$$

as a naturally corollary of the basic definition in (2.19). Eq. (2.109) is equivalent to

$$\begin{aligned} S_x(t) &= \frac{\int_\Theta S_x(t|\theta)S(x+1|\theta)dC(\theta)}{\int_\Theta S_x(x+1|\theta)dC(\theta)} \\ &= \int_\Theta S_x(t|\theta)dC(\theta|x+1)d\theta, \end{aligned}$$

where

$$\pi(\theta|x+1) = \frac{\int_0^\infty S(x+1|\theta)dC(\theta)}{\int_0^\infty S(x+1|\theta)dC(\theta)}$$

is the distribution function of Θ given $X \geq x+1$.

Apart from the hazard function and mean residual life function, higher moments of residual life can also be used for characterizing life distributions. Two such results are established in Nair et al. (1999).

Theorem 2.28. *If $E(X^2) < \infty$, the identity*

(a)

$$\mu_{(2)}(x) = E[(X-x)(X-x-1)|X > x] = Ah(x+1) + B,$$

where $A = 2p_1^{-2}p_1^{-2}(p_1p_2 - p_1 - p_2)$,

$$B = 2p_1^{-2}p_2^{-2}(p_1^2 + p_2^2 + p_1p_2(1 - p_1 - p_2)), \quad 0 < p_1, p_2 < 1,$$

holds if and only if X is distributed as geometric mixture in (2.107);

(b)

$$E[(X-x)^2|X > x] = A + Bh(x) + C(x),$$

where

$$A = -2(p_1p_2p_3)^{-1}(p_1 + p_2 + p_3),$$
$$B = 2(p_1p_2p_3)^{-1}$$

and

$$C = 2(p_1p_2p_3)^{-1}(p_1p_2 + p_2p_3 + p_3p_1),$$

if and only if X has a three-component geometric mixture

$$f(x) = \sum_{i=1}^{3} \alpha_i p_i q_i^x, \quad 0 < \alpha_i < 1, \textit{ and } \sum_{i=1}^{3} \alpha_i = 1.$$

These authors have pointed out with the help of simulated data that Theorem 2.26 is useful in model identification and inference. If the plots of the estimates $(\hat{h}(x), \hat{m}(x))$ fall along a straight line, the distribution is mixture of geometric. A quick estimate of the parameters p_1 and p_2 is obtained from the slope and intercept of the fitted line, or more accurately from a least-square fit of the line. One can estimate α by equating the sample mean with

$$E(X) = \alpha q_1 p_1^{-1} + (1-\alpha) q_2 p_2^{-1}$$

after substituting the estimate of p_1 and p_2. Nair (1983b) addressed the estimation problem by matching the factorial moments of the sample with those of the population. Denoting by $m_{(1)}$, $m_{(2)}$ and $m_{(3)}$ the first three sample factorial moments, the equations to be solved are as follows:

$$\alpha r + (1-\alpha)s = a_1 = m_{(1)}$$
$$\alpha r(r-1)^2 + (1-\alpha)s(s-1) = a_2 = \frac{m_{(2)}}{2}$$
$$\alpha r(r-1)^2 + (1-\alpha)s(s-1)^2 = a_3 = \frac{m_{(3)}}{6}$$

with $r = p_1^{-1}$ and $s = p_2^{-1}$. After some algebra, the quadratic equation

$$(a_1 + a_2 - a_1^2)s^2 + (a_1^2 + a_1a_2 - a_i - 2a_2 - a_3)s + (a_1a_2 - a_2^2) = 0$$

is obtained, which can be solved for s. Now r and α are calculated from

$$s = \frac{a_1 + a_2 - a_1 r}{a_1 - r} \text{ and } \alpha = \frac{a_1 - s}{r - s}.$$

If there are two admissible solutions for s in the above quadratic equation, the one that gives better fit is chosen as the final estimate.

Sometimes, mixtures of a more general form

$$f(x) = \sum_{i=1}^{n} \alpha_i f_i(x), \tag{2.110}$$

or

$$S(x) = \sum_{i=1}^{n} \alpha_i S_i(x),$$

where α_i are real numbers, not necessarily all positive, satisfying $\sum \alpha_i = 1$ are also considered. They are of practical significance representing distributions of order statistics, coherent systems and inference in step-stress models. Discussions of these aspects in the continuous case can be seen in Navarro and Rychlik (2007), Navarro and Shaked (2006), Balakrishnan et al. (2007, 2009), Balakrishnan and Cramer (2008), and Balakrishnan and Xie (2007). When some of the mixing constants are negative, such mixtures are called *generalized mixtures*. The representation in (2.110) can be expressed as (Navarro and Hernandez, 2004) as

$$S(x) = \alpha S_+(x) + (1 - \alpha) S_-(x)$$

with $\alpha = \sum_{\alpha_i > 0} \alpha_i > 1$,

$$S_+(x) = \sum_{\alpha_i > 0} \alpha_i S_i(x) / \sum_{\alpha_i > 0} \alpha_i \text{ and } S_+(x) = \sum_{\alpha_i < 0} \alpha_i S_i(x) / \sum_{\alpha_i < 0} \alpha_i.$$

Then,

$$S_+(x) = \alpha^{-1} S(x) + \alpha^{-1}(\alpha - 1) S_-(x)$$

and the relationship for $n = 2$ becomes

$$\min(m_1(x), m_2(x)) \le m_1(x) \le \max(m(x), m_2(x)),$$

where $m_i(x)$ is the mean residual life function of $S_i(x)$.

Recalling the formula for the reversed mean residual life of X as

$$r(x) = \frac{1}{F(x-1)} \sum_{t=1}^{x} F(t-1), \tag{2.111}$$

we see that for the two-component mixture we have

$$\begin{aligned} F(x) &= \alpha F_1(x) + (1-\alpha)F_2(x), \\ r(x) &= p_1(x-1)r_1(x) + (1-p_1(x-1))r_2(x), \end{aligned} \tag{2.112}$$

where $r_1(x)$ and $r_2(x)$ are the reversed mean residual lives of F_1 and F_2 and $p_1(\cdot)$ is as defined in (2.106). Eliminating $p_1(x)$ from

$$r(x+1) = p_1(x)r_1(x+1) + (1-p_1(x))r_2(x+1)$$

and (2.106), a relationship connecting the reversed mean residual life $r(x)$ and the reversed hazard rate $\lambda(x)$ of the mixture can be expressed in the form

$$r(x+1) = g_1(x)\lambda(x) + g_2(x), \tag{2.113}$$

where

$$g_1(x) = \frac{r_1(x) - r_2(x)}{\bar{\lambda}(x) - \bar{\lambda}_2(x)} \text{ and } g_2(x) = \frac{r_2(x)\bar{\lambda}_1(x) - r_1(x)\bar{\lambda}_2(x)}{\bar{\lambda}_1(x) - \bar{\lambda}_2(x)}.$$

Remark 2.17. The identity connecting $m(x)$ and $h(x)$ of the mixture also bears the same form

$$m(x) = l_1(x)h(x+1) + l_2(x),$$

where

$$\begin{aligned} l_1(x) &= \frac{m_2(x)\bar{h}_1(x+1) - m_1(x)\bar{h}_2(x+1)}{\bar{h}_1(x+1) - \bar{h}_2(x+1)}, \\ l_2(x) &= \frac{m_1(x) - m_2(x)}{h_1(x+1) - h_2(x+1)}. \end{aligned} \tag{2.114}$$

Example 2.25. Let X be distributed as a mixture of reversed geometric forms

$$F(x) = \alpha F_1(x) + (1-\alpha)F_2(x),$$

where

$$F_i(x) = \begin{cases} c_i^{-1}(1+c_i)^{1-b}, & x = 0 \\ (1+c_i)^{x_i - b}, & x_i = 1, 2, \dots . \end{cases}$$

Inserting

$$r_i(x) = \frac{1+c_i}{c_i} \text{ and } \bar{\lambda}_i(x) = \frac{c_i}{1+c_i}$$

in (2.113), we obtain

$$r(x+1)=\frac{c_1+c_2+2c_1c_2}{c_1c_2}-\frac{(1+c_1)(1+c_2)}{c1c_2}\lambda(x), \quad x=1,2,\dots \quad (2.115)$$

The relationship given above provides a plot of $(\lambda(x), r(x+1))$ as a decreasing straight line which is easy to verify. Notice that (2.115) is a characteristic property of this distribution.

2.12 WEIGHTED DISTRIBUTIONS

Recalling the definition of the weighted distribution in (1.20) that

$$f_W(x)=\frac{W(x)f(x)}{E[W(x)]}$$

and the expression of the survival function in (1.21) given by

$$h_W(x)=\frac{W(x)}{E[W(X)|X>x]}h_X(x),$$

the mean residual lives of X and X_W satisfy the relationship

$$\frac{m_W(x-1)-1}{m_W(x)}=\frac{p(x+1)}{p(x)}\frac{m(x-1)-1}{m(x)}, \quad p(x)=E[W(X)|X>x].$$

As special cases, we have

$$S_L(x)=\frac{(m(x)+x)S(x)}{\mu}=\frac{m^*(x)S(x)}{\mu},$$

$$h_L(x)=\frac{xh(x)}{(m(x)+x)}=\frac{xh(x)}{m^*(x)},$$

$$\frac{m_L(x-1)-1}{m_L(x)}=\frac{m(x+1)-(x+1)}{(m(x)-x)}\frac{m(x-1)-1}{m(x)}.$$

Denoting $h_n(x)$ and $m_n(x)$, for the hazard rate and the mean residual life of the equilibrium random variable Y_n (Section 1.4), we have from Nair et al. (2012b) that

$$h_n(x)=\frac{1}{m_{n-1}(x)},$$

$$m_{n-1}(x+1)=\frac{m_n(x+1)}{1+m_n(x+1)-m_n(x)}$$

and

$$h_n(x+1)=1+h_{n+1}(x+1)(1-h_{n+1}^{-1}(x)).$$

In particular, for the first-order equilibrium random variable X_E (Nair and Hitha, 1989), we have

$$h_E(x) = \frac{1}{m(x)},$$

so that the hazard rate of X_E is the reciprocal of the mean residual life of X_E. Furthermore, we have

$$h(x) = 1 + h_E(1 + h_E^{-1}(x-1)),$$
$$m(x) = m_E(x)[1 + m_E(x) + m_E(x-1)]^{-1}.$$

From the above identities, Nair and Hitha (1989) have also shown the following:

(a) a mean residual life of the form $A + Bx$ characterizes the geometric, Waring and negative hyper-geometric distributions according as $B = 0$, $B > 0$ and $B < 0$, respectively;

(b) the random variables X and X_E are such that $h(x) = C_1 h_E(x)$ ($m(x) = C_2 m_E(x)$) if and only if X is geometric for $C_1 = 1$ ($C_2 = 1$), Waring for $C_1 > 1$ ($C_2 < 1$) and negative hyper-geometric for $C_1 < 1$ ($C_2 > 1$).

More general characterizations covering the original and the nth-order equilibrium distributions have been provided by Nair et al. (2012b) as follows:

(a) $h_n(x) = (1 + Cn)h(x)$ for all x and $n = 0, 1, 2, \ldots$;
(b) $m_n(x) = (1 + Bn)^{-1} m(x)$, for all x and $n = 0, 1, 2, \ldots$;
(c) the variance residual life of Y_n,

$$\sigma_n^2(x) = C_n m_n(x)(m_n(x) - 1),$$

holds if and only if X has one of the above three distributions, the geometric ($C = 0$, $B = 0$, $C_n = 1$), Waring ($C < 0$, $B > 0$, $C_n > 1$), and negative hyper-geometric ($C > 0$, $B < 0$, $C_n < 1$).

In particular, $\sigma_n^2(x) = \sigma^2(x)$ characterizes the geometric distribution. So also the property

$$E[(X - x - 1)^{(n)} | X > x + n] = c,$$

a constant, for all x and $n = 1, 2, \ldots$.

Li (2011) obtained formulas for the equilibrium distribution of the n-fold convolution of $f(x)$ and mixtures. If f^{n*} is the n-fold convolution of f (see Section 1.5) and f_m^{n*} is the m-th order equilibrium distribution of f^{n*}, then

$$f_m^{n*}(x) = \frac{\mu_{(m)}}{\mu_{(m)}^{n*}} f_m^{(n-1)*}(x) + \frac{1}{\mu_{(m)}^{n*}} \sum_{i=1}^{m} \binom{m}{i} \mu_{(i)} \mu_{(m-i)}^{(n-1)*} f_i(x),$$

where $\mu_{(r)}^{n*}$ is the rth factorial moment of f_r^*, $m = 1, 2, \ldots, n = 2, 3, \ldots$.

When the baseline distribution of X is $f(x|\theta)$, where θ is the realization of a continuous random variable Θ, the mixture (Section 1.3)

$$g(x) = \int_{\Theta} f(x|\theta) dC(\theta)$$

has nth order equilibrium distribution

$$g_n(x) = \int_{\Theta} f_n(x|\theta) dC_n(\theta)$$

where

$$dC_n(\theta) = \frac{E(X_{(n)}|\theta) dC(\theta)}{E(X_{(n)})}.$$

Example 2.26. It has been shown in Example 1.2 that the nth order equilibrium distribution of a geometric variable X is the same as the distribution of X. Now, in this case, we have

$$\begin{aligned} dC_n(q) &= \frac{E(X_{(n)}|q) dC(q)}{E(X_{(n)})} \\ &= \left(\frac{q}{p}\right)^n \frac{dC(q)}{E(X_{(n)})}, \end{aligned}$$

showing that $g_n(x)$ is also a mixture geometric law.

Finally, we look at the relationships between various reliability measures in reversed time in the original and weighted versions. For this purpose, we note that the distribution function of the weighted law is given by

$$\begin{aligned} F_w(x) &= \sum_{t=0}^{x} W(t) f(t) \Big/ E[W(X)], \quad E[W(X)] < \infty \\ &= \frac{E[W(X)|X \le x] F(x)}{E[W(x)]}. \end{aligned}$$

Accordingly, the reversed failure rate of X_W becomes

$$\lambda_W(x) = \frac{f_W(x)}{F_W(x)} = \frac{W(x)}{E[W(x)|X \le x]} \lambda(x).$$

Similarly, from (2.64),

$$r_W(x) = \frac{1}{F_W(x-1)} \sum_{t=1}^{x} F_W(t-1)$$

gives

$$\frac{r_W(x+1)-1}{r_W(x)} = \frac{p_1(x-1)}{p_1(x)}\frac{r(x+1)-1}{r(x)}, \quad p_1(x) = E(W_1(X)|X \le x).$$

These identities, relationships and formulas are helpful while discussing properties of ageing concepts of weighted distributions and their relationships with those of the baseline distribution.

Chapter 3

Discrete Lifetime Models

3.1 INTRODUCTION

The main emphasis of this chapter is to present a discussion of the current and prospective tools that enable the representation of lifetime data through some model. A major component in such a formulation is the distribution of the underlying lifetime. Probability distributions facilitate characterization of the uncertainty prevailing in the data set by identification of the patterns of variation. By summarizing the observations into a mathematical form that contains a few unknown parameters, distributions provide a means to the best possible understanding to the basic data generating mechanism. Lifetime distributions, being a probabilistic description of the behaviour of the length of life, is to a marked extent depends on the mode of failure of the device under study. Choice of an appropriate distribution for the data set depends largely on our knowledge about the physical characteristics of the process that give rise to the observations. In some cases, the relationship between the failure mechanism and various concepts encountered in Chapter 2 can be forged. However, many situations, unlike the case of the normal distribution where the central limit theorem affords a bridge between theory and practice, the discrete distributions we confront with are difficult to justify on practical grounds. The real position is that within the range of observations we have in hand, more than one distribution may pass well established goodness-of-fit tests in many cases. Characterization properties, being unique to a specified distribution, are the only exact tools that can identify the correct model. When the sample observations provide evidence of characteristic properties exhibited by reliability functions such as hazard rate, mean residual life, etc., pertaining to a particular model with high accuracy, there is reasonable justification to choose that model. If such a model passes the goodness-of-fit test also, then we have a match between observations and the chosen distribution both in terms of comparable physical properties and statistical validity. In Chapter 2, several examples of this nature were illustrated for modelling real data with some specified distributions. The subsequent sections of this chapter will to present various discrete distributions along with their properties that can be used for modelling and analysis of lifetime data. For a detailed discussion of various aspects of reliability modelling, we refer to Blischke and Murthy (2000).

Reliability Modelling and Analysis in Discrete Time. http://dx.doi.org/10.1016/B978-0-12-801913-9.00003-8

3.2 FAMILIES OF DISTRIBUTIONS

In the presence of lack of understanding of the data generating mechanism due to insufficient details about the physical characteristics, the investigator has to be content with finding the best approximating distribution from a collection of candidate models. Based on a preliminary assessment of the features of the observations at disposal, mathematical formulation of a suitable model is initially made. A family of distributions that contains enough members with different shapes and varying characteristics can be useful aids in the process of empirical modelling described above.

3.2.1 Ord Family

The family of distributions has its origin in 1895 when Pearson introduced a system bearing his name in the continuous case. A discrete version of the Pearson system, defined by the difference equation

$$f(x) - f(x-1) = \frac{(a-x)f(x-1)}{b_0 + b_1 x + b_2 x(x-1)}, \tag{3.1}$$

was studied by Ord (1967a, 1967b), which is referred to as the *Ord family* in the sequel, where x takes values in a subset T of integers.

The members of the system with non-negative integers as the values of X are:

(i) binomial

$$f(x) = \binom{n}{x} p^x (1-p)^{n-x}, \quad x = 0, 1, 2, \ldots, n;\ 0 < p < 1,\ q = 1-p; \tag{3.2}$$

(ii) negative binomial

$$f(x) = \binom{k+x-1}{x} p^k (1-p)^x, \quad x = 0, 1, 2, \ldots; \tag{3.3}$$

(iii) Poisson

$$f(x) = e^{-\lambda}\frac{\lambda^x}{x!}, \quad x = 0, 1, 2, \ldots,\ \lambda > 0; \tag{3.4}$$

(iv) hypergeometric

$$f(x) = \binom{M}{x}\binom{N-M}{n-x} \Big/ \binom{N}{n}, \quad x = 0, 1, 2, \ldots, n; \tag{3.5}$$

(v) negative hypergeometric (beta-binomial)

$$f(x) = \binom{k+x-1}{x}\binom{N-k-x}{M-x}\Big/\binom{N}{M}, \quad x = 0, 1, \ldots, M; \qquad (3.6)$$

(vi) beta-Pascal

$$f(x) = \frac{A}{k+A}\binom{k+x-1}{x}\binom{A+B-1}{A}\Big/\binom{k+A+B+x-1}{k+A}, \qquad (3.7)$$

$x = 0, 1, 2, \ldots; A = M - 1; B = N - M.$

The first four moments of the family are given by

$$\mu_1' = (b_1 + a - 1)/(1 - 2b_2),$$
$$\mu_2 = (b_1 + a - 1)\{ab_2 + (1 - b_2)(b_1 - 2b_2)\}/(1 - 2b_2)^2(1 - 3b_2)$$
$$\mu_3 = \mu_2\{4b_2(a + 1 - b_1 - b_2) - 4b_1 - 3\}/(1 - 2b_2)(1 - 4b_2),$$

and

$$\mu_4 = 3\mu_2^2 + \frac{\mu_2}{(1 - 4b_2)(1 - 5b_2)}\Big[\{(1 - b_2)(1 - 2b_2)\}$$
$$+6\{(b_1^2 - b_2^2) + (b_2 - b_1) - 2b_2(a + b_1 - 1)\}$$
$$+b_2\mu_2\{1 - 2b_2(1 - 3b_2)^{-1}\}\Big].$$

These will be required for analysing the descriptive characteristics of the distribution, estimating parameters by the method of moments, and in computing standard errors of statistics. All distributions of the system are either J shaped or unimodal. A crucial issue of interest when families are considered for model selection, is the criteria to distinguish its members. Ord (1972) considered the quantities $I = \frac{\mu_2}{\mu}$, which is the variance mean ratio called *index of dispersion*, and $S = \frac{\mu_3}{\mu_2}$, and proposed a diagram of (S, I). In this diagram, the line $S = 2I - 1$ defines the binomial or Poisson or negative binomial distributions as $I <, =, > 1$. The other distributions are obtained by analyzing $T = S - 2I + 1$. The region $S < 2I - 1$ gives the beta-binomial, $S > 2I - 1$ corresponds to the hypergeometric or beta-Pascal according to whether $S < 1$ or > 1 in addition. Since the parameters of the family are completely determined by the first three moments, the (S, I) diagram can be used to locate a suitable distribution for a given data, by replacing the population moments with the corresponding sample moments. Ord (1972) also proposed the plot of $u_x = \frac{xp_x}{p_{x-1}}$ against x, where p_x is the sample frequency. Since successive u_x values are dependent, a smoothing $v_x = \frac{1}{2}(u_x + u_{x+1})$ is has also been suggested. For example, the binomial distribution satisfies

$$u_x = \frac{(n+1)p}{q} - \frac{p}{q}x$$

in the population, so that the plot of $(x, \frac{xp_x}{p_{x-1}})$ must show an approximate line. An interesting property of the Ord family is that the truncated version of X at either end also belongs to this family.

Various reliability functions, like the hazard rate, for the Ord family can not be expressed in simple forms. However, the members of the family admit a simple relationship between the hazard rate and the mean residual life. Ahmed (1991) has shown that X has binomial distribution in (3.2) if and only if

$$m^*(x) = np + qxh(x). \tag{3.8}$$

Since the mean residual life $m(x) = m^*(x) - x$, the relationship between $m(x)$ and $h(x)$ characterizing the binomial distribution is obvious. On similar lines, Osaki and Li (1988) have shown that X follows negative binomial distribution

$$f(x) = \binom{n-1}{x-1} p^x q^{n-x}, \quad n \geq x,\ x = 1, 2, \ldots,$$

if and only if

$$E(X|X > m) = \mu + (m+1-x)h(m+1)/p$$

for all integers $m \geq x - 1$. They have also shown that X has Poisson distribution in (3.4) if and only if

$$m^*(x) = \mu + \frac{(1+x)}{\lambda} h(x+1).$$

A more general result for the Ord family in (3.1) rewritten as

$$f(x+1) - f(x) = -\frac{(x+d)f(x)}{b_0 + b_1 x + b_2 x^2},$$

was characterized by Nair and Sankaran (1991) through the identity

$$m^*(x) = \mu + (a_0 + a_1 x + a_2 x^2)h(x+1), \tag{3.9}$$

where

$$b_i = \frac{a_i}{2a_2 + 1} \text{ and } d = \frac{a_1 - a_2 - \mu}{2a_2 + 1}, \quad i = 0, 1, 2.$$

The values of a_0, a_1 and a_2 for the distributions in (3.2)–(3.7) are listed in Table 3.1.

TABLE 3.1 Values of a_0, a_1 and a_2 in (3.9) for the Ord family

Distribution	a_0	a_1	a_2
binomial	q	q	0
Poisson	1	1	0
negative binomial	$\frac{1}{p}$	$\frac{1}{p}$	0
hypergeometric	$\frac{1}{2+2n+2M-N}$	$\frac{N+2-M}{2+2n+2M-1}$	$\frac{N-M-n}{2+2n+2M-1}$
negative hypergeometric	$\frac{N-k}{N+1-M}$	$\frac{N-k-1}{N+1-M}$	$\frac{1}{M-N-1}$
beta-Pascal	$\frac{-B}{A+2K+1}$	$\frac{-1+B}{A+2K+1}$	$\frac{-1}{A+2K+1}$

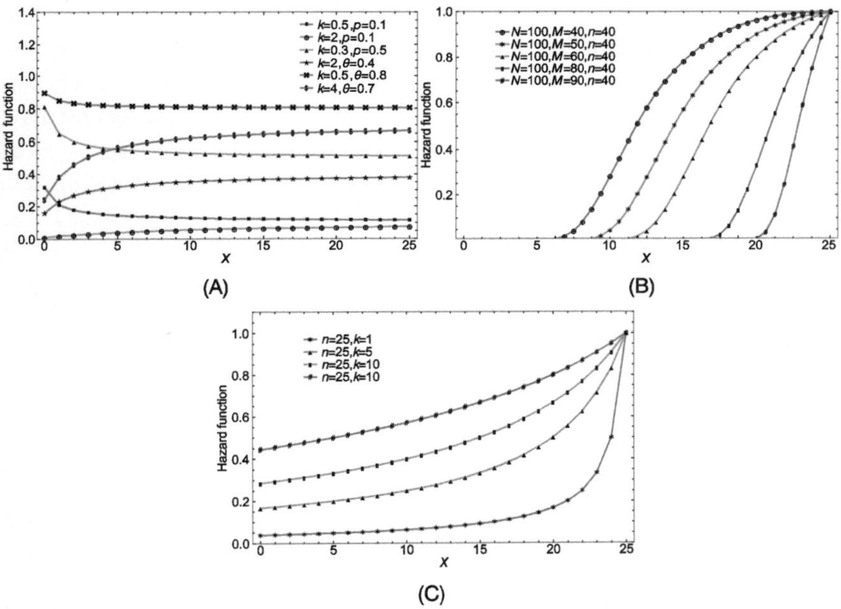

FIGURE 3.1 Hazard rate functions for (A) negative binomial distribution, (B) hypergeometric distribution, and (C) negative hypergeometric distribution.

Figs 3.1A, 3.1B and 3.1C show the hazard rates of negative binomial, hypergeometric and negative hypergeometric distributions for various values of the parameters.

Various distributional properties of the individual models along with results on inference and applications have been discussed in Johnson et al. (1992). Further, characterization problems of the family in terms of truncated moments have been addressed by Glanzel et al. (1984) and Glanzel (1987, 1991). Of particular interest is the result

$$E(X^2|X \geq x) = P(x)E(X|x \geq x) + Q(x) \tag{3.10}$$

in Glanzel (1991), where $P(x)$ and $Q(x)$ are polynomials of degree at most one, with real coefficients. Obviously, (3.10) leads to a property of the variance residual life in relation to the mean residual life.

There are two related families in connection with (3.1) that deserves special mention. One is the Katz (1965) system defined by

$$\frac{f(x+1)}{f(x)} = \frac{\alpha + \beta x}{1+x}, \quad x = 0, 1, 2 \ldots, \ \alpha > 0, \ \beta < 1. \tag{3.11}$$

It is easy to see that (3.11) is a special case of (3.1) when $b_0 = b_1$, $b_2 = 0$. This restricted model contains the binomial ($\beta < 0$), Poisson ($\beta = 0$) and negative binomial ($0 < \beta < 1$) distribution as particular members. These members can be discriminated by performing a test of hypothesis $H_0 : \mu_2 = \mu$ against $H_1 : \mu_2 > \mu$ ($\mu_2 < \mu$), with H_0 if accepted gives Poisson, otherwise $H_1 : \mu_2 > \mu$ ($\mu_2 < \mu$) gives negative binomial (binomial). The test is based on the statistic

$$Z = \frac{S^2 - \bar{x}}{\bar{x}}$$

which is closely approximated by a normal distribution $N(\frac{\beta}{1-\beta}, \frac{2}{n})$, where n is the sample size, and S^2 and $\bar{x}$ are the variance and mean of the sample. Following (3.9), for the Katz family, we see that

$$m^*(x) = \mu(1-\beta)^{-1}(1+x)h(x+1)$$

is satisfied for all x, and conversely. Sudheesh and Nair (2010) established the following characteristic properties for the Ord and Katz families, respectively:

$$\sigma^2(x) = E(c_0 + c_1 X + c_2 X^2 | X > x) + (\mu - m(x))(m(x) - x - 1), \tag{3.12}$$

where

$$c_0 = \mu + \frac{b_0 - b_1 + b_2}{1 - 2b_2}, \quad c_1 = \frac{b_1 - 1}{1 - 2b_2}, \quad c_2 = \frac{b_2}{1 - 2b_2},$$

and

$$\sigma^2(x) = \frac{\alpha + \beta m(x) + (\mu - m(x))(m(x) - x - 1)}{(1-\beta)}. \tag{3.13}$$

For example, in the Poisson case, we have

$$\sigma^2(x) = \lambda + (\lambda - m(x))(m(x) - x - 1),$$

or alternatively in terms of the variance and the hazard rate as

$$\sigma^2(x) = \lambda + (x+1)(x+1-\lambda)h(x+1) - (x+1)^2h^2(x+1).$$

Let τ be the class of real-valued functions $C(X)$ of X. Then, for the Ord family, we have

$$\inf_{C(x)\in\tau} \frac{V(C(X))}{E^2[(d_0 + d_1X + d_2X^2)\Delta C(X)]} = 1,$$

where $d_i = c_i\sigma^{-1}$ and Δ is the forward difference operator, $i = 0, 1, 2$. In particular, for the Katz family, we have

$$\inf_{C(X)\in\tau} \frac{\sigma^2(1-\beta)V[C(X)]}{E^2((\alpha+\beta X)\Delta C(X))} = 1$$

provides the lower bound to the variance of a random function $C(X)$. Under some regularity conditions, Nair and Sudheesh (2008) showed that these bounds reduce to the Cramer-Rao and Chapman-Robbins inequalities. The moments of the Katz family are

$$\mu = \frac{\alpha}{1-\beta}, \quad \mu_2 = \frac{\alpha}{(1-\beta)^2},$$

$$\mu_3 = \mu_2(1+\beta) \text{ and } \mu_4 = 3\mu_2^2 + \mu_2\frac{(\beta+2)^2}{(1-\beta)^2}.$$

It has a probability generating function

$$P(t) = \left(\frac{1-\beta t}{1-\beta}\right)^{-\frac{\alpha}{\beta}}.$$

An extension of the Ord family has been provided by Sindhu (2002) wherein the difference equation in (3.1) has the modified form

$$\frac{f(x+1)-f(x)}{f(x)} = \frac{d_0 + d_1x + d_2x^2}{k_0 + k_1x + k_2x^2} \tag{3.14}$$

with d_0, d_1, d_2, k_0, k_1 and k_2 as real constants. Besides containing all members of the Ord family, the system in (3.14) contains several other distributions such as the confluent hypergeometric distribution of Bhattacharya (1966), Borel-Tanner model (Tanner, 1953), and the Haight distribution (Haight, 1961).

Theorem 3.1. *A necessary and sufficient condition for the distribution of X to belong to the family in* (3.14) *is that*

$$d_2E(X^2|X \geq x) + (d_1 + 2k_2)E(X|X \geq x) + d_0 + k_1 - k_2 + (k_0 + k_1x + k_2x^2)h(x+1) = 0.$$

Remark 3.1. An extension of Glanzel's result in (3.10) is also possible from the above theorem. The relevant result is that X follows (3.14) if an only if

$$d_2E(X^3|X \geq x) = A(x)E(X^2|X \geq x) + B(x)E(X|X \geq x) + C(x),$$

where $A(x) = d_2x + q$ with q as a real constant and $B(x)$ and $C(x)$ are polynomials of degree at most one with real coefficients.

The approach for relationships between reliability functions in reversed time is the same and very similar results emerge. Corresponding to (3.9), the identity between $r(x)$ and $\lambda(x)$ for the Ord family is

$$r(x+1) = \mu - (a_0 + a_1x + a_2x^2)\lambda(x)$$

with $a_0 = \mu + \frac{b_0-b_1+b_2}{1-2a_2}$, $a_1 = \frac{b_1-1}{1-2b_2}$ and $a_2 = \frac{b_2}{1-2b_2}$.

Gupta et al. (1997) have given a formula for computing the hazard rate as

$$\frac{1}{h(x)} = \sum_{i=0}^{\infty} \prod_{u=x}^{x+i} \frac{f(u+1)}{f(u)} \tag{3.15}$$

when the ratio of two successive probabilities are known. By utilizing of this, for the Katz family, we have

$$h(x) = \left[1 + \frac{x!(k-1)!}{(k+x-1)!}\left\{\binom{k+x}{1+x}\beta + \cdots + \binom{k+x+j-1}{x+j}\beta^j + \ldots\right\}\right]^{-1},$$

where $k = \frac{\alpha}{\beta}$. An alternative expression is

$$h(x) = \left[{}_2F_1(1, \frac{\alpha}{\beta} + x; 1 + x; \beta)\right]^{-1}, \tag{3.16}$$

where ${}_2F_1$ is the hypergeometric function defined as

$${}_2F_1(a, b, c; x) = 1 + \tfrac{ab}{c}x + \tfrac{a(a+1)b(b+1)}{c(c+1)2!}x^2 + \ldots, \tag{3.17}$$

with $c \neq 0, -1, -2, \ldots$.

For an elaborate study on the orthogonal polynomials in the cumulative Ord family and their application to variance bounds, one may refer to the recent work of Afendras et al. (2017).

3.2.2 Power Series Family

The power series family of distributions were initially studied by Kosambi (1949) and Noack (1950). Any distribution that can be represented as

$$f(x) = \frac{a(x)\theta^x}{A(\theta)}, \quad x = 0, 1, \ldots, \theta > 0, \tag{3.18}$$

where $a(x) > 0$ and $A(\theta) = \sum a(x)\theta^x$, is said to be of *power series form*. In (3.18), $A(\theta)$ is called the series function. Sometimes, (3.18) is also referred to as the discrete linear exponential family in view of the specification of (3.18) as

$$f(x) = a(x)b(\alpha)e^{-\alpha x}.$$

Initially, Noack (1950) considered $f(x)$ with support as the whole set of non-negative integers. Patil (1962) extended the support to any arbitrary non-empty subset T of non-negative integers such that

$$A(\theta) = \sum_{x \in T} a(x)\theta^x, \tag{3.19}$$

with $a(x) > 0$, $\theta \geq 0$, so that $A(\theta)$ is positive, finite and differentiable. Then, (3.18) for $x \in T$ is called a *generalized power series distribution* (GPSD). The system in (3.19) includes the binomial, negative binomial, Poisson and logarithmic series distributions. A truncated GPSD is also a GPSD.

A further extension of the GPSD has been introduced by Gupta (1974) who defined the modified power series distributions (MPSD), specified by the probability mass function of the form

$$f(x) = \frac{a(x)(g(\theta))^x}{B(\theta)}, \quad x \in T, \tag{3.20}$$

where $a(x) > 0$, and $g(\theta)$ and $B(\theta)$ are positive finite and differentiable. When $g(\theta)$ is invertible, MPSD reduces to the GPSD. Apart from all distributions belonging to the GPSD family, the modified version contains more distributions, such as the generalized binomial distribution of Jain and Consul (1971). Being the more general form, the properties of (3.20) will be mentioned so that the results for GPSD can be deduced from them. The moments of (3.20) are

$$\mu = E(X) = \frac{B'(\theta)g(\theta)}{B(\theta)g'(\theta)} = \frac{g(\theta)}{g'(\theta)}\frac{d}{d\theta}[\log B(\theta)],$$

$$\mu_2 = \frac{g(\theta)}{g'(\theta)}\frac{d\mu}{d\theta},$$

and

$$\mu_{r+1} = \frac{g(\theta)}{g'(\theta)}\frac{d\mu_r}{d\theta} + r\mu_2\mu_{r-1}, \quad r = 1, 2, \ldots$$

The factorial moments can be calculated from the recurrence relation

$$\mu_{(r+1)} = \frac{g(\theta)}{g'(\theta)}\frac{d\mu_{(r)}}{d\theta} + \mu_{(r)}\mu - r\mu_{(r)}.$$

Consul (1995) has developed characterizations of the exponential family of distributions

$$f(x) = \exp[xp(\alpha) + q(\alpha) + s(x)] \tag{3.21}$$

which contains numerous discrete probability mass functions and probability density functions of continuous random variables for various choices of $p(\alpha)$, $q(\alpha)$ and $S(x)$. In the discrete case, the translations $a(x) = \exp[s(x)]$, $B(\theta) = \exp[-q(\alpha)]$ and $g(\theta) = \exp[p(\alpha)]$ reduce (3.21) to the form (3.20). His main result can be stated in our notation as

$$m^*(x) = \mu + \frac{g(\theta)}{g'(\theta)} \frac{\partial}{\partial\theta} (\log S(x+1)). \tag{3.22}$$

As examples, the Lagrangian Poisson distribution

$$f(x) = \frac{(1+\alpha x)^{x-1}}{x!} \theta^x \exp[-\theta(1+\alpha x)], \quad x = 0, 1, \ldots, \quad \theta > 0, \quad 0 \le \alpha \le \theta^{-1}, \tag{3.23}$$

satisfies the property

$$m^*(x) = \frac{\theta}{1-\alpha\theta} + \left(-\alpha + \frac{1}{\theta}\right)^{-1} \frac{\partial}{\partial\theta} \log S(x+1);$$

the Lagrangian negative binomial with probability function

$$f(x) = \frac{n}{n+mx} \binom{n+mx}{x} \theta^x (1-\theta)^{n+mx-x}, \quad x = 0, 1, 2, \ldots, \quad 0 < \theta < 1,\ n > 0,\ m = 0,\ 1 \le m \le \theta^{-1}, \tag{3.24}$$

is characterized by

$$m^*(x) = \frac{n\theta}{1-m\theta} + \frac{\theta(1-\theta)}{1-m\theta} \frac{\partial}{\partial\theta} \log S(x+1),\ 0 < \theta < 1. \tag{3.25}$$

Notice that (3.24) contains, as special cases, the binomial distribution ($m = 0$), negative binomial ($m = 1$), the Geeta distribution

$$f(x) = \frac{1}{\alpha x - 1} \binom{\alpha x - 1}{x} \theta^{x-1} (1-\theta)^{\alpha x - x}, \quad x = 0, 1, 2 \ldots,$$

when $n = \alpha - 1$ and x is replaced by $x - 1$ so that the identity in (3.25) is true for $x = 1, 2, \ldots, n$ (binomial), for all x (negative binomial), and $x = 2, 3, \ldots$

(Geeta). The identity in (3.22) can be used to obtain the lower bound for the variance as

$$V(C(X)) \geq E^2[g(X)\Delta C(X)],$$

when $C(x)$ is any real-valued function of x (Nair and Sudheesh, 2008) and

$$g(x) = \frac{g(\theta)}{g'(\theta)} \frac{1}{\sigma f(x)} \frac{\partial S(x+1)}{\partial \theta}.$$

3.2.3 Lerch Family

An important sub-family of the MPSD is the Hurwitz-Lerch zeta distributions (HLZD). Various properties of this class, along with their applications to reliability, have been studied by Gupta et al. (2008). The HLZD arises from the general Hurwitz-Lerch zeta function defined by

$$\phi(z, s, a) = \sum_{k=0}^{\infty} \frac{z^k}{(k+a)^s}, \; s \in C \text{ when } |z| < 1, \; a = C \backslash (0, -1, -2, \dots) \tag{3.26}$$

and $R(s) > 1$ when $|z| = 1$ and C is the set of complex numbers. Some known special cases of the function (3.26) are the Riemann zeta function

$$\zeta(s) = \sum_{k=1}^{\infty} \frac{1}{k^s} = \sum_{k=0}^{\infty} \frac{1}{(1+k)^s} = \phi(1, s, 1), \tag{3.27}$$

Hurwitz zeta function

$$\zeta(s, a) = \sum_{k=0}^{\infty} \frac{1}{(k+a)^s} = \phi(1, s, a), \tag{3.28}$$

and the polylogarithmic function

$$A(z, s) = \sum_{k=1}^{\infty} \frac{z^k}{k^s} = z \sum_{k=0}^{\infty} \frac{z^k}{(1+k)^s} = z\phi(z, s, 1). \tag{3.29}$$

For detailed properties of these functions, we refer to Erdelyi et al. (1953). Zornig and Altmann (1995) introduced the three parameter HLZD as a generalization of the Zipf, Zipf-Mandelbrot and Good distributions mentioned below. Subsequently, the family has found some applications as a model in ecology, linguistics, information sciences, statistical physics and survival analysis.

Aksenov and Savageau (2005) defined the Lerch family of distributions through the probability mass function

$$f(x) = \frac{c\,z^x}{(a+x)^s}, \quad x = 0, 1, 2, \ldots, \tag{3.30}$$

where $c^{-1} = \phi(z, s, a)$, the Lerch transcendent (Lerch, 1887), also called the Hurwitz-Lerch zeta function defined in (3.26) for $z > 0$, $a > 0$. By virtue of the relationship

$$\phi(z, s, a) = z^m \phi(z, s, m+a) + \sum_{k=0}^{m-1} \frac{z^k}{(k+a)^s}, \tag{3.31}$$

we can arrive at the distribution function

$$F(x) = 1 - z^{x+1}\phi(z, s, a+x+1)\Big/\phi(z, s, a).$$

The probability generating function and the moment generating function are, respectively,

$$P(t) = \frac{\phi(tz, s, a)}{\phi(z, s, a)}, \quad |t| \le 1,$$

and

$$M(t) = \frac{\phi(ze^t, s, a)}{\phi(z, s, a)}.$$

There is unimodality for the family if $s < 0$ and $a \ge 1$ with mode at

$$m_0 = 1 + \left[\frac{1}{z^{1/s} - 1} - a\right],$$

where $[W]$ is the integer part of W. Furthermore,

$$\mu = \frac{\phi(z, s-1, a)}{\phi(z, s, a)} - a$$

and

$$\mu_2 = (a+\mu)^2 + \frac{\phi(z, s-2, a) - 2(a+\mu)\phi(z, s-1, a)}{\phi(z, s, a)}.$$

Regarding the reliability functions, we obtain directly from the definitions that

$$h(x) = \frac{1}{(a+x)^s \phi(z, s, a+x)} \tag{3.32}$$

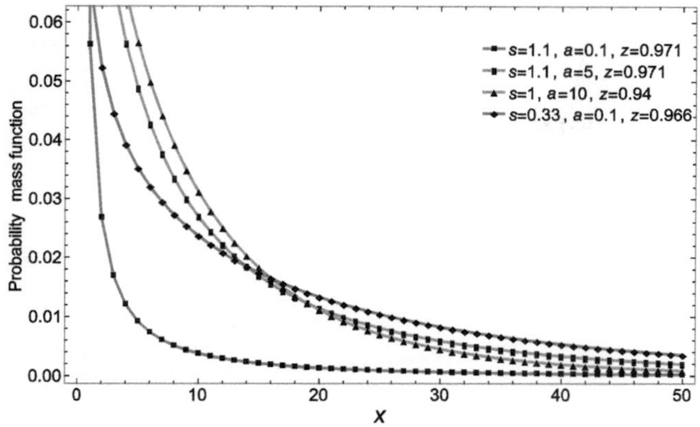

FIGURE 3.2 Probability mass functions of Hurwitz-Lerch distribution.

and

$$\lambda(x) = \frac{cz^x}{\phi(z,s,a) - z^{x+1}\phi(z,s,a+x+1)}. \tag{3.33}$$

We observe that the expressions in (3.32) leads to some new interesting recurrence relationships and formulae. Since

$$\phi(z,s,a+x) = [h(x)(a+x)^s]^{-1}$$

and

$$\phi(z,s,a+x+1) = \frac{1}{z}[\phi(z,s,a+x) - (a+x)^{-s}] = [h(x+1)(a+x+1)^s]^{-1}$$

from (3.31), we have on eliminating ϕ between the last two equations,

$$h(x+1) = z\Big(\frac{a+x}{a+x+1}\Big)^s \frac{h(x)}{1-h(x)}, \tag{3.34}$$

a recurrence relation for evaluating $h(x)$. One may use the initial value $h(0) = ca^{-s}$ in (3.34) to initiate the recurrence.

The probability mass function and the hazard rate function of the family are presented in Figs 3.2 and 3.3 respectively.

To simplify (3.33) further, we apply (Magnus et al., 1966)

$$\phi(z,s,a) - z^{x+1}\phi(z,s,ax+1) = \sum_{k=0}^{x} \frac{z^k}{(a+k)^s}$$

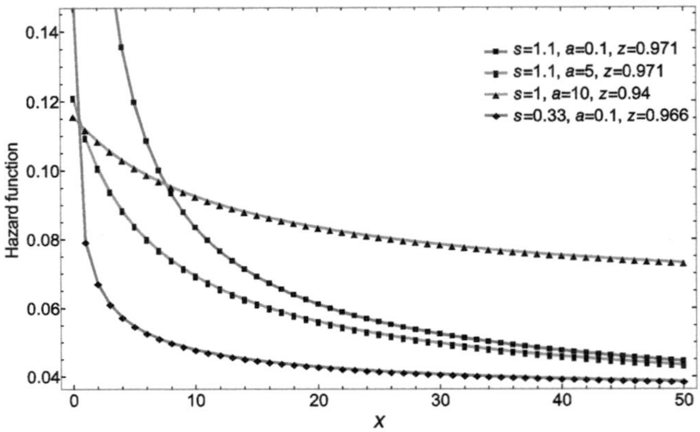

FIGURE 3.3 Hazard functions of Hurwitz-Lerch distribution.

to write a closed-form expression

$$\lambda(x) = \frac{z^x}{(a + x^s \sum_{k=0}^{x} \frac{z^k}{(a+k)^s})}$$

$$= \left[\sum_{k=0}^{x} \left(\frac{a+x}{a+x-k} \right)^s \frac{1}{z^k} \right]^{-1}. \tag{3.35}$$

The mean residual life is given by

$$m(x) = \frac{1}{z^x \phi(z, s, a+x+1)} \sum_{t=x+1}^{\infty} z^t \phi(z, s, a+t).$$

It now follows that

$$m(x+1) z \phi(z, s, a+x+2) - m(x) \phi(z, s, a+x+1) = \frac{z}{(a+x+1)^s},$$

or

$$m(x+1) - m(x) = \frac{z + m(x+1)}{(a+x+1)^s \phi(z, s, a+x+1)}. \tag{3.36}$$

Lerch family can be written in the form (3.20) with

$$B(\theta) = \phi(z, s, a),\ a(x) = (a+x)^{-s} \text{ and } g(\theta) = z$$

as a modified power series family with parameter z in place of $g(\theta) = \theta$. Hence, the moments satisfy the recurrence formula

$$\mu_{r+1} = z \frac{d\mu_r}{dz} + r \mu_z \mu_{r-1}, \quad r = 1, 2, \ldots,$$

for the central moments, and

$$\mu_{(r+1)} = z\frac{d\mu_{(r)}}{dz} + \mu_{(r)}\mu - r\mu_{(r)}$$

connecting the factorial moments. With the use of the expressions for μ and μ_2 given above, the central and factorial moments can be computed. In the context of reliability modelling, the following characterization in terms of reliability concepts are useful.

Theorem 3.2. *A discrete random variable X with support* $(0, 1, 2, \dots)$ *belongs to the Lerch family in* (3.30) *if and only if*

$$m^*(x) = \mu + z\frac{\partial \log S(x+1)}{\partial z}. \tag{3.37}$$

Proof. We have

$$S(x+1) = \sum_{t=x+1}^{\infty} \frac{z^t}{(a+t)^s \phi(z,s,a)}.$$

Differentiating with respect to z, we get

$$\frac{\partial S}{\partial z} = \sum_{t=x+1}^{\infty} \frac{tz^{t-1}}{(a+t)^s \phi(z,s,a)} - \sum_{t=x+1}^{\infty} \frac{z^t}{(a+t)^s \phi(z,s,a)} \frac{\phi'(z,s,a)}{\phi(z,s,a)}$$

and so

$$\frac{1}{S(x+1)} \frac{\partial S(x+1)}{\partial z} = \frac{1}{z}E(X|X > x) - \frac{\phi'(z,s,a)}{\phi(z,s,a)}. \tag{3.38}$$

By virtue of the identity

$$\phi(x, s-1, a) = \left(a + z\frac{\partial}{\partial z}\right)\phi(z,s,a)$$

and the fact that

$$\mu = \frac{\phi(z, s-1, a)}{\phi(z,s,a)} - a,$$

(3.38) simplifies to (3.37). To prove the sufficiency part, condition (3.7) means that

$$\sum_{t=x+1}^{\infty} tf(t) = \mu \sum_{t=x+1}^{\infty} f(t) + z \sum_{t=x+1}^{\infty} \frac{\partial f(t)}{\partial z}. \tag{3.39}$$

Changing $x+1$ to x and from the resulting equation, subtracting (3.39), yields

$$xf(x) = \mu f(x) + z\frac{\partial f(x)}{\partial z},$$

or

$$z\frac{\partial \log f(x)}{\partial z} = x - \mu = x - \frac{z}{\phi(z,s,a)}\frac{\partial \phi(z,s,a)}{\partial z}.$$

Upon integrating, we get

$$f(x) = \frac{z^x}{\phi(z,s,a)}A(x).$$

Since $\sum_{x=0}^{\infty} f(x) = 1$ and $A(x) = (a+x)^{-s}$ we have the proof. □

Theorem 3.3. *A necessary and sufficient condition for X to have Lerch distribution is that*

$$E[(a+X)^s|X>x] = \frac{(a+x+1)^s h(x+1)}{1-z}, \quad |z|<1,\ a>0, \tag{3.40}$$

for all x.

Proof. We have

$$f(x)(a+x)^s = \frac{z^x}{\phi(z,s,a)},$$

or

$$\sum_{t=x+1}^{\infty} f(t)(a+t)^s = \frac{z^{x+1}}{(1-z)\phi(z,s,a)}.$$

Thus,

$$\begin{aligned} E[(a+X)^s|X>x] &= \frac{z^{x+1}}{(1-z)S(x+1)\phi(z,s,a)} \\ &= \frac{(a+x+1)^s}{z}h(x+1), \end{aligned}$$

so that the condition is necessary. The sufficiency part follows by retracing the above steps. □

Theorem 3.4. *The random variable X has Lerch distribution if and only if*

$$E(X|X\le x) = \mu + z\frac{\partial \log F(x)}{\partial x}. \tag{3.41}$$

Proof. From Theorem 3.2 and the relationship

$$F(x)E(X|X \leq x) + S(x+1)E(X|X > x) = \mu,$$

we obtain the characterization in (3.41). □

Theorem 3.5. *The random variable X has Lerch distribution if and only if the variance residual life is given by*

$$\sigma^2(x) = z\frac{\partial \log S(x+1)}{\partial z} + z^2\frac{\partial^2 \log S(x+1)}{\partial z^2}. \tag{3.42}$$

Proof. Sudheesh and Nair (2010) have characterized the modified power series family by

$$\sigma^2(x) = \frac{g(\theta)}{g'(\theta)}\frac{\partial m^*(x,\theta)}{\partial \theta}. \tag{3.43}$$

Since the Lerch family is a special case when $g(\theta) = \theta = z$, we have

$$\sigma^2(x) = z\frac{\partial m^*(x,z)}{\partial z}. \tag{3.44}$$

Substituting (3.37) into (3.44), we obtain (3.42). □

The estimation of parameters of the Lerch family has been addressed by Aksenov and Savageau (2005), who proposed the method of moments and the method of maximum likelihood. Denoting the sample moments by

$$p_r = \frac{1}{n}\sum_{\lambda=1}^{n} x_i^r$$

from a random sample $(x_1, \ldots, x_n)$ from the distribution, the equations to be solved in the method of moments are as follows:

$$\frac{\phi(\bar{z},\bar{s}-1,\bar{a})}{\phi(\bar{z},\bar{s},\bar{a})} - \bar{a} = p_1,$$
$$\bar{a}^2 - 2\bar{a}\frac{\phi(\bar{z},\bar{s}-1,\bar{a})}{\phi(\bar{z},\bar{s},\bar{a})} + \frac{\phi(\bar{z},\bar{s}-2,\bar{a})}{\phi(\bar{z},\bar{s},\bar{a})} = p_2,$$
$$-\bar{a}^3 + 3\bar{a}^2\frac{\phi(\bar{z},\bar{s}-1,\bar{a})}{\phi(\bar{z},\bar{s},\bar{a})} - 3\bar{a}\frac{\phi(\bar{z},\bar{s}-3,\bar{a})}{\phi(\bar{z},\bar{s},\bar{a})} = p_3,$$

to get the estimates $(\bar{z},\bar{s},\bar{a})$ of (z,s,a). The above equations have to be solved numerically. Aksenov and Savageau (2005) also presented formulas for the asymptotic variances and covariances of the estimates obtained by the method of moments and the method of maximum likelihood.

Remark 3.2.

(1) The population moments given above are deduced from the general expression for the rth moment about the origin given by

$$\mu'_r = \frac{1}{\phi(z,s,a)} \sum_{j=0}^{r} \binom{r}{j} (-a)^{r-j} \phi(z, s-j, a);$$

(2) For calculation of the coefficients of skewness and kurtosis, the formula for the central moments given by

$$\mu_r = \frac{1}{\phi(z,s,a)} \sum_{j=0}^{r} (-1)^{r-j} \binom{r}{j} \phi(z, s-j, a) \left(\frac{\phi(z,s-1,a)^{r-j}}{\phi(z,s,a)} \right)^{r-j}$$

will be useful.

Gupta et al. (2008) have studied the Lerch family with positive integers as support. They considered the probability mass function of the form

$$f(x) = \frac{z^x}{T(z,s,a)(a+x)^{s+1}}, \quad \begin{array}{l} x = 1, 2, 3, \ldots, \; s \geq 0, \\ 0 \leq a \leq 1, \; 0 < z \leq 1, \end{array}$$

where

$$T(z,s,a) = \sum_{x=1}^{\infty} \frac{z^x}{(a+x)^{s+1}} = z\phi(z, s+1, a+1).$$

The basic characteristics of this distribution are evaluated from

$$\mu = \frac{T(z, s-1, a)}{T(z, s, a)} - a,$$

$$\sigma^2 = \frac{T(z, s-2, a)}{T(x, s, a)} - \left(\frac{T(z, s-1, a)}{T(z, s, a)} \right)^2,$$

and

$$\mu_{r+1} = z \frac{d\mu_r}{dz} + r\sigma^2 \mu_{r-1}, \; r = 1, 2, \ldots.$$

Other functions of interest are the probability generating function given by

$$P(t) = \frac{T(zt, s, a)}{T(z, s, a)},$$

the distribution function given by

$$F(x) = 1 - z^x \frac{T(z, s, a+x)}{T(z, s, a)},$$

the survival function given by

$$S(x) = z^{k-1} \frac{T(z, s, a + x + 1)}{T(z, s, a)},$$

the hazard rate given by

$$h(x) = \frac{z}{(a + x)^{s+1} T(z, s, a + x - 1)},$$

and the reversed hazard rate given by

$$\lambda(x) = \frac{z^k}{[T(z, s, a) - z^k T(z, s, a + x)](a + x)^{s+1}}.$$

The maximum likelihood method is proposed to estimate the parameters with the estimators solved from

$$E(X) = \bar{x},$$

$$\sum_{i=1}^{n} \frac{\log(a + x_i)}{n} = E(\log(a + X))$$

and

$$\frac{1}{n} \sum_{i=1}^{n} \frac{1}{a + x_i} = E\left(\frac{1}{a + X}\right).$$

They further observed that the likelihood equations are the same as the method of moments equations.

Zornig and Altmann (1995) and Kemp (2010) provided a list of members of the Lerch family. We now present some of these distributions and their properties. It may be noted that there is a difference in the expressions for the probability generating functions when the supports are different. For instance,

$$\begin{aligned} P(t) &= \phi(zt, s, a)/\phi(z, s, a), \quad x = 0, 1, 2, \ldots, \\ &= t\phi(zt, s, a)/\phi(z, s, a), \quad x = 0, 1, 2, \ldots, \\ &= \frac{t\phi(zt, s, a) - z^b \phi(tz, s, a + b)}{\phi(z, s, a) - z^b \phi(z, s, a + b)}, \quad x = 1, 2, \ldots, b. \end{aligned}$$

Geometric Distribution

The geometric distribution is a member of all the families discussed so far, and hence enjoys the properties of all families. In addition to some of the characteristic properties already discussed in the preceding chapter, we present a few more results here that are relevant to reliability studies. The foremost among them is the no-ageing (lack of memory) property of the geometric lifetimes. It states that X is geometric with probability mass function

$$f(x) = q^x p, \quad x = 0, 1, 2, \dots, \tag{3.45}$$

if and only if

$$P(X \geq s + t | X \geq s) = P(X \geq t) \tag{3.46}$$

for all $t, s = 0, 1, 2, \dots$, or equivalently,

$$P(X = s + t | X \geq s) = P(X = t).$$

Writing (3.46) as

$$\frac{P(X \geq s + t)}{P(X \geq s)} = P(X \geq t),$$

we can recognize the left hand side as the distribution of the residual life $(X - s | X \geq s)$. Thus, the residual life of a geometric lifetime at any age is the same as the original lifetime, thereby, justifying the name 'no-ageing' property.

Being a characteristic property, for devices that does not age with time, the geometric law provides a suitable model. This is consistent with our earlier observation that the hazard rate, mean residual life and variance residual life are constant independent of the age of the device. Further, all these properties are equivalent to one another. The notion of no-ageing, also defined as

$$S(x + y) = S(x)S(y) \tag{3.47}$$

for all x, y, forms the basis of many ageing concepts discussed in the next chapter. Also, (3.47) suitably extended to higher dimensions is employed to generate various forms of multivariate geometric distributions; see Chapter 7 for details. A simple extension of the geometric law is possible by assuming X to take values $a, a + d, a + 2d, \dots$, with

$$P(X = a + nd) = q^x p, \quad n = 0, 1, 2, \dots, \tag{3.48}$$

in which case also the no-ageing property applies when X is replaced $a + Xd$ in (3.46). More interesting results of (3.48), in connection with ageing behaviour of X, can be seen in Chapter 4. Some additional properties of the geometric distribution are as follows:

(i) If $X_1, \dots, X_n$ are independent geometric random variables with parameter p, then $X_1 + X_2 + \cdots + X_n$ is distributed as negative binomial with parameters (n, p);

(ii) If $X_1, \ldots, X_n$ are independent geometric random variables with parameters $p_1, p_2, \ldots, p_n$ respectively, then $X_{(1)} = \min(X_1, \ldots, X_n)$ is a geometric random variable with parameter $p = 1 - \prod_{i=1}^{n}(1 - p_i)$;

(iii) If X_1 and X_2 are independent random variables with same parameter p, then $X_{(2)} = \max(X_1, X_2)$ has distribution

$$
\begin{aligned}
P(X_{(2)} = x) &= P(X_1 = x, X_2 < x) \\
&\quad + P(X_2 = x, X_1 < x) \\
&\quad + P(X_1 = X_2 = x) \\
&= 2q^x p(1 - q^x) + q^{2x} p^2 \\
&= 2q^x p - q^{2x} p - q^{2x+1} p.
\end{aligned}
$$

The method of proof can be extended readily to the case of n variables.

(iv) If X is negative binomial with

$$f(x|K = k) = \binom{k + x - 1}{k - 1} q^x p^k$$

and K has geometric distribution

$$P(K = y) = p_1 q_1^{y-1}, \quad y = 1, 2, \ldots,$$

then the mixture (unconditional distribution of X) is geometric $(\frac{pp_1}{1-pq_1})$. Thus, geometric distribution arises as a mixture.

(v) Let Y be a continuous random variable with exponential distribution

$$f_1(y) = \lambda e^{-\lambda y}, \quad y > 0, \ \lambda > 0.$$

If we denote by X and U, the integer part (largest integer not exceeding Y) and the fractional part of Y, Steutel and Thiemann (1989) have shown that X and U are independent and further that X has geometric distribution in (3.45) with $p = 1 - e^{-\lambda}$; see also Deng and Chhikkara (1990). Arising from the above result, it is also shown that the order statistics $X_{k:n}$, $k = 1, 2, \ldots, n$, from the geometric distribution satisfies

$$X_{k:n} \stackrel{d}{=} \sum_{j=0}^{k-1} X_{1:n-j} + \left[\sum_{j=0}^{k-1} \frac{U_{n-j}}{n - j} \right],$$

where $\stackrel{d}{=}$ denotes equality in distribution, all variables on the right hand side are independent, and $[Y] = X$ as stated above. Also,

$$\sum_{j=0}^{k-1} X_{1:n-j} \leq X_{k:n} \leq X_{1:n-j} + n - 1$$

in which the inequalities denote the usual stochastic orderings discussed in Chapter 4.

It is more informative to pursue the above results in a more general framework. Let Y be a continuous random variable with probability density function $f_1(y)$ satisfying $P(Y \in D) = 1$, where D is an open subset of $\mathbb{R}$, the one-dimensional Euclidean space. If D can be divided into mutually exclusive regions $R_1, \ldots, R_k$ with corresponding regions $S_1, \ldots, S_k$ such that $T_i(Y) : R_i \to S_i$, $i = 1, 2, \ldots, k$, are injective and continuously differentiable. Then $U = T(Y)$ has probability density function

$$g(u) = \begin{cases} \sum_{i=1}^{k} \frac{f_1(T_i^{-1}(u))}{J(T_i^{-1}(u))}, & u \in T(R_i,) \\ 0, & u \notin T(R_i). \end{cases}$$

Here, J stands for the absolute Jacobian of transformation and $T^{-1} : T(D) \to D$ is the inverse of T (see Hoffman-Jorgensen, 1994). Applying the above transformation theorem with $D = \mathbb{R}$, $T(Y) = U$, $T(D) = [0, 1)$, then $J = 1$ for every $x = 1, 2, \ldots$. When Y is a continuous random variable with X as integer part and U as fractional part, we have the following result.

Theorem 3.6. *The distribution of U is continuous with probability density function*

$$g(u) = \sum_{x=-\infty}^{\infty} f_1(x + u), \quad 0 < u < 1.$$

Example. Starting with the exponential case

$$g(u) = \sum_{x=0}^{\infty} \lambda e^{-\lambda(x+u)} = \frac{\lambda e^{-\lambda u}}{1 - e^{-\lambda}}, \quad 0 < u < 1,$$

which is again the exponential distribution truncated at unity.

The application of the above result in reliability analysis is quite clear. On the one hand, we can convert the data on exponential to geometric and vice versa. Also by simulating exponential observations, their integer parts will be geometric. Some of the reliability functions from one distribution can be expressed in terms of the other. For example, in the exponential case, the mean residual life function is

$$\begin{aligned} m(y) &= E(Y - y|Y > y) = E(Y) \\ &= E(X + U) = E(X) + E(U) \\ &= E(X - x|X > x) + E(U) - 1 \\ &= \frac{1}{p} + \left(\frac{1}{\lambda} - \frac{e^{-\lambda}}{1 - e^{-\lambda}}\right) - 1. \end{aligned}$$

Thus, the difference between the mean residual lives of the exponential and geometric lifetimes is $\frac{1}{\lambda} - \frac{e^{-\lambda}}{1-e^{\lambda}}$. Even if failure times are observed at integer values, a very good approximation to continuous measurement can be made by virtue of the above result, from the estimated value of p, employing the equation $p = 1 - e^{-\lambda}$.

(vi) The other distributional properties are:

$$\text{probability generating function } P(t) = p(1 - qt)^{-1},$$

$$\text{mean } \mu = \frac{q}{p}, \qquad \text{variance } \mu_2 = \frac{q}{p^2},$$

and the rth factorial moment is

$$\mu_{(r)} = r!\left(\frac{q}{p}\right)^r.$$

Theorem 3.7. *For any discrete random variable X, we have*

$$E(Xh(X)) \geq \frac{2\mu^2}{E(X^2) + \mu}$$

with equality holding if and only if X is geometric.

Proof. We have

$$E(Xh(X)) = \sum_{x=0}^{\infty} \frac{xf^2(x)}{S(x)}.$$

From Cauchy-Schwarz inequality, we find

$$\left(\sum_{x=0}^{\infty} \frac{xf^2(x)}{S(x)}\right)\left(\sum_{x=0}^{\infty} xS(x)\right) \geq \left(\sum_{x=0}^{\infty} xf(x)\right)^2 = \mu^2. \tag{3.49}$$

By some straight forward calculations, we then get

$$\sum xS(x) = \frac{E(X^2) + \mu}{2}.$$

Hence, from (3.49), we have the inequality stated in the theorem. The equality sign in (3.49) holds if and only if

$$[xS(x)]^{\frac{1}{2}} = K\left(\frac{x}{S(x)}\right)^{\frac{1}{2}} f(x),$$

which gives $h(x) =$ a constant, characterizing the geometric law. □

(viii) Geometric distribution is a member of the Lerch family with $s = 0$ and $a = 1$. The reliability properties follow from the general formulas for this family and also from our discussions in Chapter 2.

Discrete Uniform Distribution

The discrete uniform distribution arises from (3.30) when $z = 1$, $s = 0$ and $a = 1$, with probability mass function

$$f(x) = \frac{1}{b}, \quad x = 1, \ldots, b. \tag{3.50}$$

It has distribution function $F(x) = \frac{x}{b}$ and survival function $S(x) = \frac{b-x+1}{b}$. Various distributional characteristics are as follows:

$$\mu = \frac{(b+1)(2b+1)}{6}, \qquad \mu_2 = \frac{b^2-1}{12},$$

$$P(t) = \frac{t(1-t^b)}{b(1-t)}, \quad t \neq 1.$$

If $X_1, \ldots, X_n$ are independent random variables with distribution in (3.50), then $X_{(1)} = \min X_i$ and $X_{(n)} = \max X_i$, $i = 1, 2, \ldots, n$ have respective probability mass functions

$$P(X_{(1)} = x) = \frac{(b-x+1)^n - (b-x)^n}{b^n}$$

and

$$P(X_{(n)} = x) = \frac{x^n - (x-1)^n}{b^n}.$$

Among the reliability functions, the hazard rate is $h(x) = \frac{1}{b-x+1}$ and the mean residual life function is $m(x) = \frac{(b-x+1)}{2}$, giving

$$h(x)m(x) = \frac{1}{2},$$

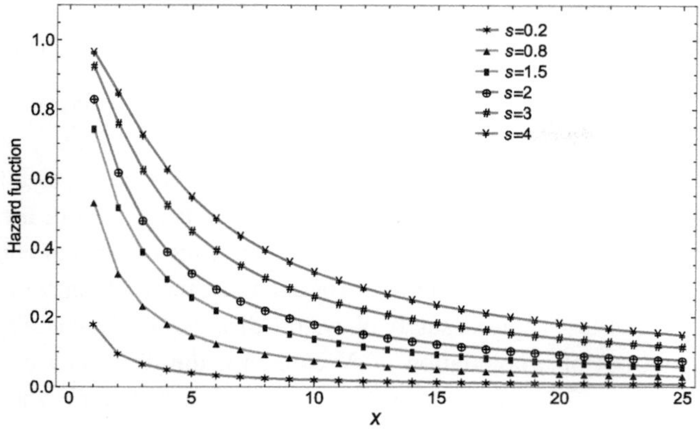

FIGURE 3.4 Hazard functions of discrete Pareto distribution.

a constant, characterizing the uniform distribution as a special case of the negative hypergeometric distribution (see Chapter 1). The variance residual life is found to be

$$\sigma^2(x) = \frac{(b-x+1)(b-x-1)}{12} = \frac{m(x)(m(x)-1)}{3}.$$

Also, the functions in reversed time turns out to be

$$\lambda(x) = \frac{1}{x}, \text{ a reciprocal linear function,}$$
$$r(x) = \frac{x}{2}, \text{ a linear function of } x,$$

with their product

$$\lambda(x)r(x) = \frac{1}{2},$$

a constant. Further, the reversed variance residual life becomes

$$v(x) = \frac{x(x-2)}{12} = \frac{r(x)(r(x)-1)}{3},$$

which is also a characterization of the discrete uniform distribution.

Discrete Pareto

The discrete Pareto distribution, also known as the Zipf distribution and as Riemann zeta distribution, is specified by the probability mass function (Fig. 3.4)

$$f(x) = Cx^{-(s+1)}, \quad x = 1, 2, \ldots, s > 0, \tag{3.51}$$

where

$$C=\left[\sum_{x=1}^{\infty}x^{-(s+1)}\right]^{-1}=\frac{1}{\zeta(s+1)}=\phi(1,s+1,1),$$

and $\zeta(\cdot)$ is the Riemann zeta function defined earlier in (3.27). As a model of random phenomenon, the distribution in (3.51) have been used in literature in different contexts. It is used to model the size or ranks of objects chosen randomly from certain type of populations, for example, the frequency of words in long sequences of text approximately obeys the discrete Pareto law. Zipf (1949) law states that out of a population of N elements, the frequency of elements of rank $(s+1)$ is (3.51) and hence the name Zipf distribution, when used in linguistics. Seal (1947) has used it to model the number of insurance policies per individual, and as the distribution of surnames by Fox and Lasker (1983); see also Good (1953) and Haight (1966). Various distributional characteristics are obtained as special cases of the corresponding results of the Lerch family. In particular, moments of (3.51) are given by

$$\mu_r'=\frac{\zeta(s-r+1)}{\zeta(s+1)},\qquad r<s,$$

and the moments are infinite for $r\geq s$. The maximum likelihood estimator $\hat{s}$ is obtained by solving

$$\frac{1}{n}\sum_{i=1}^{n}\log x_i=-\frac{\zeta'(\hat{s}+1)}{\zeta(\hat{s}+1)}.$$

Some special cases of the discrete Pareto are the Estoup model, given by

$$f(x)=\frac{c}{x},\qquad x=1,2,\ldots,b, \tag{3.52}$$

(who noted that the rank r and frequency f in a French text were related by a hyperbolic law which translates in to the form (3.52)) and the Lotka law given by

$$f(x)=\frac{c}{x^2},$$

where $c=\frac{6}{\pi^2}$. Lotka law results from the finding that the number of authors making n contributions is about n^{-a} of those making one contribution, in which empirically a is nearly equal to two. When the value of x in (3.51) is truncated at b, we obtain

$$f(x)=\frac{1}{H(b,s)x^s},\qquad x=1,2,\ldots,b, \tag{3.53}$$

which is a proper distribution with $H(b,s)=\sum_{x=1}^{b}(\frac{1}{x})^s$. In this case,

$$F(x)=\frac{H(x,s)}{H(b,s)} \text{ and } S(x)=\frac{H(b,s)-H(x-1,s)}{H(b,s)}$$

and accordingly

$$\lambda(x)=\frac{H(x,s)}{x^s} \text{ and } h(x)=\frac{1}{x^s[H(b,s)-H(x-1,s)]}.$$

The mean and variance are given by

$$\mu=\frac{H(b,s-1)}{H(b,s)} \text{ and } \mu_2=\frac{H(b,s-2)H(b,s)-H^2(b,s-1)}{H^2(b,s)},$$

and the moment generating function is given by

$$M(t)=\frac{1}{H(b,s)}\sum_{j=1}^{b}\frac{e^{jt}}{j^s}.$$

A distribution that is unrelated to the above, but bears the name 'zeta distribution' was introduced by Haight (1969) with probability mass function

$$f(x)=(2x-1)^{-\sigma}-(2x+1)^{-\sigma}, \quad x=1,2,\ldots,\ \sigma>0.$$

The distribution and survival functions are, respectively,

$$F(x)=1-(2x+1)^{-\sigma}$$

and

$$S(x)=(2x-1)^{-\sigma}.$$

Thus, the hazard and reversed hazard functions have simple closed form expressions as

$$h(x)=1-\left(\frac{2x-1}{2x+1}\right)^{\sigma}$$

and

$$\lambda(x)=\left[\left(\frac{2x+1}{2x-1}\right)-1\right]\Big/[(2x+1)^{\sigma}-1].$$

Further, the mean residual life and reversed mean residual life are found to be

$$m(x)=\mu+\frac{g(x)h(x)}{1-h(x)}-x$$

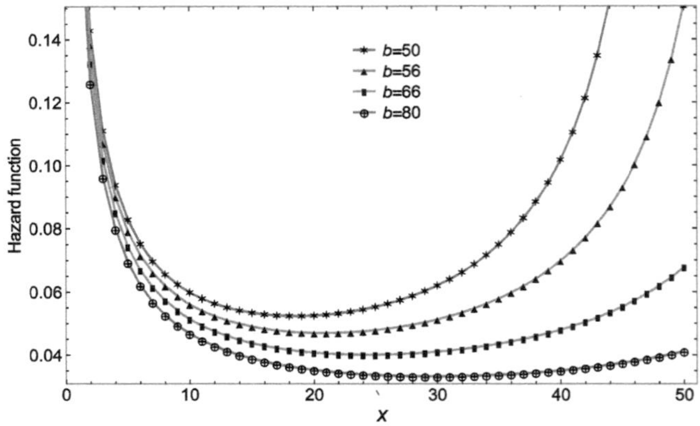

FIGURE 3.5 Hazard functions of Estoup distribution.

and

$$r(x) = x - \mu + g(x)\lambda(x),$$

where

$$g(x) = \frac{\mu(1-(2x+1)^{-\sigma}) - \zeta_x(\sigma)}{(2x-1)^{-\sigma} - (2x+1)^{-\sigma}},$$

$\zeta_x(\sigma) = \sum_{t=1}^{x} \frac{1}{t^\sigma}$ and $\mu = (1-2^{-\sigma})\zeta(\sigma)$ is the mean.

The variance of the distribution is

$$\mu_2 = (1-2^{1-\sigma})\zeta(\sigma-1) - \mu^2.$$

Notice that the mean is infinite for $\sigma \leq 1$ and the variance is infinite for $\sigma \leq 2$. Some further literature on discrete Pareto versions can be seen in Lin and Hu (2001) and Shan (2005).

Figs 3.4–3.7 show hazard rate functions of discrete Pareto, Estoup, Lotka and truncated discrete Pareto distributions.

Hurwitz-Zeta Distribution

A second major distribution belonging to the Lerch family is the Hurwitz-zeta distribution defined by

$$f(x) = \frac{c}{(a-1+x)^s}, \qquad x = 1, 2, \ldots, \ s > 1, \tag{3.54}$$

with $c^{-1} = \zeta(s,a) = \phi(1,s,a)$ being the Hurwitz-zeta function defined in (3.28). This distribution is mainly used in linguistics in connection with ranking

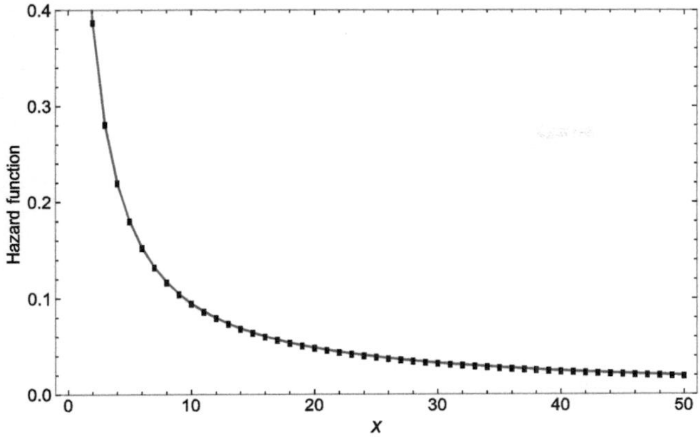

FIGURE 3.6 Hazard functions of Lotka distribution.

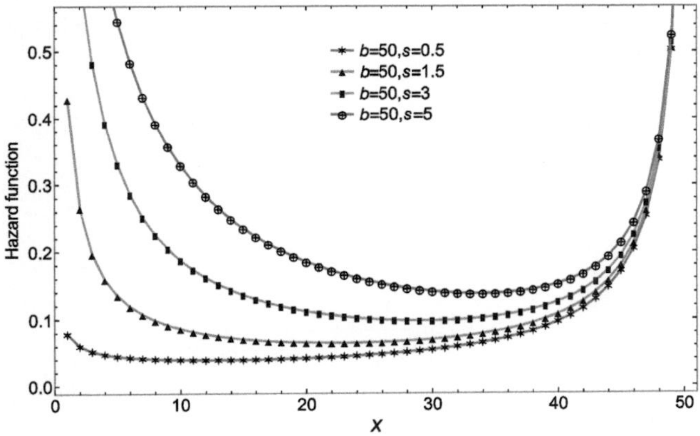

FIGURE 3.7 Hazard functions of truncated discrete Pareto.

problems. We note that

$$\zeta(s, a) - \zeta(s, a + 1) = a^{-s},$$

so that

$$F(x) = 1 - \frac{\zeta(s, a + x)}{\zeta(s, a)} \text{ and } S(x) = \frac{\zeta(s, a + x - 1)}{\zeta(s, a)}.$$

Accordingly, the reliability functions have the expressions

$$h(x) = [(a - 1 + x)^s \zeta(s, a + x - 1)]^{-1},$$
$$\lambda(x) = [(a - 1 + x)^{-s}(\zeta(s, a) - \zeta(s, a + x))]^{-1}.$$

Also,

$$\zeta(s, a+x-1) = [(a+x-1)^s h(x)]^{-1}$$
$$\zeta(s, a+x) = [(a+x)^s h(x+1)]^{-1},$$

along with

$$\zeta(s, a+x-1) - \zeta(s, a+x) = (a+x-1)^{-s},$$

leads to

$$h(x+1) = \left(\frac{a+x-1}{a+x}\right)^s \frac{h(x)}{1-h(x)}, \quad x = 1, 2, \ldots \tag{3.55}$$

Formula (3.55) enables the evaluation of successive hazard rates starting from $h(1) = \frac{1}{a^s \zeta(s,a)}$. The expression for $\lambda(x)$ given above admits further simplification on applying the relationship

$$\zeta(s, a) - \zeta(s, a+k) = \sum_{n=0}^{k-1} (a+n)^{-s},$$

which leads to a compact expression of the form

$$\lambda(x) = \frac{(a+x-1)^s}{\sum_{n=0}^{x-1} (a+n)^{-s}}.$$

Consequently, we have the recurrence relation

$$\lambda(x+1) = \lambda(x) \left[\lambda(x) + \left(\frac{a+x}{a+x+1}\right)^s\right]^{-1},$$

which can be determined iteratively by setting the initial value as $\lambda(1) = 1$. We can write the mean residual life as

$$m(x) = \frac{1}{\zeta(s, a+x)} \sum_{t=x+1}^{\infty} \zeta(s, a+t-1)$$
$$= \frac{1}{\zeta(s, a+x)} \sum_{t=x}^{\infty} \zeta(s, a+t).$$

Hence,

$$m(x)\zeta(s, a+x) - m(x+1)\zeta(s, a+x+1) = \zeta(s, a+x).$$

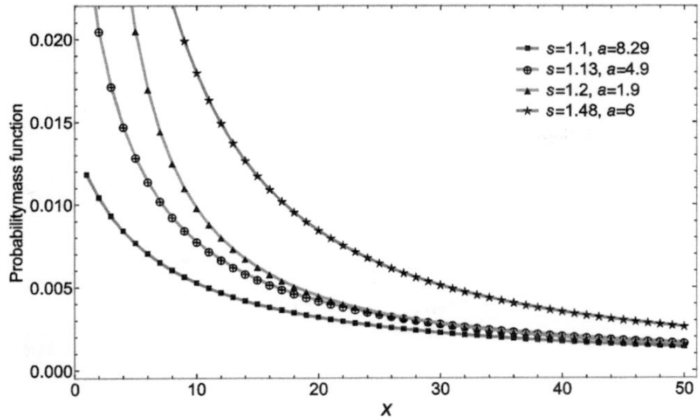

FIGURE 3.8 Probability mass functions of Hurwitz-zeta distribution.

By virtue of

$$\zeta(s, a+x+1) = \zeta(s, a+x) - (a+x)^{-s}, \tag{3.56}$$

we have a recurrence relation

$$m(x+1) = \frac{\zeta(s, a+x)(m(x)-1)}{\zeta(s, a+x) - (a+x)^{-s}}.$$

The reversed mean residual life is given by

$$r(x) = \frac{1}{\zeta(s,a) - \zeta(s, a+x-1)} \sum_{t=1}^{x-1} \zeta(s,a) - \zeta(s, a+x).$$

Further simplification yields

$$r(x+1) = 1 + \left(\frac{\sum_{t=1}^{x-1} (a+t-1)^{-s}}{\sum_{t=1}^{x} (a+t-1)^{-s}} \right) r(x).$$

The probability mass function and hazard rate function of Hurwitz-zeta distribution are presented in Figs 3.8 and 3.9, respectively.

A special case of the Hurtwitz-zeta model that is in use is the Zipf-Mandelbrot law (Mandelbrot, 1959, 1966) specified by

$$f(x) = \frac{c}{(a-1+x)}, \quad x = 1, 2, \ldots, n,$$

obtained when $s = 1$ in (3.54). See Zornig and Altmann (1995) for a discussion on the distribution.

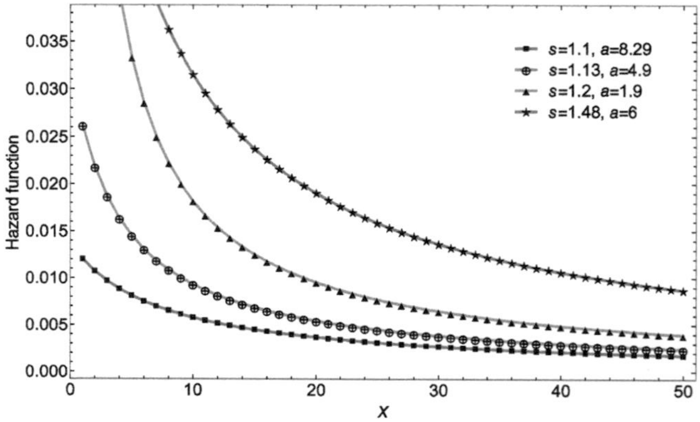

FIGURE 3.9 Hazard functions of Hurwitz-zeta distribution.

A third category of distribution in the Lerch family makes use of the general term in the polylogarithmic function $A(z, s)$ defined in (3.29).

Accordingly, the general form of the probability mass function in such cases is the Good distribution (Good, 1953) given by

$$f(x) = \frac{Cz^x}{x^s}, \quad x = 1, 2, \ldots, \ 0 < x < 1, \tag{3.57}$$

with $C = A(z, s) = z\phi(z, s, 1)$. An important member in the above form, repeatedly used in the sequel, is the logarithmic series distribution (Fig. 3.8).

Setting $s = 1$ in (3.57), we have the probability mass function of the distribution as

$$f(x) = \frac{Cz^x}{x}, \quad x = 1, 2, \ldots, \ 0 < x < 1, \tag{3.58}$$

where $C = -\log(1 - z)$. It is also member of the GPSD discussed above. It is easily seen that

$$f(x) = \frac{z(x-1)}{x} f(x-1), \quad x = 2, 3, \ldots, \ 0 < x < 1. \tag{3.59}$$

Historical remarks and the origin of the logarithmic series distribution and its properties can be found in Johnson et al. (1992). Regarding distributional properties, it may be noted that the probability generating function has the form

$$P(t) = \frac{\log(1 - zt)}{\log(1 - z)}.$$

The factorial moment of order r is

$$\mu_{(r)} = -\log(1 - z)^{-1} z^r (r - 1)!(1 - z)^{-r}$$

from which the mean and variance are obtained as

$$\mu = -\log(1-z)^{-1} z(1-z)^{-1},$$
$$\mu_2 = -\log(1-z)(1-z)^{-2}(1+z\log(1-z)).$$

Various central moments can be found from the above value of μ_2 and the recurrence relation

$$\mu_{r+1} = (1-z)\frac{d\mu_r}{dz} + r\mu_2\mu_{r-1}.$$

From the recurrence relation for $f(x)$ mentioned in (3.58), the ratio $\frac{f(x+1)}{f(x)}$ is always less than unity, so that $f(x)$ is always a decreasing function. The distribution has a long tail and it resembles the geometric distribution for large values of x. We can identify the logarithmic series distribution by plotting $\frac{xf_x}{(x-1)f_{x-1}}$ against x, where f_x is the observed frequency of x (see Eq. (3.58)). The plot is expected to give a straight line parallel to the x-axis with intercept z. In problems of estimation of z, the maximum likelihood estimator $\hat{z}$ is obtained by solving the equation

$$\bar{x} = \frac{\hat{z}}{-(1-\hat{z})\log(1-\hat{z})},$$

with $\bar{x}$ being the mean of a random sample drawn from the distribution. The reversed hazard rate function is

$$\lambda(x) = \frac{z^x}{x\sum_{t=1}^{x}\frac{z^t}{t}}$$

giving

$$\lambda(x+1) = \frac{zx\lambda(x)}{zx\lambda(x)+x+1}. \tag{3.60}$$

On the other hand, the hazard rate function is the reciprocal of an infinite series, *viz.*,

$$h(x) = \left[\sum_{t=0}^{\infty}\frac{xz^t}{x+t}\right]^{-1}.$$

Hence,

$$h(x+1) = \frac{zxh(x)}{(x+1)(1-h(x))}.$$

Fig. 3.10 gives hazard rate functions of Good distribution.

The rest of the reliability functions characterizing the logarithmic distribution are given in the following theorem.

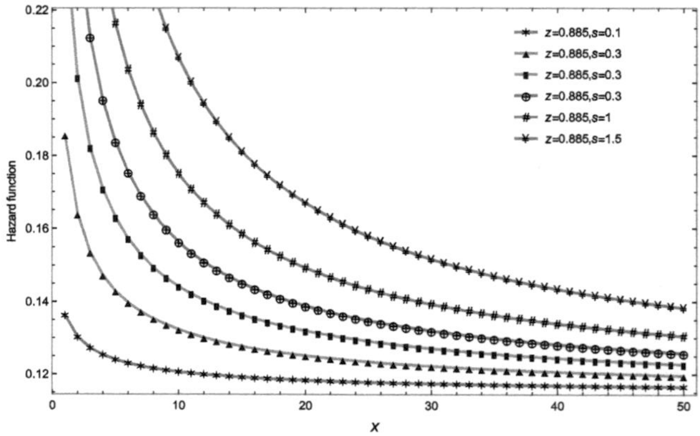

FIGURE 3.10 Hazard functions of Good distribution.

Theorem 3.8. *A discrete random variable X with the support of* $(1, 2, \ldots)$ *follows the logarithmic series distribution if and only if, for all x,*

$$m^*(x) = \frac{1+x}{1-z} h(x+1). \tag{3.61}$$

Proof. Assume that the relationship (3.61) is true for the random variable X. Then,

$$\frac{1}{S(x+1)} \sum_{t=x+1}^{\infty} t f(t) = \frac{1+x}{1-x} \frac{f(x+1)}{S(x+1)},$$

or

$$\sum_{t=x+1}^{\infty} t f(t) = \frac{1+x}{1-z} f(x+1).$$

This gives

$$x f(x) = \frac{x}{1-z} f(x) - \frac{1+x}{1-z} f(x+1)$$

and

$$\frac{f(x+1)}{f(x)} = \frac{zx}{1+x}, \quad x = 1, 2, \ldots \tag{3.62}$$

Consequently, from (3.58), X has logarithmic series distribution. The converse is proved by retracing the steps from (3.62). □

From (3.61), the mean residual life function of X is determined as

$$m(x) = \frac{1+x}{1-x}h(x+1) - x$$

with the expression for $h(x)$ given above.

Some general aspects of various distributions and their applications, other than those connected with reliability modelling, are given in the works of Vilaplana (1987), Gut (2005), Doray and Luong (1995, 1997), Kulasekera and Tonkyn (1992) and Panaretos (1989). Figs 3.1A, B, C and Fig. 3.3 provide the shapes of the hazard rate functions of the modified power series and Lerch families.

3.2.4 Abel Series Distributions

The family of Abel series distributions is named after the Norwegian mathematician Niels Henrik Abel (1802–1829) who found the series whose general term specifies the probability mass function of the family. Abel polynomials (Comptet, 1994) are defined in terms of sequence $\langle p_x(\theta) \rangle$, where

$$p_x(\theta) = \alpha_x(\theta,\lambda) = \frac{\theta(\theta + x\lambda)^{x-1}}{x!}, \quad x = 0, 1, 2, \ldots, \tag{3.63}$$

with the first term $\alpha_0 = 1$ and the second term $\alpha_1 = \theta$. Successive differentiation of (3.63) with respect to θ yields

$$\frac{d^r}{d\theta^r}\alpha_x(\theta,\lambda) = \alpha_{x-r}(\theta + r\lambda, \lambda), \quad r = 0, 1, 2, \ldots. \tag{3.64}$$

For a fixed λ, not related to θ, $\alpha_x(\theta,\lambda)$ forms the basis of a set of polynomials in θ. Hence, for a finite, positive and successively differentiable function $P(\theta) = P(\theta,\lambda)$ in θ, we have

$$P_1(\theta) = \sum_{x=0}^{\infty} \alpha_x(\theta,\lambda) g(-x\lambda), \tag{3.65}$$

where

$$g(-x\lambda) = \left[\frac{d^x}{d\theta^x} P_1(\theta)\right]_{\theta = x\lambda}.$$

Accordingly, restricting the parameter space to $(\theta,\lambda) \in \Theta \times \Lambda$ for which the terms in the expansion (3.65) are non-negative, we conclude that

$$f(x) = \frac{\alpha_x(\theta,\lambda) g(-x\lambda)}{P_1(\theta)}, \quad x = 0, 1, 2, \ldots, \tag{3.66}$$

qualifies to be a probability mass function of a discrete random variable X. The distribution in (3.66) will be designated as the *Abel series distribution* (ASD), with series function $P_1(\theta)$. It may be noted that the truncated version of the ASD is also an ASD.

Some special notation and functions are required to describe the properties of ASD, as given in Charalambides (1990). They are the shift operator E and the usual differential operator D combined to give

$$DE^{-\lambda}a(u) = \frac{da(u-\lambda)}{du}.$$

The Bell partition polynomials

$$B_n = B_n(a_1, a_2, \ldots, a_n; b_1, b_2, \ldots, b_n)$$

are defined by

$$B_n = \sum \frac{n!b_k}{k_1!\ldots k_n!}\left(\frac{a_1}{1!}\right)^{k_1}\left(\frac{a_2}{2!}\right)^{k_2}\cdots\left(\frac{a_n}{n!}\right)^{k_n}, \quad n = 1, 2, \ldots, \ B_0 = 1,$$

where the summation is extended over all partitions of n over $k_1, \ldots, k_n \geq 0$ satisfying $k_1 + 2k_2 + \cdots + nk_n = n$ and $k = k_1 + k_2 + \cdots + k_n$ is the number of parts of the partition. The special cases $f_k = 1$, $k = 1, 2, \ldots$, and $f_k = (s)_k = s(s-1)\ldots(s-k+1)$, $k = 1, 2, \ldots$, for real s of the Bell polynomials are denoted by $B'_n(a_1, a_2, \ldots a_n)$ and $B_n^{(s)}(a_1, a_2, \ldots a_n)$. Also, let

$$a(t) = \sum_{k=0}^{\infty} a_k t^k / k!$$

be the generating function of the sequence (a_k). With the above notations Charalambides (1990) has then shown that

(i) the probability generating function of ASD is

$$P(t) = [P_1(\theta)]^{-1}\{\exp[\theta q^{-1}(tDE^{-\lambda})P_1(u)]\}_{u=0},$$

where $q(w) = we^{-\lambda w}$, q^{-1} is the inverse of q and $DE^{-\lambda}a(u) = \frac{da(\mu-\lambda)}{du}$;

(ii) the factorial moments are given by

$$\mu_{(r)}(\theta) = [P_1(\theta)]^{-1}[B'_r(Q_1 \ldots Q_r)E^{\theta}P(u)]_{u=0}, \quad r = 0, 1, 2, \ldots,$$

with

$$Q_m = (-1)^{m-1}\theta\lambda^{m-1}D^m(1-\lambda D)^{-m}B_{m-1}^{(-m)}(C_1, C_2, \ldots, C_{m-1}),$$
$$m = 1, 2, \ldots,$$

where

$$C_k = (k+1-\lambda D)(1-\lambda D)^{-1}/(k+1),\ k = 1, 2, \ldots.$$

Various basic members of the ASD are obtained by the Abel series expansions of the exponential, binomial and logarithmic functions. We will discuss these distributions in some detail.

Generalized Poisson Distribution

Take

$$P_1(\theta) = e^{\theta}, \quad \frac{d^r P_1(\theta)}{d\theta^r} = e^{\theta}$$

and consequently $g(-x\lambda) = e^{-x\lambda}$. Thus, we have the probability mass function as

$$f(x;\theta,\lambda) = \frac{\theta(\theta+x\lambda)^{x-1}}{x!} e^{-(\theta+x\lambda)},$$
$$x = 0, 1, 2, \ldots,\ \theta > 0,\ 0 < \lambda < 1, \quad (3.67)$$

representing the *generalized Poisson distribution* of Consul and Jain (1973a, 1973b) who obtained it from the Lagrangian expansion. A comprehensive study of this distribution is available in Consul (1989). By way of properties of the distribution, we have:

(i) $$f(x) = \frac{(\theta+x\lambda)^{x-1}e^{-\lambda}}{(\theta+x\lambda-\lambda)^{x-2}} f(x-1); \quad (3.68)$$

(ii) $$\mu = \frac{\theta}{1-\lambda}, \qquad \mu_2 = \frac{\theta}{(1-\lambda)^3};$$

(iii) $$(1-\lambda)\mu'_{r+1} = \theta\mu'_r + \theta\frac{\partial\mu'_r}{\partial\theta} + \lambda\frac{\partial\mu'_r}{\partial\lambda}, \quad r = 0, 1, 2, \ldots,$$

and

$$\mu_{r+1} = \frac{r\theta}{(1-\lambda)^3}\mu_{r-1} + \frac{1}{1-\lambda}\left[\frac{\partial\mu_r(t)}{\partial t}\right]_{t=1}, \quad r = 1, 2, 3, \ldots,$$

where $\mu_r(t)$ is μ_r with λ and θ replaced by λt and θt, respectively.

The distribution in (3.67) is a modified power series distribution. We can estimate the parameters by the method of moments as

$$\hat{\theta} = \sqrt{\frac{m_1^3}{m_2}}$$

and

$$\hat{\alpha} = \sqrt{\frac{m_2}{m_1^3}} - \frac{1}{m_1},$$

where $\alpha = \frac{\lambda}{\theta}$ and m_i is the ith sample moment, $i = 1, 2$.

Regarding reliability characteristics, which has not been studied so far, we first observe that the hazard rate function is

$$h(x) = \frac{(\theta + x\lambda)^{x-1} e^{-\lambda x}}{x! \sum_{t=x}^{\infty} \frac{(\theta + t\lambda)^{t-1}}{t!} e^{-t\lambda}}. \tag{3.69}$$

Also, (3.69) satisfies the relationship

$$h(x+1) = \frac{(\theta + \lambda + x\lambda)^x}{(\theta + x\lambda)^{x-1}(x+1)} \frac{h(x)}{(1 - h(x))} \tag{3.70}$$

from which successive values of $h(x)$ can be calculated starting with $h(0) = e^{-\theta}$.

Secondly, the mean residual life function is obtained from the theorem due to Consul (1995) given below, with parametrization as $[\theta, \alpha)$.

Theorem 3.9. *A discrete random variable X has generalized Poisson distribution if and only if*

$$m(x) = \mu \left[1 + \frac{\partial S(x+1)}{\partial \theta} \right] - x,$$

for $x = 0, 1, 2, \ldots$.

Reliability functions in reversed time obey similar properties. We have

$$\lambda(x) = \frac{(\theta + \lambda x)^{x-1} e^{-\lambda x}}{x! \sum_{t=0}^{x} \frac{(\theta + \lambda t)^{t-1}}{t!} e^{-\lambda t}}$$

and therefrom

$$\frac{1}{\lambda(x+1)} = 1 + \frac{1}{\lambda(x)} \frac{(x+1)e^{\lambda}(\theta + \lambda x)^{x-1}}{\theta(\theta + \lambda + \lambda x)}.$$

Differentiating the distribution function

$$F(x; \theta, \alpha) = \sum_{t=0}^{x} \frac{(1 + \alpha t)^{t-1}}{t!} \theta^t e^{-\theta(1+\alpha t)}, \quad 0 \le \alpha < \frac{1}{\theta}, \ \alpha = \frac{\lambda}{\theta},$$

and simplifying the resulting expression, we obtain

$$E(X|X < x+1) = \mu \left[1 + \frac{\partial \log F}{\partial \theta} \right].$$

Thus, the reversed mean residual life has the expression

$$r(x+1)=x+1-\mu\left(1+\frac{\partial \log F}{\partial \theta}\right).$$

When $\lambda=0$ ($\alpha=0$), all the above results reduce to those of the Poisson distribution. In the Poisson process, the probability of occurrence of a single event remains constant, whereas in the generalized Poisson case the probability depends on the previous occurrence. On the basis of experimental evidence, the generalized case appears to model situations in which failures occur rarely in short periods of time where the frequency of such events are of interest over longer duration of time.

Quasi-Binomial Distribution I

Set $P_1(p)=(p+q)^n$,

$$g(-x\lambda)=n(n-1)\ldots(n-x+1)(1-p-x\lambda)^{n-x}$$

and

$$(p+q)^n=\sum_{x=0}^{n}\binom{n}{x}p(p+x\lambda)^{x-1}(1-p-x\lambda)^{x-1}.$$

When $q=1-p$, we have the probability mass function

$$f(x)=\binom{m}{x}p(p+x\lambda)^{x-1}(1-p-x\lambda)^{n-x},\quad x=0,1,2,\ldots,n,$$
$$0\le p\le 1,\ -\frac{p}{n}<\lambda<\frac{1-p}{n} \tag{3.71}$$

A random variable X having its as probability mass function as in (3.71) is said to have a *quasi-binomial distribution I* (QBD-I). The probabilities of X are computed from the recurrence relation

$$f(x+1)=\frac{(n-x)(p+\lambda x)}{(x+1)(1-p-x\lambda)}\left(1+\frac{\lambda}{p+x\lambda}\right)^{x}\left(1-\frac{\lambda}{1-p-x\lambda}\right)^{n-x-1}f(x).$$

QBD-I is unimodal and the values of the parameter λ substantially affects various characteristics of the distribution. Expressions for the mean and variance are

$$\mu=E(X)=p\sum_{k=0}^{n-1}\frac{n!}{(n-k)!}\lambda^k$$

and

$$\mu_2 = \sum_{k=0}^{n-2}(k+1)p^2\lambda^k \frac{n!}{(n-k-2)!} + \frac{1}{2}p\lambda \sum_{k=0}^{n-2}(k+1)(k+4)\lambda^4 \frac{n!}{(n-k-2)!} + \mu - \mu^2,$$

respectively. When $\lambda > (<)0$, the mean of the QBD 1 is larger (smaller) than the mean of the binomial. Further distributional aspects and estimation of the parameters have been discussed by Consul and Famoye (2006).

The hazard rate function of QBD 1 is given by

$$h(x) = \frac{\binom{n}{x} p(p+x\lambda)^{x-1}(1-p-x\lambda)^{n-x}}{\sum_{t=x}^{n}\binom{n}{t} p(p+t\lambda)^{t-1}(1-p-t\lambda)^{n-t}},$$

and there exists a recurrence formula of the form

$$\frac{1}{h(x)} = 1 + \frac{1}{h(x+1)}\left[\frac{n-x}{x+1}\left(\frac{p+x\lambda+\lambda}{1-p-x\lambda}\right)\left(1+\frac{\lambda}{p+x\lambda}\right)^{x-1}\left(1-\frac{\lambda}{1-p-x\lambda}\right)^{n-x-1}\right] \tag{3.72}$$

with $h(0) = f(0) = (1-p)^n$. Likewise, the reversed hazard rate satisfies the recurrence formula

$$\frac{1}{\lambda(x+1)} = 1 + \frac{1}{\lambda(x)}\left[\frac{x+1}{n-x}\left(\frac{1-p-x\lambda}{p+x\lambda+\lambda}\right)\left(1+\frac{\lambda}{p+x\lambda}\right)^{1-x}\left(1-\frac{\lambda}{1+p-x\lambda}\right)^{x+1-n}\right]$$

with $\lambda(0) = 1$.

The probability mass functions and hazard rate functions of quasi-binomial distribution are presented in Figs 3.11 and 3.12, respectively.

Quasi-Negative Binomial Distribution

Taking $P_1(a) = (b-a)^{-n}$, we have

$$P_1(a) = \sum_{x=0}^{\infty}\binom{n+x-1}{x} a(a+x\lambda)^{x-1}(b+x\lambda)^{-n-x}. \tag{3.73}$$

Thus,

$$f(x;a,\lambda) = \binom{n+x-1}{x} a(a+x\lambda)^{x-1}(b+x\lambda)^{-n-x},$$

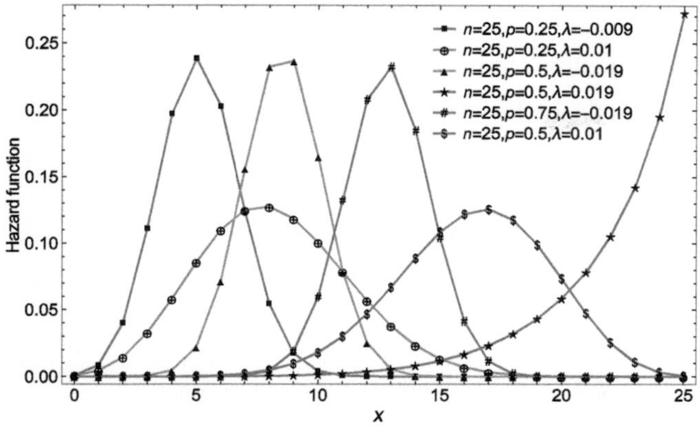

FIGURE 3.11 Probability mass functions of quasi-binomial distribution-I.

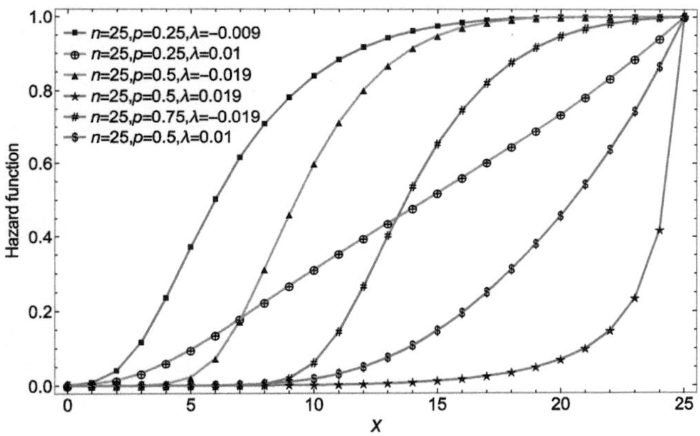

FIGURE 3.12 Hazard rate functions of quasi-binomial distribution-I.

or

$$f(x) = \binom{n+x-1}{x} p(p+x\phi)^{x-1}(q+x\phi)^{-n-x}, \tag{3.74}$$

$$x = 0, 1, 2, \ldots, \ p + x\phi < 1, \ q - p = 1,$$

defines a probability distribution, which is called the *quasi-negative binomial distribution* (QNBD). In (3.74), $p = \frac{a}{b-a}$, $q = \frac{b}{b-a}$ and $\phi = \frac{\lambda}{b-a}$ in the original expansion. Some special cases of (3.73) are as follows:

(i) the negative binomial when $\phi = 0$;

(ii) the *quasi-geometric distribution* (QGD)

$$f(x) = p(p+x\phi)^{x-1}(q+x\phi)^{-1-x}, \quad x = 0, 1, 2, \ldots,$$

when $n = 1$ and

(iii) the geometric distribution when $n = 1$ and $\phi = 0$.

The probability mass function, for successive values of x, satisfies the recursive formula

$$f(x+1) = \frac{n+x}{x+1}\left(\frac{p+x\phi+\phi}{q+x\phi}\right)\left(1+\frac{\phi}{p+x\phi}\right)^{x-1} \left(1+\frac{\phi}{q+x\phi}\right)^{-n-x} f(x),$$

and the moments are evaluated from the generating function

$$P(t) = (P_1(a))^{-1}\left[\exp aq^{-1}\left(t\frac{dP_1(\mu-\lambda)}{du}\right)\right]_{u=0}.$$

Eq. (3.74) yields the survival function as

$$S(x) = \sum_{t=x}^{\infty}\binom{n+t-1}{t}p(p+t\phi)^{t-1}(q+t\phi)^{-n-t}. \quad (3.75)$$

From (3.74) and (3.75), we find that the hazard rate function can be computed from the recursive formula

$$\frac{1}{h(x)} = 1+\frac{n+x}{x}\frac{(1+\frac{\phi}{p+x\phi})^{x-1}}{(1+\frac{\phi}{1+p+x\phi})^{n+x}}\left(1+\frac{1}{p+x\phi+\phi}\right)^{-1}\frac{1}{h(x+1)}.$$

Similarly, the reversed hazard rate function satisfies the recursive formula

$$\frac{1}{\lambda(x+1)} = 1+\frac{1}{\lambda(x)}\frac{(x+1)}{(n+x)}\left(\frac{q+x\phi+\phi}{p+x\phi+\phi}\right)\frac{(1+\frac{\phi}{q+x\phi})^{x+n}}{(1+\frac{\phi}{p+x\phi})^{x-1}}.$$

When $n = 1$, we have the results for the quasi-geometric law. Models that lead to QNBD as well as various properties of QNBD have been discussed by Bilal and Hassan (2006) and Hassan and Bilal (2008). They have found the mean and variance to be

$$\mu = np\,{}_2F_0[1, n+1, -; \phi]$$

and

$$\begin{aligned}\mu_2 =& np\,{}_2F_0[1, n+1, -; \phi] + p(p+2\phi)n(n+1)\,{}_2F_0[2, n+2, -; \phi] \\ &+ p\phi^2 n(n+1)(n+2)\,{}_2F_0[3, n+3, -; \phi] - \mu^2,\end{aligned}$$

respectively, where

$$_2F_0[1, n+1, -; \phi] = \sum_{j=0}^{\infty} i^{(j)} (n+1)^{(j)} \frac{\phi^j}{j!}.$$

Quasi-Logarithmic Series Distribution

In this case, an extension of the usual logarithmic series distribution is obtained by taking $P_1(a) = -\log(1-a)$. Then, the Abel series expansion yields

$$-\log(1-a) = \sum_{x=1}^{\infty} \frac{(1-\lambda x)^x}{x} a(a+x\lambda)^{x-1}$$

so that

$$f(x) = \frac{\theta a(a+x\lambda)^{x-1}}{x}(1-x\lambda)^{-x},\ 0<a<1,\ \theta^{-1} = -\log(1-a),$$

$x = 1, 2, 3, \ldots$, is a probability mass function, representing the *quasi-logarithmic distribution.* The hazard rate function becomes

$$h(x) = \frac{(a+x\lambda)^{x-1}}{x(1-x\lambda)^x} \Big/ \sum_{t=x}^{\infty} \frac{(a+t\lambda)^{t-1}}{t(1-t\lambda)^t}.$$

Recurrence relations for $h(x)$ and $\lambda(x)$ can be obtained in the usual manner. For instance, a recursive relation for the hazard function is given by

$$\frac{1}{h(x)} = 1 + \frac{x}{\lambda+1}\left(\frac{a+\lambda+x\lambda}{1-\lambda-x\lambda}\right)\left(1+\frac{\lambda}{a+x\lambda}\right)^{x-1} \left(1+\frac{\lambda}{a+x\lambda}\right)^{-x}\frac{1}{h(x+1)}.$$

Quasi-Binomial Distribution II

The *quasi-binomial distribution II* (QBD II) can be derived in different ways. We obtain it here as a property of Abel polynomials and the corresponding series expansion. A sequence $\langle s_n(a)\rangle$ is of the binomial type if

$$s_n(a+b) = \sum_{x=0}^{n} \binom{n}{x} s_x(a) s_{n-x}(b)$$

for all b. With the choice of

$$s_n(\theta) = \theta(\theta+n\lambda)^{n-1} = n!\alpha_n(\theta,\lambda), \tag{3.76}$$

we have

$$\frac{d^x}{d\theta^x}\alpha_n(\theta+b\lambda)=\alpha_{n-x}(\theta+b+x\lambda,\lambda)$$

from (3.64).

Hence,

$$g(-x\lambda)=\alpha_{n-x}(b,\lambda).$$

Using the expansion in (3.66), we have

$$\alpha_n(\theta+b,\lambda)=\sum_{x=0}^{n}\alpha_x(\theta,\lambda)\alpha_{n-x}(b,\lambda)$$

so that

$$\frac{1}{s_n(\theta+b,\lambda)}\sum_{x=0}^{n}\binom{n}{x}s_a(\theta,\lambda)s_{n-x}(b,\lambda)=1.$$

Substituting for s_n from (3.76), we see that

$$f(x)=\binom{n}{x}\frac{\theta b}{\theta+b}\frac{(\theta+x\lambda)^{x-1}[b+(n-x)\lambda]^{n-x-1}}{(\theta+b+n\lambda)^{n-1}}, \tag{3.77}$$

$x=0,1,2,\ldots,n,\ \theta>0,\ b>0$ and $\lambda>-\frac{\theta}{n}$. The distribution defined by (3.77) is called QBD II. When $\lambda=0$, it reduces to the binomial model. A reparametrization of $p=\theta(\theta+b+n\lambda)$ and $\alpha=\lambda(\theta+b+n\lambda)^{-1}$ provide another QBD II version

$$f(x)=\binom{n}{x}\frac{(1-p-n\alpha)p}{(1-n\alpha)}(p+\alpha x)^{x-1}(1-p-\alpha x)^{n-x-1}, \tag{3.78}$$
$$0<p<1,\ -\frac{p}{n}<\alpha<\frac{(1-p)}{n},\ x=0,1,2,\ldots,n.$$

The mean and variance of the form in (3.77) are

$$\mu=\frac{n\theta}{\theta+b}$$

and

$$\mu_2=\frac{n^2\theta b}{(\theta+b)^2}-\frac{n(n-1)\theta b}{\theta+b}\sum_{i=0}^{n-2}\frac{(n-2)_{(i)}\lambda^i}{(\theta+b+n\alpha)^{i+1}},$$

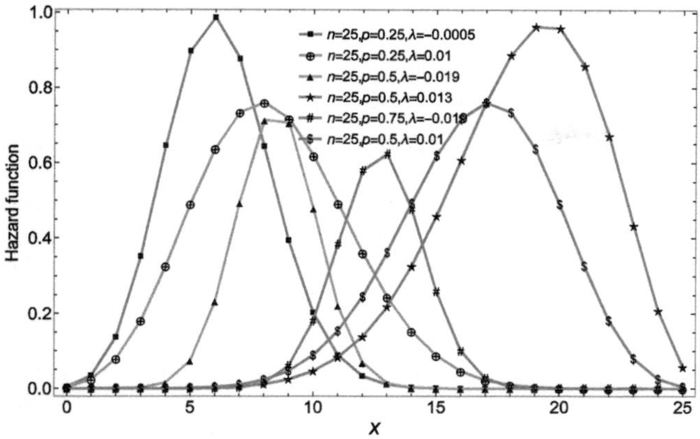

FIGURE 3.13 Probability mass functions of quasi-binomial distribution-II.

respectively. From (3.78), we can write expressions for the hazard rate and reversed hazard rate functions, respectively, as

$$h(x) = \frac{\binom{n}{x}(p+\alpha x)^{x-1}(1-p-\alpha x)^{n-x-1}}{\sum_{t=x}^{n}\binom{n}{t}(p+\alpha t)^{t-1}(1-p-\alpha t)^{n-t-1}}$$

and

$$\lambda(x) = \frac{\binom{n}{x}(p+\alpha x)^{x-1}(1-p-\alpha x)^{n-x-1}}{\sum_{t=0}^{n}\binom{n}{t}(p+\alpha t)^{t-1}(1-p-\alpha t)^{m-t-1}}.$$

They satisfy the recursive relationships

$$\frac{1}{h(x)} = 1 + \frac{1}{h(x+1)}\frac{n-x}{x+1}\left(\frac{p+\alpha+\alpha x}{1-p-\alpha x}\right)\left(1+\frac{\alpha}{p+\alpha x}\right)^{x-1}\left(1-\frac{\alpha}{1-p-\alpha x}\right)^{n-x-2},$$

and

$$\frac{1}{\lambda(x+1)} = 1 + \frac{1}{\lambda(x)}\frac{x+1}{n-x}\left(\frac{1-p-\alpha-\alpha x}{p+\alpha x}\right)\left(1+\frac{\alpha}{p+\alpha x}\right)^{-x}\left(1-\frac{\alpha}{1-p-\alpha x}\right)^{x+1-n}.$$

Figs 3.13 and 3.14 provide probability mass functions and hazard rate functions of quasi-binomial distribution II.

The Abel series distributions presented above have been obtained by Nandi and Das (1994). They have proposed estimates of the parameters of the distributions by equating the proportion of zeros and ones in the sample with $f(0)$ and $f(1)$ in the case of QBD I, QNBD and the generalized Poisson, and equating the sample mean and proportion of zeros with the population mean and $f(0)$ for QBD II. With real data, it has also been shown that various distributions become useful in practice as satisfactory models. Chapter 4 of Consul and Famoye

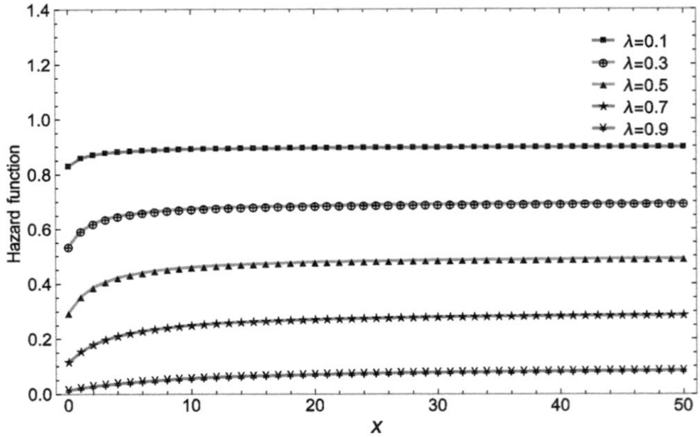

FIGURE 3.14 Hazard rate functions of quasi-binomial distribution-II.

(2006) and the references therein provide more information on distributional properties, characterizations and application of the Abel series distributions in the general framework of Lagrangian expansions.

3.2.5 Lagrangian Family

Consider a function $g(z)$ which is successively differentiable satisfying $g(0) \neq 0$ and $g(1) = 1$. Then, the numerically smallest root $z = l(t)$ of the transformation $z = tg(z)$ defines a probability generating function with Lagrangian expansion

$$z = P(t) = \sum_{x=1}^{\infty} \frac{t^x}{x!} \left[\frac{d^{x-1}}{dt^{x-1}} g(z)^x \right]_{z=0} \tag{3.79}$$

provided that the derivative inside the square brackets is non-negative for all $x \geq 0$. The corresponding probability mass function is

$$f(x) = \frac{1}{x!} \left[\frac{d^{x-1}}{dz^{x-1}} (g(z))^x \right]_{z=0} \tag{3.80}$$

which constitutes the basic *Lagrangian distributions*. Detailed study of the properties, applications and estimation problems connected with this family is available in works of Mohanty (1966), Consul and Shenton (1972, 1973, 1975), and the books by Consul (1989) and Consul and Famoye (2006). We, therefore, represent here only these results that are most pertinent to reliability modelling and analysis. A large number of distributions belong to the Lagrangian family, of which we discuss only those for which the support is appropriate and where some useful and simple results on various reliability functions can be presented.

Chronologically, the earliest distribution is that of Borel specified by

$$f(x) = \frac{(x\lambda)^{x-1}}{x!} e^{-x\lambda}, x = 1, 2, \ldots. \tag{3.81}$$

By transferring the random variable X in (3.81) to $Y = X - 1$, we have the generalized Poisson distribution presented earlier in (3.67). Since details of the Borel distribution can be evaluated in terms of these of the generalized Poisson, we abstain from further discussions here.

Haight Distribution

The Haight (1961) distribution has probability mass function

$$f(x) = \frac{1}{2s-1}\binom{2x-1}{x} q^x p^{x-1}, \quad x = 1, 2, \ldots, \quad 0 < q = 1 - p < 1. \tag{3.82}$$

From the probability generating function

$$P(t) = q(1 - pt)^{-1},$$

the basic characteristics such as the mean, variance and higher moments can all be easily derived. The hazard rate function is

$$h(x) = \frac{\binom{2x-1}{x} q^x p^{x-1}}{(2x-1)\sum_{t=x}^{\infty} \frac{1}{2t-1}\binom{2t-1}{t} q^t p^{t-1}}, \qquad x = 1, 2, \ldots .$$

As usual, $h(x)$ can be evaluated recursively as

$$\frac{1}{h(a)} = 1 + \frac{(4x-2)pq}{(x+1)h(x+1)}, \qquad x = 1, 2, \ldots, \tag{3.83}$$

with $h(1) = q$. Differentiating the survival function

$$S(x+1) = \sum_{t=1}^{x+1} \frac{1}{2t-1}\binom{2t-1}{t} q^t p^{t-1}$$

with respect to p and rearranging the terms, we get

$$E(X|X > x) = \frac{pq}{q-p}\left[\frac{1}{p} + \frac{\partial \log S(x+1)}{\partial p}\right]. \tag{3.84}$$

Accordingly, the mean residual life function is

$$m(x) = \frac{pq}{q-p}\left[\frac{1}{p} + \frac{\partial \log S(x+1)}{\partial}\right] - x, \qquad x = 1, 2, \ldots. \tag{3.85}$$

Differentiating $S(x+1)$ twice with respect to p, we get

$$\frac{\partial^2 S(x+1)}{\partial p^2} = \sum_{t=x+1}^{\infty} \left\{ \frac{1}{2t-1}\binom{2t-1}{t}\left[\frac{(t-1)(t-2)q^t p^{t-1}}{p^2}\right] - t(t-1)\left[\frac{2}{pq} - \frac{1}{q^2}\right]p^{t-1}q^t \right\}$$

which leads to

$$\frac{1}{S(x+1)}\frac{\partial^2 S(x+1)}{\partial p^2} = \frac{2}{p^2} - \frac{2q^2+(q-p)^2}{p^2q^2}E(X|X>x) + \frac{(q-p)^2}{p^2q^2}E(X^2|X>x)$$

and to

$$\begin{aligned} &E(X^2|X>x) \\ &= \frac{p^2q^2}{(q-p)^2}\left[\frac{1}{S(x+1)}\frac{\partial^2 S(x+1)}{\partial p^2} + \frac{2q^2+(q-p)^2}{p^2q^2}E(X|X>x)\right]. \end{aligned} \quad (3.86)$$

Now, the variance residual life can be computed from (3.84) and (3.86), as

$$\sigma^2(x) = E(X^2|X>x) - E^2(X|X>x).$$

With the aid of the distribution function

$$F(x) = \sum_{t=1}^{x} \frac{1}{2t-1}\binom{2t-1}{t} q^t p^{t-1},$$

some similar manipulations yield reliability functions in reversed time. The expressions are

$$\frac{1}{\lambda(x+1)} = 1 + \frac{x+1}{2(2x-1)xqp}\frac{1}{\lambda(x)},$$

$$r(x) = x - \frac{pq}{q-p}\left[\frac{1}{p} + \frac{\partial \log F(x)}{\partial p}\right], \quad x = 1, 2, \ldots$$

and

$$v(x) = E(X^2|X \le x) - E^2(X|X \le x),$$

where

$$E(X|X \le x) = \frac{pq}{q-p}\left[\frac{1}{p} + \frac{\partial \log F(x)}{\partial}\right]$$

and

$$E(X^2|X \le x) = \frac{p^2q^2}{(q-p)^2}\left[\frac{\partial^2 F(x)}{\partial p^2} - \frac{2}{p^2} + \frac{2q^2+(q-p)^2}{p^2q^2}E(X|X \le x)\right].$$

Geeta Distribution

A discrete random variable X is said to have *Geeta distribution* with parameters θ and α if

$$f(x)=\frac{1}{\alpha x-1}\binom{\alpha x-1}{x}\theta^{x-1}(1-\theta)^{\alpha x-x},$$
$$x=1,2,\ldots,\ 0<\theta<1,\ 1<\alpha<\theta^{-1}.$$

The distribution is L-shaped and unimodal with

$$\mu=(1-\hat{\theta})(1-\alpha\theta)^{-1}$$

and

$$\mu_2=(\alpha-1)\theta(1-\theta)(1-\alpha\theta)^{-3}.$$

A recurrence relation is satisfied by the central moments of the form

$$\mu_{r+1}=\theta\mu\frac{d\mu_r}{d\theta}+r\mu_2\mu_{r-1},\quad r=1,2,\ldots.$$

Estimation of the parameters can be done by the method of moments or by the maximum likelihood method. Moment estimates are

$$\hat{\mu}=\bar{x}\text{ and }\hat{\alpha}=\frac{S^2-\bar{x}(\bar{x}-1)}{S^2-\bar{x}^2(\bar{x}-1)}$$

with $\bar{x}$ and s^2 being the sample mean and variance. The maximum likelihood estimates are

$$\hat{\mu}_M=\bar{x}$$

and $\hat{\alpha}$ is iteratively obtained by solving the equation

$$\frac{(\alpha-1)\bar{x}}{\alpha\bar{x}-1}=\exp\left[-\frac{1}{n\bar{x}}\sum_{x=2}^{k}\sum_{i=2}^{x}\frac{xf_x}{x-i}\right]$$

with $n=\sum_{x=1}^{k}f_x$ as the sample size and f_x is the frequency of $x=1,2,\ldots,k$. For details, see Consul (1990). As the Geeta model is a member of the MPSD, the reliability properties can be readily obtained from those of the MPSD discussed in Section 3.2.2.

Generalized Geometric Distribution II

Famoye (1997) obtained the model

$$f(x) = \frac{1}{x}\binom{mx}{x-1}\theta^{x-1}(1-\theta)^{mx-x+1},$$
$$x = 1, 2, \ldots, \ 0 < \theta < 1, \ 1 \le m \le \theta^{-1}, \quad (3.87)$$

which is a generalization of the geometric distribution, as (3.87) becomes the geometric law when $m = 1$. It is obtained from the Lagrangian expansion of the generating function of the geometric distribution. The mean and variance are

$$\mu = (1-\theta m)^{-1} \text{ and } \mu_2 = m\theta(1-\theta)(1-\theta m)^{-3}.$$

Other distributional properties are derived from the central moments that satisfy the recurrence formula

$$\mu_{r+1} = \theta(1-\theta)(1-\theta m)^{-1}\frac{d\mu_r}{d\theta} + r\mu_{r-1}\mu_2, \quad r = 1, 2, \ldots,$$

using the value of μ_2 given above. For any given set of values of x, the hazard rate function satisfies from the recursive relationship

$$\frac{1}{h(x)} = 1 + \frac{\theta(1-\theta)^{m-1}}{(x+1)h(x+1)}\frac{\Gamma(mx+m+1)\Gamma(mx-x+2)}{\Gamma(nx+1)\Gamma(mx+m-x+1)}, \ x = 1, 2, \ldots,$$

with $h(1) = (1-\theta)^m$. Working as in the earlier cases, the mean residual life satisfies the relationship

$$m(x) = \mu\left[\frac{\partial \log S(x+1)}{\partial\theta} + \frac{1}{\theta(1-\theta)}\right] - x.$$

Also,

$$\frac{1}{S(x+1)}\frac{\partial^2 S(x+1)}{\partial\theta^2} = \frac{E(X-1)(X-2)}{\theta^2} - 2\frac{E(X-1)(mX-X+1)}{\theta(1-\theta)} + \frac{E(mX-x)(mX-x+1)}{(1-\theta)^2}.$$

Simplifying, we obtain

$$E(X^2|X > x) = \frac{1}{S(x+1)}(1-\theta)^{-2}$$
$$\times\left[\frac{\partial^2 S(x+1)}{\partial\theta^2} - (3\theta^2 + 2\theta m - 8\theta^2 + 2\theta - 3)E(X|X > x) - 2(1-\theta)\right].$$

This, along with

$$E(X|X>x)=\mu\left[\frac{\partial \log S}{\partial\theta}+\frac{1}{\theta(1-\theta)}\right],$$

gives an expression for the variance residual life. The reversed hazard rate is derived from the recurrence relation

$$\frac{1}{\lambda(x+1)}=1+\lambda(x)\frac{(x+1)\Gamma(mx+m-x+1)\Gamma(mx+1)}{\Gamma(mx+m+1)\Gamma(mx+2-x)\theta(1-\theta)^{m-1}}.$$

By adopting calculations similar to those of the mean residual life, we have the reversed mean residual life as

$$r(x)=x-\mu\left[\frac{\partial \log F(x)}{\partial x}+\frac{1}{\theta(1-\theta)}\right].$$

Finally, the estimates of the parameters can be obtained by the moment method, or the maximum likelihood method, or by equating the sample mean and frequency at $x=1$ with the mean and $nf(1)$, where n is the sample size.

There are several other families of discrete distributions discussed in the literature like the generalized hypergeometric family (Kemp, 1968a, 1968b), Gould series distributions (Charalambides, 1986), factorial series distributions (Berg, 1974) and their further extensions in various directions. Many of the members of these families overlap with those discussed here already. Furthermore, the reliability functions in other cases are of complicated forms, but can be obtained by some of the methods already detailed here.

3.3 DISCRETE ANALOGUES OF CONTINUOUS DISTRIBUTIONS

Several discrete distributions that are members of various families have been presented in the preceding section. They can serve as black-box models in the sense that their derivations were not based on the physical properties of the failure mechanism resulting in lifetimes. In a slightly different scenario, in this section, we consider various forms of discrete models that are obtained from popular continuous life distributions. This can be accomplished in different ways. Let Y be a continuous random variable representing lifetime with survival function $\bar{F}(x)=P(Y\geq x)$. If times are grouped into unit intervals, the discrete variable $X=[Y]$, the largest integer less than or equal to Y, has its probability mass function as

$$\begin{aligned} f(x)&=P(X=x)=P(x\leq Y<x+1)\\ &=\bar{F}(x)-\bar{F}(x+1),\quad x=0,1,2,\ldots \end{aligned} \tag{3.88}$$

The survival function $S(x)$ of X and $\bar{F}(x)$ of Y will be the same at all integer points. As a simple example, when Y is exponential with $\bar{F}(x) = \exp[-\lambda x]$, we have

$$\begin{aligned} f(x) &= \exp(-\lambda x) - \exp(-\lambda(x+1)) \\ &= q^x - q^{x+1} = pq^x, \quad p = 1-q,\ q = e^{-\lambda}, \end{aligned}$$

which represents the geometric distribution.

Discrete Weibull Distribution

Taking $\bar{F}(x) = \exp[-\left(\frac{x}{\alpha}\right)]^\beta$ and applying (3.88), we have

$$S(x) = q^{x^\beta}, \quad q = \exp[-\frac{1}{\alpha^\beta}], \quad x = 0, 1, 2, \ldots, \ 0 < q < 1. \tag{3.89}$$

This will be referred to as the *discrete Weibull distribution I*, first proposed by Nakagawa and Osaki (1975). For the model in (3.89), we readily find

$$f(x) = q^{x^\beta} - q^{(x+1)^\beta}$$

and

$$h(x) = 1 - q^{(x+1)^\beta - x^\beta}, \quad x = 0, 1, 2, \ldots.$$

The reversed hazard rate is

$$\lambda(x) = \frac{q^{x^\beta} - q^{(x+1)^\beta}}{1 - q^{(x+1)^\beta}} = \frac{1 - q^{x^\beta}}{1 - q^{(x+1)^\beta}} - 1.$$

Two special cases of interest are the geometric when $\beta = 1$ and the discrete Rayleigh when $\beta = 2$. There are no closed-form expressions for the mean, variance, and higher-order moments. Alikhan et al. (1989) noted that

$$P(X = 1) = 1 - q \text{ and } P(X = 2) = q - q^{2^\beta},$$

and then proposed estimating the parameters q and β by equating the theoretical and estimated probabilities from the data of the observations $X = 1$ and $X = 2$. Thus, these estimates are given by

$$\hat{q} = S_n(1) \text{ and } \hat{\beta} = \frac{1}{\log 2} \frac{\log S_n(2)}{\log S_n(1)},$$

where $S_n(x)$ is the empirical survival function. One can use two other values of X instead of 1 and 2, depending on the number of observations in the sample that are equal to the chosen integer values. The larger the frequencies of the numbers chosen, the better will be the estimates. Bracquemond and Gaudoin (2003) proposed the method of maximum likelihood for the estimation of q and β and observed that the estimates are biased, but of good quality for large values of q. One can use the distribution as an approximation to the continuous counterpart. Further, a shock model interpretation is also available, with X denoting the number of shocks survived by a device and q is the probability of surviving more than one shock. It can also be noted that (3.89) satisfies the extension of the lack of memory property that

$$P(X \geq (x^{\beta} + y^{\beta})^{\frac{1}{\beta}} | X \geq x) = P(X \geq y).$$

Discrete Half-Logistic Distribution

The half-logistic model in the continuous case has survival function (Balakrishnan, 1985)

$$\bar{F}(x) = \frac{2}{1+\exp(\frac{x}{\sigma})}, \qquad x > 0,\ \sigma > 0.$$

Accordingly, the *discrete half-logistic distribution* is specified by the survival function

$$S(x) = \frac{2}{1+\exp(\frac{x}{\sigma})}, \qquad x = 0, 1, 2, \ldots, \tag{3.90}$$

or by the probability mass function

$$f(x) = \frac{2(e^{\frac{1}{\sigma}} - 1)e^{\frac{x}{\sigma}}}{(1+e^{\frac{x}{\sigma}})(e^{\frac{x+1}{\sigma}} - 1)}, \qquad x = 0, 1, 2, \ldots.$$

The mean and variance of the distribution in (3.90) are

$$\mu = 2\sum_{n=1}^{\infty}(1+e^{\frac{n}{\sigma}})^{-1} \text{ and } \mu_2 = \frac{2(e^{\frac{1}{\sigma}} - 1)}{(1+e^{\frac{1}{\sigma}})^2} + 2\sum_{n=1}^{\infty}\frac{2n+1}{1+e^{\frac{n+1}{\sigma}}} - \mu^2.$$

The hazard rate and reversed hazard rate functions of the distribution in (3.90) can be derived as follows:

$$h(x) = \frac{(e^{\frac{1}{\sigma}} - 1)e^{\frac{x}{\sigma}}}{1+e^{\frac{x+1}{\sigma}}},$$

$$\lambda(x) = \frac{2(e^{\frac{1}{\sigma}} - 1)e^{\frac{x}{\sigma}}}{(1+e^{\frac{x}{\sigma}})(e^{\frac{x+1}{\sigma}} - 1)}, \qquad x = 0, 1, \ldots.$$

There is only one parameter (σ) involved in (3.90) which can be estimated by the method of maximum likelihood.

Geometric Weibull Distribution

This is a three-parameter model based on the continuous exponential Weibull distribution

$$\bar{F}(x) = \exp[-\lambda x - \{\lambda(x-a)^+\}^\beta]$$

introduced by Zacks (1984), where $(x-a)^+ = \max(0, x-a)$. Of the three parameters, $\lambda > 0$ represents scale, $\beta > 0$ the shape and $a > 0$ is the change-point. The model was proposed in a practical situation when the hazard rate is constant until time a and thereafter it is increasing like a Weibull hazard rate. In the discrete analogue also, the same physical meaning prevails with

$$S(x) = \exp[-\lambda x - \{\lambda(x-a)^+\}^\beta] \tag{3.91}$$

and

$$f(x) = \exp[-\lambda x - \{\lambda(x-a)^+\}^\beta] - \exp[-\lambda(x+1) - \{\lambda(x+1-a)^+\}^\beta].$$

We then have

$$h(x) = 1 - \exp[-\lambda + \lambda^\beta(\{(x-a)^+\}^\beta - \{(x-a+1)^+\}^\beta)]$$

as the hazard rate.

Telescopic Distributions

A more general family of distributions result if we consider the discrete analogue of Y with survival function

$$\bar{F}(x) = \exp[-\alpha k_\theta(x)], \quad x > 0,$$

when $k_\theta(x)$ is a strictly increasing function with $k_\theta(0) = 0$ and $k_\theta(x) \to \infty$ and $x \to \infty$. It includes the exponential, Rayleigh, Weibull, linear-exponential, and Gompertz distributions as special cases. Corresponding to $\bar{F}(x)$ above, we have the distribution of X as

$$f(x) = q^{k_\theta(x)} - q^{k_\theta(x+1)}, \quad x = 0, 1, 2, \ldots, \; 0 < q < 1, \; q = e^{-\alpha}. \tag{3.92}$$

For the model in (3.92), it follows (Roknabadi et al., 2009) that

$$\begin{aligned}
S(x) &= q^{k_\theta(x)}, \\
h(x) &= 1 - q^{k_\theta(x+1) - k_\theta(x)}, \\
\lambda(x) &= q^{k_\theta(x) - k_\theta(x+1)}/1 - q^{k_\theta(x+1)},
\end{aligned}$$

$$m(x) = \sum_{i=x}^{\infty} q^{k_\theta(i+1)-k_\theta(x)},$$

$$r(x) = \sum_{i=1}^{x} \frac{1-q^{k_\theta(i)}}{1-q^{k_\theta(x+1)}},$$

and the alternative hazard rate is

$$h_1(x) = (k_\theta(x-1) - k_\theta(x)) \log q.$$

Discrete Inverse Weibull Distribution

Jazi et al. (2010) considered the inverse Weibull law

$$\bar{F}(x) = 1 - \exp[-ax^{-b}], \quad x > 0,$$

to derive the *discrete inverse Weibull distribution* with survival function

$$S(x) = 1 - q^{x^{-b}}, \quad x = 1, 2, \ldots,$$

and probability mass function

$$f(x) = \begin{cases} q, & x = 1, \\ q^{x^{-b}} - q^{(x-1)^{-b}}, & x = 2, 3, \ldots, \; 0 < q < 1, \; b > 0. \end{cases} \tag{3.93}$$

The distribution in (3.93) has decreasing $f(x)$ for $q \geq \frac{\log 2}{2^{-b}-1}$ and unimodal with mode at 2, otherwise. When β becomes small, the tail becomes longer. The mean and variance do not have closed-form expressions. The hazard rate function and the alternative hazard rate function are, respectively,

$$h(x) = \frac{q^{x^{-b}} - q^{(x-1)^{-b}}}{1 - q^{(x-1)^{-b}}}$$

and

$$h_1(x) = \log\left(\frac{1 - q^{(x-1)^{-b}}}{1 - q^{x^{-b}}}\right), \quad x = 1, 2, \ldots$$

In order to estimate the parameters Jazi et al. (2010) considered the method of proportions, method of moments, a heuristic algorithm and an inverse Weibull probability plots. The discrete version of the probability plot is given by

$$y = -\beta x + \log(-\log q),$$

which is a straight line and accordingly β and q can be estimated by a simple linear regression model fitted to the data.

Discrete Generalized Exponential Distribution

Recalling that the generalized exponential distribution has its survival function as

$$\bar{F}(x) = 1 - (1 - e^{-\lambda x})^{\alpha}, \quad x > 0,$$

Nekoukhou et al. (2013) derived its discretized version as

$$\begin{aligned} f(x) &= (1 - p^{x+1})^{\alpha} - (1 - p^{x})^{\alpha}, \quad \alpha > 0,\ 0 < p < 1 \\ &= \sum_{j=1}^{\infty} (-1)^{j+1} \binom{\alpha}{j} p^{jx}(1 - p^{j}) \end{aligned} \tag{3.94}$$

with support $(0, 1, 2, \ldots)$. For integer values of α, the sum in (3.94) terminates at α. The distribution is unimodal for all α and p. There is a closed form expression for the moments given by

$$E(X^r) = \sum_{j=1}^{\infty} (-1)^{j+1} \binom{\alpha}{j} \sum_{t=1}^{r} S(r,t)t! \left(\frac{p^j}{1 - p^j}\right)^t, \quad r = 1, 2, \ldots,$$

where

$$S(r,t) = \frac{1}{t!} \sum_{j=0}^{t-1} (-1)^j \binom{t}{j} (t - j)^r$$

is the Stirling number of the second kind. In particular, we have

$$\mu = \sum_{j=1}^{\infty} (-1)^{j+1} \binom{\alpha}{j} \frac{p^j}{1 - p^j}$$

and

$$\mu_2 = \sum_{j=1}^{\infty} (-1)^{j+1} \binom{\alpha}{j} \frac{p^j(1 + p^j)}{(1 - p^j)^2} - \mu^2.$$

The parameters are estimated by the method of maximum likelihood by solving the equations

$$\sum_{j=1}^{n} \frac{(1 - p^{x_j+1})^{\alpha} \log(1 - p^{x_j+1}) - (1 - p^{x_j})^{\alpha} \log(1 - p^{x_j})}{(1 - p^{x_j+1})^{\alpha} - (1 - p^{x_j})^{\alpha}} = 0$$

and

$$\alpha \sum_{j=1}^{n} (1 - p^{x_j})^{\alpha - 1} \frac{p^{x_j - 1} x_j - (1 - p^{x_j+1})^{\alpha} p^{x_j}}{(1 - p^{x_j+1})^{\alpha} - (1 - p^{x_j})^{\alpha}} (x_j + 1) = 0.$$

Using the survival function

$$S(x) = 1 - (1 - p^x)^\alpha,$$

we can write the hazard rate and reversed hazard rate functions as

$$h(x) = \frac{(1 - p^{x+1})^\alpha - (1 - p^x)^\alpha}{1 - (1 - p^{x+1})^\alpha}$$

and

$$\lambda(x) = \frac{(1 - p^{x+1})^\alpha - (1 - p^x)^\alpha}{(1 - p^{x+1})^\alpha} = 1 - \left(\frac{1 - p^x}{1 - p^{x+1}}\right)^\alpha, \quad x = 0, 1, \ldots .$$

The alternative hazard rate becomes

$$h_1(x) = \log\left(\frac{1 - (1 - p^x)^\alpha}{1 - (1 - p^{x+1})^\alpha}\right), \quad x = 0, 1, \ldots .$$

Discrete Gamma Distribution

Chakraborty and Chakravarty (2012) considered the well-known two-parameter gamma distribution with survival function

$$\bar{F}(x) = \frac{1}{\theta^k \Gamma(k)} \int_x^\infty u^{k-1} e^{-\frac{u}{\theta}} du = \frac{\Gamma(k, \frac{x}{\theta})}{\Gamma(k)},$$

where

$$\Gamma\left(k, \frac{x}{\theta}\right) = \int_{\frac{x}{\theta}}^\infty \frac{1}{\theta^k} u^{k-1} e^{-u} du$$

is the upper incomplete gamma function, to derive the discrete gamma model with probability mass function

$$f(x) = \frac{\Gamma(k, \frac{x}{\theta}) - \Gamma(k, \frac{x+1}{\theta})}{\Gamma(k)}, \quad k, \theta > 0, \ x = 0, 1, \ldots . \qquad (3.95)$$

As particular cases, we have the one-parameter discrete gamma when $\theta = 1$ and the geometric when $\theta = 1$, $k = 1$. The distributional characteristics can be ascertained from the moments satisfying the recursive formula

$$\mu'_r = \sum_{j=1}^{r-1} (-1)^{j+1} \binom{r-1}{j} g^{r-j}(k, \frac{x}{\theta}) + g^{r-1}\left(k, \frac{x}{\theta}\right) - \mu'_{r-1}$$

where

$$g^r(k, \frac{x}{\theta}) = \sum_{x=1}^\infty x^r \Gamma(k, \frac{x}{\theta}) \Big/ \Gamma(k).$$

In particular, we have

$$\mu = g^0\left(k, \frac{x}{\theta}\right)$$

and

$$\mu_2 = 2g^1\left(k, \frac{x}{\theta}\right) - g^0\left(k, \frac{x}{\theta}\right) - \left[g^0\left(k, \frac{x}{\theta}\right)\right]^2.$$

From a random sample of failure times $(x_1, \ldots, x_n)$, estimates of θ and k can be obtained by solving the likelihood equations

$$\sum_{j=1}^{n} \frac{\int_{x_j/\theta}^{x_{j+1}/\theta} u^{k-1} e^{-u} \log u du}{\Gamma(k, \frac{x_j}{\theta}) - \Gamma(k, \frac{x_{j+1}}{\theta})} = n\psi(k)$$

and

$$\sum_{j=1}^{n} \frac{\Gamma(k+1, \frac{x_j}{\theta}) - \Gamma(k+1, \frac{x_{j+1}}{\theta})}{\Gamma(k, \frac{x_j}{\theta}) - \Gamma(k, \frac{x_{j+1}}{\theta})} = nk$$

with $\psi(k) = \frac{d \log \Gamma(k)}{dk}$, being the digamma function.

The reliability functions of interest are the survival function

$$S(x) = \frac{\Gamma(k, \frac{x}{\theta})}{\Gamma(k)},$$

the hazard rate function

$$h(x) = 1 - \frac{\Gamma(k, \frac{x+1}{\theta})}{\Gamma(k, \frac{x}{\theta})},$$

the alternative hazard rate function

$$h_1(x) = \log \frac{\Gamma(k, \frac{x}{\theta})}{\Gamma(k, \frac{x+1}{\theta})},$$

and the reversed hazard rate function

$$\lambda(x) = \frac{\Gamma(k, \frac{x}{\theta}) - \Gamma(k, \frac{x+1}{\theta})}{\Gamma(k) - \Gamma(k, \frac{x}{\theta})}.$$

The hazard rate functions of discrete gamma distribution are presented in Fig. 3.15.

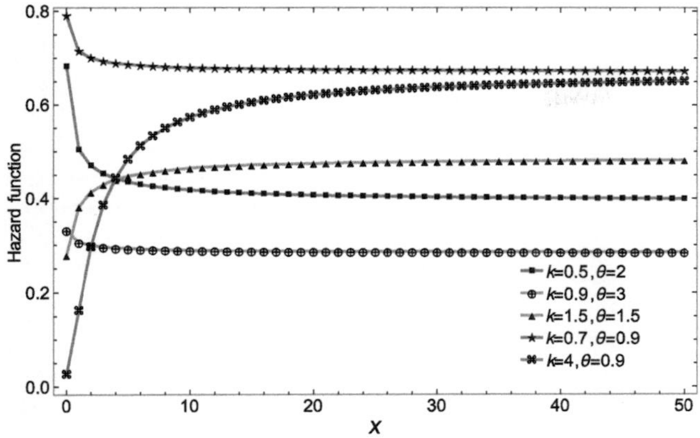

FIGURE 3.15 Hazard rate functions of discrete gamma distribution.

Discrete Lindley Distribution

The Lindley distribution, in the continuous case, has its survival function as

$$\bar{F}(x) = \frac{e^{-\theta x}(1+\theta+\theta x)}{1+\theta}, \quad x > 0, \theta > 0.$$

Accordingly, the discrete version takes on the form

$$f(x) = \frac{\lambda^x}{1-\log\lambda}[\lambda\log\lambda + (1-\lambda)(1-\log\lambda^{x+1})], \quad x = 0, 1, 2, \ldots \tag{3.96}$$

and $\lambda = e^{-\theta}$. From the probability generating function

$$P(t) = \frac{(\lambda-1)(t\lambda-1)+(1-2\lambda+t\lambda^2)\log\lambda}{(1-t\lambda)^2(1-\log\lambda)},$$

it is easy to obtain

$$\mu = \frac{\lambda[(\lambda-2)\log\lambda+1-\lambda]}{(1-\lambda)^2(1-\log\lambda)}$$

and

$$\mu_2 = \frac{\lambda[(1-\lambda)^2-(3-4\lambda+\lambda^2)\log\lambda+(2-3\lambda)\log^2\lambda]}{(1-\lambda)^4(1-\log\lambda)^2}.$$

We also have

$$S(x) = \frac{\lambda^x(1-(1+x)\log\lambda)}{1-\log\lambda}, \tag{3.97}$$

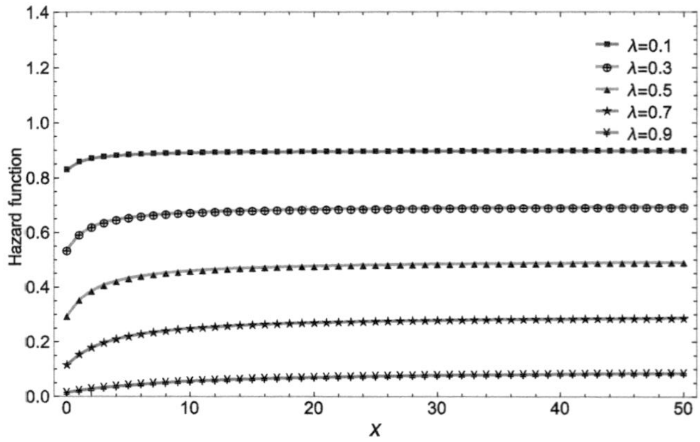

FIGURE 3.16 Hazard rate functions of discrete Lindley distribution.

$$F(x) = \frac{1 - \lambda^{x+1} + ((2+x)\lambda^{x+1} - 1)\log\lambda}{1 - \log\lambda}, \tag{3.98}$$

so that

$$h(x) = \frac{\lambda\log\lambda + (\lambda - 1)(\log\lambda^{x+1} - 1)}{1 - (1+x)\log\lambda}, \tag{3.99}$$

$$\lambda(x) = \frac{\lambda^x[(\lambda\log\lambda + (\lambda-1)(\log\lambda^{x+1} - 1))]}{1 - \lambda^{x+1}[(2+x)\lambda^{x+1} - 1]\log\lambda},$$

and

$$h_1(x) = \log\frac{(2+x)\log\lambda - 1}{\lambda(3+x)\log\lambda - 1}, \quad x = 0, 1, \ldots. \tag{3.100}$$

Fig. 3.16 presents hazard rate functions of discrete Lindley distribution.

The maximum likelihood estimate of λ can be obtained as the solution of the equation

$$n\bar{x} + \frac{n}{1 - \log\lambda} + \sum_{j=1}^{n} \frac{(\lambda + x_j)\lambda\log\lambda - (x_j+1)(1-\lambda)}{\lambda\log\lambda + (1-\lambda)(1 - (x_j+1)\log\lambda)} = 0,$$

where $\bar{x}$ is the mean of a random sample of size n from the discrete Lindley distribution.

Besides the above results on the discrete Lindley distribution described in Gomez-Deniz and Calderin-Ojeda (2011), there are some other results on the reliability characteristics. We note that

$$\sum_{t=x}^{\infty} S(t) = \frac{\lambda^x}{(1-\lambda)^2}[(1-\lambda)(1-x\log\lambda)-\log\lambda].$$

Hence, the mean residual life function can be expressed in a simple form as

$$m(x) = \frac{1-\log\lambda}{(1-\lambda)^2}\left[\frac{(1-\lambda)(1-\log\lambda - x\log\lambda)-\log\lambda}{1-\log\lambda-\lambda\log\lambda - x\lambda\log\lambda}\right] \qquad (3.101)$$

which is a homographic function of x. Moreover, as a corollary to Eq. (3.9) characterizing the Ord family, we have the following theorem.

Theorem 3.10. *A random variable X, taking values in* $(0, 1, \ldots)$, *will be distributed as discrete Lindley distribution if and only if*

$$m^*(x) = \mu(\lambda\log\lambda + (1-\lambda)(1-\log\lambda) - x(1-\lambda)\log\lambda)h(x+1)$$

for all x.

Unlike in the case of most discrete models, the variance residual life has also a closed-form expression given by

$$\sigma^2(x) = \frac{2(1-\log\lambda)}{(1-\lambda)^4}\,\frac{(1-\lambda)(1-\lambda-2\log\lambda)-\log\lambda - x(1-\lambda)\log\lambda}{1-(x+2)\log\lambda} - m(x)(m(x)-1) \qquad (3.102)$$

with $m(x)$ as given in (3.101).

The reversed mean residual life has the expression

$$r(x) = x - \frac{\lambda(1-\lambda^x)}{1-\lambda} - \frac{\log\lambda}{(1-\lambda)^2}[(1-\lambda^x)\lambda(2-\lambda) + x(1-\lambda)\lambda^x].$$

A characterization of the distribution by relationship between $r(x)$ and $\lambda(x)$ can be derived along the lines of Theorem 3.3.

Theorem 3.11. *The random variable X, taking values in* $(0, 1, \ldots)$, *will have discrete Lindley distribution if and only if*

$$r(x) = \mu - [\lambda\log\lambda + (1-\lambda)(1-\log\lambda) - x(1-\lambda)\log\lambda]\lambda(x)$$

for all $x = 0, 1, 2, \ldots$.

Further, the odds functions are

$$w(x) = \frac{\lambda^{x+1}(1-(2+x)\log\lambda)}{1-\lambda^{x+1}+[(2+x)\lambda^{x+1}-1]}\log\lambda$$

and $\bar{\omega}(x) = [\omega(x)]^{-1}$.

A second approach to construct discrete analogues of continuous distributions is to use the formula

$$P(X=x)=\frac{g(x)}{\sum_{x=0}^{\infty}g(t)}, \tag{3.103}$$

where $g(x)$ is the probability density function of a continuous random variable. When $g(x)=Ce^{-\frac{x}{\theta}}x^{\alpha}$, for example, (3.103) results in

$$f(x)=\frac{q^{x}x^{\alpha}}{\sum_{t=1}^{\infty}q_t^{t\alpha}}, \quad x=1,2,\ldots,\ q=e^{-\frac{1}{\theta}},$$

the Good distribution in (3.57). For a detail study of this distribution, we refer to Kulasekera and Tonkyn (1992). Obviously, the geometric distribution can also be derived from the exponential density function in this manner. See Sato et al. (1999) for an application of such a model, and also the convolution of their discrete exponential in the form of a negative binomial law. Lai and Wang (1995) obtained an analogue of the power distribution with density function

$$g(x)=\frac{\alpha x^{\alpha-1}}{\beta^{\alpha}}, \quad 0\le x\le\beta,\ \beta>0,$$

in the form

$$f(x)=\frac{x^{\alpha}}{\sum_{x=0}^{b}x^{\alpha}},\ x=0,1,\ldots,b,$$

with hazard rate

$$h(x)=\frac{x^{\alpha}}{\sum_{t=x}^{b}t^{\alpha}}$$

and reversed hazard rte

$$\lambda(x)=\frac{x^{\alpha}}{\sum_{t=0}^{x}t^{\alpha}}.$$

Lai (2013) has also suggested construction of new models by discretizing the continuous hazard rate function or the alternative hazard rate function. Further discussion on the above methods and models derived therefrom can be seen in Kemp (2008) and Krishna and Pundir (2009).

3.4 SOME OTHER MODELS

In this section, we present some other models arising from a variety of considerations.

Discrete Weibull Distribution II

Stein and Dattero (1984) introduced a second form of Weibull distribution by specifying its hazard rate function as

$$h(x) = \begin{cases} (\frac{x}{m})^{\beta-1}, & x = 1, 2, \ldots, m, \\ 0, & x = 0 \text{ or } x > m. \end{cases}$$

The probability mass function and survival function are derived from $h(x)$ using the formulas in Chapter 2 to be

$$f(x) = \left(\frac{x}{m}\right)^{\beta-1} \prod_{i=1}^{x-1} \left[1 - \left(\frac{i}{m}\right)^{\beta-1}\right], \quad x = 1, 2, \ldots, m,$$

and

$$S(x) = \prod_{i=1}^{\min(x,m)} \left[1 - \left(\frac{x}{m}\right)^{\alpha-1}\right]. \tag{3.104}$$

By comparison, the discrete Weibull I has survival function of the same form as the continuous counterpart, while discrete Weibull II has the same form for the hazard rate function. Stein and Dattero (1984) have pointed out that a series system with two components that are independent and identically distributed have a distribution of the form in (3.104).

Discrete Weibull Distribution III

A third type of Weibull distribution proposed by Padgett and Spurrier (1985) is specified by

$$f(x) = (1 - \exp[-c(x+1)^{\alpha}]) \exp\left[-c \sum_{j=0}^{x} j^{\alpha}\right],$$
$$x = 0, 1, 2, \ldots, \quad c > 0, \quad -\infty < \alpha < \infty, \tag{3.105}$$

with

$$S(x) = \exp\left[-c \sum_{j=0}^{x} j^{\alpha}\right].$$

Accordingly, the hazard function is

$$h(x) = 1 - \exp[-cx^{\alpha}].$$

For $\alpha = 0$, (3.105) reduces to the geometric case. An important advantage of this model is that the hazard rate is flexible, in the sense that it can assume different shapes. In view of the complex nature of the probability mass function, the maximum likelihood estimates becomes computationally tedious and intensive. Bracquemond and Gaudoin (2003) have pointed out that the quality of the maximum likelihood estimate of c, as regards bias, increases with c, while the bias for α is small except for very small samples.

"S" Distribution

Bracquemond and Gaudoin (2003) derived the "S" distribution based on some physical characteristics of the failure pattern through a shock-model interpretation. They have assumed a system in which on each demand a shock can occur with probability p and not occur with probability $(1-p)$. The number of shocks N_x at the xth demand is such that the hazard rate is an increasing function of N_x satisfying

$$h_N(x) = 1 - \pi^{N_x}, \quad 0 < \pi \leq 1.$$

Then, the survival function, given N_x, is

$$\begin{aligned} S(x) &= P(X > x) = E((P(X > x)|N_x, x \geq 1) \\ &= E[\pi^{\sum_{i=1}^{\infty} N_i}]. \end{aligned}$$

Further, if $U_x = N_x - N_{x-1}$, the U_x's are independent Bernoulli (p) random variables, so that

$$S(x) = \prod_{i=1}^{x} E(\pi^{(x-i+1)U_i}).$$

This leads to the "S" distribution specified by the probability mass function

$$f(x) = p(1 - \pi^x) \prod_{i=1}^{x-1} (1 - p + p\pi^i)$$

and the survival function

$$S(x) = \prod_{i=1}^{x-1} (1 - p + p\pi^i).$$

The interpretation given to the parameters is that p is the probability of a shock and π is the probability of surviving such a shock. We then have

$$h(x) = p(1 - \pi^x).$$

TABLE 3.2 Hazard rate functions

Distribution	$f(x)$	$h(x)$
binomial	$\binom{n}{x} p^x q^{n-x}$	$\{1 + \frac{1}{\binom{n}{x}\theta^x}[(1+\theta)^n - \sum_{j=0}^{x} \binom{n}{x}\theta^j]\}^{-1}$, $\theta = \frac{p}{q}, q = 1 - p$
Poisson	$e^{-\lambda}\frac{\lambda^x}{x!}$	$\{1 + \frac{x!}{\lambda^x}(e^\lambda - 1 - \sum_{j=1}^{x} \frac{\lambda^j}{j!})\}^{-1}$
negative binomial	$\binom{k+x-1}{x} p^k q^x$	$\{1 + \binom{k+x-1}{x}^{-1} q^x p^k [1 - \sum_{j=0}^{x} \binom{k+j-1}{j} p^k q^j]\}^{-1}$
Haight zeta	$(2x-1)^{-\sigma} - (2x+1)^{-\sigma}$	$\frac{(2x-1)^{-\sigma} - (2x+1)^{-\sigma}}{1-(2x-1)^{\sigma}}$, $x = 2, 3, \ldots$
geometric	$q^x p$	p
Waring	$\frac{(a-b)(b)_x}{(a)_{x+1}}$	$\frac{a-b}{a+x}$
negative hypergeometric	$\frac{\binom{-1}{x}\binom{-k}{n-x}}{\binom{-1-k}{n}}$	$\frac{k}{k+n-x}$
uniform	$f(x) = \frac{1}{b}, x = 1, \ldots, b$	$\frac{1}{b-x+1}$
arithmetic	$\frac{x}{b(b+1)}, x = 1, 2, \ldots, b$	$\frac{(b+x)(b-x+1)}{b(b+1)}$
S	$p(1-\pi^2)\prod_{i=1}^{x-1}(q + p\pi^i)$	$p(1-\pi^x)$
Weibull I	$q^{x^\beta} - q^{(x+1)^\beta}$	$1 - q^{(x+1)^\beta - x^\beta}$
Weibull II	$(\frac{x}{m})^{\beta-1}\prod_{i=1}^{x-1}(1 - (\frac{i}{n})^{\beta-1})$	$(\frac{x}{m})^{\beta-1}$
Weibull III	$(1 - \exp(-c(x+1)^\alpha))$ $\exp[-\sum_{j=0}^{x} j^\alpha]$	$\exp[-cx^\alpha]$

continued on next page

TABLE 3.2 *(continued)*

Distribution	$f(x)$	$h(x)$
power	$\frac{x^\alpha}{\sum_{x=0}^{b} x^\alpha}$, $x = 0, 1, \ldots, b$	$\frac{x^\alpha}{\sum_{t=x}^{b} t^\alpha}$
Lindley	$\frac{\lambda^x}{1-\log\lambda}[\lambda\log\lambda + (1-\lambda)(1-(x+1)\log\lambda)]$	$\frac{\lambda\log\lambda+(\lambda-1)((x+1)\log\lambda-1)}{1-(x+1)\log\lambda}$
gamma	$\frac{\Gamma(k,\frac{x}{\theta})-\Gamma(k,\frac{x+1}{\theta})}{\Gamma(k)}$	$1-\frac{\Gamma(k,\frac{x+1}{\theta})}{\Gamma(k,\frac{x}{\theta})}$
quasi binomial II	$\binom{n}{x}\frac{(1-p-n\alpha)p}{1-n\alpha}(p+\alpha x)^{x-1}(q-\alpha x)^{n-x-1}$	$\frac{\binom{n}{x}(p+\alpha x)^{x-1}(q-\alpha x)^{n-x-1}}{\sum_{k=x}^{n}\binom{n}{t}(p+\alpha t)^{t-1}(q-\alpha t)^{n-t-1}}$
quasi logarithmic	$\frac{\theta a(a+x\lambda)^{x-1}(1-x\lambda)^{-x}}{x}$	$\frac{(a+x\lambda)^{x-1}}{x(1-x\lambda)^x\sum_{t=x}^{\infty}\frac{(a+t\lambda)^{t-1}}{t(1-t\lambda)^t}}$
generalized Poisson	$\frac{\theta(\theta+x\lambda)^{x-1}}{x!}e^{-(\theta+x\lambda)}$	$\frac{(\theta+x\lambda)^{x-1}e^{-\lambda x}}{x\sum_{t=x}^{\infty}\frac{(\theta+t\lambda)^{t-1}}{t!}e^{-\lambda t}}$
Hurtwitz-zeta	$\frac{C}{(a-1+x)^s}$	$[(a-1+x)^s\zeta(s,a+x-1)]^{-1}$
Lerch	$\frac{Cz^x}{(a+x)^s}$	$[(a+x)^s\phi(z,s,a+x)]^{-1}$
linear hazard rate	$\begin{cases}(1-a)\ldots(1-a-b(x-1))(a+bx) & x=1,2,\ldots,n \\ a & x=0,\ \alpha+\beta n=1\end{cases}$	$a+bx$

Various properties of the distribution as well as the estimation issues have not been studied yet.

There are two other distributions proposed by Salvia and Bollinger (1982) and their generalizations by Padgett and Spurrier (1985), which are essentially particular cases of the models already discussed. Alzaatreh et al. (2012) provided a general method for deriving new distributions from continuous or discrete models. Let $G(x)$ be the distribution function of a random variable Y which may be continuous or discrete and $a(x)$ be the probability density function of a continuous random variable T taking values in $[0, \infty)$. Then, a new distribution can be defined by the distribution function

$$F(x) = \int_0^{-\log(1-G(x))} a(t)dt = A(-\log(1 - G(x))),$$

where $A(\cdot)$ is the distribution function of T. When Y is discrete, the new distribution has probability mass function

$$f(x) = A(-\log(1 - G(x))) - A(-\log(1 - G(x - 1))).$$

For instance, when X has a geometric (p) distribution, the corresponding distribution arising from a continuous distribution with $G(\cdot)$ as the distribution function is

$$f(x) = G(c(x + 1)) - G(cx),$$

where $c = -\log p$. Another category of models arise when they are required to satisfy certain specific properties for their reliability characteristics, such as bathtub shaped hazard rate functions. Such distributions will be taken up later on in Chapter 5. Expressions for hazard rate function for some distributions are presented in Table 3.2 for ease reference.

Chapter 4

Discrete Ageing Concepts

4.1 INTRODUCTION

The reliability of a unit or system is thought of as the probability that it carries out the intended function over a specific period of time. In this context, the age of the unit is reckoned as the time during which it functions in this manner before failing to do so. Also by the term ageing, we mean the phenomenon by which the life remaining to the unit is affected by its current age in some probabilistic sense. Generally, ageing is classified into positive ageing, negative ageing and no ageing. If the residual lifetime is decreasing when the age is increasing, we say that the unit is ageing positively. For instance, various equipments in common use or mechanical systems decrease its efficiency due to wear and tear as a result of prolonged use. Naturally, the remaining lifespan decreases when the time for which they are used increases. Thus, in short, the effect of ageing is to decrease the reliability in the case of positive ageing. On the other hand, there are situations in which the performance of a unit improves with increasing age. A classical example is that of human being whose remaining lifetime increases once they pass infancy. So also is the case of equipments that undergo efficient preventive maintenance. In all such cases, we say that the unit has negative ageing, which in technical terms is explained as the increase in residual lifetime as the age increases. In contrast to the above two ageing categories, there are situations in which the residual lifetime remains the same at all ages. This is described as no-ageing property. We have seen in Chapter 2 that only geometric distribution processes such a property among all discrete lifetime models. The basic reliability concepts such as survival (reliability) function, hazard rate, mean residual life and others form the fabric with which the ageing characteristics are built upon.

Various ageing criteria discussed below play a fundamental role in the development of reliability theory and practice. To begin with, they give us an indication of the pattern of failure exhibited by the unit and the behaviour of the lifetime as the unit undergoes ageing. Secondly, life distributions can be classified according to the ageing concepts, which would enable choice of the appropriate model easier. For example, if the equipment exhibits positive ageing, one needs to consider only the subclass of distributions that possesses such a characteristic to choose the model instead of the entire set of models that can

Reliability Modelling and Analysis in Discrete Time. http://dx.doi.org/10.1016/B978-0-12-801913-9.00004-X

be proposed as candidate models. Further, information about the particular category of ageing can be used as an additional input in non-parametric inference. Many of the ageing criteria considered here can be expressed in terms of certain geometric properties such as convexity, star-shapedness and stochastic orders. This will serve as additional information to be employed in modelling and analysis. Finally, there are various properties possessed by the ageing classes that can be successfully employed in analysis and prediction.

4.2 STOCHASTIC ORDERS

In many practical problems, it becomes necessary to compare two life distributions with reference to some of their characteristics. For example, two manufacturers may produce devices intended for the same purpose using different technologies resulting in non-identical life distributions. It will then be of interest for a customer to know which of the devices have a larger mean residual life at all ages. Descriptive measures like the averages can give a global picture of comparison, though they may not be as informative as time-dependent measures in revealing the inherent reliability aspects. Stochastic orders provide the necessary tools in such situations.

To define stochastic orders, we need some basic notions from set theory. A binary relation R on a non-empty set A is said to be a pre-order if (i) $a\ R\ a$ (reflexivity) for all a in A and (ii) for a, b, c in A, $a\ R\ b$ and $b\ R\ c \Rightarrow a\ R\ c$ (transitivity). If, in addition to (i) and (ii), we also have the property that $a\ R\ b$ and $b\ R\ a \Rightarrow a = b$, then R is called a partial order. In formulating stochastic order relations, we include pre-order and partial orders.

Let C be a characteristic of interest in a distribution. Further, let X and Y be random variables with distribution functions $F_X(x)$ and $F_Y(x)$, respectively. When X has lesser (greater) C than Y, we say that X is stochastically smaller (greater) than Y in C (or equivalently, F_X is smaller (greater) than F_Y) and write it as $X \leq_C (\geq_C) Y$ (equivalently $F_X \leq_C (\geq_C) F_Y$). In defining such stochastic orders, it should be the case that there should be a measure k of C that satisfies $k_X(C) \leq (\geq) k_Y(C)$. The definitions of various stochastic orders employed in the sequel are given below. Shaked and Shanthikumar (2007) have covered definitions (a) through (e), (h) and (j) and the others are discrete analogues of the continuous cases discussed by them.

We say that the random variable X is smaller than Y in

- **(a)** usual stochastic order, denoted by $X \leq_{\text{st}} Y$, iff $S_X(x) \leq S_Y(x)$ for all $x \geq 0$;
- **(b)** hazard rate order denoted by $X \leq_{\text{hr}} Y$, iff $h_X(x) \geq h_Y(x)$, $x \geq 0$, $x < \max(U_x, U_y)$, where U_x and U_y are the right end points of the supports of X and Y;

(c) likelihood ratio order, denoted by $X \leq_{\mathrm{lr}} Y$, iff $\frac{f_X(x)}{f_Y(x)}$ is non-increasing in x over the union of supports of X and Y;

(d) mean residual life order, denoted by $X \leq_{\mathrm{mrl}} Y$, iff $m_X(x) \leq m_Y(x)$, all $x \leq \max(U_x, U_y)$;

(e) harmonic mean residual life order, denoted by $X \leq_{\mathrm{hmrl}} Y$, iff $\frac{S_X(x)}{E(X)} \leq \frac{S_Y(x)}{E(Y)}$;

(f) increasing convex order, denoted by $X \leq_{\mathrm{icx}} Y$, iff $\sum_{t=x}^{\infty} S_X(t) \leq \sum_{t=x}^{\infty} S_Y(t)$;

(g) increasing concave order, denoted by $X \leq_{\mathrm{icv}} Y$, iff $\sum_{t=0}^{x} S_X(t) \geq \sum_{t=0}^{\delta x} S_Y(t)$;

(h) probability generating function order, denoted by $X \leq_{\mathrm{pgf}} Y$, iff $E(s^X) \geq E(s^Y), 0 < s < 1$;

(i) variance residual life order, denoted by $X \leq_{\mathrm{vrl}} Y$, iff $\sigma_X^2(x) \leq \sigma_Y^2(x)$;

(j) reversed mean inactivity time order, denoted by $X \leq_{\mathrm{MIT}} Y$, iff $r_X(x) \geq r_Y(x)$;

(k) concave transform order, denoted by $X \leq_c Y$ if $G^{-1}F(x)$ is concave in the support of X, where F and G are respectively the distribution functions of X and Y;

(l) star order, denoted by $X \leq_* Y$, if $G^{-1}F(x)$ is star-shaped in x, or $\frac{G^{-1}F(x)}{x}$ is decreasing in $x = 0, 1, 2, \ldots$;

(m) sub-additive (super-additive), written as $X \leq_{\mathrm{sa}} Y$ ($X \leq_{\mathrm{su}} Y$), if $G^{-1}F(x)$ is sub-additive (super-additive) in x or $G^{-1}F(x+y) \leq (\geq) G^{-1}F(x) + G^{-1}F(y)$.

Some of the partial orders given above are also used in defining ageing classes. As will be exposed in the sequel, such definitions involve two types of comparisons. One is to compare the characteristic of a life distribution with that of the geometric distribution and the other is a stochastic order involving the probability distribution of the residual life at different ages. The concept of equilibrium distribution also plays a predominant role in interpreting ageing classes in terms of partial orders between the original distribution and its equilibrium form. All the above orderings are not independent of one another, as there exists several implications and equivalence under additional conditions. These often provide simple proofs in establishing the relative strengths of various ageing concepts. Some implications among the stochastic orders are shown in Fig. 4.1. Some of the orders in the diagram defined subsequently in the appropriate contexts are also included in Fig. 4.1 for the sake of completeness.

In the following sections of this chapter, we discuss various classes of life distributions induced by the behaviour of the reliability functions explained in the preceding chapter. Around forty different classes and their properties have been studied in the continuous case. However, discrete counterparts of many of

$$
\begin{array}{ccccc}
 & & X \leq_{\text{hmrl}} Y & \Rightarrow & X \leq_{\text{icx}} Y \\
 & & \Uparrow & & \\
X \leq_{\bar{m}} Y & \Rightarrow & X \leq_{\text{mrl}} Y & & \\
 & & \Uparrow & & \\
X \leq_{\text{lr}} Y & \Rightarrow & X \leq_{\text{hr}} Y & \Rightarrow & X \leq_{\text{st}} Y \\
\Downarrow & & & & \\
X \leq_{\text{rhr}} Y & & & & \\
\Downarrow & & & & \\
X \leq_{\text{MIT}} Y & & & & \\
\Downarrow & & & & \\
X \leq_{\text{VIT}} Y & & & &
\end{array}
$$

$$X \leq_{c} Y \Rightarrow X \leq_{*} Y \Rightarrow X \leq_{\text{su}} Y.$$

FIGURE 4.1 Implications among different ageing concepts.

them have not found a place in the literature yet. Accordingly, besides presenting a review of the existing classes, we also define and discuss the properties of some important classes that have been left out. A redeeming feature of these classes is that all of them have physical interpretations and therefore become appropriate when information about the failure mechanism exhibited by the data matches with that reflected in the class. Secondly, in many practical problems, the data may shed some light on the behaviour of the reliability functions so that in such situations, the analyst can restrict the choice of the model from the corresponding class. The parametric families already discussed, generally belonging to one of the sub-classes, makes it easier to use them once the class can be identified.

Life distributions are classified on the basis of ageing concepts which, in some sense, describe the pattern in which a device or system deteriorates or improves with age. One important topic considered in a studying a particular class is to examine whether the class preserves various reliability operations like the formation of coherent systems, mixtures, convolution, etc. Secondly, bounds on reliability functions and similar characteristics are found as specific properties of various classes. Various tests of hypothesis on whether a given sample of lifetimes indicate the presence of an ageing concept have also been considered in the literature for the continuous case. In the discrete case, most of these problems are still open for investigation. While discussing different ageing classes, we group together those based on hazard rate, functions of residual life, survival functions, and reliability functions in reversed time.

Eventhough one often examines whether a device is positively or negatively ageing by means of classes to be discussed shortly, no less important is the concept of no-ageing. Recall from Eq. (3.46) that a lifetime distribution satisfies the no-ageing property if and only if

$$P(X \geq s+t|X \geq s) = P(X \geq t)$$

for all s and t, which says that the original lifetime and residual lifetime have the same probability at all ages. This is also referred to as the *lack of memory property* that characterizes the geometric law. There has been some attempts to define the memory of a distribution in the continuous case by Muth (1977) and in the discrete case by Nair (1983a). Since the memory classification of distributions is related to ageing, there are some points worth noting in this context.

Definition 4.1. A distribution is said to have no memory in $[0, t]$, full memory, positive memory, negative memory iff

$$\bar{r}(t) = \bar{r}(0)(\bar{r}(t) = \bar{r}(0) - 1, \bar{r}(t) < \bar{r}(0), \bar{r}(t) > \bar{r}(0)),$$

where $\bar{r}(t) = E(X - t|X \geq t)$. Nair (1983a) obtained a measure of memory of a distribution as

$$M = \sum_{t=1}^{\infty} w_t(\bar{r}(t-1) - \bar{r}(t)),$$

with w_t as the weight function $w_t = \sum_{s=t}^{\infty} S(s)/\sum_{u=0}^{\infty} S(u)\bar{r}(u)$.

Accordingly, for a discrete distribution, memory is measured by

$$M = \frac{2\bar{r}^2(0) + \bar{r}(0) - E(X^2)}{E(X^2) + \bar{r}(0)}. \tag{4.1}$$

The degenerate distribution has full memory ($M = 1$), geometric distribution has no memory ($M = 0$), Poisson distribution has positive memory ($0 < M < 1$), and Waring distribution has negative memory. Obviously, the ageing of a device is positive (negative) iff its memory at a point t, $\bar{r}(t) - \bar{r}(t+1) > (<)0$ for all t in the support of its lifetime X. It is to be noted that $M = 0$ does not mean that X is geometric, nor $M > 0$ guarantees that the memory is positive at all points in the support. With reference to the concepts introduced, $m(x) = 1 + \bar{r}(x+1)$ and $\bar{r}(0) = \mu$. Sandhya and Rajasekharan (2012b) have shown that the measure of memory is independent of change of origin, but not of scale, and also that it can be extended to cover random variables $Z = a + k + X$, where $a \geq 0$ and $k \geq 1$ are integers.

It will be of some interest to reliability analysts to have a criterion by which the better one among two devices can be chosen in terms of a known reliability characteristic. The notion of memory can be employed as a tool for this purpose by invoking stochastic orders in terms of memory.

Definition 4.2. Let X and Y be two discrete lifetimes. Then, X is smaller than Y in memory order, written as $x \leq_{\bar{m}} Y$, iff $\mu_X \geq \mu_Y$ and

$$\bar{r}_X(t) - \bar{r}_X(t+1) \geq \bar{r}_Y(t) - \bar{r}_Y(t+1), \quad t = 0, 1, 2, \ldots \tag{4.2}$$

Denoting by $\bar{m}(t) = \bar{r}(t) - \bar{r}(t+1)$, we have for $\mu_X > \mu_Y$, $X \leq_{\bar{m}} Y \Leftrightarrow \bar{m}_X(t) \geq \bar{m}_Y(t)$. In terms of the mean residual life function $m(x)$, $\bar{m}(t) = m(t-1) - m(t)$, $t = 0, 1, 2, \ldots$. Thus, $X \leq_{\bar{m}} Y$ has the physical interpretation that $\bar{r}_X(t) - \bar{r}_Y(t)$ is non-decreasing, or $m_X - m_Y$ is non-increasing, so that the amount of change in $m_X(x)$ is less than the corresponding change in $m_Y(x)$. The memory order $\leq_{\bar{m}}$ satisfies the following properties:

1. If $X \leq_{\bar{m}} Y$, then $X \geq_{\text{mrl}} Y$.

Proof. We have

$$\begin{aligned}
X \leq_{\bar{m}} Y &\Rightarrow m_X(t-1) - m_Y(t+h) \leq m_Y(t-1) - m_Y(t+h) \\
&\Rightarrow m_X(-1) - m_X(h) \leq m_Y(-1) - m_Y(h) \\
&\Rightarrow m_X(h) \geq m_X(-1) - m_Y(-1) + m_Y(h) \\
&\Rightarrow m_X(h) \geq (\mu_X - \mu_Y) + m_Y(h) \\
&\Rightarrow m_X(h) \geq m_Y(h), \quad h = 0, 1, 2, \ldots \\
&\Rightarrow X \geq_{\text{mrl}} Y;
\end{aligned}$$

2. The memory order is stronger than the mean residual life order. In other words, $X \leq_{\text{mrl}} Y$ does not imply $X \geq_{\bar{m}} Y$. To see this, let X be a geometric variable with $p = \frac{3}{4}$ so that $m_X(x) = \frac{1}{p} = \frac{4}{3}$ for $x = -1, 0, 1, \ldots$. Suppose Y follows discrete Lindley distribution in (3.95) with $\lambda = \frac{1}{2}$. Then, from (3.100), we have

$$m_Y(x) = \frac{2 + 2x\log 2 + 3\log 2}{1 + x\log 2 + 2\log 2}.$$

It is easy to see that $m_Y(x) > m_X(x)$ for all x, and so $X \leq_{\text{mrl}} Y$. We also have

$$\begin{aligned}
m_Y(-1) - m_Y(0) &= \frac{2 + \log 2}{1 + \log 2} - \frac{2 + 3\log 2}{1 + 2\log 2} = -(\log 2)^2 \\
&< 0 = m_X(-1) - m_X(0)
\end{aligned}$$

showing that $X \leq_{\bar{m}} Y$; □

3. If $X \leq_{\bar{m}} Y$, then $E(X^r) \geq E(Y^r)$ for $r \geq 1$, provided the expectations are finite.

Proof. We have

$$\begin{aligned} X \leq_{\bar{m}} Y &\Rightarrow X \geq_{\text{mrl}} Y \text{ by Property 1 above} \\ &\Rightarrow Ea(X) \geq E(a(Y)) \text{ for all increasing convex functions } a(\cdot) \\ &\Rightarrow E(X^r) \geq E(Y^r) \text{ by taking } a(t) = t^k; \end{aligned}$$

□

4. If $X \leq_{\bar{m}} Y$, then $M_X \geq M_Y$, where M is the measure of memory obtained in (4.1), provided that $E(X) = E(Y)$.

Proof. From (4.1), we have

$$\begin{aligned} M_X - M_Y &= \frac{2\mu^2 + \mu - E(X^2)}{E(X^2) + \mu} - \frac{2\mu^2 + \mu - E(Y^2)}{E(Y^2) + \mu}, \quad \mu = E(X) = E(Y) \\ &= \frac{[E(Y^2) - E(X^2)]}{(E(X^2) + \mu)(E(Y^2) + \mu)}(2\mu^2 + 2\mu). \end{aligned}$$

Since $X \leq_{\bar{m}} Y$, $E(X^2) \geq E(Y^2)$, and hence $M_X \leq M_Y$. □

4.3 CLASSES BASED ON HAZARD RATE

4.3.1 Monotone Hazard Rates

Definition 4.3. A discrete lifetime X is said to have an increasing (decreasing) hazard rate, IHR (DHR), iff the sequence $\langle h(x) \rangle$ is increasing. In other words, X is IHR (DHR) according as $h(x+1) \geq (\leq) h(x)$ for all $x = 0, 1, \ldots$. We note that there are several equivalent conditions for X to be IHR (DHR).

Theorem 4.1. *X has IHR (DHR) iff*

(a) $S_x(t) = \frac{S(x+t+1)}{S(x+1)}$ *is decreasing (increasing) in x for all $t = 0, 1, \ldots$. This is interpreted by saying that residual life distributions are stochastically decreasing (increasing) in x;*

(b) $\langle \log S(x) \rangle$ *is a concave (convex) sequence, or* $\langle S(x) \rangle$ *is log-concave (log-convex);*

(c) *the odds function of the residual life $w_x(t)$ is increasing (decreasing) in t;*

(d) $S(x)$ *is PF_2 (the determinant*

$$\begin{vmatrix} S(x_1 + t_1) & S(x_1 + t_2) \\ S(x_2 + t_1) & S(x_2 + t_2) \end{vmatrix} \geq 0$$

for $t_1 \leq t_2$, $x_1 \leq x_2$, or in other words, $S(x+y)$ is TP_2, in x and y);

(e) $[S(x+1)]^2 \geq (\leq) S(x)S(x+2)$ *for all $x = 0, 1, 2, \ldots$;*

(f) $X \leq_c (\geq_c) Y$*, where Y is a geometric random variable with distribution function $F_Y(x) = 1 - q^x$, $x = 0, 1, 2, \ldots$;*

(g) $x \geq_{lr} (\leq_{lr}) X_E$, *where* X_E *is the equilibrium random variable given in* (1.23);

(h) $X \geq_{lr} (\leq_{lr})(X_E - x|X_E > x)$ *for all* x;

(i) $(X - x|X > x) \geq_{lr} (\leq_{lr})(X_E - y|X_E > y)$ *for all* $x, y, x \leq y$;

(j) $X_E \geq_{lr} (\leq_{lr})(X_E - x|X_E > x)$ *for all* x;

(k) $(X - x|X > x) \geq_{st} (\leq_{st})(X - y|X > y)$, *for all* x, y $x \leq y$;

(l) $(X - x|X > x) \geq_{hr} (\leq_{hr})(X - y|X > y)$, *for all* $x, y, x \leq y$;

(m) $X \geq_{hr} (\leq_{hr})(X - x|X > x)$ *for all* x.

Proof. In establishing the above characterizations, only the proof for the IHR case will be given as the corresponding result for DHR can be proved quite similarly.

(a) We observe that

$$S_x(t) = \frac{S(x+t+1)}{S(x+1)} = \frac{\prod_{u=0}^{x+t}(1-h(u))}{\prod_{u=0}^{x}(1-h(u))} = \prod_{u=x}^{x+t}(1-h(u)).$$

When X is IHR, each term in the product is decreasing and so $S_x(t)$ is decreasing. Conversely, when $t = 1$,

$$\frac{S(x+2)}{S(x+1)} = 1 - h(x+1),$$

so that when the left hand side is decreasing, $h(x)$ is increasing or X is IHR;

(b) $\log S(x)$ is concave $\Leftrightarrow \log S(x+1) - \log S(x) = \log(1-h(x))$ is decreasing, and X is IHR and conversely;

(c) The proof follows from the identity $\bar{w}_x(t) = \frac{1}{S_x(t+1)} - 1$ and (a) above;

(d) Since $S_x(t)$ is decreasing, we see from condition (d) in the definition of PF_2 in Chapter 1 and (b) above that $S_x(t)$ is PF_2 iff X is IHR;

(e) This is a restatement of (a);

(f) When Y is geometric, $F_Y^{-1}(u) = \frac{\log(1-u)}{\log q}$. Thus,

$$\begin{aligned} X \leq_c Y &\Leftrightarrow F_Y^{-1}F_X(x) \text{ is concave} \\ &\Leftrightarrow \frac{1-F_X(x)}{\log q} \text{ is concave} \\ &\Leftrightarrow \log(1-F_X(x)) \text{ is concave} \\ &\Leftrightarrow \log\frac{S(x+2)}{S(x+1)} \text{ is decreasing} \\ &\Leftrightarrow X \text{ is IHR;} \end{aligned}$$

(g) From (1.23), the probability mass function of X_E is

$$f_E(x) = \frac{S(x+1)}{\mu}.$$

Hence,

$$\begin{aligned} X \geq_{\text{lr}} X_E &\Leftrightarrow \frac{S(x+1)}{\mu f(x)} \text{ is decreasing in } x \\ &\Leftrightarrow \frac{S(x+1)}{S(x)} \frac{S(x)}{f(x)} \text{ is decreasing in } x \\ &\Leftrightarrow \frac{1-h(x)}{h(x)} \text{ is decreasing in } x \\ &\Leftrightarrow X \text{ is IHR;} \end{aligned}$$

(h) Since the probability mass function of X_E is $f_E(x) = \frac{S(x+1)}{\mu}$, the corresponding survival function is

$$S_E(x) = \sum_{u=x}^{\infty} \frac{S(u+1)}{\mu}. \tag{4.3}$$

From (4.3), the survival function of $(X_E - x|X > x)$ then turns out to be

$$S_{E,x}(t) = \frac{\sum_{u=x+t+1}^{\infty} S(u+1)}{\sum_{u=x+1}^{\infty} S(u+1)}. \tag{4.4}$$

Hence, the probability mass function of (4.4) becomes

$$f_{E,x}(t) = \frac{S(x+t+2)}{\sum_{u=x+1}^{\infty} S(u+1)}.$$

Now,

$$\begin{aligned} X \geq_{\text{lr}} (X_E - x|X_E > x) &\Leftrightarrow \frac{f_{E,x}(t)}{f_X(t)} \text{ is decreasing in } t \\ &\Leftrightarrow \frac{S(x+t+2)}{S(t) - S(t-1)} \text{ is decreasing in } t \\ &\Leftrightarrow \frac{\prod_{u=t}^{x+t+1}(1-h(u))}{h(t)} \text{ is decreasing in } t \\ &\Leftrightarrow h(t) \text{ is increasing in } t \Leftrightarrow X \text{ is IHR;} \end{aligned}$$

(i) We have

$$(X-x|X>x)\geq_{\mathrm{lr}}(X_E-y|X_E>y)\Leftrightarrow \frac{S(x)}{f(t+x)}\frac{S(t+y+2)}{\sum_{u=y+1}^{\infty}S(u+1)}$$

$$\text{is decreasing in } t$$

$$\Leftrightarrow \frac{S(t+y+2)}{f(t+x)} \text{ is decreasing in } t$$

$$\Leftrightarrow \prod_{u=x+t}^{t+y+1}\frac{(1-h(u))}{h(t+x)} \text{ is decreasing}$$

$$\Leftrightarrow X \text{ is IHR;}$$

(j) We have

$$X_E\geq_{\mathrm{lr}}(X_E-x|X_E>x)\Leftrightarrow \frac{S(x+t+2)}{\sum_{u=x+1}^{\infty}S(u)}\Big/(S(t+1)) \text{ is decreasing in } t$$

$$\Leftrightarrow \frac{S(x+t+2)}{S(t+1)} \text{ is decreasing in } t$$

$$\Leftrightarrow X \text{ is IHR;}$$

(k) The proof is obvious from (a) above;

(l) The probability mass function of $(X-x|X>x)$ is

$$\frac{f(t+x+1)}{S(x+1)},$$

and the survival function is $\frac{S(t+x+1)}{S(x+1)}$. Accordingly, the hazard rate of $(X-x|X>x)$ becomes

$$h_x(t)=\frac{f(t+x+1)}{S(t+x+1)}=h(t+x+1).$$

Similarly, for $(X-y|X>y)$ is

$$h_y(t)=h(t+y+1)$$

$$(X-x|X>x)\geq_{\mathrm{hr}}(X-y|X>y)\Leftrightarrow h(t+y+1)$$

$$\geq h(t+x+1) \text{ for } x\leq y$$

$$\Leftrightarrow X \text{ is IHR;}$$

(m) We have

$$(X-x|X>x)\geq_{\mathrm{hr}}(X-x|X\geq x)\Leftrightarrow h(t)\leq h(t+x+1) \text{ for all } t \text{ and } x$$

$$\Leftrightarrow X \text{ is IHR.}$$

□

Apart from providing vital information about the ageing pattern, the IHR/DHR family of life distributions possess several interesting properties in their own right. Some such properties are listed below.

The important properties associated with discrete models in insurance, finance, reliability, and queueing are convolution and mixing; see Chapter 1 for details. Pavlova et al. (2006) have discussed these aspects with necessary proofs and counter examples.

(i) If F_1 and F_2 are distribution functions of IHR random variables X_1 and X_2 then the convolution $F = F_1 * F_2$ is IHR. (Notice the difference of terminology used in Pavlova et al., 2006, Hu et al. (2003) and in the present work. The above authors define strongly IHR (S-IHR) notion whenever $h(x)$ is increasing for $x = 0, 1, 2, \dots$, and IHR notion whenever $h(x)$ is increasing for $x = 1, 2, \dots$.) When the support is $x = 1, 2, \dots$, $F = F_1 * F_2$ need not be IHR;

(ii) The DHR class of life distributions remain to be DHR under the formation of mixtures. However, the class of IHR distributions is not closed under formation of mixtures;

(iii) Gupta et al. (1997) defined

$$\eta(x) = \frac{f(x) - f(x+1)}{f(x)}$$

and have shown that

(a) if $\Delta\eta(x) > 0$, then X is IHR,

(b) if $\Delta\eta(x) < 0$, then X is DHR,

(c) if $\Delta\eta(x) = 0$, the distribution of X satisfies $f(x) = C^x f(0)$, where C is a constant and accordingly X is distributed either as geometric, or as uniform with $f(x) = f(0)$ or as

$$f(x) = \frac{C^x}{1 + C + \cdots + C^m}, \qquad x = 0, 1, \dots, m.$$

In the last two cases, X is IHR. Notice that $\Delta\eta(x) > 0$ is not a sufficient condition for IHR;

(iv) Since the alternative hazard rate $h_1(x)$ is related to $h(x)$ through the identity

$$h(x) = 1 - e^{-h_1(x)}.$$

$h(x)$ is increasing (decreasing) if and only if $h_1(x)$ increasing (decreasing). Hence, all the properties of IHR (DHR) mentioned above and hereafter hold good irrespective of whether h_1 or h is used to define the hazard rate;

(v) If X is DHR, then its distribution function is concave. This follows from the fact that the cumulative hazard rate $H_1(x)$ defined in (2.12) is concave whenever X is DHR and $S(x) = \exp[-H_1(x)]$ which is decreasing. Also, if H_1 is concave, so F;

(vi) If F is concave, then it is log concave. Further if the probability mass function is log-convex, the distribution is DHR and hence F is log-concave. Log-concavity of $S(x)$ corresponds to IHR class. It becomes apparent that the log-concave class of distributions contains the concave class which in turn contains the DHR distributions;

(vii) If X is IHR, then $S(x) - q^x$, $x = 1, 2, \ldots$, changes in sign at-most once and the change is from positive to negative;

(viii) If X has an increasing hazard rate, then the residual life X_x also has increasing hazard rate. This is evident from the fact that if $h(t)$ is hazard rate of X, the corresponding rate for X_x is $h(t + x + 1)$;

(ix) Let Y_n be the random variable possessing equilibrium distribution of order n of X (see (1.25)). If Y_n is IHR (DHR), then Y_{n+1} is IHR (DHR) for every n. But, the converse need not be true. Consequently, $Y_0 = X_E$ is IHR (DHR) whenever X is IHR (DHR);

(x) The mixture of geometric distributions has a decreasing hazard rate. From (1.16), we find that mixture of geometric distributions, when p has a beta distribution, is Waring which has decreasing hazard rate;

(xi) We say that X has increasing (decreasing) likelihood ratio, ILR (DLR), if $f(x)$ is log-concave (log-convex). Sometimes, ILR (DLR) is also viewed as an ageing concept. We then have the implication ILR (DLR) $\Rightarrow$ IHR (DHR).
To prove this assertion, we note that X is ILR implies $\frac{f(x+1)}{f(x)}$ is decreasing or $g(x) - g(x+1) \geq 0$, $g(x) = \frac{f(x+1)}{f(x)}$. Now,

$$
\begin{aligned}
\frac{1}{h(x+1)} - \frac{1}{h(x)} &= \frac{S(x+1)}{f(x+1)} - \frac{S(x)}{f(x)} \\
&= [g(x+1) - g(x)] + g(x+1)[g(x+2) - g(x)] + \ldots \\
&\leq 0
\end{aligned}
$$

whenever X is ILR. Thus, $h(x+1) \geq h(x)$ and X is IHR. The converse of the proposition need not be true. Hence, ILR is a stronger positive ageing concept than IHR and is useful in situations where IHR is too stringent a condition;

(xii) A sufficient condition for two discrete lifetimes X and Y to satisfy $X \geq_{\text{hr}} Y$ is that $X \geq_{\text{lr}} Y$; that is, $X \geq_{\text{lr}} Y \Rightarrow X \geq_{\text{hr}} Y$. Assuming $X \geq_{\text{lr}} Y$, by definition, $\frac{f_X(x)}{f_Y(x)}$ is increasing. This means that

$$\frac{f_X(x+1)}{f_Y(x+1)} \geq \frac{f_X(x)}{f_Y(x)} \text{ or } \frac{f_X(x+1)}{f_X(x)} \geq \frac{f_Y(x+1)}{f_Y(x)}.$$

Adding 1 to both sides, we get

$$\frac{f_X(x)+f_X(x+1)}{f_X(x)} \geq \frac{f_Y(x)+f_Y(x+1)}{f_Y(x)},$$

or

$$\frac{f_X(x)+f_X(x+1)}{f_X(x)+f_Y(x+1)} \geq \frac{f_X(x)}{f_Y(x)}. \tag{4.5}$$

Likewise,

$$\frac{f_X(x+2)+f_X(x+3)}{f_X(x+2)+f_Y(x+3)} \geq \frac{f_X(x+2)}{f_Y(x+2)} \geq \frac{f_X(x)}{f_Y(x)}. \tag{4.6}$$

Inequalities in (4.5) and (4.6) together yield

$$\frac{\sum_{i=0}^{3} f_X(x+i)}{f_Y(x+i)} \geq \frac{f_X(x)}{f_Y(x)}.$$

Proceeding similarly, as the number of terms on the left hand side tends to infinity, we get

$$\frac{S_X(x)}{S_Y(x)} \geq \frac{f_X(x)}{f_Y(x)} \text{ or } h_Y(x) \geq h_X(x),$$

which implies $X \geq_{\text{hr}} Y$;

(xiii) $X \geq_{\text{hr}} Y \Rightarrow X \geq_{\text{st}} Y$.
First, we note that $X \geq_{\text{hr}} Y$ iff $\frac{S_X(x)}{S_Y(x)}$, is increasing in x or

$$\frac{S_X(x+1)}{S_Y(x+1)} \geq \frac{S_X(x)}{S_Y(x)} \geq \cdots \geq \frac{S_X(0)}{S_Y(0)} = 1.$$

Thus, $S_X(x) \geq S_Y(x)$ or $X \geq_{st} Y$;

(xiv) Sometimes, it is enough to know the values of hazard rate at time zero to ascertain whether X is IHR (DHR), by employing the following result that involves a new ordering. We say that X is less than Y in the *initial hazard rate*, written as $X \leq_{h(0)} Y$, iff $h_X(0) \geq h_Y(0)$. The required result is that

$$X \text{ is IHR (DHR)} \Leftrightarrow X_{x_2} \leq_{h(0)} X_{x_1}, \quad 0 \leq x_1 \leq x_2;$$

(xv) If X and Y have the same mean,

$$X \leq_{\text{hr}} Y \Leftrightarrow X_E \leq_{\text{lr}} Y_E.$$

(xvi) The IHR property is not preserved for coherent systems.

The reliability bounds for IHR and DHR classes have been discussed by Sengupta et al. (1995). The nature of the hazard rates of some of the distributions discussed in Chapter 3 has been displayed in Table 4.1. Hazard rates and probability mass functions of these distributions can be seen from Table 3.2.

4.3.2 Increasing Hazard Rate (2)

The concept of increasing (decreasing) hazard rate of order 2-IHR (2) (DHR (2)) was introduced in the continuous case using the notion of stochastic dominance. Corresponding to the continuous case, we say that among two discrete random variables X_1 and X_2 with distribution functions $F_1(x)$ and $F_2(x)$, respectively, X_1 has stochastic dominance of the first order (SD_1) over X_2 if $F_1(x) \leq F_2(x)$ for every x. Similarly, stochastic dominance of order 2, SD_2, is defined as

$$\sum_{t=0}^{x} F_1(t) \leq \sum_{t=0}^{x} F_2(t)$$

and that of order 3, SD_3 as

$$\sum_{t=0}^{x} \sum_{u=0}^{t} F_1(u) \leq \sum_{t=0}^{x} \sum_{u=0}^{t} F_2(u).$$

It is easy to see that,

$$\text{SD}_1 \Rightarrow \text{SD}_2 \Rightarrow \text{SD}_3.$$

Sometimes, instead of using the distribution functions, the survival functions are used to define $\overline{SD}_2$ as

$$\sum_{t=x}^{\infty} S_1(t) \geq \sum_{t=x}^{\infty} S_2(t)$$

and $\overline{SD}_3$ as

$$\sum_{t=x}^{\infty} \sum_{u=0}^{t} S_1(u) \geq \sum_{t=x}^{\infty} \sum_{u=0}^{\infty} S_2(u),$$

where S_1 and S_2 are the survival functions of X_1 and X_2, respectively. It follows that

TABLE 4.1 Monotonicity of hazard rates of discrete life distributions

	Distribution	Nature of hazard rate
1.	binomial	IHR
2.	Poisson	IHR
3.	negative binomial	IHR for $k > 1$, DHR for $k < 1$, constant for $k = 1$
4.	Haight zeta	DHR
5.	Waring	DHR
6.	negative hyper-geometric	IHR
7.	uniform	IHR
8.	arithmetic	DHR
9.	S	IHR
10.	Weibull	IHR for $\beta > 1$, DHR for $\beta < 1$, constant for $\beta = 1$
11.	Weibull II	IHR for $\beta > 1$, DHR for $\beta < 1$
12.	Weibull III	IHR for $\alpha > 0$, DHR for $\alpha < 0$, constant for $\alpha = 0$
13.	power	IHR
14.	Lindley	IHR
15.	gamma	IHR for $k > 1$, DHR for $k < 1$ constant for $k = 1$
16.	Hurwitz-zeta	DHR
17.	Lerch family	DHR
18.	linear hazard rate	IHR for $b > 0$, DHR for $b < 0$
19.	logarithmic series	DHR
20.	geometric	constant
21.	Burr	$h(x) = 1 - \theta^{\log \frac{(1+x)^\alpha}{1+x^\alpha}}$ DHR in $0 < d < c$, $d > c$; $h(x)$ takes a jump at $x = 1$ and then decreases; $\alpha = c$, $h(0) = h(1)$ where $c = \frac{\log 3}{\log 2}$
22.	discrete generalized exponential	IHR $\alpha \geq 1$, DHR $\alpha \leq 1$
23.	inverse Weibull	DHR for $(1 - q)^{-1} \geq (1 - q^{2^{-\beta}})^{1/2}$

$$SD_1 \Rightarrow \overline{SD}_2 \Rightarrow \overline{SD}_3,$$

since SD_1 is equivalent to $S_1(x) \geq S_2(x)$.

In the sequel, we denote the various types of dominance relations by $X_1\ SD_1\ X_2$, $X_1\ SD_2\ X_2$, $X_1\ \overline{SD}_2\ X_2$, and so on.

In defining IHR (2) (DHR (2)), the requirement is that the residual life X_{x_1} has stochastic dominance of order 2 over X_{x_2}, $x_1 > x_2$.

Definition 4.4. We say that X has increasing (decreasing) hazard rate of order 2, IHR (2) (DHR (2)), if and only if, for every fixed x, the sum

$$a_x(t) = \sum_{u=0}^{x} \frac{S(t+u)}{S(x)} = \sum_{u=x}^{x+t} \frac{S(u)}{S(x)}$$

is decreasing (increasing) in t.

From condition (k) in Theorem 4.1,

$$(X - x|X > x) \geq_{\text{st}} (X - y|X > y) \text{ for all } x, y, x \leq y \Leftrightarrow \text{IHR},$$

$X_x SD_1 X_y \Leftrightarrow X_x \geq_{\text{st}} X_y$ and $X_x SD_1 X_y \Rightarrow X_x SD_2 X_y$, and so we have

$$X \text{ has IHR } \Rightarrow X \text{ has IHR (2)}.$$

Thus, the notion of IHR (2) (DHR (2)) is weaker than that of IHR (DHR). The properties of the IHR (2) class and its applications have not been studied yet.

4.3.3 Increasing Hazard Rate Average

It is known that the IHR class of distributions is not closed under the formation of coherent systems. The increasing hazard rate average concept was introduced in an effort to find the class of distributions that preserves this closure property. In contrast to the continuous case, there are two definitions of the new class depending on whether $h(x)$ or $h_1(x)$ is used as the definition of the hazard rate.

Definition 4.5. The random variable X (or its distribution) belongs to the increasing (decreasing) hazard rate average-1 class, IHRA (DHRA), if the hazard rate average

$$A(x) = \frac{H(x)}{x} = \frac{1}{x} \sum_{t=0}^{x-1} h(t)$$

is increasing (decreasing).

The IHRA-class satisfies the following properties:

(i) If X has IHR (DHR) property, then X has IHRA (DHRA) property, but the converse need not be true;

(ii) A probability mass function $f(x)$ belongs to both IHRA and DHRA-class if and only if

$$f(x) = \begin{cases} (1-p)^x p, & x = 0, 1, 2, \ldots, k-1, \\ (1-p)^k, & x = k. \end{cases} \tag{4.7}$$

When $k \to \infty$ we have the geometric law. Results (i) and (ii) are from Bracquemond et al. (2001);

(iii) X is IHRA if and only if $X \leq_* X_G$, where X_G is a geometric variable with parameter p;

(iv) Y_n has IHR (DHR) $\Rightarrow$ Y_n has IHRA (DHRA).

Definition 4.6. We say that X belongs to the IHRA-1 (DHRA-1) class if the alternative hazard rate average $A_1(x) = \frac{H_1(x)}{x} = \frac{1}{x}\sum_{t=0}^{x-1} h_1(t)$ is increasing (decreasing) in x.

The IHRA-1 (DHRA-1) distributions have the following properties:

(a) If X has IHR (DHR), then X has IHRA-1 (DHRA-1);

(b) IHRA-1 (DHRA-1) $\Rightarrow$ IHRA (DHRA), but the converse need not be true;

(c) A distribution has both IHRA-1 and DHRA-1, then its probability mass function has the form in (4.7);

(d) IHRA-1 class is closed under the formation of coherent systems;

(e) The IHRA-1 class is characterized by the property that for each $0 < q < 1$, $S(x) - q^x$, $x = 1, 2, \ldots$ has at most one change of sign and this change occurs from positive to negative;

(f) $(\frac{\mu_{(r)}}{r!})^{\frac{1}{r}}$ is decreasing in r and $\mu_{(r)} \leq r!\mu^r$, where $\mu_{(r)}$ is the rth factorial moment;

(g) The convolution of two IHRA-1 distributions is IHRA-1;

(h) The DHRA-1 class is closed under mixing;

(i) Let $\{X_n, n \geq 0\}$ be a stochastically monotone Markov chain with monotone paths on a partially ordered countable set S. Also, let C be an increasing set with its complement $\bar{C}$ finite. Then, the first passage time $T_i(i, c)$ from state i to C is IHRA-1. Conversely, every IHRA-1 distribution is either the first passage time distribution to an increasing set for a stochastically monotone Markov chain with monotone paths on a partially ordered finite set or else is the limit of such distributions (Brown and Rao, 1980);

(j) The bounds for distributions in the IHRA-1 class are given by

$$S(x) \geq \begin{cases} \delta_x, & x < \mu_r^{\frac{1}{r}}, \\ 0, & x \geq \mu_r^{\frac{1}{r}}, \end{cases}$$

where δ_x is given by

$$(\delta_x^{-\frac{1}{x}} - 1) \sum_{t=1}^{\infty} t^r \delta_x^{\frac{1}{x}} = \mu_r$$

and

$$S(x) \leq \begin{cases} 1, & x \leq \mu_r^{\frac{1}{r}} - 1, \\ \delta_x, & x \geq \mu_r^{\frac{1}{r}} - 1, \end{cases}$$

in which case

$$(\delta_x^{-\frac{1}{x}} - 1) \sum_{t=1}^{x} t^r \delta_x^{\frac{1}{x}} = \mu_r. - (x+1)^r$$

In both cases, the bound is sharp. For details, see Sengupta et al. (1995).

4.3.4 Now Better Than Used in Hazard Rate

The idea used in the new better (worse) than used in hazard rate, NBUHR (NWUHR), as an indicator of positive ageing is that at the initial age the hazard rate will be less than that of a used one (at any age).

Definition 4.7. A discrete lifetime X is NBUHR (NWUHR) iff $h(x) \geq (\leq) h(0)$ for all $x \geq 0$.

Remark 4.1.

(1) The above definition is equivalent to saying that $S(x+1) \leq S(x)S(1)$ since this will mean

$$\frac{S(x+1)}{S(x)} \leq 1 - f(0) \Leftrightarrow 1 - h(x) \leq 1 - h(0);$$

(2) $h(x) \geq (\leq) h(0) \Leftrightarrow (h_1(x) \geq (\leq) h_1(0)$. Hence, both $h(x)$ and $h_1(x)$ produce the same class of NBUHR (NWUHR) life distributions.

We have the following properties for the NBUHR class.

(i) the convolution of two distributions belonging to the NBUHR class need not be NBUHR;

(ii) mixtures of NWUHR are also NWUHR. To prove this assertion, let X be distributed as $S(x;\theta)$, where θ is continuous with distribution function $C(\theta)$ so that the mixture distribution has survival function $S(x)$, where

$$S(x) = \int_{\Theta} S(x|\theta)dC(\theta), \quad \theta \in \Theta.$$

Assume X to be NWUHR. Then,

$$\begin{aligned} S(x)S(1) &= \left[\int_{\Theta} S(x|\theta)dC(\theta)\right]\left[\int_{\Theta} S(1|\theta)dC(\theta)\right] \\ &\leq \int_{\Theta} S(x|\theta)S(1|\theta)dC(\theta) \\ &= \int_{\Theta} S(x+1|\theta)dC(\theta) = S(x+1), \end{aligned}$$

so that $S(x)$ is NWUHR;

(iii) $\mu_{(r+1)} \geq (\leq) \frac{\mu_{(r+1)}}{(r+1)} f(0)$ according as X is HBUHR (NWUHR);

(iv) The survival function of a shock model is given by (see Chapter 1)

$$S(x) = \sum_{k=0}^{\infty} \bar{P}_K P(N(x) = k), \tag{4.8}$$

where $(N(x), x \geq 0)$ is the number of shocks received by the component in $[0, x]$ and $\bar{P}_k$ is the probability that the component survives k shocks. Then, we have from Abouammoh et al. (1988) that

(a) $S(x)$ is NBUHR if $\{\bar{P}_k\}_{k=0}^{\infty}$ has the discrete NBUHR property if

$$P(N(x) = k) = \frac{e^{-\lambda t}(\lambda t)^k}{k'},$$

a homogeneous Poisson process;

(b) $S(x)$ is NBUHR if $\{\bar{P}_k\}_{k=0}$ has discrete NBUHR property if $N(x)$ is a non-homogeneous Poisson process

$$P(N(x) = k) = e^{\Lambda(t)} \frac{[\Lambda(t)]^k}{k!}$$

satisfying $\Lambda(0) = 0$ and $\Lambda'(t) > \Lambda'(0)$;

(v) IHRA (DHRA) $\Rightarrow$ NBUHR (NWUHR);

(vi) There are two different ways in which NBUHR can be interpreted as a partial order. First, we say that two random variables X and Y are ordered with respect to initial failure rate $h(0)$, $X \leq_{h(0)} Y$, iff $h_X(0) \geq h_Y(0)$. The discrete version of this order proposed in the continuous case by

Deshpande et al. (1990) reads $X_x \leq_{h(0)} X$ which means that $h_{X_x}(0) \geq h_X(0)$ or $h(t) \geq h(0)$ for all t, since $h_{X_x} = h(x+t)$. Thus, $X_x \leq_{h(0)} X \Leftrightarrow X$ is NBUHR. The second is in terms of an 'S-order' (Abouammoh, 1990) which says that $X \leq_S Y$ iff $\Delta G^{-1}F(x) \leq \Delta G^{-1}[F(y)]_{y=0}$. Taking G as the geometric distribution, it is easy to see that $X \leq_S Y \Leftrightarrow X$ is NBUHR. Of these, the $\leq_{h(0)}$ offers a comparison between the hazard rate of residual life distribution with the original distribution, while the second compares the baseline distribution with the geometric law. Reversing the inequalities, we have

$$X_x \geq_{h(0)} X \Leftrightarrow \text{NWUHR and } X \geq_S Y \Leftrightarrow \text{NWUHR}.$$

A weaker class of life distributions can be obtained if we use the averages of the hazard rates.

Definition 4.8. The lifetime X is said to belong to the new better than used in hazard rate average, NBUHRA (new worse than used in hazard rate average, NWUHRA), iff

$$h(0) \leq \frac{1}{x}\sum_{t=0}^{x-1} h(t)(h(0) \geq \frac{1}{x}\sum_{t=0}^{x-1} h(t)).$$

Remark 4.2. It may be noted that NBUHRA $\Leftrightarrow H(0) \leq \frac{H(x)}{x}$, where $H(x)$ is the cumulative hazard rate in $[0, x-1]$. When the alternative hazard rate $h_1(x)$ is used, we have a different class of life distributions.

Definition 4.9. We say that X has new better (worse) than used in hazard rate average-2, NBUHRA-1 (NWUHRA-1), iff $h_1(0) \leq (\geq)\frac{1}{x}\sum_{t=0}^{x-1} h_1(t)$ (Khalique, 1989), or equivalently $[S(1)]^x \geq S(x)$ for all x.

Theorem 4.2. *X has NBUHRA-1* $\Rightarrow$ *X has NBUHRA.*

Proof. We have

$$X \text{ has NBUHRA-1 } \Leftrightarrow h_1(0) - \frac{H_1(x)}{x} \leq 0. \tag{4.9}$$

Since the arithmetic mean is not less than the geometric mean, we get

$$\log\left[\frac{1}{x}\sum_{t=0}^{x-1}(1-h(t))\right] \geq \frac{1}{x}\sum_{t=0}^{x-1}\log(1-h_1(t))$$
$$\Rightarrow \log\left(1 - \frac{H(x)}{x}\right) \geq -\frac{H_1(x)}{x}. \tag{4.10}$$

Also,

$$\log(1 - H(0) = -h_1(0). \tag{4.11}$$

From (4.10) and (4.11), we find

$$\log(1 - h(0)) - \log\left(1 - \frac{H(x)}{x}\right) \geq \frac{H_1(x)}{x} - h_1(0) \geq 0$$

by (4.9). The last inequality means that $\frac{H(x)}{x} \geq h(0)$, and so X has NBUHRA. □

In terms of implications, we note that

$$\text{NBUHR} \Rightarrow \text{NBUHRA},$$
$$\text{IHRA} \Rightarrow \text{NBUHRA} \Rightarrow \text{NBUHRA-1}$$

4.4 CLASSES BASED ON RESIDUAL LIFE

Th importance of residual life in reliability and survival analysis has already been pointed out in Section 1.3. In this section, we study classes of life distributions arising from the monotonicity of mean residual life, variance residual life and harmonic mean residual life.

4.4.1 Decreasing Mean Residual Life

Recall that the mean residual life function (MRL) of X is given by

$$m(x) = E(X - x|X > x) = \frac{1}{S(x+1)} \sum_{t=x+1}^{\infty} S(t), \; x = -1, 0, 1, \ldots.$$

Definition 4.10. We say that the random variable X has decreasing (increasing) mean residual life, DMRL (IMRL), if the sequence $\langle m(x) \rangle$ is decreasing (increasing) in x.

Theorem 4.3. *(i) X is DMRL (IMRL) iff*

$$\frac{1}{S(x+1)} \sum_{t=x+1}^{\infty} S(t)$$

is decreasing (increasing) in x;

(ii) X is DMRL (IMRL) iff there exists a decreasing (increasing) non-negative sequence $\langle a(x) \rangle$ such that, for all $x = 0, 1, 2, \ldots,$

$$h(x) = \frac{a(x) - a(x-1) + 1}{1 + a(x)};$$

(iii) *X is DMRL (IMRL) iff X_E is IHR (DHR), provided $E(X) < \infty$;*

(iv) *If X has finite mean, then X is DMRL (IMRL) iff $X \geq_{hr} X_E$;*

(v) *X is DMRL (IMRL) iff $X_E \geq_{hr} (X_E - t|X_E > t)$;*

(vi) *X is DMRL (IMRL) iff $Y_{x_1} \geq Y_{x_2}$, $0 \leq x_1 \leq x_2$, where Y_x stands for the residual life random variable corresponding to X_E;*

(vii) *X is DMRL (IMRL) iff*

(a) $X_x \geq_{mrl} (\leq_{mrl}) X_{x'}$, $x' \geq x$,

(b) $X \geq_{mrl} (\leq_{mrl}) X_{x'}$,

(c) $X + x \leq_{mrl} X + x'$;

(viii) *X is DMRL (IMRL) iff $\sum_{t=0}^{\infty} S_x(t)$ is decreasing (increasing);*

(ix) *X is DMRL (IMRL) iff $E[\phi(X - x)|X > x]$ is decreasing (increasing) in x for all increasing convex functions $\phi(\cdot)$.*

Proof. Property (i) follows from the definition of $m(x)$. The proof of (ii) is available in Ebrahimi (1986) with suitable modifications in the definitions adopted there and in the present work for the mean residual life. In proving (iii), we note that the result is obvious from the fact that

$$h_{X_E}(x) = \frac{1}{m_X(x)}. \tag{4.12}$$

The result in (iv) is established from

$$\begin{aligned} X \geq_{\text{hr}} X_E &\Leftrightarrow h_X(x) \leq \frac{1}{m_X(x)} \text{ from (4.12)} \\ &\Leftrightarrow h_X(x) m_X(x) \leq 1 \\ &\Leftrightarrow 1 + m(x) - m(x-1) \leq 1 \\ &\Leftrightarrow m(x) \text{ is decreasing.} \end{aligned}$$

When X is such that

$$X_E \geq_{\text{hr}} (X_E - t|X_E > t) \Leftrightarrow \frac{1}{m(x)} \leq \frac{1}{m(x+1)} \Leftrightarrow X \text{ is DMRL},$$

as stated in (v). The proof of (vi) is similar. The results in (vii) follow from the definition of $m(x)$ and the fact that $m_{X_x}(t) = m(x+t)$. To prove (vii), observe that

$$\sum_{t=0}^{\infty} S_x(t) = \sum_{t=0}^{\infty} \frac{S(x+t+1)}{S(x+1)} = \frac{1}{S(x+1)} \sum_{t=x+1}^{\infty} S(t) = m(x).$$

Finally, in establishing (vii), we have

$$
\begin{aligned}
E[\phi(X-x)|X>x] &= \sum_{t=x+1}^{\infty} \phi(t-x)\frac{f(t)}{S(x+1)} \\
&= \sum_{t=x+1}^{\infty} \frac{\phi(t-x)}{S(x+1)}[S(t)-S(t+1)] \\
&= \frac{1}{S(x+1)}\Big[\sum_{t=x+1}^{\infty} \phi(t-x)S(t) - \sum_{t=x+2}^{\infty} \phi(t-x-1)S(t)\Big] \\
&= \phi(1) + \frac{1}{S(x+1)}[\sum_{t=x+2}^{\infty} [\phi(t-x)-\phi(t-x-1)]S(t) \\
&= \phi(1) + m(x)E[\phi(X_E-x)-\phi(X_E-x-1)|X>x+1],
\end{aligned} \tag{4.13}
$$

since

$$
E[\phi(X_E-x)-\phi(X_E-x-1)|X_E>x+1] = \frac{1}{S(t)}\sum_{t=x+2}^{\infty}[\phi(t-x)-\phi(t-x-1)]S(t).
$$

Now, assume that X is DMRL. The hazard rate of X_E is decreasing and so

$$
(X_E - x|X_E > x) \leq_{\text{st}} (X_E - x'|X_E > x'), \quad x' < x.
$$

Also, $X \leq_{\text{st}} Y$ is equivalent $E\phi(X) \leq E\phi(Y)$ for all increasing convex functions. Thus, $E\phi(X_E - x|X_E > x)$ is a decreasing function of X and consequently from (4.13), $E\phi(X - x|X > x)$ is a decreasing function for all increasing convex function, which proves the necessary part. To prove the sufficiency part, take $\phi(x) = x$ which is convex and increasing. □

There are several results that help in analyzing the DMRL (IMRL) class of life distributions and these are as follows:

(a) If X is IHR (DHR), then X is DMRL (IMRL), assuming the existence of $E(X) = \mu$.

Proof. If X has increasing hazard rate, $S(x)$ is log-concave or $f_E(x) = \frac{S(x+1)}{\mu}$ is log-concave which means that $S_E(x)$ is log-concave or X_E has increasing hazard rate. From the identity $h_{X_E}(x) = \frac{1}{m(x)}$, X is then DMRL; □

(b) If $\sum_{t=x}^{\infty} S(t)$ is log-concave, then X is DMRL.

Proof. When $\sum_{t=x}^{\infty} S(t)$ is log-concave, the survival function of X_E is log-concave and therefore X_E has increasing hazard rate. Hence, from (a), X is DMRL; □

(c) We have

$$m(x) - m(x-1) = h(x)m(x) - 1.$$

Hence, irrespective of the monotonicity of $h(x)$, X is DMRL (IMRL) if and only if $h(x)m(x) \leq (\geq 1)$. However, if X is IHR, we can write

$$m(x) \leq \frac{1}{h(x)} \leq \frac{1}{h(0)} = \frac{1}{f(0)}$$

which provides an upper bound for $m(x)$;

(d) If $X \leq_{\bar{m}} Y$, then Y is IMRL $\Rightarrow$ X is IMRL and X is DMRL $\Rightarrow$ Y is DMRL. This follows from the fact that

$$X \leq_{\bar{m}} Y \Leftrightarrow m_X(t-1) - m_X(t) \leq m_Y(t-1) - m_Y(t);$$

(e) If X and Y are DMRL, then

$$X \leq_{\bar{m}} Y \Rightarrow \frac{m_X(x)}{m_Y(x)}$$

is increasing in x.

Proof. Since Y is DMRL, $\frac{1}{m_Y(x)}$ is increasing, and

$$\begin{aligned} X \leq_{\bar{m}} Y &\Rightarrow m_X(x) - m_Y(x) \text{ is increasing} \\ &\Rightarrow \frac{m_X(x) - m_Y(x)}{m_Y(x)} \text{ is increasing} \\ &\Rightarrow \frac{m_X(x)}{m_Y(x)} \text{ is increasing;} \end{aligned}$$ □

(f) If X and Y are IMRL, the ratio $\frac{m_X(x)}{m_Y(x)}$ is increasing and $X \geq_{\text{mrl}} Y$, then $X \leq_{\bar{m}} Y$.

Proof. Since $X \geq_{\text{mrl}} Y$ we have $m_X(x) \geq m_Y(x)$ for all x. Thus,

$$m_X(0) \geq m_Y(0) \Rightarrow 1 + \mu_X \geq 1 + \mu_Y \Rightarrow \mu_X \geq \mu_Y. \tag{4.14}$$

Further,

$$
\begin{aligned}
&\frac{m_X(x)}{m_Y(x)} \text{ is increasing} \\
&\Rightarrow \frac{m_X(x-1)}{m_X(x)} \geq \frac{m_Y(x-1)}{m_Y(x)} \Rightarrow \frac{m_X(x-1) - m_X(x)}{m_X(x)} \\
&\qquad \geq \frac{m_Y(x-1) - m_Y(x)}{m_X(x)}
\end{aligned} \tag{4.15}
$$

$$
\Rightarrow m_X(x-1) - m_X(x) \geq m_Y(x-1) - m_Y(x). \tag{4.16}
$$

From (4.14) and (4.16), the conclusion $X \leq_{\bar{m}} Y$ follows; □

(g) $X \leq_{\bar{m}} (\geq_{\bar{m}}) Y$ if and only if $X(Y)$ is IMRL (DMRL), where Y is geometrically distributed.

Proof. When Y is geometric, $m_Y(x) = p^{-1}$ for all $x = 0, 1, 2, \ldots$ Then,

$$
\begin{aligned}
X \leq_{\bar{m}} Y &\Leftrightarrow m_X(x-1) - m_X(x) \geq m_Y(x-1) - m_Y(x) = 0 \; \forall x \\
&\Leftrightarrow m_X(x-1) \geq m_X(x) \; \forall x \\
&\Leftrightarrow X \text{ is DMRL;}
\end{aligned}
$$

□

Remark. The above result expresses the DMRL (IMRL) class in terms of ordering $\leq_{\bar{m}}$. In other words, if X has positive (negative) memory at all points of the support of X, then X is DMRL (IMRL);

(h) X is DMRL (IMRL) if and only if

$$
(X_E - x | X_E > x) \geq_{hr} (\leq_{hr}) (X_E - y | X_E > y) \; \forall y \geq x.
$$

The above statement is equivalent to $h_{X_E}(x+t) \leq (\geq) h_E(y+t)$ for $y \geq x$ and all $t \geq 0$. Hence, X_E is IHR (DHR) and so X is DMRL (IMRL);

(i) X is DMRL (IMRL) if and only if

$$
(X - x | X > x) \geq_{\text{st}} (\leq_{\text{st}}) (X_E - y | X_E > y).
$$

Proof. We have

$$
\begin{aligned}
&(X - x | X > x) \geq_{\text{st}} (X_E - y | X_E > y) \\
&\qquad \Leftrightarrow \frac{S(t+x+1)}{S(x+1)} \geq \frac{\sum_{u=x+t+1}^{\infty} S(u+1)}{\sum_{u=x+1}^{\infty} S(x+1)} \\
&\qquad \Leftrightarrow \frac{\sum_{u=x+1}^{\infty} S(x+1)}{S(x+1)} \geq \frac{\sum_{u=x+t+1}^{\infty} S(u+1)}{S(t+x+1)} \text{ from (4.4)} \\
&\qquad \Leftrightarrow m(x) \geq m(x+t) \Leftrightarrow X \text{ is DMRL.}
\end{aligned}
$$

The IMRL case is obtained by reversing the inequalities; □

(j) X is DMRL (IMRL) if and only if

$$(X_E - x|X_E > x) \geq_{\text{st}} (X_E - y|X_E > y), \ y \geq x.$$

Proof. The above statement, when X is DMRL, means that $\frac{S_{X_E}(t+x)}{S_{X_E}(x)}$ is stochastically decreasing. In other words X_E has an increasing hazard rate which is equivalent to saying that X has DMRL; □

(k) X is DMRL (IMRL)

$$\Leftrightarrow (X - x|X > x) \geq_{\text{icx}} (\leq_{\text{icx}})(X - y|X > y), \quad y > x.$$

Proof. Recall that $X \leq_{icx} Y$ iff $\sum_{t=x}^{\infty} S_X(t) \leq \sum_{t=x}^{\infty} S_Y(t)$. Hence,

$$\begin{aligned}
(X - x|X > x) &\geq_{\text{icx}} (X - y|X > y) \\
&\Leftrightarrow \sum_{t=x}^{\infty} \frac{S(t+x+1)}{S(x+1)} \geq \sum_{t=y}^{\infty} \frac{S(t+y+1)}{S(y+1)} \\
&\Leftrightarrow \frac{\sum_{u=t+x+1}^{\infty} S(u)}{S(x+1)} \geq \frac{\sum_{u=t+y+1} S(u)}{S(y+1)} \\
&\Leftrightarrow m(x) - t \geq m(y) - t \Leftrightarrow m(x) \geq m(y), \ y > x \\
&\Leftrightarrow X \text{ is DMRL;}
\end{aligned}$$

□

(l) X is DMRL $\Leftrightarrow X \geq_{\text{icx}} (\leq_{\text{icx}})(X - x|X > x)$,
IHR (DHR) $\Rightarrow$ DMRL (IMRL);

(m) Thus, the DMRL class is weaker than the IHR class. A proof of this result is given by Lai and Xie (2006) with a slight change of definition of the mean residual life in terms of $\bar{r}(x)$. But, the converse of this result need not be true; see Ebrahimi (1986). It is of some interest to know what additional condition is required for the reverse implication DMRL $\Rightarrow$ IHR to hold.

Theorem 4.4. *If $m^*(x) = E(X|X > x)$ is convex, then DMRL $\Rightarrow$ IHR.*

Proof. Since $m^*(x)$ is convex,

$$m^*(x) - m^*(x-1) \geq m^*(x-1) - m^*(x-2).$$

But $m^*(x) = m(x) + x$, and so we have

$$m(x) - m(x-1) \geq m(x-1) - m(x-2)$$

which shows that $m(x) - m(x-1)$ is an increasing function. The identity

$$h(x) = \frac{m(x) - m(x-1) + 1}{m(x)}$$

is the product of $\frac{1}{m(x)}$ which is increasing since X is DMRL and the term $m(x) - m(x-1) + 1$ which is also increasing. Hence, X is IFR; □

(n) $X \leq_{\text{hr}} Y \Rightarrow X \leq_{\text{mrl}} Y$.

Proof. We have

$$\begin{aligned}
X \leq_{\text{hr}} Y &\Leftrightarrow \frac{S_X(x)}{S_Y(x)} \text{ is decreasing in } x \\
&\Leftrightarrow \frac{S_X(x+1)}{S_X(x)} \leq \frac{S_Y(x+1)}{S_Y(x+2)} \\
&\Leftrightarrow \frac{S_X(x+k+1)}{S_X(x+1)} \leq \frac{S_Y(x+k+1)}{S_Y(x+1)} \\
&\Rightarrow \sum_{t=x+1}^{\infty} \frac{S_X(t)}{S_X(x+1)} \leq \frac{\sum_{x+1}^{\infty} S_Y(t)}{S_Y(x+1)} \\
&\Leftrightarrow m_X(x) \leq m_Y(x) \Leftrightarrow X \leq_{\text{mrl}} Y.
\end{aligned}$$

□

The converse of this result is true only with an additional condition as seen from the following theorem.

Theorem 4.5. *Given that $\frac{m_X(x)}{m_Y(x)}$ is increasing in x, $X \leq_{mrl} Y \Rightarrow X \leq_{hr} Y$.*

Proof. $\frac{m_X(x)}{m_Y(x)}$ is increasing in x

$$\begin{aligned}
&\Leftrightarrow \frac{m_X(x)}{m_Y(x)} \geq \frac{m_X(x-1)}{m_Y(x-1)} \\
&\Leftrightarrow \frac{m_Y(x-1)}{m_Y(x)} \geq \frac{m_X(x-1)}{m_X(x)} \\
&\Leftrightarrow \frac{m_Y(x-1) - m_Y(x)}{m_Y(x)} \geq \frac{m_X(x-1) - m_X(x)}{m_X(x)} \\
&\Leftrightarrow \frac{m_X(x) - m_X(x-1)}{m_X(x)} \geq \frac{m_Y(x) - m_Y(x-1)}{m_Y(x)} \\
&\Leftrightarrow \frac{m_X(x) - m_X(x-1)}{m(x)} + \frac{1}{m_X(x)} \geq \frac{m_Y(x) - m_Y(x-1)}{m_Y(x)} + \frac{1}{m_Y(x)}
\end{aligned} \tag{4.17}$$

since X is DMRL implies $m_X(x) \leq m_Y(x)$. Thus, (4.17) is equivalent to $h_X(x) \geq h_Y(x)$, and so $X \leq_{\text{hr}} Y$. □

Since $X \leq_{\text{hr}} Y \Rightarrow X \leq_{\text{st}} Y$, the above theorem provides a condition for which $X \leq_{\text{mrl}} Y \Rightarrow X \leq_{\text{st}} Y$. An alternative condition for the same implication is given in the following theorem.

Theorem 4.6. *If* $\frac{m_X(x)}{m_Y(x)} \geq \frac{1+\mu_X}{1+\mu_Y}$, *then* $X \leq_{mrl} Y \Rightarrow X \leq_{st} Y$.

Proof. Since

$$S_X(x) = \prod_{t=0}^{x-1} \frac{m_X(t-1)-1}{m_X(t)}$$
$$= \frac{m_X(-1)}{m_X(x-1)} - \prod_{t=0}^{x-1} \frac{1}{m_x(t)},$$

we have

$$S_X(x) - S_Y(x) = \frac{m_X(-1)}{m_X(x-1)} - \frac{m_Y(-1)}{m_Y(x-1)}$$
$$- \left[\prod_{t=0}^{x-1}\left(\frac{1}{m_X(t)} - \frac{1}{m_Y(t)}\right)\right].$$

Since $X \leq_{\text{mrl}} Y$, the terms inside the square brackets is positive and so

$$S_X(x) - S_Y(x) \leq \frac{1+\mu_X}{m_X(x-1)} - \frac{1+\mu_Y}{m_Y(x-1)} \leq 0.$$

Thus, $X \leq_{st} Y$; □

(o) The convolution of two DMRL (IMRL) random variables need not be DMRL.

(p) The mixture of two IMRL variables is IMRL. We refer to Pavlova et al. (2006) for a proof of the propositions in (0) and (p).

Table 4.2 contains a list of discrete models, their mean residual life, and the classes to which they belong to.

Remark 4.3. There are no closed-form expressions for the mean residual life in the case of several distributions. However, the nature of their mean residual lives are given in brackets. Binomial (DMRL), hyper-geometric (DMRL), logarithmic series (IMRL), Poisson (DMRL), Weibull I (IDMRL for $0 < \beta < 1$, constant for $\beta = 1$ and DIMRL for $\beta > 1$), Weibull 2 (IMRL for $0 < \beta < 1$, constant $\beta = 1$, DMRL for $\beta > 1$), and Weibull III (IMRL for $\alpha < 0$, DMRL for $\alpha > 0$ and constant for $\alpha = 0$).

4.4.2 Decreasing Mean Residual Life in Harmonic Average

The decreasing (increasing) mean residual life in harmonic average, DMRLHA (IMRLHA), as the name suggests, is based on the nature of the monotonicity of the harmonic mean of the mean residual life.

TABLE 4.2 Mean residual life of functions of some discrete models

Distribution	Mean residual life function	Class
Height zeta	$(1-2^{-\sigma})\xi(\sigma)$ $[(2x+1)^{\sigma}(1-\zeta_x(\sigma))]-x$	IMRL for $\sigma > 1$
geometric	$\frac{1}{p}$	constant MAL
Waring	$\frac{a+x}{a-b-1}$	IMRL
negative hyper geometric	$\frac{k+n-x}{n+1}$	DMRL
uniform	$\frac{b-x+1}{2}$	DMRL
arithmetic	$\frac{b-x+1}{b-x}$	IMRL
Lindley	$\frac{(1-\lambda)(1-(1+x)\log\lambda)-\log\lambda}{(1-\lambda)^2[1-(x+2)\log\lambda]}$	DMRL
Lerch family	$\frac{\phi(z,s-1,a)}{\phi(z,s,a)}-a$ $+x\frac{\partial \log}{\partial z}(z^{x+1}\frac{\phi(z,s,a+x+1)}{\phi(z,s,a)})$	IMRL
Logarithmic series	$\frac{1+x}{1-z}[\sum_{t=0}^{\infty}\frac{(x+1)z^t}{(x+1+t)}]^{-1}$	IMRL

Definition 4.11. The discrete lifetime X is said to be DMRLHA (IMRLHA) iff

$$B(x) = \left[\frac{1}{x}\sum_{t=-1}^{x-2}\frac{1}{m(t)}\right]^{-1} \tag{4.18}$$

is decreasing (increasing) in x.

The above definition can be given an interpretation in terms of the reliability characteristic of the equilibrium distribution of X. There are several such criteria of ageing introduced in reliability analysis, through a comparison of the distributions of X and X_E. Note that a unit with lifetime X which ages more rapidly will become worse in the case of X_E. This scenario will become much predominant when we consider other ageing concepts in the sequel. Since the mean residual life $m(x)$ of X is the reciprocal of the hazard rate $h_{X_E}(x)$ of X_E, (4.18) is equivalent to

$$B(x) = \left[\frac{1}{x}\sum_{t=0}^{x-1}h(t)\right]^{-1}. \tag{4.19}$$

The right hand side of (4.19) is the reciprocal of the average of the hazard rate so that

$$X \text{ is DMRLHA} \Leftrightarrow X_E \text{is IFRA}.$$

We further have

$$X \text{ is DMRL } \Leftrightarrow X_E \text{ is IFR } \Rightarrow X_E \text{ is IFRA } \Leftrightarrow X \text{ is DMRLHA.}$$

Thus, DMRLHA is a milder positive ageing criterion than DMRL. There is also an inter-connection between the ageing concepts and the hmrl order defined earlier in Section 4.2. Observe that

$$X \leq_{\text{hmrl}} Y \Leftrightarrow B_X(x) \leq B_Y(x).$$

Theorem 4.7. $(X - x|X > x) \geq_{hmrl} (X - y|X > y) \Leftrightarrow X$ *is DMRL,* $y \geq x$.

Proof. We have

$$\begin{aligned}
(X - x|X > x) &\geq_{\text{hmrl}} (X - y|X > y) \\
&\Leftrightarrow \frac{\sum_{u=t+1}^{\infty} S(u+x)}{S(x)m(x)} \geq \frac{\sum_{u=t+1}^{\infty} S(u+y)}{S(y)m(y)} \\
&\Leftrightarrow \frac{\sum_{u=x+t+1}^{\infty} S(u)}{\sum_{u=y+t+1}^{\infty} S(u)} \geq \frac{\sum_{u=x}^{\infty} S(u)}{\sum_{u=y}^{\infty} S(u)} \\
&\Leftrightarrow \frac{\sum_{u=x}^{\infty} S(u)}{\sum_{u=y}^{\infty} S(y)} \text{ is increasing in } x \text{ for } y \geq x \\
&\Leftrightarrow m(x) \geq m(y) \Leftrightarrow X \text{ is DMRL.}
\end{aligned}$$

□

Theorem 4.8.

$$X_E \leq_{hmrl} Y \Leftrightarrow X_E \; SD\, 1 \; Y_E,$$

where SD 1 is the stochastic dominance of order 1.

Proof. We have

$$\begin{aligned}
X_E \text{ SD } 1\; Y_E &\Leftrightarrow \frac{1}{\mu_X} \sum_{t=x+1}^{\infty} S(t) \leq \frac{1}{\mu_Y} \sum_{t=x+1}^{\infty} S_Y(t) \\
&\Leftrightarrow X \leq_{\text{hmrl}} Y.
\end{aligned}$$

□

Theorem 4.9. $X \leq_{mrl} Y \Rightarrow X \leq_{hmrl} Y$.

Proof. We have

$$\begin{aligned}
X \leq_{\text{mrl}} Y &\Rightarrow m_X(x) \leq m_Y(x) \text{ for } x \geq -1 \\
&\Leftrightarrow \frac{1}{m_X(x)} \geq \frac{1}{m_Y(x)} \\
&\Leftrightarrow h_{X_E}(x) \geq h_{Y_E}(x)
\end{aligned}$$

$$\Leftrightarrow S_{X_E}(x) \le S_{Y_E}(x)$$
$$\Leftrightarrow X_E \text{ SD } 1\ Y_E$$
$$\Leftrightarrow X \le_{\text{hmrl}} Y \qquad \square$$

Some further properties of DMRLHA and $\le_{\text{hmrl}}$ will be discussed in connection with other ageing classes. Notice that the dual class IMRLHA has similar properties obtained by reversing the inequalities in the results for DMRLHA and also by replacing the ageing concept by its dual.

4.4.3 Used Better Than Aged

When a unit is working with unknown age, to assess its future ageing behaviour, used better than aged concept can be employed.

Definition 4.12. The distribution of X, for which $1 < m(\infty) < \infty$, is used better (worse) than aged, UBA (UWA), iff

$$S(x+1) \ge (\le) \frac{m(\infty)-1}{m(\infty)} S(x) \quad \text{for } x = 0, 1, 2, \ldots$$

From the definition, we see that

$$\frac{S(x+1)}{S(x)} \ge \frac{m(\infty)-1}{m(\infty)} \Leftrightarrow 1 - h(x) \ge 1 - h(\infty)$$
$$\Leftrightarrow h(x) \le h(\infty)$$

for all x, provided $h(\infty)$ is finite. If $P(t)$ denotes the probability generating function given by

$$P(t) = \sum_{x=0}^{\infty} t^x f(x), \ |t| < t_0$$

with the radius of convergence, $t_0 \ge 1$, Willmot and Lin (1994) have shown that

$$\lim_{x\to\infty} \frac{S(x+1)}{S(x)} = \frac{1}{t_0}$$

provided the limit exists. It has been further established by Willmot and Cai (2001) that if X is DMRL (IMRL), then

$$S(x+1) \ge (\le) \frac{S(x)}{t_0}.$$

Identifying t_0 with $\frac{m(\infty)}{m(\infty)-1}$, we have the result that if X is DMRL (IMRL) with $1 < m(\infty) < \infty$,

$$\text{DMRL (IMRL)} \Rightarrow \text{UBA (UWA)}.$$

Willmot and Cai (2000) have also shown that in the mixed Poisson distribution with

$$f(x) = \int_0^{\infty} \frac{(\lambda t)^x e^{-\lambda t}}{x!} dG(t) \tag{4.20}$$

where G is a distribution on $[0, \infty)$, the ageing property UBA (UWA) holds for F iff the mixed Poisson distribution is UBA (UWA). Eq. (4.20) also provides many examples of UBA (UWA) distributions. We further observe that

$$S(x+y) \geq (\leq) t_0^y S(x), \quad x = 0, 1, 2, \ldots, \ y = 1, 2, \ldots \tag{4.21}$$

A still weaker class of life distributions can be obtained by considering the following definition.

Definition 4.13. We say that X is used better (worse) than aged in expectation, UBAE (UWAE), if $m(x) \geq m(\infty)$ for all x and $1 < m(\infty) < \infty$.

When X is UBA, the inequality in (4.21) gives

$$\frac{1}{S(x+1)} \sum_{t=x+1}^{\infty} S(t) \geq \sum_{y=1}^{\infty} \left(\frac{m(\infty) - 1}{m(\infty)} \right)^y,$$

which is the same as $m(x) \geq m(\infty)$, thus justifying the averaging implied in the definition of UBAE. Thus,

$$\text{UBA} \Rightarrow \text{UBAE}.$$

Theorem 4.10. *A necessary and sufficient condition that X is UBAE (UWAE) is that X_E is UBA (UWA).*

Proof. We have

$$\begin{aligned} X \text{ is UBAE} &\Leftrightarrow m(x) \geq m(\infty) \\ &\Leftrightarrow \frac{1}{h_{X_E}(x)} \geq \frac{1}{h_{X_E}(\infty)} \\ &\Leftrightarrow h_{X_E}(x) \leq h_{X_E}(\infty) \\ &\Leftrightarrow X_E \text{ is UBA.} \end{aligned}$$

□

4.4.4 Decreasing Variance Residual Life

In this section, we discuss classes of discrete life distributions based on the monotone behaviour of the variance residual life $\sigma^2(x)$ defined in Section 2.4.

Definition 4.14. We say that X has decreasing (increasing) variance residual life, DVRL (IVRL), iff $\sigma^2(x)$ is decreasing (increasing) in x.

Recalling from (2.42) that

$$\sigma^2(x+1)-\sigma^2(x)=h(x+1)[\sigma^2(x+1)-m(x+1)(m(x)-1)],$$

we have

(i) X is DVRL iff $\sigma^2(x+1)\le m(x+1)(m(x)-1)$,
(ii) X is IVRL iff $\sigma^2(x+1)\ge m(x+1)(m(x)-1)$.

Theorem 4.11. *DMRL (IMRL) ⇒ DVRL (IVRL).*

Proof. We prove the case of DMRL from which the IMRL case is obvious. We have

$$\begin{aligned}
&\sigma^2(x+1)-m(x+1)(m(x)-1)\\
&\quad=\frac{2}{S(x+2)}\sum_{t=x+2}^{\infty}\sum_{u=t+1}^{\infty}S(u)-m(x+1)(m(x+1)-1)\\
&\qquad\quad -m(x+1)(m(x)-1)\\
&\quad=\frac{2}{S(x+2)}\sum_{t=x+2}^{\infty}m(t)S(t+1)-m(x+1)(m(x+1)+m(x)-2)\\
&\quad=\frac{2}{S(x+2)}\sum_{t=x+2}^{\infty}[m(t)-m(x+1)]\\
&\qquad\quad +\frac{2m(x+1)}{S(x+2)}\left[\sum_{t=x+2}^{\infty}S(t)-S(x+2)\right]\\
&\qquad\quad -m(x+1)(m(x+1)+m(x)-2)\\
&\quad=\frac{2}{S(x+2)}\sum_{t=x+2}^{\infty}[m(t)-m(x+1)]S(t+1)+2m(x+1)(m(x+1)-1)\\
&\qquad m(x+1)(m(x+1)+m(x)-2)\\
&\quad=\frac{2}{S(x+2)}\sum_{t=x+2}^{\infty}[m(t)-m(x+1)]S(t+1)\\
&\qquad +m(x+1)(m(x+1)-m(x)).
\end{aligned}\tag{4.22}$$

When X is DMRL, $m(t)\le m(x+1)$ for $t=x+2,x+3,\ldots$ and $m(x+1)\le m(x)$. Thus, the expression on the right hand side of (4.22) is negative, and so X is DVRL. □

Theorem 4.12. *If $S(x)$ is strictly decreasing, X is DVRL (IVRL) if and only if X_E is DMRL.*

Proof. We have

$$\begin{aligned}
m_E(x) &= \sum_{t=x+1}^{\infty}\sum_{u=t+1}^{\infty} S(u) \Big/ \sum_{t=x+2}^{\infty} S(t) \\
&= \frac{\mu_{(2)}(x)S(x+1)}{2S(x+2)m(x+1)} \qquad \text{from (2.38)} \\
&= \frac{\mu_{(2)}(x)}{2m_2(x+1)(1-h(x+1))} \\
&= \frac{\mu_{(2)}(x)}{2(m(x)-1)}. \qquad (4.23)
\end{aligned}$$

□

We also have

$$\sigma^2(x) = \mu_{(2)}(x) + m(x)(m(x)-1), \qquad (4.24)$$

and from (4.23) that

$$\frac{\mu_{(2)}(x)}{m(x)-1} = 2m_E(x). \qquad (4.25)$$

So,

$$\begin{aligned}
\frac{\mu_{(2)}(x+1)}{m(x+1)-1} - \frac{\mu_{(2)}(x)}{m(x)-1} &= 2(m_E(x+1) - m_E(x)) \\
&= 2(m_E(x+1)h_E(x+1) - 1) \\
&= 2\left(\frac{m_E(x+1)}{m(x+1)} - 1\right) \\
&= \frac{\mu_2(x+1)}{(m(x+1)-1)m(x+1)} - 2. \qquad (4.26)
\end{aligned}$$

Using (4.24), we get

$$\begin{aligned}
2(m_E(x+1) - m_E(x)) &= \frac{\sigma^2(x+1) + m(x+1)(m(x+1)-1)}{m(x+1)(m(x+1)-1)} - 2 \\
&= \frac{\sigma^2(x+1)}{m(x+1)(m(x+1)-1)} - 1. \qquad (4.27)
\end{aligned}$$

Now,

$$\frac{\sigma^2(x+1)}{m(x+1)(m(x)-1)} = \frac{(m(x+1)-1)(1+m_E(x+1)-m_E(x))}{m(x)-1} \qquad (4.28)$$

and

$$\frac{\sigma^2(x)}{m(x)(m(x)-1)} = 2(m_E(x) - m_E(x-1)) + 1 \text{ from (4.27).} \tag{4.29}$$

Using (4.28) and (4.29), we then obtain

$$\begin{aligned}\sigma^2(x) - \sigma^2(x+1) &= m(x+1)h(x+1)[m(x) - 1 - (m(x+1) - 1)] \\ &\quad [1 + m_E(x+1) - m_E(x)] \\ &= \frac{m(x+1)h(x+1)}{1 + m_E(x) - m_E(x-1)}[m_E(x-1) - m_E(x)] \\ &= m(x+1)h(x+1)(m(x) - m_E(x)) \qquad (4.30) \\ &= m(x+1)m(x)h(x+1)[m_E(x-1) - m_E(x)] \qquad (4.31)\end{aligned}$$

upon utilizing the identity $m(x) = m_E(x)[1 + m_E(x) - m_E(x-1)]^{-1}$ in Section 2.11. Thus, the sign of $\sigma^2(x) - \sigma^2(x+1)$ is the same as that of $m_E(x-1) - m_E(x)$, and so X is DVRL (IVRL) iff X_E is DMRL (IMRL).

Remark 4.4.

(1) Since

$$\frac{\mu_{(2)}(x+1)}{m(x+1)-1} - \frac{\mu_{(2)}(x)}{m(x)-1} = 2(m_E(x+1) - m_E(x)),$$

as a corollary to the above theorem, we have that X is DVRL (IVRL) if and only if $\frac{\mu_{(2)}(x)}{m(x)-1}$ is decreasing (increasing);

(2) If $S(\cdot)$ is strictly decreasing we have as a result of (4.36) that X is DVRL (IVRL) if and only if $m(x) \le (\ge) m_E(x)$. This is also equivalent to the statement that

$$X \text{ is DVRL (IVRL)} \Leftrightarrow X \le_{\text{mrl}} (\ge_{\text{mrl}}) X_E.$$

Regarding the closure properties of the DVRL class with respect to formation of mixtures and convolutions is that the DVRL class is not closed with respect to either of these two properties (see Al-Zahrani et al., 2013).

Theorem 4.13. *If X DVRL (IVRL) with $1 < m(\infty) < \infty$, then X is UBAE.*

Proof. We have

$$\begin{aligned}X \text{ is DVRL (IVRL)} &\Rightarrow X_E \text{ is DMRL (IMRL)} \\ &\Rightarrow X_E \text{ is UBA} \Rightarrow X \text{ is UBAE.}\end{aligned}$$

□

In order to compare the variance residual lives of two distributions, we require an ordering which is defined as follows.

Definition 4.15. We say that X is smaller than Y in variance residual life order, $X \leq_{\text{vrl}} Y$, iff $\sigma_X^2(x) \leq \sigma_Y^2(x)$ for all x.

Example 4.1. Let X be uniformly distributed with $f_X(x) = \frac{1}{n+1}$, $x = 0, 1, 2, \dots, n$, and Y follows the distribution

$$f_Y(x) = \frac{2(n-x+1)}{(n+1)(n+2)}, \quad x = 0, 1, \dots, n.$$

Then,

$$V_X(x) = \frac{(n-x+1)(n-x-1)}{12} \text{ and } V_Y(x) = \frac{(2+n-x)(n-x-1)}{18}$$

so that

$$V_X(x) - V_Y(x) = -\frac{(n-x-1)^2}{36} < 0.$$

Hence, $X \leq_{\text{vrl}} Y$.

4.5 CLASSES BASED ON SURVIVAL FUNCTION

The survival function, being the probability that a device is functioning properly beyond age x can give an indication of ageing when compared at different points of survival. Some ageing concepts based on the survival function by different kinds of comparisons will be discussed in this section.

4.5.1 New Better Than Used

Definition 4.16. A lifetime random variable X is said to be *new better than used* (NBU) or *new worse than used* (NWU) accordingly as $S(x+t) \leq (\geq) S(x)S(t)$ (Esary et al., 1973).

The definition means that X is NBU (NWU) if the conditional probability of survival at age x for a device that has survived to age t is less (greater) than the survival probability at age x of a new item. Notice that when the equality $S(x+t) = S(x)S(t)$ holds, we have the no-ageing property characterizing the geometric law. It is easy to see that X is NBU (NWU) if and only if $X \geq_{\text{st}} (\leq)(X - x | X > x)$. The class NBU (NWU) satisfies the following properties:

(i) X is NBU (NWU) iff $X \leq_{\text{su}} (\leq_{\text{sa}}) Y$, where Y is a geometric random variable;

(ii) IFRA $\Rightarrow$ NBU;

(iii) Consider a renewal sequence $\langle U_n \rangle$ defined by

$$U(t) = \sum_{x=0}^{\infty} U_x t^x = \left[1 - \sum_{x=1}^{\infty} t^x f(x)\right]^{-1},$$

where $f(x)$ is the probability mass function of a random variable X so that $\sum t^x f(x) = P(t)$ is the probability generating function with radius of convergence t_0. Then (Berenhaut and Lund, 2002), if X is NBU,

$$|U_x - U_\infty| \le (\frac{t_0}{\alpha} - 1)^{-1} (\frac{t_0}{c})^{-x}, \quad x \ge 0,$$

where $c = \frac{1}{2}(1+5)^{\frac{1}{2}}$;

(iv)

$$S(x) \ge \begin{cases} \delta_k, & k < (\mu_r')^{\frac{1}{r}}, \\ 0, & k < (\mu_r')^{\frac{1}{r}}, \text{ with } \mu_r' = E(X^r), \end{cases}$$

where δ_k satisfies

$$(\delta_k^{-1} - 1) \sum t^r \delta_k^t = \frac{\mu_r'}{k^r}$$

and the bound given above is sharp (Sengupta et al., 1995);

(v) NBU $\Rightarrow$ NBUHR.

Proof. We have

$$\begin{aligned} \text{NBU} &\Rightarrow S(x+t) \le S(x)S(t) \text{ for all } x, t \\ &\Rightarrow S(x+1) \le S(x)S(1) \\ &\Rightarrow \frac{S(x+1)}{S(x)} \le 1 - f(0) \\ &\Rightarrow 1 - h(x) \le 1 - h(0) \Rightarrow X \text{ is NBUHR;} \end{aligned}$$

□

(vi) X is IFR $\Leftrightarrow X_x$ is NBU.

Proof. We have

$$\begin{aligned} X \text{ is NBU} &\Leftrightarrow \frac{S(x+1)}{S(x)} \le S(t) \\ &\Leftrightarrow \frac{S(x+t)}{S(x+t-1)} \frac{S(x+t-1)}{S(x+t-2)} \cdots \frac{S(x+1)}{S(x)} \le S(t) \\ &\Leftrightarrow \prod_{u=0}^{t-1} (1 - h(x+u)) \le \prod_{u=0}^{t-1} (1 - h(u)) \end{aligned}$$

$$\Leftrightarrow \prod_{u=0}^{t-1}(h(u) - h(x+u)) \leq 0 \quad \text{for all } t = 1, 2, \ldots$$

$$\Leftrightarrow X \text{ is IFR;}$$

□

(vii) We have

$$S(ux) \leq [S(x)]^u, \quad u = 1, 2, \ldots, \quad x \geq 0,$$

obtained after setting $t = x$ in the definition of NBU, and then by employing induction;

(viii) An NBU life distribution possesses finite moments of all orders;

Proof. Let t be a finite positive integer such that $x = tk$. Then, by (vi) above, we have

$$S(x) = S(tk) \leq [S(t)]^k = [S(t)]^{\frac{x}{t}}.$$

Now,

$$\begin{aligned} E(X^{(r)}) \sum_{x=0}^{\infty} x^{(r)} f(x) &= \sum_{x=0}^{\infty} x^{(r)}[S(x) - S(x+1)], \\ &= \sum_{x=0}^{\infty}[x^{(r)} - (x-1)^{(r)}]S(x) \\ &= r\sum_{x=0}^{\infty}(x-1)^{(r-1)}S(x) \\ &\leq r\sum_{x=1}^{\infty}(x-1)^{(r-1)}[S(t)^{\frac{1}{t}}]^x \\ &= r\sum_{x=1}^{\infty}(x-1)^{(r-1)}p^x, \end{aligned}$$

where $p = [S(t)]^t$, so that $0 < p < 1$. Thus,

$$E(X^{(r)}) \leq rp\sum_{x=0}^{\infty} x^{(r-1)}p^x = rpqE(X_G^{(r-1)}) < \infty,$$

where X_G is a geometric random variable; □

(ix) X is NBU if and only if $\bar{\omega}_x(t) \geq \bar{\omega}(x)$. We refer to Section 2.9 for the definitions of the odds functions $\bar{\omega}(t)$ and $\bar{\omega}(x)$ from which it is easy to verify the above property. The result gives an alternative methodology to check the NBU property;

(x) If x is NBU, then the odds ratio for survival satisfies

$$w(x+t-1) \leq w(x-1)w(t-1), \quad x,t=0,1,2,\ldots$$

Proof. X is NBU $\Rightarrow S(x+t) \leq S(x)S(t)$

$$\Rightarrow 1-F(x+t-1) \leq (1-F(x-1))(1-F(t-1))$$
$$\Rightarrow F(x+t-1) \geq F(x-1)+F(t-1)-F(x-1)F(t-1)$$
$$= F(x-1)F(t-1), \tag{4.32}$$

since $F(x-1)+F(t-1) \geq 2\sqrt{F(x-1)F(t-1)} \geq 2F(x-1)F(t-1)$. Thus,

$$\Rightarrow \frac{1}{F(x+t-1)} \leq \frac{1}{F(x-1)F(t-1)}. \tag{4.33}$$

Also, since X is NBU,

$$S(x+1) \leq S(x)S(t). \tag{4.34}$$

Upon multiplying (4.33) and (4.34), we get the desired result; □

(xi) If X_1 and X_2 are NBU, their convolution is also NBU. Let $S_i(x)$ be a mixture of $S_i(x)$, $i=1,2,\ldots$, with each S_i NWU and no two distributions S_i and S_j cross. Then, $S(x)$ is NWU. For a proof see Pavlova et al. (2006).

(xii) Let $X_1, X_2, \ldots$ be a sequence of independent lifetimes each with NWU distributions. Assume N to be a positive integer-valued random variable. Then, if N is independent of X_i's, $S_N = X_1+X_2+\cdots+X_N$ is NWU.

Proof. Let $G(\cdot)$ be the survival function of S_N. We can write

$$\begin{aligned} G(x+t) &= P(S_N \geq x+t) \\ &= \sum_{n=1}^{\infty} P(S_n \geq x+t)P(N=n) \\ &\geq \sum_{n=1}^{\infty} P(S_n \geq x)P(S_n \geq t)P(N=n), \text{ by NWU property} \\ &\geq \sum_{n=1}^{\infty} P(S_n \geq x)P(N=n)\sum_{n=1}^{\infty} P(S_n \geq t)P(N=n) \\ &= G(x)G(t), \end{aligned}$$

so that S_n is NWU. □

There are some variants and modifications to the NBU (NWU) concept. The first of these stems from the motivation to obtain certain properties analogous to those in the continuous case.

Definition 4.17. A random variable X is NBU-1 (NWU-1) iff

$$H(x+t) \geq H(x) + H(t),$$

where $H(x) = \sum_{t=0}^{x-1} h(t)$ is the cumulative hazard rate.

By way of properties of NBU-1, we have

(i) IFRA-1 $\Rightarrow$ NBU-1;
(ii) neither NBU (NWU) $\Rightarrow$ NBU-1 (NWU-1) nor NBU-1 (NWU-1) $\Rightarrow$ NBU (NWU).

For a counter example in this regard, one may refer to Shaked et al. (1995).

Using the alternative failure rate $h_1(x)$ instead of $h(x)$ and applying the same technique, we can have another definition for NBU as

$$H_1(x+t) \geq H_1(x) + H_1(t), \quad x = 1, 2, \ldots, \tag{4.35}$$

where $H_1(x) = \sum_{t=0}^{x-1} h_1(t)$ is the alternative cumulative hazard rate. Since $H_1(x) = \log S(x)$, this leads to the NBU class. Thus, the above equation provides a characterization of NBU. It is easy to see that (4.35) can be expressed in terms of the alternative hazard rate as

$$\sum_{t=0}^{x-1} h_1(t) \leq \sum_{t=k+1}^{k+x-1} h_1(t) \quad \text{for all } k \geq 0,\ x > 0.$$

Eq. (4.35) also means that the function $H_1(\cdot)$ is super addictive.

Another associated concept is NBU-t_0 (NWU-t_0) which satisfies

$$S(t_0 + x) \leq S(t_0)S(x).$$

When a unit deteriorates over time, say upto t_0, to make the system more effective, replacements or repairs are undertaken. But such operations may not revert the unit to the original state of effectiveness or performance. The NBU-t_0 requires that the lifetime after t_0 is smaller compared to the original one in probability sense. Obviously,

$$\text{NBU} \Rightarrow \text{NBU-}t_0.$$

Instead of keeping t_0 fixed, we can think of the above behaviour beyond t_0 giving rise to the NBU*t_0 class defined by

$$S(x+t) \leq S(x)S(t), \quad x = 0, 1, \ldots, \ t \geq t_0.$$

We then have

$$\text{NBU} \Rightarrow \text{NBU} * t_0 \Rightarrow \text{NBU-}t_0.$$

A second approach is to extend the ordering implied in NBU. Recall that X is NBU (NWU) is equivalent to $X \leq_{\text{st}} (\geq_{\text{st}})(X - x|X > x)$. This means that

$$S(t) \geq \frac{S(x+t+1}{S(t+1)} = S_x(t),$$

which implies stochastic dominance of order one between X and X_x; see Section 4.4. A still larger class can be achieved by invoking stochastic dominance of order 2 of X over X_x.

Definition 4.18. We say that X is new better (worse) than used of order 2-NBU (2) (NWU (2)) iff

$$\sum_{y=0}^{t} S(y) \geq \sum_{y=0}^{t} S_x(y). \tag{4.36}$$

In terms of stochastic orders, Definition 4.18 is equivalent to $X \leq_{\text{icv}} X_x$. Thus, in NBU (2), the ordering $\leq_{\text{st}}$ in NBU is replaced by $\leq_{\text{icv}}$. Since $X \leq_{\text{st}} Y \Rightarrow \leq_{\text{icv}} Y$, we have

$$\text{NBU} \Rightarrow \text{NBU (2)}.$$

Using the survival function of X, (4.36) can also be written as

$$S(x+1)\sum_{y=0}^{t} S(y) \geq \sum_{y=x+1}^{t+x+1} S(y).$$

A second extension of the usual stochastic order in NBU similar to the above provides another class of life distributions called *new better than used in convex order* (NBUC).

Definition 4.19. The lifetime X belongs to the NBUC class iff

$$X \geq_{\text{icx}} (X - x|X > x) \text{ for all } x = 0, 1, 2, \ldots \tag{4.37}$$

When the inequality is reversed, we have the *new worse than used in convex order* or NWUC.

The NBUC (NWUC) concept involves a comparison of the lifetimes of a new unit with that of a used unit at time x with the same life distribution and says that the new unit performs better than an old unit in the sense of convex order. Thus, NBUC (NWUC) indicates positive (negative) ageing. Using the survival functions, we can write (4.37) as

$$\sum_{y=t}^{\infty} S(y) \geq \sum_{y=t}^{\infty} S_x(y) = \sum_{y=t}^{\infty} \frac{S(x+y+1)}{S(x+1)},$$

or

$$S(x+1) \sum_{y=t}^{\infty} S(y) \geq \sum_{y=t+x+1}^{\infty} S(y). \tag{4.38}$$

We can also interpret (4.38) as $X \geq_{\mathrm{st}} (X_E - x|X_E > x)$, a stochastic comparison between the original lifetime and the residual life of the equilibrium random variable. In this sense, it is a usual stochastic order, designating a first order stochastic dominance SD_1. On the other hand, (4.37) is a stochastic dominance $S\bar{D}_2$ of order two. Some properties of the NBUC class are as follows:

(1) DMRL $\Rightarrow$ NBUC

Proof. First, we show that

$$X \leq_{\mathrm{mrl}} Y \Leftrightarrow (X - x|X > x) \leq_{\mathrm{icx}} (Y - x|Y > x).$$

We have

$$\begin{aligned}(X - x|X > x) \leq_{\mathrm{icx}} (Y - x|Y > x) \Leftrightarrow & \frac{1}{S_X(x+1)} \sum_{y=t+x+1}^{\infty} S_X(y) \\ & \leq \frac{1}{S_Y(x+1)} \sum_{y=t+x+1}^{\infty} S_X(y) \\ & \Leftrightarrow X \leq_{\mathrm{mrl}} Y\end{aligned}$$

When $x = -1$,

$$X \leq_{\mathrm{mrl}} Y \Rightarrow X \leq_{\mathrm{icx}} Y \text{ and } X \geq_{\mathrm{mrl}} Y \Rightarrow X \geq_{\mathrm{icx}} Y.$$

Taking $Y = X - x|X > x$, we get

$$\begin{aligned} X \text{ DMRL } & \Leftrightarrow X \geq_{\mathrm{mrl}} (X - x|X > x) \\ & \Rightarrow X \geq_{\mathrm{icx}} (X - x|X > x) \Rightarrow X, \text{ is NBUC.} \end{aligned}$$ □

The converse need not be true, for example, the modified geometric law is DMRL, but not NBUC;

(2) NBU $\Rightarrow$ NBUC, since the usual stochastic order implies increasing convex order;

(3) Pavlova et al. (2006) have shown that if X_1 and X_2 are NBUE, their convolution is also NBUC. Further, a mixture of NWUC distributions is also NWUC provided that no two of the constituent distributions cross each other.

The next extension of the NBU class requires the notion of the probability generating function order. Recall from the definition in Section 4.2 that $X \le_{\text{pgf}} Y$ iff

$$\sum_{x=0}^{\infty} s^x f_X(x) \le \sum_{x=0}^{\infty} s^x f_Y(x), \quad 0 < s < 1.$$

Writing $f(x) = S(x) - S(x+1)$, it can be seen that the above definition is equivalent to

$$\sum_{x=1}^{\infty} s^x S_X(x) \le \sum_{x=1}^{\infty} s^x S_Y(x).$$

Elabatal and Ahsanullah (2012) have investigated the properties of the class NBU_{pg} (NWU_{pg}), based on $\le_{\text{pgf}}$ order.

Definition 4.20. The lifetime variable X belongs to the new better (worse) than used in pgf order, NBU_{pg} (NWU_{pg}) iff $X_x \le_{\text{pg}} (\ge_{\text{pg}})\ X$.

The above definition means that

$$\sum_{t=0}^{\infty} s^x S(x+t+1) \le S(x+1) \sum_{t=0}^{\infty} s^t S(t), \quad 0 < s < 1.$$

The properties of this class are summarized as follows:

(1) Let X and Y be independent lifetimes with $P(X > Y) > 0$. Then, the random residual life $X_Y = (X - Y | X > Y)$ satisfies

$$X \text{ is } \text{NBU}_{\text{pg}} \Leftrightarrow X_Y \le_{\text{pg}} X;$$

(2) Let $Y = g(X)$ be an increasing function and $p(X)$ be the inverse of $g : N \to N$, $N = (0, 1, 2, \ldots)$. If X is NBU_{pg} (NWU_{pg}) and $p(\cdot)$ is star-shaped (anti star-shaped), then Y is NBU_{pg} (NWU_{pg});

(3) Consider a renewal process of independent and identically distributed random variables with inter-arrival times X_i with distribution function $F(\cdot)$ with $F(0) = 0$, $S_k = \sum_{i=1}^{k} X_i$, $S_0 = 0$. The excess lifetime at t

$$r(t) = S_{N(t)+1} - t,$$

where $N(t)$ is the renewal counting process given by $N(t) = \sup\{n | S_n \leq t\}$, is such that $r(0)$ has distribution function $F(\cdot)$ or $r(0) =_{\text{st}} X$. Then, if $r(i)$ is decreasing in i for all $i = 0, 1, 2, \ldots, m$ in the pgf order,

(a) F is NBU_{pg},

(b) F is $\text{NBU}_{\text{pg}} \Rightarrow r(i) \leq_{\text{pg}} r(0)$.

In this connection, we note two results that involves stochastic comparison of X_Y and X through other stochastic orders, and these results are as follows.

Theorem 4.14. *X is NBU if and only if $X_Y \leq_{st} X$, where Y is any non negative discrete random variable.*

Proof. Observe that

$$S_{X_Y}(t) = \frac{\sum_{y=0}^{\infty} S_X(y+t+1) f_Y(y)}{\sum_{y=0}^{\infty} S_X(y+t+1) f_Y(y)}, \quad t > -1.$$

Also,

$$\begin{aligned} X \text{ is NBU} &\Leftrightarrow S_X(x+y) \leq S_X(x) S_X(y) \\ &\Rightarrow \sum_{y=0}^{\infty} S_X(x+y) f_Y(y) \leq \sum_{y=0}^{\infty} S_X(x) S_Y(y) f_Y(y) \\ &\Rightarrow \frac{\sum_{y=0}^{\infty} S_X(x+y) f_Y(y)}{\sum_{y=0}^{\infty} S_X(y) f_Y(y)} \leq S_X(x) \\ &\Rightarrow S_{X_Y}(x) \leq S_X(x) \Rightarrow X_Y \leq_{\text{st}} X. \end{aligned}$$

Conversely, if $X_Y \leq_{\text{st}} X_x$ for all Y, taking Y identically equal to t, $S_x(t) \leq S_x(x) \Rightarrow X$ is NBU. □

Theorem 4.15. *X is NBUC $\Leftrightarrow X_Y \leq_{icx} X$ for any Y.*

The proof is quite similar to that of Theorem 4.14.

4.5.2 New Better Than Used in Expectation

The concept of new better (worse) than used in expectation, NBUE (NWUE), requires that a used device at any age has a mean residual life smaller than that of a new device of the same kind. Thus, NBUE (NWUE) refers to positive (negative) ageing.

Definition 4.21. A discrete lifetime X belongs to the NBUE (NWUE) class if $m(x) \leq (\geq) \mu$.

The NBUE life distributions possess the following properties:

(1) If X is NBUE then $X_E \leq_{\text{st}} X$.

Proof. X is NBUE $\Rightarrow \frac{1}{S(x+1)} \sum_{t=x+1}^{\infty} S(t) \leq \mu$

$$\Rightarrow \sum_{t=x}^{\infty} \frac{S(t+1)}{\mu} \leq S(x+1) \leq S(x) \Rightarrow X_E \leq_{\text{st}} X.$$

One can also write the result as $X \geq_{st} X_E + 1$ if the second inequality in the last line is dispensed with; □

(2) X is NBUE $\Rightarrow X_E \leq_{\text{cx}} X_G$, where X_G is a geometric random variable with the same mean as X. The $\leq_{\text{cx}}$ order implies that $E(X) = E(Y)$ and $X \leq_{\text{icx}} Y$.

Proof.

$$\begin{aligned} X \text{ is NBUE} &\Leftrightarrow S_E(x) \leq S(x+1) = \mu f_E(x), \quad \mu = E(X) \\ &\Leftrightarrow h_E(x) \geq \frac{1}{\mu}. \end{aligned}$$

Thus,

$$S_E(x+1) = \prod_{t=0}^{x} (1 - h_E(t)) \leq \left(\frac{\mu - 1}{\mu} \right)^{x+1},$$

which is the same as $X_E \leq_{\text{cx}} X_G$, where X_G is geometric with mean $E(X_G) = \mu = E(X)$. □

Remark 4.5. Properties (1) and (2) offer different kinds of comparisons. Notice that the converse of (1) may not be true. For example if

$$f(x) = \begin{cases} \frac{2}{3}, & x = 0, \\ \frac{1}{3}, & x = 1, \end{cases}$$

$X_E \leq_{\text{st}} X$. But, $m(0) = \frac{2}{3} > \mu = \frac{1}{3}$, so that X is not NBUE;

(3) **(a)** X_E is NBU $\Rightarrow X$ is NBUC,
(b) NBUC $\Rightarrow$ NBUE.

Proof.

$$\begin{aligned} X_E \text{ is NBU } &\Rightarrow S_E(x+y) \leq S_E(x) S_E(y) \leq S(x+1) S_E(y) \\ &\Rightarrow X \text{ is NBUC}. \end{aligned}$$

This proves (a). To prove (b), we recall that X is NBUE if

$$S(x+1)\sum_{y=t}^{\infty} S(y) \geq \sum_{y=t+x+1}^{\infty} S(y),$$

or

$$S(x+1)S_E(t-1) \geq S_E(t+x).$$

Setting $t = 1$,

$$S(x+1) \geq S(x+1) \text{ for all } x,$$

which means that X is NBUE. This proves the result in (b); □

(4) Whenever X is NBUE, $\frac{\sigma^2}{\mu(\mu-1)} \leq 1$ (Nair et al., 2012b);

(5) The convolution of two NBUE distributions is NBUE while the mixture of NWUE distributions is NWUE provided that no two of them cross each other;

(6) There are some useful inequalities for the NBUE family of life distributions. We first observe that

$$\begin{aligned}
\mu_n = E(X^n) &= \sum_{0}^{\infty} x^n f(x) \\
&= \sum_{0}^{x} t^n f(t) + \sum_{x+1}^{\infty} t^n f(t) \\
&= F(x)E(X^n|X \leq x) + S(x+1)E(X^n|X > x) \\
&= (1 - S(x+1))r_n^*(x) + S(x+1)m_n^*(x),
\end{aligned} \tag{4.39}$$

where $r_n^*(x) = E(X^n|X \leq x)$ and $m_n^*(x) = E(X^n|X > x)$. Thus,

$$S(x+1) = \frac{\mu_n - r_n^*(x)}{m_n^*(x) - r_n^*(x)}. \tag{4.40}$$

From (4.39), we have

$$S(x+1) \leq \frac{\mu_n}{m_n^*(x)}, \quad x = 0, 1, 2, \ldots, \tag{4.41}$$

an upper bound for reliability. Similarly we can also have a lower bound. Since $S(x+1) \leq 1$,

$$(\mu_n - r_n^*(x)) \leq (m_n^*(x) - r_n^*(x)),$$

or

$$(\mu_n - r_n^*(x)) + r_n^*(x) - (m_n^*(x) \le 0.$$

Hence, we have

$$(\mu_n - x^n)(m_n^*(x) - r_n^*(x)) \le (\mu_n - r_n^*(x))(m_n^*(x) - x^n),$$

or

$$\frac{\mu_n - x^n}{m_n^*(x) - x^n} \le \frac{\mu_n - r_n^*(x)}{m_n^*(x) - r_n^*(x)} \quad \text{for } \mu_n \ge x^n.$$

This leads to

$$S(x+1) \ge \frac{\mu_n - x^n}{m_n^*(x) - x^n}, \quad x \le \mu_n^{\frac{1}{n}}. \tag{4.42}$$

If we work in a similar manner, a sharper inequality than (4.41), *viz.*,

$$S(x+1) \le \frac{\mu_n - r_n^*(x)}{x^n - m_n^*(x)}, \quad x > \mu_n^{\frac{1}{n}},$$

is obtained. In terms of the mean, (4.41) gives

$$S(x+1) \ge \frac{\mu - x}{m_1^*(x) - x} \ge \frac{\mu - x}{\mu}, \quad x \le \mu,$$

on using $m(x) = m_1^*(x) - x \le \mu$, since X is NBUE. Thus, we have

$$S(x+1) \ge \begin{cases} \frac{\mu - x}{\mu}, & x \le \mu, \\ 0, & x > \mu. \end{cases}$$

Similarly, when X is NWUE,

$$S(x+1) \le \frac{\mu}{\mu + x}, \quad x \ge 0.$$

Also when $n = 1$ in (4.41), for $x = 0, 1, 2, \ldots$, we have

$$S(x+1) \ge \frac{\mu}{m(x) - x};$$

(7) NBU (2) $\Rightarrow$ NBUE;
(8) DMRL $\Rightarrow$ NBUE;
(9) Neither DVRL $\Rightarrow$ NBUE nor NBUE $\Rightarrow$ DVRL, but both share the common property that $\frac{V(X)}{\mu(\mu-1)} \le 1$.

Another associated notion is NBU-(3) (NWU-3) defined by

$$\sum_{x=0}^{\infty}\sum_{y=0}^{x} S(x+y) \le (\ge) \sum_{x=0}^{\infty}\sum_{y=0}^{x} S(x)S(y)$$

which is a more general class than NBU-(2), so that NBU-(2) $\Rightarrow$ NBU-(3). Nofal (2012) has shown that the convolution of two independent NBU-(3) variables is again NBU-(3).

4.5.3 Harmonically New Better That Used in Expectation

The harmonically new better (worse) than used in expectation, HNBUE (HNWUE), class of distributions have the property that the harmonic mean residual life of a unit of age x is not better than the harmonic mean of a new unit.

Definition 4.22. We say that X has HNBUE property iff

$$\sum_{x=0}^{n}[E(X-x)|X>x]^{-1} \ge \frac{n}{E(X)}.$$

Also, X is HNWUE if the above inequality is reversed.

Suppose X_G is a geometric variable with parameter $p = \frac{1}{1+\mu}$. Then, the notion of HNBUE property defined above is equivalent to X being convex ordered with equal mean relative to X_G (Bhattacharjee, 2002). This leads to characterization of the geometric law as well.

Theorem 4.16. *Suppose X is HNBUE with mean μ. Then, X is geometric if and only if $C_X(x) = C_{X_G}(x)$, where $C(\cdot)$ stands for the coefficient of variation.*

The other useful characteristics of the HNBUE property are as follows:

(a) NBUE $\Rightarrow$ HNBUE;

(b) the survival function

$$\bar{H}(x) = \sum_{k=0}^{\infty} e^{-\lambda x}\frac{(\lambda x)^k}{k!}\bar{P}_k$$

is HNBUE if $\bar{P}_k$ has discrete HNBUE property;

(c) If $\bar{P}_k$ is discrete HNBUE and $\Lambda(x)$ is anti-starshaped, then

$$\bar{H}(x) = \sum_{k=0}^{\infty} e^{-\Lambda x}\frac{(\Lambda x)^k}{k!}\bar{P}_k$$

is HNBUE, where $\Lambda(x)$ is the mean value function of a non-homogeneous Poisson process. The results (b) and (c) are from Klefsjo (1981);

(d) The HNBUE class is preserved under the operation of convolution and HNBUE class is closed under the formation of mixtures of HNWUE distributions no two of which shall cross;

(e)

$$\text{HNBUE} \Leftrightarrow \sum_{t=x}^{\infty} S(t+1) \leq \mu \left(\frac{\mu-1}{\mu}\right)^{x+1}; \tag{4.43}$$

(f) Some moment relations for HNBUE distributions can be obtained. The factorial moments of the geometric distribution are

$$\mu_{(r)} = r!\frac{q}{p^r} = r!\mu(1+\mu)^{r-1}.$$

Also, defining the binomial moments

$$\beta_r = E\left[\binom{X+r}{r}\right] = \sum_{x=0}^{\infty} \binom{x+r}{r} f(x)$$

and the characteristic index

$$\alpha_r = 1 - \frac{\beta_r}{(1+\mu)^r},$$

we have in the geometric case

$$\beta_r^G = (1+\mu)^r \text{ and } \alpha_r^G = 0.$$

When X is HNBUE,

$$\beta_r \leq \beta_r^G,\ \mu_{(r)} \leq \mu_{(r)}^G,\ \alpha_r = 1 - \frac{\beta_r}{(1+\mu)^r},$$

and when X is HNWUE,

$$\beta_r \geq \beta_r^G,\ \mu_{(r)} \geq \mu_{(r)}^G,\ \alpha_r = \frac{\beta_r}{(1+\mu)^r} - 1.$$

Also, if X is HNWUE,

$$\alpha_r = \frac{\alpha_{r-1}}{1+\mu} \quad \text{or} \quad \alpha_r \geq \alpha_{r-1},\ r \geq 2.$$

For the sake of completion of the present review, we mention three more ageing classes which have continuous analogues. We say that X is

Definition 4.23. k-HNBUE (k-HNWUE) iff

$$\sum_{x=0}^{n}[E(X-x)^k|X>x]^{-1} \geq (\leq)\frac{n}{E(X^k)}.$$

It is easy to see that $k=1$ corresponds to the usual HNBUE.

Definition 4.24. HNBUE (3) (HNWUE (3)) iff

$$\sum_{u=x}^{\infty}\sum_{t=u}^{\infty} S(t+1) \leq (\geq)\mu^2\left(\frac{\mu-1}{\mu}\right)^{x+1}.$$

Being a stochastic dominance relation $\overline{SD}_3$ over $\overline{SD}_2$ given in (4.43) above, we have HNBUE $\Rightarrow$ HNBUE (3).

Definition 4.25. NBUCA (NWUCA) iff

$$\sum_{t=0}^{\infty}\sum_{y=t}^{\infty} S(y) \geq (\leq)\sum_{t=0}^{\infty}\sum_{y=t+x+1}^{\infty} S(y).$$

Obviously, NBUC $\Rightarrow$ NBUCA.

4.6 CLASSES BASED ON RELIABILITY FUNCTIONS IN REVERSED TIME

In view of the similarity between the usual reliability functions and those in reversed time, several studies have been made on the monotonicity of the latter. These include the nature of the reversed hazard rate, reversed mean residual life (mean inactivity time) and the reversed variance residual life (variance inactivity time) and some associated functions. Unlike the usual reliability functions, the reversed ones generally do not exhibit monotonicity in both directions (increasing, decreasing) and hence the functions in reversed time are not useful in describing ageing of devices. However, these classes of life distributions are employed in other areas to good effect, as will be seen in the sequel.

4.6.1 Decreasing Reversed Hazard Rate

Definition 4.26. A discrete lifetime X has decreasing reversed hazard rate (DRHR) if the sequence $\langle\lambda(x)\rangle$ is decreasing in x.

Some properties of DRHR models have been studied by Nanda and Sengupta (2005). They have pointed out that grouped data arising from samples of log-concave distributions are DRHR. In stress-strength models where the cause for failure is the stress accumulated through independent and identically distributed exponentials and the strength has log-concave distribution, the number of shocks resulting in failure is DRHR. Further instances where DRHR distributions arise in practice have also been given by Nanda and Sengupta (2005). Various equivalent conditions for X to be DRHR and other properties are presented below.

Theorem 4.17. *The distribution of X belongs to the DRHR class if and only if any one of the following conditions are satisfied:*

(a) $\frac{F(x+k)}{F(x)}$ *is decreasing for all* $x = 0, 1, \ldots,$

(b) *The distribution function* ${}_xF(t)$ *of* $(x - X|X \le x)$ *is stochastically increasing in* x.

The proofs of this and the next two theorems are given by Nanda and Sengupta (1999).

Theorem 4.18. *For* $0 < x < y$, *the following statements are equivalent:*

(a) *X DRHR,*

(b) $(x - X|X \le x) \le_{st} (y - X|X \le y)$,

(c) $(x - X|X \le x) \le_{hr} (y - X|X \le y)$,

(d) $F(x)$ *is log-concave,*

(e) $F^{p+q}(x) \ge F^p(\frac{q}{p}x)F^q(\frac{p}{q}x)$, *p, q being positive integers.*

Theorem 4.19. **(1)** *If* $f(x)$ *is decreasing in* x, *then X is DRHR;*

(2) *If X is DHR then X is DRHR;*

(3) *If* $\frac{f(x+t)}{f(x)}$ *is monotone in* x *for* $t = 0, 1, 2, \ldots,$ $x = 1, 2, \ldots,$ *then X is DRHR;*

(4) *If* $f(x)$ *has a finite number of modes, then there is an age* x_0 *which is less than or equal to the right most mode such that the distribution of the residual life at age greater than* x_0 *is DRHR;*

(5) *If a sequence of DRHR distributions converges to a limiting distribution, then that distribution in DRHR;*

(6) *If a parallel system consists of components with independent lives that are DRHR, then the system lifetime is DRHR;*

(7) *If a k-out-of-n system has components that have independent and identically distributed DRHR lifetimes, then the system lifetime is also DRHR;*

(8) *The DRHR class of distributions is closed with respect to convolution;*

(9) *The lifetime of a series system with independent DRHR component lives need not be DRHR. Similarly, the DRHR property is not preserved under the formation of mixtures;*

(10) *If X is DRHR, then*

$$S(x+1) \leq \begin{cases} 1 & \text{if } x < [\mu], \\ \max_{i=x,x+1,\dots}(1-a_i^{i-x}) & x \geq [\mu], \end{cases}$$

when $[\mu]$ *is the integer part of* μ *and* a_i *is the unique solution of*

$$\frac{(1-a_i^{i+1})}{1-a_i} = i+1-\mu.$$

From our earlier discussions on reserved hazard rate function $\lambda(x)$ in Section 2.6, we may recall that $\lambda(x)$ is a conditional probability and also that $\lambda(0) = 1$ irrespective of the distribution of X. Thus, it is evident that $\lambda(x)$ cannot be an increasing function when the support of X is set of all non-negative integers. However, if we exclude the value $X = 0$ and the support of X is finite, $\lambda(x)$ can be increasing. The distribution in (2.63) has $\lambda(x)$ non-decreasing for $x = 1, 2, \dots, b$; see Remark 2.10. A random variable X_b taking values in $(0, 1, 2, \dots, b)$, $b < \infty$, has increasing reversed hazard rate (IRHR) if $\langle\lambda(x)\rangle$, $x = 1, 2, \dots$, is a non-decreasing sequence such that $\lambda(x+1) \geq \lambda(x)$ for $x \geq 1$.

4.6.2 Increasing Mean Inactivity Time

Recall the definition of the mean inactivity time (reversed mean residual lifetime) as

$$r(x) = E(x-X|X<x) = \sum_{t=1}^{x} F(t-1)/F(x-1).$$

The inactivity time ${}_xX = (x-X|X<x)$ has distribution function

$$\begin{aligned} {}_xF(t) &= (x-X \leq t|X<x) \\ &= \frac{F(x-1)-F(x-t-1)}{F(x-1)}, \end{aligned}$$

and the corresponding probability mass function is

$${}_xf(t) = \frac{f(x-t)}{F(x-1)}.$$

Also,

$$P(x-X \geq t|X<x) = \frac{S(t-x)}{F(x-1)}.$$

Thus, the hazard rate of ${}_xX$ is $h(t-x)$.

Definition 4.27. A discrete lifetime X is said to have increasing mean inactivity time (IMIT) iff $\langle r(x) \rangle$ is an increasing sequence for $x = 1, 2, \ldots$.

Example 4.2. From Examples 2.18 and 2.19, it is clear that the geometric and uniform distributions are IMIT.

Theorem 4.20. *We have*

$$DRHR \Rightarrow IMIT.$$

Proof. Consider

$$r(x) = \frac{1}{F(x-1)} \sum_{t=1}^{x} F(t).$$

Since X is DRHR, $\frac{F(t-x)}{F(t)}$ is an increasing function of t, and so it follows that $r(x)$ is increasing in x and hence X is IMIT. □

The converse of the above result need not be true, without some conditions on X. The next theorem provides such a condition.

Theorem 4.21. *Assume that $r^*(x) = E(X|X < x)$ is concave. Then, IMIT ⇒ DRHR.*

Proof. When $r^*(x)$ is concave,

$$r^*(x+1) - 2r^*(x) + r^*(x-1) \leq 0.$$

Since $r(x) = x - r^*(x)$, the above condition simplifies to

$$r(x) - r(x-1) \leq r(x+1) - r(x),$$

or $r(x+1) - r(x)$ is an increasing function. Further, X is IMIT implies that $\frac{1}{r(x)}$ is decreasing. Thus,

$$\lambda(x) = \frac{1}{r(x)}(1 + r(x) - r(x+1))$$

is decreasing, and the proof is completed. □

Stochastic orders involving functions in reversed time are quite useful in establishing several theoretical results. We now present some important theorems in this connection.

Theorem 4.22.

$$\textit{(i)} \quad X \geq_{lr} Y \Rightarrow X \geq_{rhr} Y; \tag{4.44}$$

$$\textit{(ii)} \quad X \geq_{rhr} Y \Rightarrow X \geq_{st} Y. \tag{4.45}$$

Theorem 4.23.

$$X \leq_{rhr} Y \Rightarrow X \leq_{MIT} Y.$$

Proof. First, we show that

$$\frac{F_X(x)}{F_Y(x)} \text{ is decreasing } \Leftrightarrow X \leq_{\text{rhr}} Y. \tag{4.46}$$

We have

$$\begin{aligned}
\frac{F_X(x)}{F_Y(x)} \text{ decreasing } &\Leftrightarrow \frac{F_X(x+1)}{F_Y(x)} \geq \frac{F_X(x+1)}{F_Y(x+1)} \\
&\Leftrightarrow \frac{F_Y(x+1)}{F_Y(x)} - 1 \geq \frac{F_X(x+1)}{F_X(x)} - 1 \\
&\Leftrightarrow \frac{f_Y(x+1)}{F_Y(x)} \geq \frac{f_X(x+1)}{F_X(x)} \\
&\Leftrightarrow \lambda_Y(x+1)\frac{F_Y(x+1)}{F_Y(x)} \geq \lambda_X(x+1)\frac{F_X(x+1)}{F_X(x)} \\
&\Leftrightarrow \frac{\lambda_X(x+1)}{1-\lambda_X(x+1)} \geq \frac{\lambda_Y(x+1)}{1-\lambda_Y(x+1)} \\
&\Leftrightarrow \lambda_X(x+1) \geq \lambda_Y(x+1) \\
&\Leftrightarrow X \geq_{\text{rhr}} Y.
\end{aligned}$$

Now,

$$\begin{aligned}
X \leq_{\text{rhr}} Y &\Leftrightarrow \frac{F_Y(x)}{F_X(x)} \text{ is decreasing in } x \\
&\Leftrightarrow \frac{F_X(x+1)}{F_X(x)} \geq \frac{F_Y(x+1)}{F_Y(x)} \\
&\Leftrightarrow \frac{\sum_{t=1}^{x} F_X(t-1)}{F_X(x-1)} \geq \frac{\sum_{t=1}^{x} F_Y(t-1)}{F_Y(x-1)} \\
&\Leftrightarrow r_X(x) \geq r_Y(x) \Leftrightarrow X \leq_{\text{MIT}} Y.
\end{aligned}$$

□

Theorem 4.24. *If $\frac{r_X(x)}{r_Y(x)}$ is an increasing function of x, then $X \leq_{MIT} Y \Rightarrow X \leq_{rhr} Y$.*

Proof. Since

$$X \leq_{\text{MIT}} Y, \text{ we have } \frac{1}{r_X(x)} \leq \frac{1}{r_Y(x)} \tag{4.47}$$

TABLE 4.3 Increasing reversed hazard function

X	0	1	2	3	4
$f(x)$	$\frac{3}{16}$	$\frac{1}{16}$	$\frac{2}{16}$	$\frac{3}{16}$	$\frac{7}{16}$
$\lambda(x)$	1	$\frac{1}{4}$	$\frac{1}{3}$	$\frac{1}{3}$	$\frac{7}{16}$
$r(x)$	0	1	$\frac{7}{4}$	$\frac{13}{6}$	$\frac{22}{9}$

TABLE 4.4 Decreasing reversed mean residual life

X	0	1	2	3
$f(x)$	$\frac{36}{324}$	$\frac{4}{324}$	$\frac{91}{324}$	$\frac{193}{324}$
$g(x)$	0	1	$\frac{9}{5}$	$\frac{55}{36}$
$\lambda(x)$	1	$\frac{1}{10}$	$\frac{91}{131}$	$\frac{193}{324}$

Then, $\frac{r_X(x)}{r_Y(x)}$ is increasing

$$\Leftrightarrow \frac{r_X(x+1)}{r_X(x)} \geq \frac{r_Y(x+1)}{r_Y(x)}$$
$$\Leftrightarrow \frac{r_X(x)-r_X(x+1)}{r_X(x)} \leq \frac{r_Y(x)-r_Y(x+1)}{r_Y(x)}$$
$$\Leftrightarrow \frac{r_X(x)-r_X(x+1)+1}{r_X(x)} \leq \frac{r_Y(x)-r_Y(x+1)+1}{r_Y(x)} \quad \text{using (4.47)}$$
$$\Leftrightarrow \lambda_X(x) \leq \lambda_Y(y) \Leftrightarrow X \leq_{rhr} Y.$$

□

Theorem 4.25. *We have*

$$X \leq_{hr\,x} X \Leftrightarrow X \text{ is IFR.}$$

To examine whether MIT can be decreasing, we observe that $r(1) = 1$ for all distributions and

$$r(x) = \frac{1}{F(x-1)}[F(0) + \cdots + F(x-1)] > 1$$

for all $x = 2, 3, \ldots$. Hence, $r(x)$ cannot be a decreasing function for all x, but can be decreasing for $x = 2, 3, \ldots, b$. Thus, we define X to have decreasing mean inactivity time (DMIT) iff $\langle r(x) \rangle$ is decreasing for $x = 2, 3, \ldots, b < \infty$. See the following tables that illustrate this fact. Tables 4.3 and 4.4 give examples of increasing reversed hazard and decreasing reversed mean residual life, respectively. Table 4.5 exhibits some distributions for which these two reliability functions have simple closed-forms that can be used for modelling as well as for constructing examples and counter examples.

Remark. It is informative to note from the above tables that the functions in reversed time do not manifest the same behaviour as the hazard rate and mean residual life. For example, when $h(x)$ is increasing (decreasing) $m(x)$ is decreasing (increasing) whereas in Table 4.3, both $\lambda(x)$ and $r(x)$ are increasing. Likewise, when $r(x)$ is decreasing (excluding the points $x = 0$ and $x = 1$, which has the same value for all distributions), $\lambda(x)$ does not have monotone behaviour.

4.6.3 Increasing Variance Inactivity Time

The reversed variance residual life or the variance inactivity time was defined in Section 2.8 as

$$\begin{aligned} v(x) &= E[(x-X)^2|X<x] - r^2(x) \\ &= \frac{2}{F(x-1)}\sum_{t=1}^{x}\sum_{u=1}^{t} F(u-1) - r(x)(r(x)+1). \end{aligned}$$

In this section, we discuss classes of life distributions for which $v(x)$ is monotonic.

Definition 4.28. A lifetime random variable X is said to have increasing variance inactivity time (IVIT) iff $\langle v(x) \rangle$ is an increasing sequence.

The uniform distribution, arithmetic distribution and the distribution in (2.80) belong to the IVIT class. Some properties of the IVIT class are as follows:

(1) X is IVIT iff $v(x) \le r(x)(r(x+1)-1)$. This result is an immediate consequence of (2.73);

(2) IMIT $\Rightarrow$ IVIT.

Proof. We have

$$\begin{aligned} & v(x) - r(x)(r(x+1)-1) \\ &= \frac{2}{F(x-1)}\sum_{t=1}^{x}\sum_{u=1}^{t} F(t-1) - r(x)(r(x)+1) - r(x)(r(x+1)-1) \end{aligned} \tag{4.48}$$

$$= \frac{2}{F(x-1)}\sum_{t=1}^{x} F(t-1)r(t-1) - r(x)(r(x)+r(x+1))$$

TABLE 4.5 Reversed hazard and reversed mean residual life functions

Distribution	$F(x)$	$\lambda(x)$	$r(x)$
reversed geometric	$(1+c)^{x-b},\ x=0,1,\ldots,b$	$\frac{c}{1+c}$	$\frac{c}{1+c}[1-(1+c)^{-x}]$
geometric	$1-q^{x+1},\ x=0,1,\ldots$	$\frac{pq^x}{1-q^{x+1}}$	$\frac{xp-q(1-q^{x+1})}{p(1-q^x)}$
uniform	$\frac{x}{b},\ x=1,2,\ldots,b$	$\frac{1}{x}$	$\frac{x}{2}$
constant RMRL	$\begin{cases} c^{-1}(1+c)^{1-b}, & x=0 \\ (1+c)^{x-b}, & x=1,2,\ldots,b \end{cases}$	$\begin{cases} 1, & x=0 \\ \frac{c-1}{c}, & x=1 \\ \frac{c}{c+1}, & x=2,\ldots,b \end{cases}$	$\frac{c}{1+c}$
arithmetic (negative skew)	$\frac{x(x+1)}{b(b+1)},\ x=1,2,\ldots,b$	$\frac{2}{x+1}$	$\frac{x+1}{3}$
arithmetic (positive skew)	$\frac{x(2b-x+1)}{b(b+1)},\ x=1,2,\ldots,b$ $\frac{\binom{\theta+x-1}{x-1}}{\binom{\theta+b-1}{x-1}},\ x=1,2,\ldots,b$	$\frac{2(b-x+1)}{2b-x+1}$ $\frac{\theta}{x+\theta-1}$	$\frac{2(x-1)(b-x+1)}{3}$ $\frac{x+\theta-1}{\theta}$
Lindley	$1-\frac{\lambda^{x+1}(1-(x+2)\log\lambda)}{1-\log\lambda}$	$\frac{\lambda^x[\lambda\log\lambda+(\lambda-1)\log\lambda-1]}{1-\frac{\lambda^{x+1}[1-(x+2)\log\lambda]}{1-\log\lambda}}$	$x-\frac{\lambda(1-\lambda^x)}{1-\lambda}$ $-\frac{\log\lambda}{(1-\lambda^2)}[(1-\lambda^x)$ $\lambda(2-\lambda)+x(1-\lambda)\lambda^x$

$$= \frac{2}{F(x-1)} \sum_{t=1}^{x} F(t-1)(r(t-1) - r(x))$$

$$+ \frac{2}{F(x-1)} \sum_{t=1}^{x} F(t-1)r(x) - r(x)(r(x+1) + r(x))$$

$$= \frac{2}{F(x-1)} \sum_{t=1}^{x} F(t-1)(r(t-1) - r(x)) + 2r^2(x)$$

$$- r(x)(r(x) + r(x-1))$$

$$= \frac{2}{F(x-1)} \sum F(t-1)(r(t-1) - r(x)) + r(x)(r(x) - r(x+1)).$$

When X is IMIT, both terms on the right hand side are negative and hence by (4.48) above, X is IVIT; □

(3) If X and Y are two lifetimes, X is said to be smaller than Y in variance inactivity time order, $X \leq_{\mathrm{VIT}} Y$, iff $v_X(x) \geq v_Y(x)$ for all x. This order is connected to the MIT order by

$$X \leq_{\mathrm{MIT}} Y \Rightarrow X \leq_{\mathrm{VIT}} Y.$$

In the spirit of our discussions so far, the classes of life distributions and stochastic orders using functions of reversed time, other classes and stochastic orders may be defined. Since the methodology for working interrelationships and properties are similar, we confine ourselves to giving just a few definitions and implications only.

- We say that X has decreasing reversed hazard rate average (DRHRA) iff $\frac{1}{x}\sum_{t=0}^{x-1} \lambda(t)$ is decreasing in x.

Theorem 4.26. *We have*

$$DRHR \Rightarrow DRHRA.$$

Proof. Let $\bar{A}(x) = \frac{1}{x}\sum_{t=0}^{x-1} \lambda(t)$. Then,

$$A(x+1) - A(x) = \frac{1}{x+1} \sum_{0}^{x} \Lambda(t) - \frac{1}{x} \sum_{0}^{x-1} \lambda(t)$$

$$= \frac{1}{x(x+1)} \left[x\lambda(x) - \sum_{0}^{x-1} \lambda(t) \right]$$

$$= \frac{1}{x(x+1)} \sum_{t=0}^{x-1} [\lambda(x) - \lambda(t)].$$

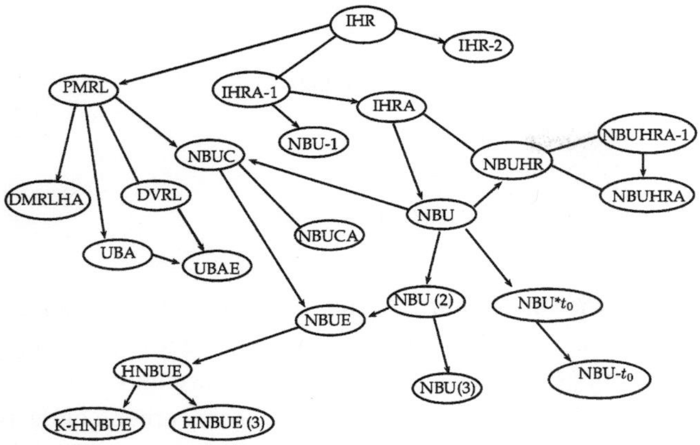

FIGURE 4.2 Implications among the ageing concepts.

Since X is DRHR, $\lambda(x) \leq \lambda(t)$ for $t = 0, 1, \ldots, x-1$ and so $A(x+1) \leq A(x)$; □

- X has decreasing reversed hazard rate average-1 (DRHRA-1) iff $\bar{A}_1(x) = \frac{1}{x}\sum_{t=0}^{x-1} \lambda_1(t)$ is decreasing in x. We have

$$\text{DRHR} \Rightarrow \text{DRHRA-1}$$

and

$$\text{DRHRA-1} \Rightarrow \text{DRHRA};$$

- X has reversed NWU property (RNBU) iff

$$F(x+y)F(0) \leq F(x)F(y), \; x, y = 0, 1, 2, \ldots \tag{4.49}$$

It means that the inactivity time at the beginning is better than that at the end. The random variable X satisfies the reserved NBU (RNBU) property if the inequality in (4.48) is reversed. Both RNBU and RNWU properties are satisfied by the reversed geometric distribution, and

$$\text{DRHRA} \Rightarrow \text{RNWU}.$$

Various implications we have noticed in the discussions so far are consolidated and presented in Figs 4.2 and 4.3 for ease of reference.

4.7 AGEING PROPERTIES FOR WEIGHTED DISTRIBUTIONS

The concept of weighted distributions and their characteristics with special reference to two important special cases viz. length-biased and equilibrium models,

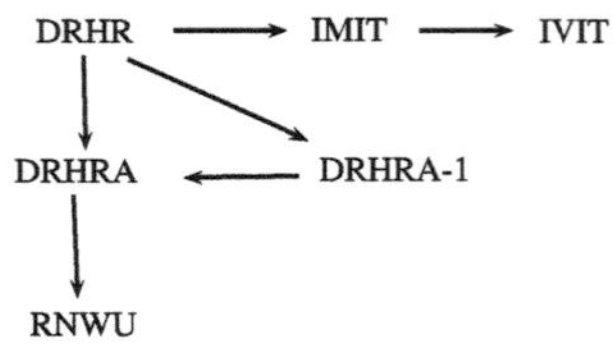

FIGURE 4.3 Implications among concepts in reverse time.

were discussed earlier in Section 1.9. This was followed in Section 2.11 by the derivation of various reliability functions and their inter relationships for such models. The relationships between the original distribution and its weighted version give interesting connection between the ageing criteria of the two distributions. Retaining our notations in Section 2.11, we have the following results.

Theorem 4.27. *If the weight function $w(x)$ is increasing, then $X \leq_{st} X_w$, $X \geq_{hr} X_w$, and also $X \leq_{mrl} X_w$, and the reverse inequalities hold when $w(x)$ is decreasing when the monotonicity condition on $w(x)$ is dropped,*

(a) $\frac{S(x)}{S_w(x)}$ *is non-decreasing* $\Leftrightarrow X \leq_{hr} X_w$;

(b) $\frac{\sum S_w(t)}{\sum S(t)}$ *is non-decreasing* $\Leftrightarrow X \leq_{mrl} X_w$.

Proof. The probability mass function of X_W is

$$f_w(x) = \frac{w(x)f(x)}{E\,w(X)},$$

and the survival function is

$$\begin{aligned} S_w(x) &= \sum_{t=x}^{\infty} \frac{w(x)f(x)}{Ew(x)} \\ &= \frac{w(x)S(x)}{Ew(X)} + \sum_{t=x+1}^{\infty} \frac{(w(t)-w(t-1))}{E\,w(x)} S(t) \\ &= \frac{S(x)(w(x)+M(x))}{\mu_w},\ M(x) = \sum_{t=x+1}^{\infty} (w(t)-w(t-1))S(t) \end{aligned} \tag{4.50}$$

and $\mu_w = Ew(X)$. The hazard rate function of X_w is

$$h_w = \frac{w(x)h(x)}{w(x)+M(x)}. \tag{4.51}$$

The relationships $S_w(x) \geq S(x)$ and $h_w(x) \leq h(x)$ are evident from (4.50) and (4.51). Observe that when $w(\cdot)$ is increasing, $M(x) \geq 0$. When $\frac{S_w}{S}$ is increasing

$w(x) + M(x)$ is increasing, so that $h_w(x) \le h(x)$ which proves (a). The mean residual life function of the weighted distribution is

$$m_w(x) = \sum_{t=x+1}^{\infty} \frac{S(t)(w(t) + M(t))}{S(x+1)(w(x+1) + M(x+1))} \tag{4.52}$$

From (4.52), we obtain (b). □

Theorem 4.28. *If X is IFR (DFR) and $w(x)$ is increasing and concave (decreasing and convex), then X_w is IFR (DFR).*

Proof. From (4.51), we have

$$\begin{aligned}
h_w(x+1) - h_w(x) &= \frac{w(x+1)h(x+1)}{w(x+1) + M(x+1)} - \frac{w(x)h(x)}{w(x) + M(x)} \\
&= [(w(x+1) + M(x+1))(w(x) + M(x))]^{-1} \\
&\quad [(M(x) - M(x+1))(h(w(x+1)h(x+1) + w(x)h(x))] \\
&\quad + (h(x+1) - h(x))w(x)w(x+1) \\
&\quad - w(x+1)h(x+1)M(x+1) - w(x)h(x)M(x) \\
&\ge 0,
\end{aligned}$$

by the assumptions of the theorem. □

Theorem 4.29. *(a) If X has increasing likelihood ratio, then X_w also has increasing likelihood ratio;*
(b) X is DMRL $\Rightarrow X_w$ is DMRL;
(c) $X \le_{hr} Y$ and $w(\cdot)$ is increasing in x, $X_w \le_{hr} Y_w$ and

$$B_{X_w}(x) \le B_{Y_w}(x),$$

where $B(\cdot)$ is the mean residual life in harmonic average defined in (4.18).

There are more elaborate results in the special case of equilibrium distributions and their higher orders. As before, let Y_n denote the equilibrium random variable of order n so that $Y_0 = X$ and $Y_1 = X_E$. Then the following ageing properties have been obtained by Nair et al. (2012b):

(a) If Y_n is IFR (DFR) then Y_{n+1} is IFR (DFR) for every n. But, the converse need not be true;
(b) X is DMRL implies that Y_n is DMRL;
(c) Y_n is IFR (2) $\Rightarrow Y_{n+1}$ is IFR $\Rightarrow Y_n$ is DMRL;
(d) Y_n is IFR $\Rightarrow Y_{n-1}$ is DMRL $\Leftrightarrow Y_{n-2}$ is DVRL, $n = 2, 3, \ldots$;

(e) X is DVRL $\Rightarrow Y_n$ is DVRL;
(f) X is DMRL $\Rightarrow Y_n$ is UBA; Y_n is UBA $\Rightarrow Y_{n+1}$ is UBA;
(g) Y_n is UBA $\Leftrightarrow Y_{n-1}$ is UBAE;
(h) Y_n is IFR $\Rightarrow Y_{n-1}$ is DMRL $\Rightarrow Y_n$ is UBA;
(i) Y_n is NBU $\Rightarrow Y_{n-1}$ is NBUC;
(j) Y_n is NBU $\Rightarrow Y_{n-1}$ is NBUE;
(k) Y_{n-1} NBUE $\Rightarrow Y_n$ is NBUFR;
(l) Y_{n-1} NBUC $\Rightarrow Y_n$ is NBUFR;
(m) If X is IFRA, then $S_E(x)/S(x)$ is decreasing;
(n) If X is NBUE and $E(X_E) = E(X)$, then X is geometric and conversely;
(o) If the distribution of X belongs to the modified power series family and $E(X_E) = E(X)$, then X is geometric.

The length-biased models are straightforward special case of X_w when $w(x) = x$ and therefore inherits all the general properties of the weighted distribution. Finally, we view the residual life distribution of $X_x = (X - x|X > x)$ also as a weighted model with $w(x) = \frac{f(x+t)}{f(x)}$. As noted earlier, the hazard and mean residual life functions of X_x are

$$h_x(t) = h(x+t+1) \text{ and } m_x(t) = m(x+t+1).$$

Comparison of ageing concepts of X and X_x are of interest in reliability analysis. We note that X is IFR (DMRL) if X_x is NBU (NBUE) and both IFR and DMRL of X ensures that $\sigma_x^2 = V(X_x)$ is increasing. When the equilibrium distributions of two random variables X and Y need a comparison, it is helpful to note that

(i) $X \geq_{\text{hr}} Y \Rightarrow X_E \geq_{\text{hr}} Y_E \Rightarrow X_E \geq_{\text{st}} Y_E$;
(ii) $X \leq_{\text{mrl}} Y \Rightarrow X_E \geq_{\text{mrl}} Y_E$ and if $\frac{m_{E,x}(x)}{m_{E,y}(x)}$ is increasing, $X_E \leq_{\text{mrl}} Y_E \Rightarrow X \leq_{\text{mrl}} Y$;
(iii) $X_E \leq_{\text{st}} Y_E \Leftrightarrow X \leq_{\text{hmrl}} Y$.

Similar results exist in the case of functions in reversed time although they do not have interpretations in terms of ageing. We state some of these below:

(1) If X is DRHR and $w(x)$ is decreasing and log-concave then X_w is DRHR;
(2) $\leq_{\text{rhr}} Y \Rightarrow X_{w_1} \leq_{\text{rhr}} Y_{w_2}$ provided that $w_1(x)$ and $w_2(x)$ the weight functions in respect to X and Y, are such that $w_2(x)/w_1(x)$ is decreasing;
(3) X is IMIT $\Rightarrow X_w$ is IMIT if $E[w(X)|X < x]$ is decreasing and log-convex;
(4) X is IMIT $\Rightarrow X_w$ is DRHR if $w(x)$ is decreasing and log-concave and $r(x)$ is log-convex;
(5) $X \leq_{\text{MIT}} Y \Rightarrow X_{w_1} \leq_{\text{MIT}} Y_{w_2}$ if $w_1(\cdot)$ is decreasing and $w_2(\cdot)$ is increasing.

4.8 RELATIVE AGEING

There are many practical situations when one has to compare the reliabilities of more than one device. For example, when the same kind of device is produced by several manufactures, the choice has to be made with reference to the ageing patterns of the competing devices. Relative ageing concepts specify which of two devices age faster by comparing the two on the basis of some ageing criterion. Also there are models in which the nature of ageing depends on the parameter values, for example, the Weibull distribution. Thus an analysis that reveals the relationship between the ageing property and the model parameters is required. This becomes more important in cases where there are critical values of the parameters which partitions one ageing feature into another which is distinctly different.

4.8.1 Ordering by Ageing Concepts

When we deal with two devices to ascertain which one ageing more positively than the other, ageing criteria like hazard rate, mean residual life, etc. come into the picture. Let X and Y denote the lifetimes of two devices where ageing behaviours need comparison.

Definition 4.29. The random variable X is ageing faster than Y in hazard rate written as $X \leq_{\text{IHR}} Y$ iff $\frac{h_X(x)}{h_Y(x)}$ is increasing in x, provided $h_Y(x) \neq 0$.

If the hazard rates of X and Y are such that $\frac{h_X(x)}{h_Y(x)}$ is a constant (increasing, decreasing), we may say that the device with lifetime X age at the same rate (faster, slower) as the device with lifetime Y. While studying $X \leq_{\text{IHR}} Y$, three types of phenomenon can occur:
(a) $h_X(x) \geq h_Y(X)$ for all x,
(b) $h_X(x)$ crosses $h_Y(x)$ from below, and
(c) $h_X(x) \leq h_Y(x)$ for all x.

Example 4.3.

(1) X has geometric distribution $G(\frac{3}{4})$ and Y has Waring distribution with hazard rate $h_Y(x) = \frac{2}{3+x}$, $x = 0, 1, 2, \ldots$. Then, $X \leq_{\text{IHR}} Y$ and $h_X(x) \geq h_Y(x)$;

(2) X has negative hypergeometric law with $h_X(x) = \frac{k}{k+n-x}$ and $h_Y(x) = p$, the geometric hazard rate. Then, $\frac{h_X(x)}{h_Y(x)} = \frac{k}{(k+n-x)p}$ is increasing for all k and $0 < p < 1$. Further,

$$h_X(x) - h_Y(x) = \frac{k}{k+n-x} - p = k(1-p) + (n-x) \leq 0$$

holds for $x \leq n$, the support of X;

(3) Finally, $h_X = \frac{2}{3+x}$ as in Example (1) and $h_Y(x) = \frac{1}{6-x}$, corresponding to the uniform distribution in $[0,5]$. In this case, h_X and h_Y crosses at $x = 3$ and $X \leq_{\text{IHR}} Y$.

If we look at the binary relation $\leq_{\text{IHR}}$ more closely, it says that the lifetime X is less than lifetime Y if X is more IHR than Y. Also,

$$X \leq_{\text{IHR}} Y \Leftrightarrow \frac{h_X(x+1)}{h_Y(x)} \geq \frac{h_X(x)}{h_Y(x)} \Leftrightarrow \frac{h_X(x+1)}{h_X(x)} \geq \frac{h_Y(x+1)}{h_Y(x)}$$

and so,

$$X \leq_{\text{IHR}} Y \text{ and } Y \leq_{\text{IHR}} Z \Rightarrow \frac{h_X(x+1)}{h_Y(x)} \geq \frac{h_Z(x+1)}{h_Z(x)}$$
$$\Rightarrow X \leq_{\text{IHR}} Z.$$

Moreover,

$$X \leq_{\text{IHR}} Y \text{ and } Y \leq_{\text{IHR}} X \Rightarrow h_X(x) = c h_Y(x)$$

for some constant $c > 0$. Thus, the ordering $\leq_{\text{IHR}}$ defines a partial order among equivalence classes generated by the ratio $\frac{h_Y}{h_X}$. Further, $X =_{\text{IHR}} Z$ corresponds to the equivalence class in which the proportional hazard rate model for discrete lifetimes hold. The partial order $\leq_{\text{IHR}}$ can be used to define the IHR concept:

$$X \text{ is IHR(DHR) } \Leftrightarrow X \leq_{\text{IHR}} X_G,$$

where X_G is the geometric random variable. When $x f_X(x)$ is increasing and $x f_Y(x)$ is decreasing, the ordering $X \leq_{\text{IHR}} Y$ holds.

When the hazard rate is replaced by other ageing concepts, we can have stochastic orders representing relative ageing. We consider only some important representations in what follows.

Definition 4.30. A random lifetime X is ageing faster than another random lifetime Y in hazard rate average, written as $X \leq_{\text{IHRA}} Y$

$$\text{iff } \frac{\sum_0^{x-1} h_X(t)}{\sum_0^{x-1} h_Y(t)}$$

is increasing in x.

We interpret $X \leq_{\text{IHRA}} Y$ by saying that a device with, lifetime X has lesser life than the device with, lifetime Y iff X is more IHRA than Y. Since X has greater IHRA, it ages more positively than Y. Also

$$X \text{ is IHRA (DHRA)} \Leftrightarrow X \leq_{\text{IHRA}} X_G,$$
$$X \leq_{\text{IHR}} Y \Rightarrow X \leq_{\text{IHRA}} Y.$$

Definition 4.31. We say that X is ageing faster than Y in mean residual life, $X \leq_{\text{DMRL}} Y$, iff $\frac{m_X(x)}{m_Y(x)}$ is decreasing in x.

As before, one can show that $\leq_{\text{DMRL}}$ defines a partial order which means that X has less decreasing mean residual life than Y so that X is more positively ageing than Y. Like the other orders, $\leq_{\text{DMRL}}$ satisfies

$$X \text{ is DMRL (IMRL)} \Leftrightarrow X \leq_{\text{DMRL}} X_G.$$

Similar orders can be defined using the concepts of NBU, NBUE, NBUHR, and so on.

4.8.2 Specific Ageing Factor

Consider two devices or systems whose lifetimes follow the same distribution with survival function $S(x)$. One of the devices is new and the other is aged x units. Then, $S(x)$ is the probability that the new device worked at least age x and $\frac{S(x+y)}{S(x)}$ is the probability that the older device aged y survives the same duration. Following Bryson and Siddiqui (1969) in the continuous case, we define the specific ageing factor

$$A(x, y) = \frac{S(x)S(y)}{S(x+y)}$$

which compares the survival probabilities of the two devices. It seems that the potential of the specific ageing factor in describing ageing patterns and relative ageing have not been fully exploited. Note that $A(x, y) \geq 1$ (≤ 1) if and only if the older system is more (less) aged in the sense that it has lesser (greater) chance of surviving the duration x than the new one. The relationships of $A(x, y)$ with some of the ageing criteria are detailed below.

Theorem 4.30. *The random variable X has increasing (decreasing) hazard rate if and only if $A(x, y)$ is increasing (decreasing) in y.*

Proof. We have

$$\begin{aligned}
A(x, y) \text{ is increasing} &\Leftrightarrow A(x, y+1) - A(x, y) \geq 0 \\
&\Leftrightarrow S(y+1)S(x+y) \geq S(y)S(x+y+1) \\
&\Leftrightarrow \frac{S(y+1)}{S(y)} \geq \frac{S(x+y+1)}{S(x+y)} \\
&\Leftrightarrow 1 - h(x) \geq 1 - h(x+y) \\
&\Leftrightarrow X \text{ is IHR.}
\end{aligned}$$

By reversing the above steps, we have the converse. The proof of the DHR case is analogous. □

Theorem 4.31.

(i) *X is NBU (NWU)* $\Leftrightarrow A(x, y) \geq (\leq) 1$ *for all* $y = 0, 1, 2, \ldots$;
(ii) *X is NBU-*y_0 *(NWU-*y_0*)* $\Leftrightarrow A(x, y_0) \geq (\leq) 1$;
(iii) *X is NBU**y_0 *(NWU**y_0*)* $\Leftrightarrow A(x, y) \geq (\leq 1)$, $x = 0, 1, 2, \ldots, y \geq y_0$.

The implications to IHRA, DMRL, etc., are evident from Theorem 4.29 and to NBUE, HNBUE, etc., are evident from Theorem 4.31.

A second approach to quantify relative ageing of a new device and an old device with the same life distribution is to consider the identity

$$P(X \geq g(x, y) + x | X \geq x) = P(X \geq y) \tag{4.53}$$

for all $x, y = 0, 1, 2, \ldots$. In the above representation, $g(x, y)$ represents the expected lifelength of an old unit that has survived till x which has the same probability of survival (reliability) as a new unit of age y. Eq. (4.53) can be written as

$$S(x + g(x, y)) = S(x)S(y),$$

or

$$H_1(x + g(x, y)) = H_1(x) + H_1(y), \tag{4.54}$$

where $H_1(\cdot)$ is the alternative cumulative hazard rate defined in (2.12). We can now write

$$M(x, y) = \frac{g(x, y)}{y}$$

and interpret it as the rate at which an old unit is losing or gaining life in relation to a new unit with identical life distribution. Thus, M provides a measure of the effect of ageing of a device. We call $M(x, y)$ as the *relative ageing factor*.

The expressions of $M(x, y)$ of some discrete life distributions are presented in Table 4.6.

We now discuss the properties of $M(x, y)$ and its role as a measure of relative ageing of a device:

(i) Since $H_1(0) = 0$,

$$H_1(g(x, 0)) = H_1(x) + H_1(0)$$

implies $g(x, 0) = x$ and similarly $g(0, y) = y$. Also, $g(0, 0) = 0$;

(ii) When $M(x, y) = 1$, $g(x, y) = y$ so that (4.52) becomes the lack of memory property. Thus, $M(x, y) = 1$ represents the non-ageing property and conversely, characterizing the geometric law;

TABLE 4.6 Relative ageing factors for some life distributions

Distribution	$S(x)$	$g(x, y)$	$M(x, y)$
geometric	q^x	y	1
Weibull I	q^{x^β}	$(x^\beta + y^\beta)^{\frac{1}{\beta}} - x$	$\frac{(x^\beta + y^\beta)^{\frac{1}{\beta}} - x}{y}$
Lomax	$(1 + \frac{x}{\alpha})^c$, $\alpha > 0, c > 0$	$y(1 + \frac{x}{\alpha})$	$1 + \frac{x}{\alpha}$
Burr	$(1 + x^\lambda)^{-1}$	$(x^\lambda + y^\lambda + x^\lambda y^\lambda)^{\frac{1}{\lambda}} - x$	$\frac{(x^\lambda + y^\lambda + x^\alpha > 0 y^\lambda)^{\frac{1}{\lambda}} - x}{y}$
half-logistic	$2(1 + e^{\frac{x}{\sigma}})^{-1}$ $x = 0, 1, \ldots, \sigma > 0$	$\sigma \log[\frac{1}{2}(1 + e^{\frac{x}{\sigma}})(1 + e^{\frac{y}{\sigma}}) - 1]$	$y^{-\frac{1}{\sigma}} \log[\frac{1}{2}(1 + e^{\frac{x}{\sigma}})(1 + e^{\frac{y}{\sigma}}) - 1] - x$
re-scaled beta	$(1 - \frac{x}{R})^c$, $x = 0, 1, \ldots, R$	$y(1 - \frac{x}{R})$	$(1 - \frac{x}{R})$
power (uniform $\alpha = 1$)	$1 - (\frac{x}{\beta})^\alpha$, $x = 1, 2, \ldots, \beta$	$[[x^\alpha + y^\alpha - (\frac{x}{\beta})^\alpha y^\alpha]^{\frac{1}{\alpha}} - x]y^{-1}$ $(1 - \frac{x}{\beta})$	

(iii) $M(x, y)$ provides a measure of the extent of ageing of a device at different ages. For example, assume that X has Weibull distribution in Table 4.6 with shape parameter $\beta = 2$ and that a device with lifetime X has survived a lifetime of 3 units. To compare the extent of ageing of this device with a new one with the same life distribution for a time period of 4 units, taking $x = 3$ and $y = 4$, we have

$$g(x, y) = 2 \text{ and } M(x, y) = \frac{5-3}{4} = \frac{1}{2}.$$

Thus, the old unit has to work twice the time compared to the new one to produce the same output. The former is clearly wearing out and is therefore subject to positive ageing. In general, $g(x, y) \leq (\geq) x + y$ for any x and a fixed y indicates the effect of spent life on the remaining life y in relation to the device;

(iv) There are some properties of $M(x, y)$ that relate to ageing concepts as detailed below.

Theorem 4.32. *The lifetime X is NBU (NWU) if and only if*

$$M(x, y) \leq (\geq) 1.$$

Proof. Since $H_1(x)$ is an increasing function, $g(x, y) \leq y$ implies

$$H_1 g(x, y + x) \leq H_1(x + y).$$

So, from (4.54), we have

$$H_1(x) + H_1(y) \leq H_1(x + y),$$

which is a necessary and sufficient condition for X to be NBU, by Eq. (2.22). Also,

$$g(x, y) \leq y \Leftrightarrow M(x, y) \leq 1$$

and the theorem is thus proved. The proof of NWU case is obtained by reversing the inequalities. □

Remark. A criterion for positive (negative) ageing emerges from Theorem 4.32. When $M(x, y) \geq (\leq 1)$, X is ageing positively (negatively) and when $M(x, y) = 1$, X satisfies the no-ageing property.

Remark. As in Theorem 2.9,

$$M(x, y_0) \leq (\geq) 1 \Leftrightarrow X \text{ is NBU-}y_0 \text{ (NWU-}y_0\text{)}.$$

For all x and $y \geq y_0$,

$$M(x, y) \leq (\geq) 1 \Leftrightarrow X \text{ is NBU}^* y_0 \text{ (NWU}^* y_0).$$

(v) $M(x, y)$ does not depend on the lifelength y in the case of the geometric, Lomax and re-scaled beta distributions.

Since the specific ageing factor $A(x, y)$ and relative ageing factor $M(x, y)$ are similar in their purposes, it seems that a comparison of the two is natural. First, we observe that

$$\begin{aligned}\log A(x, y) &= -\log S(x + y) + \log S(x) + \log S(y) \\ &= H_1(x + y) - H_1(x) - H_1(y)\end{aligned}$$

is different from $M(x, y)$. For instance, the Weibull distribution has

$$M(x, y) = \frac{(x^\beta + y^\beta)^{\frac{1}{\beta}} - x}{y}$$

whereas

$$A(x, y) = \frac{q^{x^\beta + y^\beta}}{q^{(x+y)\beta}}.$$

Generally, the expressions of $A(x, y)$ are algebraically more complex than those of $M(x, y)$.

Theorem 4.33. *$M(x, y)$ is decreasing $\Leftrightarrow$ X is IHR $\Leftrightarrow$ $H_1(x)$ is convex $\Leftrightarrow$ $A(x, y)$ is increasing in x for every y.*

Proof. Define

$$H_1^{-1}(t) = \inf[t : H_1(x) \geq t]$$

and take the largest integer contained in t' as $H^{-1}(t)$. Since $H(x)$ is monotone, the inverse is unique. Then,

$$\begin{aligned}&M(x, y) \text{ is decreasing in } x \\ &\quad \Leftrightarrow g(x, y) \text{ is decreasing} \\ &\quad \Leftrightarrow H_1^{-1}(H_1(x) + H_1(y)) \text{ is decreasing from (4.54)} \\ &\quad \Leftrightarrow H_1^{-1}(t + s) - H_1^{-1}(t) \text{ is decreasing in } t \text{ for every } s = H_1^{-1}(y) \\ &\quad \Leftrightarrow H_1^{-1}(t) \text{ is concave} \\ &\quad \Leftrightarrow H_1(t) \text{ is convex} \\ &\quad \Leftrightarrow X \text{ is IHR.}\end{aligned}$$

The rest of the implications follow from Theorem 4.30. □

The $M(x, y)$ function has applications in ordering life distributions on the basis of the NBU property.

Definition 4.32. The random variable X is less NBU than the random variable Y, denoted by $X \leq_M Y$, iff $M_X(x, y) \leq M_Y(x, y)$ for all $x, y = 0, 1, 2, \ldots$.

Our next result relates the order $\leq_M$ and a superadditive order in $H_{1,X}(x)H_{1,Y}^{-1}(x)$, where $H_{1,X}(\cdot)$ is the alternative cumulative hazard rate of X.

Theorem 4.34.

$$X \leq_M Y \Leftrightarrow H_{1,X}H_{1,Y}^{-1}(x+y) \geq H_{1,X}H_{1,Y}^{-1}(x) + H_{1,X}H_{1,Y}^{-1}(y). \tag{4.55}$$

Proof. We have

$$\begin{aligned} X \leq_M Y &\Leftrightarrow M_X(x, y) \leq M_Y(x, y) \\ &\Leftrightarrow g_X(x, y) \leq g_Y(x, y) \\ &\Leftrightarrow H_{1,X}^{-1}(H_{1,X}(x) + H_{1,X}(y)) \leq H_{1,Y}(H_{1,Y}(x) + H_{1,Y}(y)) \\ &\Leftrightarrow H_{1,X}(x) + H_{1,X}(y) \leq H_{1,X}[H_{1,Y}^{-1}(H_{1,Y}(x) + H_{1,Y}(y))] \\ &\Leftrightarrow H_{1,X}(H_{1,Y}^{-1}(t)) + H_{1,X}(H_{1,Y}(s)) \leq H_{1,X}(H_{1,Y}^{-1}(t+s)), \end{aligned}$$

which proves the required result. □

Sengupta and Deshpande (1994) stated that X is ageing faster than Y in the superadditive sense if the inequality on the right hand side of (4.55) is satisfied. Thus, our notion of $M(x, y)$ provides a stochastic order that is equivalent to the above definition of Sengupta and Deshpande (1994).

Remark. If Y is NBU, $X \leq_{\text{NBU}} Y \Rightarrow X$ is NBU.

Finally, the relative ageing factor is useful in comparing life distributions by saying which is more positive ageing (or negative ageing by reversing the inequality in (4.55), so that $X \geq_M Y$ is equivalent to the subadditivity of $H_{1,X}\ H_{1,Y}^{-1}(\cdot)$ leading to NWU class) in the sense of NBU. For example, the Weibull distribution is less NBU than the Lomax distribution since

$$x^\beta + y^\beta \leq (x + y + \frac{xy}{\alpha})^\beta$$

for $\beta > 1$. Thus, the Lomax distribution is more positive ageing than the Weibull.

4.8.3 Ageing Intensity Function

In order to evaluate the ageing phenomenon quantitatively, Jiang et al. (2003) proposed the ageing intensity function for continuous lifetimes. It is the ratio of the hazard rate to a baseline hazard rate, which has been chosen by the authors as the average of $h(x)$ in $[0, x]$. In the discrete case, the corresponding rate becomes

$$A_1(x) = \frac{h_1(x)}{H_1(x)/x} = \frac{h_1(x)}{H_1(x)}. \tag{4.56}$$

The intensity function is unity when $h_1(x)$ is a constant (geometric case), $A_1(x) > 1$ when X is IHR, and $A_1(x) < 1$ when X is DHR. Thus, large (small) values of $A_1(x)$ indicate positive (negative) ageing. When $h(x)$ is used to define the hazard rate, we have a slightly different form as

$$A_1(x) = \frac{xh(x)}{H(x)}. \tag{4.57}$$

Writing $A_1(x)$ in terms of $h(x)$, we have

$$A_1(x) = \frac{x \log(1 - h(x))}{\sum_{t=0}^{x-1} \log(! - H(x) + H(x-1))}.$$

Example 4.4. Let X be distributed in the S-form (Section 3.3) with

$$h(x) = p(1 - \pi^x), \quad 0 < p < 1,\ 0 \leq \pi < 1,\ x = 1, 2, \ldots,$$

$$H(x) = \sum_1^x h(t) = px - \frac{p\pi}{1-\pi}(1 - \pi^x),$$

so that

$$A(x) = \frac{(1-\pi)(1-\pi^x)xp}{(1-\pi)xp - p\pi(1-\pi^x)}.$$

Similarly,

$$A_1(x) = \frac{x \log(q + p\pi^x)}{\log \prod_{t=1}^x (q + \pi^t).}$$

As seen from the above example, $A(x)$ and $A_1(x)$ are not equal generally. However, in the geometric case, we have $A(x) = A_1(x) = 1$.

Example 4.5. For the Pareto distribution, we have

$$S(x) = \left(\frac{\alpha}{x + \alpha - 1}\right)^\beta \quad \text{(see Example 2.4)},$$

$$A_1(x) = \frac{x(\log(x+\alpha) - \log(x+\alpha-1))}{\log(x+\alpha-1) - \log x}.$$

It can be verified that $A_1(x)$ is a decreasing function and therefore Pareto lifetimes exhibit negative ageing.

Reducing (4.56) recursively after writing it as

$$A_1(x) = x\left(\frac{\log S(x+1)}{\log S(x)} - 1\right),$$

and

$$S(x) = S(1)^{(A(1)+1)(\frac{A_1(2)}{2}+1)+\cdots+(\frac{A_1(x-1)}{x-1}-1)}.$$

The probability mass function can be determined by using $\sum_{x=0}^{\infty} f(x) = 1$. Thus, the ageing intensity function determines the distribution uniquely. This facilitates comparison of distributions vis-a-vis their ageing intensity functions. See Sankaran et al. (2016a) for details.

Chapter 5

Bathtub Distributions

5.1 INTRODUCTION

In the last chapter, we considered models in which the hazard rates have one of the three forms; constant, increasing or decreasing represented by the geometric, IHR and DHR distributions. However, there are several data sets in which none of the above classes of models provide a satisfactory fit. See for example the data in Davis (1952) on bus motor failures and the subsequent works by Bain (1978) and Smith and Bain (1975) on non-monotone hazard rate models. There are several instances where such models find applicability in other disciplines like biology and ecology, as pointed out by Paranjpe et al. (1985) and Paranjpe and Rajarshi (1986). Among non-monotonic hazard rate models, the bathtub shaped (BT) ones have generated considerable interest among researchers resulting in a large volume of literature, especially in the continuous case. There appears to be some justification in the growing interest on bathtub shaped hazard rate functions in reliability and survival analysis. Instances can be drawn in the lifetime of a device or organism in which the initial phase of life shows a high rate of failure or death when organisms die early due to genetic disorders, hereditary defects, hostile environment, etc., while manufactured items fail due to reasons like inferior raw material, production defects and errors in quality assurance. In the second phase of life, failure is rather few that happens due to chance and therefore the hazard rate remains more or less constant. Finally, we have period of life in which the hazard rate increases rapidly leading to the ultimate failure of the device. These three phases are some times called the 'infant mortality', 'useful life' and 'wear out' periods, respectively. Identification of the time intervals during which these phases happen is of great value for reliability analysis to prepare and plan to enhance the lifetimes of devices. To avoid infant mortality, often 'burn in' procedures are employed, while replacement policies are envisaged by eliminating components with short lives before wear out phase reaches an advanced stage. When the nature of the hazard function in bathtub shaped models is reversed, that is, initially an increasing hazard rate, then a constant, and finally a decreasing hazard rate, we have an upside-down bathtub shaped (UBT) model.

In this chapter, we discuss the shapes of non-monotonic reliability functions such as hazard rate and mean residual life. Among these, the bathtub-shaped

Reliability Modelling and Analysis in Discrete Time. http://dx.doi.org/10.1016/B978-0-12-801913-9.00005-1

hazard functions and the corresponding life distributions play a prominent role. Other forms such as roller-coaster curves and periodic functions also form part of our discussion. Definitions and criteria for identifying each variety of shape and mechanisms that generate such models are some of the key topics in this connection. Various models, their analysis and methods of construction of new models are also presented in the sequel.

5.2 DEFINITIONS AND TECHNIQUES FOR IDENTIFICATION OF BATHTUB DISTRIBUTIONS

Let $\langle a_n \rangle$ be a sequence of real numbers. It has a bathtub shape, BT (upside-down bathtub shape, UBT), if there exist integers $1 \le n_1 \le n_2 \le \infty$ such that

$$a_1 > a_2 > \cdots > a_{n_1-1} > a_{n_1} = \cdots = a_{n_2} < a_{n_2+1} < \ldots$$

$(a_1 < a_2 < \cdots < a_{n_1-1} < a_{n_1} = \cdots = a_{n_2} > a_{n_2+1} > \ldots)$. Hence, n_1 and n_2 are called the change points of the sequence $\langle a_n \rangle$ (Mi, 1993). Using this definition, a lifetime random variable X has bathtub-shaped hazard rate or bathtub-shaped distribution if $h(x)$ is strictly decreasing for $x = 0, 1, \ldots, x_0$, constant for $x_0, \ldots, x_1$, and strictly increasing for $x = x_1, x_1 + 1, \ldots$. We call x_0 and x_1 as the change points. Frequently, it is enough to consider models in which there is only one change point. In such cases, we can say that the hazard rate function $h(x)$ is such that $h(x) \le h(x_0)$ for $x = 0, 1, 2, \ldots, x_0$ and $h(x) \ge h(x_0)$ for $x = x_0, x_0 - 1, \ldots$.

A similar definition can be provided when we consider the mean residual life function $m(x)$. A life distribution is known as increasing (decreasing) initially and then increasing (decreasing) mean residual life, IDMRL (DIMRL), if $m(x) \le (\ge) m(x + 1)$ for $x = 0, \ldots, x_0 - 1$ and $m(x) \ge (\le) m(x + 1)$ for $x = x_0, x_0 + 1, \ldots$. The mean residual life function is maximized (minimized) at time x_0, the change point. Guess and Park (1988) have shown that life distribution is DIMRL (IDMRL) at x_0 if and only if there exists a sequence $\langle a(x) \rangle$, $a(x) > 0$ for all x, such that

$$\begin{aligned} a(x) \le (\ge) a(x+1), \quad x &= 0, 1, \ldots, x_0 - 1, \ a(x) \ge (\le) a(x+1), \\ x &= x_0, x_0 + 1, \ldots \end{aligned}$$

and

$$1 - h(x) = \frac{a(x)}{1 + a(x+1)}, \qquad x = 0, 1, 2, \ldots$$

Notice that, in the above case of $a(x)$,

$$S(0) = 1, \quad S(x) = \prod_{t=0}^{x-1} \frac{a(t)}{1 + a(t+1)}, \qquad x = 1, 2, \ldots,$$

and

$$a(x) = E(X - x|X \geq x).$$

Mi (1996) has defined X to have a BT hazard rate average if the sequence $\{F(x)\}^{1/x}$, $x \geq 1$, has a bathtub-shape. He has shown that

(i) if X has support set $(1, 2, \ldots)$ and BT hazard rate $h(x)$, $x \geq 1$, with change points $x_0 > 1$ and $x_1 > x_0$, then it has a BT hazard rate average with either a unique change point $y_0 \geq x_1$ or two change points y_0 and $y_0 + 1$;

(ii) if $\langle h(x), x \geq 1\rangle$ has a BT shape with change points x_0 and $x_1 < \infty$, then for the sequence $\{a(x) = E(X - x|X \geq x), x \geq 1\}$,

(a) $\langle a(x)\rangle$ strictly decreases if $f(1) < (1 + \mu)^{-1}$,
(b) $a(1) = a(2) > a(3) > \ldots$ if $f(1) = (1 + \mu)^{-1}$,
(c) $\langle a(x)\rangle$ has UBT shape with a unique change point $y_0 \leq x_0$ or two change points $y_0 - 1$ and y_0 if $f(1) > (1 + \mu)^{-1}$;

(iii) if $\langle a(x), x \geq 1)$ has a UBT shape with a unique change point y_0 and $\langle \Delta a(x), x \geq 1\rangle$ has a BT shape with change point $y_0 + 1$, then $\langle h(x), x \geq 1\rangle$ has BT shape with unique change point y_0 or $y_0 + 1$ or two change points x_0 and $x_0 + 1$.

Remark 5.1. Note that (iii) is like a converse of (ii) and spells out the mutual relationship between the shapes of $a(x)$ and $h(x)$. The relationship between $a(x)$ and the mean residual life $m(x)$ considered here is

$$\frac{m(x)}{m(x-1) - 1} = \frac{1 + a(x+1)}{a(x)}.$$

Remark 5.2. When the support of the distribution is $(0, 1, 2, \ldots)$, the results in (ii) (a), (b), (c) above becomes the following: if $\langle h(x)\rangle$ has UBT hazard rate with a change point $x_0 > 0$, and if

(1) $f(0) < (1 + \mu)^{-1}$, then $\langle m(x)\rangle$ has either BT shape with unique change point $y_0 = m_0 + 1 \leq x_0$, or two change points y_0 and $y_1 \leq x_0$;
(2) $f(0) > (1 + \mu)^{-1}$ then $\langle m(x)\rangle$ is strictly increasing;
(3) $f(0 = (1 + \mu)^{-1}$, then $m(0) = m(1) < m(2) < \ldots$ (Tang et al., 1999).

The above results have application to 'burn-in' strategies. It is customary to test certain devices, which have high initial hazard rates under conditions of field operation, to eliminate or reduce such early failures before shipping them to customers. Such an operation of screening equipments to eliminate the short lived ones is called *burn-in*. The burn-in is carried out on the products for a fixed time b and only those that survive the procedure are put into service. If the burn-in time is excessive, it will result in a loss to the manufacturer in terms of several costs. On the other hand, if the burn-in is for a short time, the problem of early failures may still persist among a percentage of products ending up with return costs. So, an important problem in the testing process is the determination of the optimal time point b upto which the test has to be conducted. The choice of b depends on the particular criterion to be used for which the hazard rate, mean residual life, coefficient of variation of residual life, etc., form the basis.

If the criterion is longest mean life without consideration of costs, we determine b such that the mean residual life is maximal, that is, to find $b^* = \max_{b \geq 0} m(b)$, which is the optimal burning time. If $h(x)$ is BT shaped with change point x_0 and $x_0 = 0$, there is no need to burn-in. Suppose $h(x)$ has BT shape with change points $x_0 > 0$ and $x_1 = \infty$, then $b^* = x_1$. If X is DFR we need to consider the cost and if $x_0 \leq x_1 < \infty$, then b^* is the unique change point of $m(x)$ or $a(x)$.

Guess and Park (1988) provided an example of an IDMRL model with

$$a(x) = \frac{\theta^2}{\alpha|x - x_0| + \theta} + \gamma,\ x = 0, 1, 2, \ldots,\ \theta > 0,\ \alpha \geq 0, \gamma \geq 0. \qquad (5.1)$$

Choosing at $\alpha = 1$, $x_0 = 40$, $\theta = 62.71$ and $\gamma = 61.71$, they showed that $a(x)$ is IDMRL at $x_0 = 40$. The hazard rate function is not BT-shaped. In fact, both the mean residual life function and the hazard rate function increase for $x = 0, 1, \ldots, 39$. One key point to be noted is that minimizing the hazard rate is not equivalent to minimizing the mean residual life. Guess and Park (1988) opines that although for many situations minimizing the hazard rate is the best, for many others maximizing mean residual life would be preferable for reducing the average cost. It is of interest to note the hazard rate corresponding to (5.1) is given by

$$1 - h(x) = \frac{\alpha(\theta + |x - x_0 + 1|)(\theta^2 + \gamma(\alpha|x - x_0| + \theta))}{(\alpha|x - x_0| + \theta)[(1 + \gamma)(\alpha|x - x_0 + 1| + \theta^2)]}$$

and the survival function is

$$S(x) = \prod_{t=0}^{x-1} (1 - h(t)).$$

5.3 MODELS

In this section, we present lifetime distributions possessing BT (UBT) shaped reliability functions. Compared to the continuous case, the number of such models in discrete time are far less. The behaviour of the hazard rate function or the mean residual life function, as the case may be, is analyzed in each case.

5.3.1 Inverse Weibull Model

Jazi et al. (2010) proposed a discrete inverse Weibull distribution which has a unimodal or decreasing hazard rate function. A random variable X follows the discrete inverse Weibull distribution if it has a distribution function of the form

$$F(x) = q^{x^{-\beta}}, \; x = 1, 2, \ldots \tag{5.2}$$

and probability mass function

$$f(x) = \begin{cases} q, & x = 1, \\ q^{x^{-\beta}} - q^{(x-1)^{-\beta}}, & x = 2, 3, \ldots, \; 0 < q < 1, \; \beta > 0. \end{cases}$$

The mass function is decreasing for $\log q > \frac{\log 2}{2^{-\beta}-1}$ and unimodal otherwise with mode at 2. Moments of the distributions do not have closed-forms, and can be expressed as infinite series. For the model in (5.2), the hazard rate functions are

$$h(x) = \frac{q^{x^{-\beta}} - q^{(x-1)^{-\beta}}}{1 - -q^{(x-1)^{-\beta}}}$$

and

$$h_1(x) = \log\left(\frac{1 - -q^{(x-1)^{-\beta}}}{1 - q^{x^{-\beta}}}\right), \quad x = 1, 2, \ldots . \tag{5.3}$$

It can be seen that $h_1(x)$ is decreasing if $(1-q)^{-1} \geq (1 - q^{2^{-\beta}})$ and is unimodal otherwise. Thus, we have the case of a UBT hazard rate. Setting $t = \log x$ and $y = \log(-\log F(x))$ in (5.2), we get

$$y = \beta t - \log(-\log q), \tag{5.4}$$

giving a straight line plot. If the data points lie approximately on a straight line in the (t, y) plot in (5.4), the inverse Weibull model is a plausible one. Jazi et al. (2010) have recommended four methods for estimating the parameters. These are the method of proportions, in which the proportions of 1 and 2 in the sample and in the population are equated and solved for q and β; method of moments, a heuristic algorithm, and by the simple linear regression model in (5.4) fitted to the data.

In the special case $\beta = 2$, the distribution in (5.2) is called the *discrete inverse Rayleigh model*. Hussain and Ahmad (2014) have discussed the properties of this model which they specify through the probability mass function

$$f(x) = q^{(x+1)^{-2}} - q^{x^{-2}}, \quad 0 < q < 1, \; x = 0, 1, 2, \ldots, \tag{5.5}$$

by including the point $x = 0$. They have pointed out that X following (5.5) is IHR for $x < 1.0694543(\sqrt{-\log q})$, DHR when $x > 1.0694543(\sqrt{-\log q})$ at integer values, and is UBT with change point at either $x = 1$ or $x = 2$ for $0 < q < 0.75$ and change point $x = 2$ as $q \to 0$.

A closer look at the distributions in (5.2) and (5.5) reveals that they have more properties useful in lifetime data analysis. The model in (5.2) is characterized by the property

$$P(X \leq x | X \leq (x^{-\beta} + y^{-\beta})^{-\frac{1}{\beta}}) = P(X \leq y), \quad x = 1, 2, \ldots, \ \beta > 0,$$

or by

$$F[(x^{-\beta} + y^{-\beta})^{-\frac{1}{\beta}}] = F(x)F(y), \tag{5.6}$$

Eq. (5.6) accounts for the departure of the inverse Weibull distribution from the reversed lack of memory property defined in (2.58). Other functions of interest are the reversed hazard rates

$$\lambda(x) = 1 - q^{(x-1)^{-\beta}} \text{ and } \lambda_1(x) = \left(\frac{1}{(x-1)^\beta} - \frac{1}{x^\beta}\right)(-\log q).$$

Hence, for the inverse Rayleigh, we have

$$\lambda_1(x) = \left(\frac{1}{(x-1)^2} - \frac{1}{x^2}\right)(-\log q) = \frac{2x-1}{x^2(x-1)^2}(-\log q), \ x = 1, 2, \ldots$$

When $\beta = 1$ in (5.2), we have the inverse geometric distribution defined by

$$F(x) = q^{\frac{1}{x}}, \ x = 1, 2, \ldots, \tag{5.7}$$

(or $F(x) = q^{\frac{1}{x+1}}, x = 0, 1, 2, \ldots$) as a member of the family of inverse Weibull. Fig. 5.1 shows the nature of $f(x)$ for different values of β. In the case of (5.7), $S(x) = 1 - q^{\frac{1}{x-1}}$ and the probability mass function is $f(x) = q^{\frac{1}{x}} - q^{\frac{1}{x-1}}$ so that

$$1 - h(x) = \frac{1 - q^{\frac{1}{x}}}{1 - q^{\frac{1}{x-1}}}. \tag{5.8}$$

Hence,

$$h(x+1) - h(x) = \frac{1 - q^{\frac{1}{x}}}{1 - q^{\frac{1}{x-1}}} - \frac{1 - q^{\frac{1}{x+1}}}{1 - q^{\frac{1}{x}}}$$

whose sign depends on

$$(1 - q^{\frac{1}{x}})^2 - (1 - q^{\frac{1}{x-1}})(1 - q^{\frac{1}{x+1}}).$$

The hazard rate function is increasing.

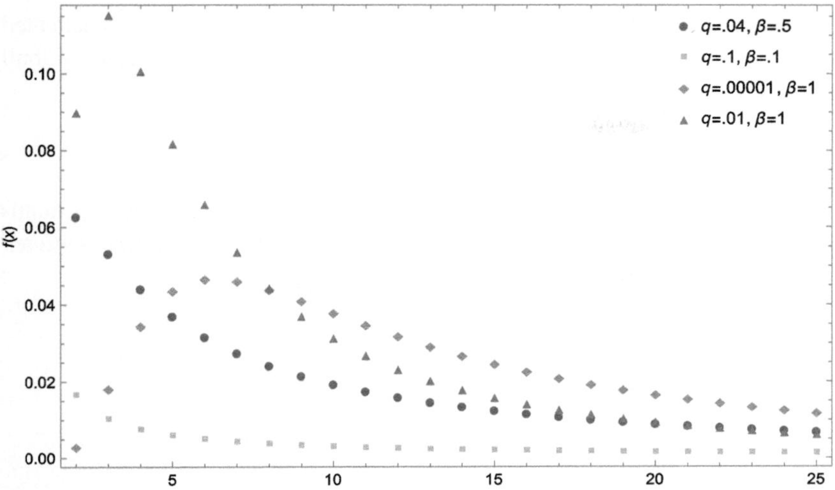

FIGURE 5.1 Probability mass functions of discrete inverse Weibull distribution for different parameter values.

Also, the mean residual life function can be expressed in terms of the recurrence relation

$$m(x) = \left(\frac{1 - q^{\frac{1}{x-1}}}{1 - q^{\frac{1}{x}}} \right) (m(x-1) - 1), \ x = 2, 3, \ldots .$$

More convenient expressions are available for the reversed hazard rates as

$$\lambda(x) = 1 - \frac{q^{\frac{1}{x-1}}}{q^{\frac{1}{x}}}$$

and

$$\lambda_1(x) = \log \frac{q^{\frac{1}{x}}}{q^{\frac{1}{x-1}}} = \frac{1}{x(x+1)}(-\log q).$$

Thus, the inverse geometric distribution is characterized by an alternative hazard function of the form $\frac{k}{x(x+1)}$, $k > 0$. Further,

$$-\log F(x) = \frac{(-\log q)}{x}$$

which is of the form $y = \frac{k}{x}$, a rectangular hyperbola. Therefore, the sample plots lie approximately on the curve $xy = k$, and inverse geometric distribution can be a candidate distribution for such data. Further, an estimate of k and hence q can be derived graphically which can provide an initial value for more sophisticated means of estimation already available in connection with the inverse Weibull law specialized to $\beta = 1$.

5.3.2 Competing Risk Models

Jiang (2010) presented two types of discrete distributions, one is the exponentiated Poisson model and the other is a competing risk model. The *exponentiated Poisson distribution* has distribution function

$$F(x) = \left[e^{-\lambda} \sum_{t=0}^{x} \frac{\lambda^t}{t!} \right]^{\theta}, \qquad x = 0, 1, 2, \ldots, \ \theta > 0, \ \lambda > 0,$$

and probability mass function

$$f(x) = \left[e^{\lambda} \sum_{t=0}^{x} \frac{\lambda^t}{t!} \right]^{\theta} - \left[e^{\lambda} \sum_{t=0}^{x-1} \frac{\lambda^t}{t!} \right]^{\theta}.$$

It has an increasing or BT-shaped hazard function. The competing risk model takes the form of the distribution function as

$$F(x) = 1 - (1 - F_1(x))(1 - F_2(x)),$$

where F_1 represents a geometric law and F_2 conforms to an exponentiated Poisson distribution. In this case also, the hazard rate is either increasing or BT shaped, which has the form

$$h(x) = p + (1 - p)h_1(x),$$

where $h_1(x)$ is the hazard rate of $F(x)$.

5.3.3 Modified Weibull Distribution

Nooghabi et al. (2011) considered the discretized version of the modified Weibull law with survival function

$$\bar{F}(x) = P(X > x) = \exp[-ax^b \exp(\lambda x)], \quad x > 0, \ a, b, \lambda > 0,$$

to arrive at the discrete survival function

$$S(x) = q^{x^b c^x}, \quad x = 0, 1, 2, \ldots, \ 0 < q < 1, \ b > 0, \ c \geq 1. \tag{5.9}$$

The hazard rate function is

$$h(x) = 1 - \frac{q^{(x+1)^b c^{x+1}}}{q^{x^b c^x}}.$$

Special cases of this distribution are the discrete Weibull distribution for $c = 1$, discrete Rayleigh ($c = 1$, $b = 2$), and geometric ($b = 1$, $c = 1$). For $0 < b < 1$, $c2^{p-1} < 1$, the hazard rate function is BT-shaped; otherwise, it is increasing. The parameters of the model can be estimated by either the method of maximum likelihood or by fitting a linear regression model of the form

$$y = a + bx_1 + dx_2,$$

where $a = \log(-\log q)$, $d = \log c$, $x_1 = x$ and $x_2 = e^x$. The alternative hazard rate function

$$\begin{aligned} h_1(x) &= \log \frac{q^{x^b c^x}}{q^{(x+1)^b c^{x+1}}} = c^x((x+1)^b c - x^b)(-\log q) \\ &= Ac^x(c(x+1)^b - x^b), \quad A > 0, \end{aligned}$$

has a more compact and tractable form.

5.3.4 Additive Weibull Distribution

The discrete additive Weibull distribution originates as the discrete version of the continuous additive Weibull law defined by

$$\bar{F}(x) = P(X > x) = \exp[-\lambda_1 x^\alpha - \lambda_2 x^\beta], \quad x > 0,\ \lambda_1, \lambda_2, \beta > 0.$$

Thus, the discrete survival function $S(x)$ also have the form

$$S(x) = q_1^{x^\alpha} q_2^{x^\beta}, \quad q_1 = e^{-\lambda_1},\ q_2 = e^{-\lambda_2},\ x = 0, 1, 2, \ldots, \tag{5.10}$$

with

$$f(x) = q_1^{x^\alpha} q_2^{x^\beta} - q_1^{(x+1)^\alpha} q_2^{(x+1)^\beta}.$$

Thus, the hazard rate function is of the form

$$h(x) = 1 - q_1^{(x+1)^\alpha - x^\alpha} q_2^{(x+1)^\beta - x^\beta}.$$

Bebbington et al. (2012) have studied the shape of this hazard function and observed the following:

(i) if either $\alpha \geq 1$ and $\beta > 1$, or $\alpha > 1$ and $\beta \geq 1$ then $h(x)$ is strictly increasing;

(ii) if $\alpha = 1$ and $\beta = 1$, then $h(x)$ is constant;

(iii) if $\alpha \leq 1$ and $\beta < 1$ or $\alpha < 1$ and $\beta \leq 1$, then $h(x)$ is strictly decreasing;

(iv) if $\alpha < 1 < \beta$, then $h(x)$ is bathtub shaped with minimum achieved at one of the three points $[t_{\alpha,\beta}]$, $1+[t_{\alpha,\beta}]$ and $2+[t_{\alpha,\beta}]$, where $[t_{\alpha,\beta}]$ is the largest integer in $t_{\alpha,\beta}$ and

$$t_{\alpha,\beta} = \left(\frac{\alpha(1-\alpha)\lambda_1}{\beta(\beta-1)\lambda_2}\right)^{\frac{1}{\beta-\alpha}}.$$

By symmetry, the same conclusion holds for $\beta < 1 < \alpha$. The alternative hazard rate function

$$h_1(x) = \lambda_1((x+1)^\alpha - x^\alpha) + \lambda_2((x+1)^\beta - x^\beta)$$

also has the same shape as $h(x)$. These above authors have also applied the results to several data sets after estimating the parameters by the method of maximum likelihood.

5.3.5 Modified Weibull Extension

Noughabi et al. (2013) have proposed the discrete modified Weibull extension model specified by

$$S(x) = q^{\alpha(e^{(\frac{x}{\alpha})^\beta}-1)}, \quad x = 0, 1, 2, \ldots, \; 0 < q < 1, \; \beta > 0, \; \alpha > 0, \tag{5.11}$$

to model bathtub shaped hazard rate data. The distribution has probability mass function as

$$f(x) = q^{-\alpha}\left(q^{\alpha e^{(\frac{x}{\alpha})^\beta}} - q^{\alpha e^{(\frac{x+1}{\alpha})^\beta}}\right). \tag{5.12}$$

Eq. (5.11) is the discrete version of the continuous modified Weibull extension

$$\bar{F}(x) = \exp[\lambda\alpha(1 - e^{-(\frac{x}{\alpha})^\beta}], \quad \lambda, \alpha, \beta > 0, x \geq 0,$$

of Xie et al. (2002b). The hazard rate, mean residual life and α-percentile residual life function are, respectively,

$$h(x) = 1 - q^{\alpha(e^{(\frac{x+1}{\alpha})^\beta} - e^{(\frac{x}{\alpha})^\beta})},$$

$$m(x) = \sum_{x=1}^{\infty} q^\alpha \left[e^{(\frac{x+1}{\alpha})^\beta} - E^{-(\frac{x}{\alpha})^\beta}\right]$$

and

$$q_\alpha(x) = \inf\{x : S(x) \leq 1 - \alpha\}.$$

These authors have also considered a special case of the additive Weibull model.

5.3.6 Discrete Reduced Modified Weibull Distribution

The discrete version of the reduced modified distribution introduced by Almalki and Nadarajah (2014) possess a bathtub-shaped hazard rate function. Their distribution has survival function

$$S(x) = q^{\sqrt{x}(1+bc^x)}, \quad x = 0, 1, \ldots, \; 0 < q < 1, \; b > 0, \; c \geq 1. \tag{5.13}$$

When $b = 0$, it is a special case of the discrete Weibull distribution. The probability mass function is

$$f(x) = q^{\sqrt{x}(1+bc^x)} - q^{\sqrt{x+1}(1+bc^{x+1})},$$

and so the hazard rate function has the form

$$h(x) = 1 - q^{\sqrt{x+1}(1+bc^{x+1}) - \sqrt{x}(1+bc^x)}.$$

It has been shown that the hazard rate function is increasing if $bc(c - \sqrt{2}) < \sqrt{2} - 1$ and has BT shape otherwise. The method of maximum likelihood has been employed to estimate the parameters. Almalki and Nadarajah (2014) also compared the model competency with the bathtub-shaped modified Weibull and the additive Weibull distributions with the aid of the Akaike information Criterion (Akaike, 1974), the Bayesian information criterion (Schwarz, 1978), and the consistent Akaike information criterion (Hurvich and Tsai, 1989). The fits obtained by them were found to be superior than these competing models.

5.4 METHODS OF CONSTRUCTING BATHTUB MODELS

A variety of data sets that display bathtub-shaped hazard rates are available and this has resulted in a large number of continuous distributions possessing such a property, each to fit into a specific data set. In the discrete case also, there is abundance of data sets but only a few distributions so far developed to model them as could be seen from the review of existing distributions in the last section. In the absence of physical conditions that suggest a form of some reliability function, often one has to construct a model that matches the characteristics of the observations. This calls for some general methods of construction of distributions with bathtub hazard rates. However, it needs to be mentioned that some of the methods explained here may result in bathtub models, but they need not always end up with such distributions.

5.4.1 Discretizing Continuous Distributions

A general method of construction is to verify whether the discrete version of a continuous distribution with bathtub hazard rate inherits the same property. All the models we have seen in Section 5.2 belong to this category. As a further example, let us consider the quadratic failure rate distribution introduced by Bain (1974, 1978) with hazard rate

$$h(x) = a + bx + cx^2, \quad x > 0,\ b > 0,\ c > 0,$$

and survival function

$$\bar{F}(x) = \exp[-ax - \frac{bx^2}{2} - \frac{cx^3}{3}],\ x \geq 0.$$

Accordingly, the discrete model is specified by

$$S(x) = q^{ax+\frac{bx^2}{2}+\frac{cx^3}{3}} \tag{5.14}$$

with probability mass function

$$f(x) = q^{ax+\frac{bx^2}{2}+\frac{cx^3}{3}} - q^{a(x+1)+\frac{b}{2}(x+1)^2+\frac{c(x+1)^3}{3}}. \tag{5.15}$$

From (5.14) and (5.15), we find

$$\begin{aligned} h(x) &= 1 - \frac{q^{a(x+1)+\frac{b}{2}(x+1)^2+\frac{c}{3}(x+1)^3}}{q^{ax+\frac{b}{2}x^2+\frac{c}{3}x^3}} \\ &= 1 - q^{a+\frac{b}{2}(2x+1)+\frac{c}{3}(3x^2+3x+1)}. \end{aligned}$$

Also,

$$h(x+1) - h(x) = q^{a+\frac{b}{2}(2x+1)+\frac{c}{3}(3x^2+3x+1)}(1 - q^{b+2c(x+1)}).$$

Thus, the zero of the above equation becomes

$$x_0 = -\frac{b}{2c} - 1, \quad b < 0, c > 0.$$

Further, $h(x_0+1) - h(x_0) = 1 - q > 0$ and $h(x_0) - h(x_0 - 1) = 1 - q^{-3c} < 0$ so that $h(x)$ is decreasing in $[0, x_0 - 1)$ and increasing in $[x_0 + 1, \infty)$. Thus, $h(x)$ is bathtub-shaped with change point $[x_0]$, with $[d]$ representing the integer part of d. We refer to (5.15) as the *discretized quadratic hazard model* (DQHM).

To estimate the parameters of the model, we may proceed as follows. In a random sample of n lifetimes, let $S_n(x)$ be the empirical survival function. Then, $S_n(x)$ is the sample proportion of devices that survive age x. We estimate the parameters a, b, c by minimizing

$$L(a, b, c) = \sum_x (S_n(x) - S(x))^2.$$

TABLE 5.1 Observed and expected frequencies for quadratic hazard rate model

Age interval	Observed frequency	Expected frequency
0–4	9	8
5–18	9	10
19–50	8	7
51–67	8	8
68–84	9	10
> 84	7	7

The estimating equations are obtained as $\frac{\partial L}{\partial a} = 0$, $\frac{\partial L}{\partial b} = 0$ and $\frac{\partial L}{\partial c} = 0$. Solving these equations, the least-square estimates of a, b and c can be obtained.

To illustrate the methodology, we apply it to the Aarset (1987) data pertaining to 50 lifetimes of devices by taking the first two observations 0.1 and 0.2 as zeros. Using numerical methods, L is seen to attain the minimum at

$$\hat{a} = 0.0379894,\ \hat{b} = -0.00190263,\ \hat{c} = 0.0000274169,$$

with $\hat{L} = L(\hat{a}, \hat{b}, \hat{c}) = 0.0517965$, a small value. The change point x_0 is estimated as $[x_0] = 33$. It is further seen that $h(x)$ is non-increasing on the set $A = (0, 1, 2, \ldots, 33)$ and non-decreasing on A^c. With the above estimates, the observed and estimated frequencies were determined and are presented in Table 5.1.

The chi-square value of 0.468 as against the tabled value $\chi_2^2 = 5.991$ suggests that the model is not rejected at 5% level of significance. Also, expected and observed frequencies are so close seen to be quite each other.

Nooghabi et al. (2011), fitted the modified Weibull distribution (MWD) in (5.9), Bebbington et al. (2012) fitted the additive Weibull (AW) in (5.10), and Almalki and Nadarajah (2014) fitted the reduced modified Weibull model (RMW) in (5.13) to the same set of observations. Table 5.2 contains the observed and expected frequencies under these models. A straightforward comparison reveals that the discretized quadratic hazard rate model provides a better fit than other models in the sense of the chi-square test. All the models shown in the table provide reasonable fits for the data, however.

The survival function and the hazard rate function are presented in Figs 5.2A and 5.2B, respectively.

Since this distribution has not appeared in our discussions so far, some of its important properties are given below. The moments and other related character-

TABLE 5.2 Comparison of observed and expected frequencies under different models

Class	Observed frequency	Expected frequency			
		DQHM	MWD	AW	RMW
0–4	9	8	8	10	9
5–18	9	10	7	7	7
19–50	8	7	13	7	8
51–67	8	8	7	6	6
68–84	9	10	5	14	14
> 84	7	7	10	6	6
Expected frequency		0.47	2.04	3.44	2.70

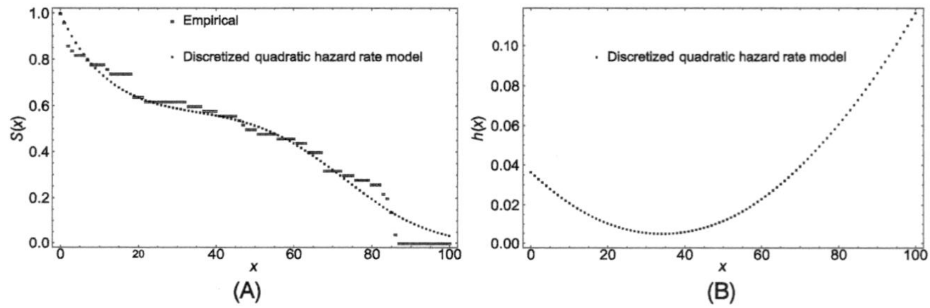

FIGURE 5.2 Plots of survival and hazard functions. (A) Survival function. (B) Hazard function.

istics of (5.15) are difficult to evaluate analytically, since

$$E(X^r) = \sum_{x=0}^{\infty} x^r [q^{ax+\frac{b}{2}x^2+\frac{c}{3}x^3} - q^{a(x+1)+\frac{b}{2}(x+1)^2+\frac{c}{3}(x+1)^3}].$$

The quantile-based measures are much easier to evaluate by employing

$$F(x) = 1 - e^{-[a(x+1)+\frac{b}{2}(x+1)^2+\frac{c}{3}(x+1)^3]}.$$

We have as measure of location

$$\text{median} = M,$$

as measure of dispersion the

$$\text{quartile deviation } = Q_3 - Q_1,$$

and as measures of skewness and kurtosis,

$$\text{Coefficient of skewness, } S = \frac{Q_1 + Q_3 - 2M}{Q_3 - Q_1},$$

where M, Q_1 and Q_3 are the solutions of $F(M)=\frac{1}{2}$, $F(Q_1)=\frac{1}{4}$, $F(Q_3)=\frac{3}{4}$ representing the median, first quartile and third quartile, respectively, and

$$\text{Coefficient of kurtosis, } T = O_7 - O_5 + O_3 - O_1/Q_3 - Q_1$$

where

$$F(O_t)=\frac{t}{8}, \quad t=1,3,5,7.$$

The model in (5.14) has two important special cases. One is the geometric law when $b=0=c$ and $p=^{e^{-a}}$, and the second is the discrete version of the linear hazard rate distribution when $c=0$ in $\bar{F}(x)$ above with survival function

$$S(x)=q^{ax+\frac{b}{2}x^2}, \quad q=e^{-1},\ a,b>0. \tag{5.16}$$

Eventhough (5.16) does not have a BT shape, in view of the importance of the distribution as seen in the continuous case, we discuss some properties here of the discrete version.

From (5.16), the probability mass function is

$$\begin{aligned} f(x) &= q^{ax+\frac{b}{2}x^2} q^{a(x+1)+\frac{b}{2}(x+1)^2} \\ &= q^{ax+\frac{b}{2}x^2}(1-q^{a+\frac{b}{2}(2x+1)}). \end{aligned}$$

Thus, the hazard rate is

$$h(x)=1-q^{a+\frac{b}{2}(2x+1)}$$

which is IHR. Using the definition

$$m(x)=\frac{1}{q^{ax+\frac{b}{2}x^2}}\sum_{t=x+1}^{\infty} q^{at+\frac{b}{2}t^2}$$

of the mean residual life function, we have a recurrence relation

$$m(x)=m(x-1)[q^{-(a+\frac{b}{2}(2x-1))}-1], \quad x=0,1,\ldots.$$

Obviously, X is DMRL in this case.

5.4.2 Using General Conditions

In this section, we derive some general conditions for a distribution to possess bathtub-shaped hazard rate function. Choice of a probability mass function $f(x)$ that meets these requirements will lead to a BT or UBT hazard rate. The theorems presented below will also help to check the shape of $h(x)$ especially in cases where the survival function or hazard rate cannot be obtained in closed-form to analytically determine its monotonicity. In the following discussions, the symbols I, D, B and U, respectively, stand for situations in which $h(x)$ is strictly increasing, strictly decreasing, strictly decreasing, and then strictly decreasing and strictly increasing and then strictly decreasing. The theorems considered below are the discrete versions of the results in Glaser (1980) and Ghitany (2004).

Theorem 5.1. *Let X be a discrete random variable taking values* $0, 1, 2, \dots b$*,* $b \le \infty$*. Define*

$$\eta(x) = \frac{f(x) - f(x+1)}{f(x)},$$

where $f(\infty) = 0$ *when* $b = \infty$*.*

Then:

(i) *If* $\eta(x)$ *is D, then* $h(x)$ *is D;*
(ii) *If* $\eta(x)$ *is I, then* $h(x)$ *is I;*
(iii) *If* $\eta(x)$ *is B and* $f(0) = 0$ $(f(0) = \infty)$*, then* $h(x)$ *is* $I(B)$*;*
(iv) *If* $\eta(x)$ *is U and* $f(0) = 0$ $(f(0) = \infty)$*, then* $h(x)$ *is* $U(D)$*.*

Proof. Let $g(x) = \frac{1}{h(x)}$.

Then,

$$\Delta g(x) = g(x+1) - g(x) = \frac{h(x) - h(x+1)}{h(x)h(x+1)} \tag{5.17}$$

and also

$$\begin{aligned}
\Delta g(x) &= \frac{S(x+1)}{f(x+1)} - \frac{S(x)}{f(x)} \\
&= \frac{S(x+1)}{f(x+1)} - \frac{S(x+1) + f(x)}{f(x)} \\
&= S(x+1)\left(\frac{f(x) - f(x+1)}{f(x)}\right) - 1 \\
&= \frac{\eta(x)}{f(x+1)} \sum_{t=x+1}^{\infty} f(t) - 1 \\
&= \frac{1}{f(x+1)}\left[\sum_{t=x+1}^{\infty} (\eta(x) - \eta(t)) f(t) + \sum_{t=x+1}^{\infty} f(t)\eta(t)\right] - 1 \\
&= \frac{1}{f(x+1)} \sum_{t=x+1}^{\infty} (\eta(x) - \eta(t)) f(t)
\end{aligned}$$

$$+\frac{1}{f(x+1)}(f(x+t)-f(b))-1$$

$$=\frac{1}{f(x)}\sum_{t=x+1}^{\infty}(\eta(x)-\eta(t))f(t)-f(b), \tag{5.18}$$

irrespective of whether b is finite or infinite. We distinguish these different cases.

(i) When $\eta(x)$ is D, $\eta(x)-\eta(t)<0$ and $f(b)=f(\infty)=0$. Hence, from (5.18), $\Delta g(x)>0$ which in turn implies $h(x)>h(x+1)$, and hence $h(x)$ is D.

The proof of (ii) is similar.

(iii) Writing

$$\Delta g(x)=\frac{a(x)}{f(x+1)}, \tag{5.19}$$

we have

$$\begin{aligned}
a(x)&=\sum_{t=x+1}^{\infty}(\eta(x)-\eta(t))f(t)-f(b)\\
&=\eta(x)S(x+1)-[f(x+1)-f(b)]-f(b)\\
&=\eta(x)S(x+1)-f(x+1)
\end{aligned}$$

irrespective of whether b is finite or infinite. Then,

$$\begin{aligned}
\Delta a(x)&=\eta(x+1)[S(x+1)-f(x+1)]\\
&\qquad-\eta(x)S(x+1)-[f(x+2)-f(x+1)]\\
&=\Delta\eta(x)S(x+1).
\end{aligned} \tag{5.20}$$

Now,

$$\begin{aligned}
\Delta h(x)&=(g(x)-g(x+1))h(x)h(x+1)\\
&=-\frac{\Delta g(x)}{g(x)g(x+1)}=-\frac{a(x)}{f(x)g(x)g(x+1)}.
\end{aligned} \tag{5.21}$$

To prove (iii), assume that $\eta(x)$ is B and has exactly one zero, say x_0. Since $\eta(x)$ is I in $[x_0,b]=x_0,x_0+1,\ldots,b$, we must have

$$\eta(x)-\eta(t)<0,\quad x_0\le x<t\le b,$$

and so

$$\Delta g(x)<0\Rightarrow h(x)<h(x+1).$$

Thus, $h(x)$ is I in $[x_0,b]$. This fact will be referred to as property P.

When $f(0)=0$, $\eta(x)$ being B, it is D in $[1, x_0-1]$ and so is $\Delta a(x)$ by virtue of (5.19). Also, $f(0)=0$ and $f(1)>0$ means that $a(1)<0$ and this together with the decreasing nature of $\Delta a(x)$ shows that $a(x)<0$ in $[1, x_0)$. Therefore, from (5.20), $\Delta h(x)>0$ in $[1, x_0)$. Combining this with property P, $h(x)$ is I in $[1, b)$ and the first part of (iii) is proved.

Now, take $f(0)=\infty$ so that $g(0)=0$ and $\Delta g(1)>0$. Hence, from (5.19), $a(1)>0$. Since $a(x)$ is D in $[1, x_0)$, $a(x)>0$ if $a(x_0)>0$. This gives $\Delta h(x)<0$ by (5.21) or $h(x)$ is D in $[1, x_0)$. Along with P, this means that $h(x)$ is B. On the other hand, if $a(x_0)<0$ there exists an x_1 in $[1, x_0)$ such that $\alpha(x)>0$ in $[1, x_1)$, $a(x_1)=0$ and $a(x)<0$ in (x_1, x_0). Thus, from (5.20), $h(x)$ is B in $[1, x_0)$ and along with P this guarantees that $h(x)$ is B in $[1, b)$. The proof of (iv) is quite similar. □

The case when b is finite follows the same line of proof with $f(b)=0$ and is given below.

Theorem 5.2. *Let X be the random variable as in Theorem 5.1 with $b<\infty$. Then:*

(i) *If $\eta(x)$ is D and $f(b)=0$, then $h(x)$ is D;*
(ii) *If $\eta(x)$ is I, then $h(x)$ is I;*
(iii) *If $\eta(x)$ is B and $f(0)=0$ $(f(0)=\infty)$, then $h(x)$ is $I(B)$;*
(iv) *If $\eta(x)$ is U and $f(0)=0$ $(f(0)=\infty)$, and $f(b)=0$, then $h(x)$ is $U(I)$.*

Another set of sufficient conditions which are perhaps more easier to work with, is the following.

Theorem 5.3. *When $b<\infty$, (i) if $\eta(x)\le 0$ and $f(b)>0$, then $h(x)$ is I; and (ii) if $\eta(x)$ is D and $f(0)=f(b)=\infty$, then $h(x)$ is B.*

Proof. When $\eta(x)\le 0$, $\Delta g(x)=S(x+1)\eta(x)-1\le 0$ and so $h(x)$ is I, proving (i). In the second case, $f(0)=\infty$ and

$$\eta(0)=1-\frac{f(1)}{f(0)}=1.$$

Hence, from (5.19), $a(0)>0$. Similarly, $f(b)=\infty$ implies $a(b)<0$. Since $\eta(x)$ is strictly decreasing, we can find an x_2 such that $a(x)>0$ in $[0, x_2]$, $a(x_2)=0$ and $a(x)<0$ in $[x_2, b]$. From (5.21), the zeros of $a(x)$ is the same as those of $\Delta h(x)$ and therefore $h(x)$ is B. □

Remark 5.3. Gupta et al. (1997) considered the shape of $\eta(x)$ and showed that $\Delta\eta(x)>(<,=)0$, then $h(x)$ is increasing (decreasing, increasing). It covers the case $\Delta\eta(x)=0$ for all x only. The result here extends the behaviour of $\eta(x)$ to cover BT and UBT distributions.

One main problem with the above theorems is that they require the computation of $\Delta\eta(x)$ which is algebraically complicated for many distributions having bathtub-shaped hazard rate. To overcome this difficulty, we have the next result which needs only the function $\eta(x)$ in the calculations.

Theorem 5.4. *The random variable X has*

(i) *IHR (DHR) if and only if* $h(x+1) \geq (\leq)\eta(x)$;

(ii) *a BT (UBT)-shaped hazard rate if and only if* $h(x+1) = \eta(x)$ *possesses a unique zero* x_0 *such that* $h(x-1) \geq (\leq h(x))$ *in* $[0, x_0)$ *and* $h(x-1) \leq (\geq)$ $h(x)$ *in* $[x_0, b)$.

Proof. We have

$$\begin{aligned} h(x+1) - h(x) &= \frac{f(x+1)}{S(x+1)} - \frac{f(x)}{S(x)} \\ &= \frac{f(x+1)(S(x+1) + f(x))}{S(x)S(x+1)} - f(x)S(x+1) \\ &= \frac{[f(x+1) - f(x)]S(x+1) + f(x)f(x+1)}{S(x)S(x+1)} \\ &= -\eta(x)h(x) + h(x)h(x+1), \end{aligned}$$

$$\text{or } \frac{h(x+1) - h(x)}{h(x)} = h(x+1) - \eta(x). \tag{5.22}$$

Thus, $h(x+1) - h(x)$ has the same sign as $h(x+1) - \eta(x)$ which proves (i), and also $h(x+1) - h(x)$ has the same zero as $h(x+1) - \eta(x)$ which proves (ii). □

Example 5.1. Let X be distributed with survival function

$$S(x) = \frac{1}{1+cx^2}, \quad x = 0, 1, 2, \ldots, \; c > 0.$$

Then,

$$\begin{aligned} f(x) &= (1+cx^2)^{-1} - (1+c(x+1)^2)^{-1}, \\ h(x) &= 1 - (1+cx^2)(1+c(x+1)^2)^{-1}, \end{aligned}$$

and

$$\eta(x) = 1 - \frac{[1+c(x+1)^2]^{-1} - [1+c(x+1)^2]^{-1}}{(1+cx^2)^{-1} - [1+c(x+1)^2]^{-1}}.$$

Thus,

$$\eta(x) - h(x+1) = \frac{2x+3}{2x+1}\frac{1+c(x+1)^2}{1+c(x+2)^2} - 1$$

leads to the equation

$$2cx^2 + 4cx + c - 2 = 0$$

whose zeros are

$$x = -1 \pm \frac{1}{2}\left(4 - 2\left(1 - \frac{2}{c}\right)\right)^{1/2}.$$

A positive solution greater than zero exists only when $x \leq 2$. As an illustration, when $c = \frac{2}{31}$, the change point of $h(x)$ is $x_0 = 3$. We further see that $h(x)$ is increasing for $x = 0, 1, 2$ and decreasing for x in $[3, \infty)$. Thus, X has a UBT-shaped hazard rate when $c < 2$ and X is DHR when c is in $[2, \infty)$.

Other than being useful in the analysis of the shape of the hazard rate function, $\eta(x)$, which is called the score function, also possess several other properties; see, Nair et al. (2016). Some simple relations between $\eta(x)$ and $h(x)$ can be employed to derive life distributions with monotone and non-monotone hazard rates. We give two examples in this regard.

Example 5.2. Assume that $\eta(x) = ch(x+1)$ or $h(x+1) - \eta(x) = (1-c)$. Then, using (5.22), we arrive at the recurrence relation

$$h(x+1) = \frac{h(x)}{1 + ch(x)}, \quad x = 0, 1, 2, \ldots.$$

Successive reduction yields the solution

$$h(x) = \frac{h(0)}{1 + (c-1)h(0)x}$$

which is of the form

$$h(x) = \frac{1}{\alpha x + \beta}, \quad \beta = \frac{1}{h(0)} > 0 \text{ and } \alpha = (c-1) \text{ real.} \tag{5.23}$$

From Xekalaki (1983), the hazard function is of the form (5.23) if and only if X is

(i) geometric with

$$f(x) = \frac{1}{\beta}\left(\frac{\beta - 1}{\beta}\right)^x, \quad x = 0, 1, 2, \ldots, \ (c = 1);$$

(ii) Waring with

$$f(x) = \left(\frac{\beta - 1}{\alpha}\right) x/\beta \left(\frac{\beta}{\alpha} + 1\right)_x, \quad x = 0, 1, 2, \ldots, \ (c > 1);$$

(iii) negative hyper geometric with

$$f(x) = \binom{-1}{x}\binom{\frac{1}{x}}{n-x}\Big/\binom{-1+\frac{1}{\alpha}}{n},$$

$x = 0, 1, \ldots, n, n = \frac{1-\beta}{\alpha}, (c < 1)$.

Further special cases are when $\alpha = -1$ in (ii), X has uniform distribution on $(0, 1, \ldots, \beta - 1)$. In this case, $h(x) = \frac{1}{\beta - x}$ and $f(x) = \frac{1}{\beta}, x = 0, 1, 2, \ldots \beta - 1$. Also, when $\beta = \alpha + 1$ in (ii), we have the Yule distribution.

Example 5.3. Consider the relationship

$$h(x+1) - \eta(x) = \frac{\alpha x + \beta}{h(x)}.$$

Then, from (5.22), we have the recurrence relation

$$h(x+1) = h(x) + \alpha x + \beta$$

which gives

$$h(x) = \frac{\alpha x(x-1)}{2} + \beta(x-1) + h(0) \tag{5.24}$$

Since (5.24) can be written as

$$h(x) = a_1 x^2 + b_1 x + c_1,$$

we have a quadratic hazard rate in which the conditions $a_1 < 0$ and $b_1^2 - 4a_1(c_1 - 1) < 0$ are imposed. Thus, $h(x)$ provides a bathtub-shape with change point $x_0 = -\frac{1}{2}(1 + \frac{b_1}{a_1})$, where x_0 is the integer part of the right hand side which is positive. The random variable X has survival function

$$S(x) = \prod_{t=0}^{x-1}[1 - (a_1 t^2 + b_1 x + c_1)]. \tag{5.25}$$

Eq. (5.25) represents a family of distributions consisting of the geometric distribution when $a_1 = b_1 = 0$, the linear hazard rate distribution when $a_1 = 0$, $b_1 > 0$, $c_1 > 0$, and the linear hazard rate distribution with bounded support on $(0, 1, \ldots, -\frac{c_1}{b_1})$, where $-\frac{c_1}{b_1}$ is a positive integer. Distribution in (5.25) will be referred to as the *quadratic hazard rate model* (QHM).

Remark 5.4. In the discussions from Theorem 5.4 to Example 5.2, the hazard function $h(x)$ is not an additional input as it can also be written in terms of $\eta(x)$

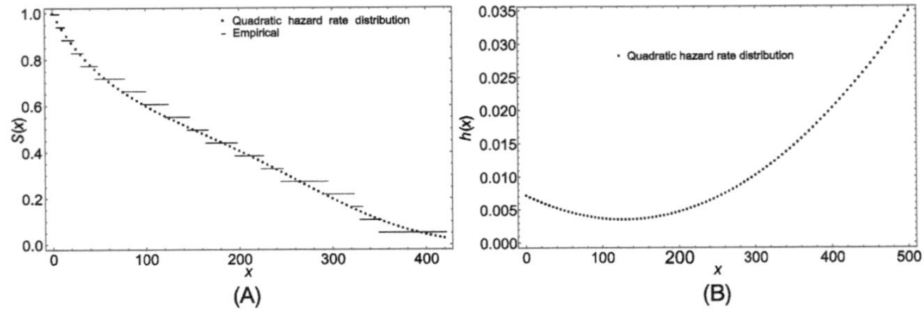

FIGURE 5.3 Plots of survival and hazard functions. (A) Survival function. (B) Hazard function.

as

$$h(x)=\frac{\prod_{t=1}^{x-1}(1-\eta(t))f(0)}{1-f(0)\sum_{t=1}^{x}(1-\eta(1))\dots(1-\eta(t-1))}.$$

Thus, the application of Theorem 5.4 requires only the probability mass function.

The quadratic hazard model was fitted to the data on the lifetimes of 18 electronic devices reported in Wang (2000). To estimate the parameter of the model, the method of least-squares, minimizing the deviation of the theoretical hazard rate from the empirical counterpart, was applied. This means that

$$L(a_1,b_1,c_1)=\sum\left(a_1x^2+b_1x+c_1-\frac{\hat{S}(x)-\hat{S}(x+1)}{\hat{S}(x)}\right)^2$$

is minimized over the 18 sample values in the data. The estimates obtained were

$$\hat{a}_1=2.25484\times 10^{-7},\ \hat{b}_1=-0.000057436,\ \text{and}\ \hat{c}_1=0.00720904$$

for a minimum L value of $L_{\min}=0.13483$. The goodness-of-fit ascertained through the K-S statistic with value 0.0641 does not reject the model. Almalki and Nadarajah (2014) also employed the same data for modelling with their reduced modified Weibull distribution. Their K-S statistic value 0.084 which is slightly higher than that of the above model. The graphs of the survival function and hazard function are presented in Fig. 5.3A and 5.3B.

Example 5.4. In this example, we assume that

$$h(x+1)-\eta(x)=\frac{1}{h(x)}\left[\alpha-\frac{\theta\beta}{(1+\beta x)(1+\beta(x+1))}\right]. \tag{5.26}$$

The identity the (5.26) leads to the solution

$$h(x) = \alpha x + \frac{\theta}{1+\beta x}, \quad \theta = h(0). \tag{5.27}$$

For (5.27) to be hazard rate, the conditions to be satisfied are $0 < \theta < 1$ and $\alpha > 0$, $\beta > 0$. The survival function is

$$S(x) = \begin{cases} \prod_{t=1}^{x-1}(1 - \alpha t - \frac{\theta}{1+\beta t}), & x = 1, 2, 3, \ldots, \\ 1, & x = 0. \end{cases} \tag{5.28}$$

We note that (5.27) corresponds to the discrete version of the Hjorth (1980) model for the hazard rate. But, the survival function (5.28) is not the discretized form of the Hjorth survival function in the continuous case.

A second application is the use of $\eta(x)$ function in comparing life distributions based on stochastic orders involving reliability functions. We say that X is less than Y in score function order iff, $\eta_X(x) \geq \eta_Y(x)$, where X and Y are discrete lifetime random variables, and we write this order as $X \leq_\eta Y$.

Theorem 5.5. *The score function order $X \leq_\eta Y$ is equivalent to the likelihood ratio order $\leq_{lr}$.*

Proof. We have

$$\begin{aligned} X \leq_\eta (Y) &\Leftrightarrow 1 - \frac{f_X(x+1)}{f_X(x)} \geq 1 - \frac{f_Y(x+1)}{f_Y(x)} \\ &\Leftrightarrow \frac{f_Y(x+1)}{f_X(x+1)} \geq \frac{f_Y(x)}{f_X(x)}, \end{aligned}$$

so that $\frac{f_Y(x)}{f_X(x)}$ is an increasing function, and hence $X \leq_\eta Y \Leftrightarrow X \leq_{\text{lr}} Y$. □

Remark 5.5. Since the likelihood ratio order implies both hazard rate order and reversed hazard rate order, we have

$$X \leq_\eta (Y) \Rightarrow X \leq_{\text{hr}} Y \text{ and } X \leq_\eta (Y) \Rightarrow X \leq_{\text{rh}} Y.$$

Thus, $X \leq_\eta Y$ is a sufficient condition to establish both $\leq_{\text{hr}}$ and $\leq_{\text{rh}}$. All other properties of the likelihood ratio order stated in Shaked and Shanthikumar (2007) hold for $\leq_\eta$ as well.

5.4.3 Construction From Mixtures

Recalling the discussion of mixtures of distributions in Section 2.10, we consider the mixture of distribution with lifetimes X_1 and X_2, written in the form

$$f(x) = \alpha f_1(x) + (1-\alpha) f_2(x),$$

where $f_1(x)$ and $f_2(x)$ are probability mass functions of X_1 and X_2, respectively, and the hazard rate function of the mixture distribution $f(x)$ is

$$h(x) = p(x)h_1(x) + (1 - p(x))h_2(x).$$

In the above formulation, $h_1(x)$ and $h_2(x)$ are the hazard rates of X_1 and X_2, while

$$p(x) = \frac{\alpha S_1(x)}{\alpha S_1(x) + (1 - \alpha)S_2(x)}$$

which survival functions $S_1(x)$ and $S_2(x)$ corresponding to $f_1(x)$ and $f_2(x)$.

A mixture of two (or more) distributions of which one has decreasing hazard rate and the other (or at least one of the others) has increasing hazard rate may lead to a BT distribution.

Example 5.5. Let X be a geometric variable with $S_1(x) = q^x$ and hazard rate $h_1(x) = p$, and X_2 have a survival function

$$S_2(x) = (1 + \theta x)q^{\theta x}, \quad \theta > 0,\ q = e^{-1}.$$

Then,

$$h_2(x) = \frac{1 + \theta x - (1 + \theta + \theta x)q^{\theta}}{1 + \theta x}$$

and so

$$h(x) = \frac{\alpha p + (1 - \alpha)(1 + \theta x - (1 + \theta + \theta x)q^{\theta})}{\alpha + (1 - \alpha)(1 + \theta x)q^{\theta(x-1)}}$$

has bathtub shape.

5.4.4 Convex Functions

A strictly convex function $h(x)$ satisfies the conditions for a hazard rate function producing a BT distribution.

Example 5.6. Consider the function

$$h(x) = 1 - e^{-(ax^2+bx+c)}, \quad b < 0,\ a < 0,$$

which satisfies $0 \le h(x) \le 1$ and $\sum h(x)$ is divergent. The function is also convex. The survival function

$$\begin{aligned} S(x) &= \prod_{t=0}^{x-1} (1 - h(t)) \\ &= q^{\frac{ax^3}{3} + \frac{a-b}{2}x^2 + \frac{a-3b-6c}{6}x} \end{aligned} \tag{5.29}$$

TABLE 5.3 Goodness of fit for Aarset data

Class	0–4	5–18	19–50	51–67	68–84	> 84
Observed	9	9	8	8	9	7
Expected	7	10	7	8	10	8

admits a BT hazard rate. In fact,

$$h(x+1) - h(x) = q^{ax^2+bx+c}(1 - q^{a(2x+1)+b})$$

which admits the zero

$$x_0 = -\frac{(b+a)}{2a}$$

providing a bathtub-shape whenever $b > -a$ with change point x_0. The model parameters can be estimated by minimizing

$$L(a,b,c) = \sum (ax_i^2 + bx_i + c + \log(1 - \hat{h}(x)))^2,$$

where $\hat{h}(x) = \frac{\hat{S}(x) - \hat{S}(x+1)}{\hat{S}(x)}$, the empirical hazard rate. These results were used on the Aarset data considered earlier in Table 5.3. The estimates obtained were

$$\hat{a} = 0.0000227975, \; \hat{b} = -0.00156645, \; \hat{c} = 0.326186.$$

For the same classes taken, the χ^2 value of 1.03929 does not reject the model. The observed and expected frequencies used in χ^2 test are given in Table 5.3.

Thus, (5.29) represents a BT model with change point $x_0 = 34$.

There are some other methods prescribed for continuous lifetimes that can be adapted to the discrete case as in the previous sections. We refer interested readers to Nair et al. (2013, 2017a) for details.

5.5 PROPERTIES OF BT MODELS

Just as in the case of monotonic reliability functions discussed in Chapter 4, there are many useful properties enjoyed by the classes of BT and UBT distributions. These include closure properties with regard to reliability operations as well as some bounds and moment characteristics. A detailed exposition of the properties and examples are given in Nair et al. (2017a).

Series Systems

Let us consider a two-component system in which the lifetimes of the components are X_1 and X_2. If the system remains in operation only when both components are functioning, we say that the system has series structure. In such a structure, the minimum lifetime, $\min(X_1, X_2)$, is the lifetime of the system. It has survival function

$$S(x) = S_{X_1}(x)S_{X_2}(x).$$

Using the definition of the alternative hazard rate $h_1(x)$, we can write the system hazard rate as

$$h_1(x) = h_{1,X_1}(x) + h_{1,X_2}(x),$$

and so

$$h_1(x+1) - h_1(x) = (h_{1,X_1}(x+1) - h_{1,X_1}(x)) + (h_{1,X_2}(x+1) - h_{1,X_2}(x)).$$

If the hazard rates of X_1 and X_2 are BT (UBT) with a common change point x_0, then

$$h_1(x_0+1) - h_1(x_0) = 0$$

so that x_0 is a change point of $h_1(x)$. Thus, if the component hazard rates are BT (UBT) with a common change point, a series system is also BT (UBT) with the same change point as that of the components.

Parallel Systems

In a parallel system with identical components, the system will work only if at least one of the component functions. Thus, the lifetime of the system is given by maximum (X_1, X_2) and its survival function is

$$\begin{aligned} S_1(x) &= 1 - [1 - S(x)]^2 \\ &= 2S(x) - S^2(x), \end{aligned}$$

where $S(x)$ is the common survival function of X_1 and X_2. When X_1 and X_2 are BT (UBT), $S_1(x)$ need not be BT (UBT).

Convolution

The convolution of two BT (UBT) distributions need not be BT (UBT).

Example 5.7. The discretized quadratic hazard rate model

$$S(x) = q^{ax + \frac{b}{2}x^2 + \frac{c}{3}x^3}$$

encountered in (5.14) is BT when $a = 0.005674$, $b = -0.00003678$ and $c = 1.5557666 \times 10^{-7}$. But the convolution of the distribution with itself has an increasing hazard rate at the above parameter values.

Example 5.8. Consider the hazard rate

$$h(0) = 0.01,\ h(1) = 0.02,\ h(2) = 0.6 \text{ and } h(x) = 0.4,\ x \geq 3.$$

It is UBT. But, the convolution of the distribution with itself has a hazard rate that is initially increasing, then decreasing, and finally increasing.

Mixtures

The mixture of two BT (UBT) models may not be BT (UBT).

Example 5.9. Take $S_1(x)$ as in Example 5.7 and

$$S_2(x) = \frac{\exp[-\frac{a}{2}x^2]}{(1+bx)^{c/b}}, \quad x = 0, 1, 2, \ldots.$$

Both have BT hazard rates, the latter is taken at $a = 0.00032741$, $b = 5.05530198$ and $c = 0.2812267$. Their mixture

$$S(x) = 0.692 S_1(x) + 0.308 S_2(x)$$

has initially increasing, then decreasing, and again increasing hazard rate.

Example 5.10. The distribution

$$S_1(x) = \left(1 + c\left(\frac{x}{a}\right)^b\right)^{\frac{1}{c}}, \quad x = 0, 1, 2, \ldots,\ a, b > 0,$$

has UBT hazard at $a = 1.81$, $b = 1.184$ and $c = -0.35$. The same shape is obtained for the distribution

$$S_2(x) = (1 + cx^{\alpha})^{-1}$$

at $\alpha = 2.330086$ and $c = 2.78613758 \times 10^{-6}$. When $\alpha = 0.698$, their mixture has an initially decreasing hazard rate.

Reliability Bounds

If X is BT (UBT), then $X \leq (\geq) X_G$, where G is a geometric random variable with parameter $h(x_0)$ with x_0 being the change point of the hazard rate of X.

Proof. Assume that X is BT with change point x_0. Then, $h(x)$ is decreasing in $[0, x_0)$ and therefore $1 - h(x) \leq 1 - h(x_0)$. Thus,

$$\begin{aligned} S_X(x) &= \prod_{t=0}^{x-1}(1 - h(t)) \leq \prod_{t=0}^{x-1}(1 - h(x_0)) \\ &= q^x = S_G(x),\ q = 1 - h(x_0). \end{aligned} \tag{5.30}$$

So, $X \le X_G$. The proof of the UBT case is obtained by reversing the inequalities.

A BT distribution has moments of all orders $r > 0$. This follows from the fact that

$$X \le X_G \Rightarrow E(X^{(r)}) \le E(X_G^{(r)}) \tag{5.31}$$

$$= r!\left(\frac{1-h(x_0)}{h(x_0)}\right)^r, \text{ for every } r > 0. \tag{5.32}$$

On the other hand, if $E(X^r) < \infty$, the lower bound for the rth factorial moment is

$$E(X^{(r)}) \ge r!\left(\frac{1-h(x_0)}{h(x_0)}\right).$$

□

Remark 5.6. The result in (5.30) gives the bound on the reliability function of a BT distribution, while (5.32) gives the upper bound of the factorial moments of the BT model.

Residual Life Distribution

If X has BT (UBT) distribution, the corresponding residual life distribution has BT (UBT) hazard rate.

Proof. When X has BT (UBT) hazard rate, there exist a point x_0 such that

$$h(x_0+1) - h(x_0) = 0.$$

The distribution of the residual life $(X - t|X > x)$ has survival function $\frac{S(t+x+1)}{S(x+1)}$ and hazard rate $h(x+t+1)$. Hence,

$$h(y+1) - h(y) = 0, \quad y = x+t,$$

also has a zero, say $y_0 = x + t_0$. Thus, the residual life distribution is BT (UBT) with change point $x_0 - t_0$ according as $h(x)$ is BT (UBT), provided $x_0 > t$. □

Equilibrium Distribution

If Y_n is the nth order equilibrium random variable corresponding to X, then Y_n has BT (UBT) distribution if and only if the mean residual life function of Y_{n-1} is BT (UBT) with the same change point as the hazard rate function of Y_n.

The proof follows from the fact $h_n(x) = \frac{1}{m_{n-1}(x)}$; see the identity in (2.112).

Shape of the Probability Mass Function

When X has a BT hazard rate, then the probability mass function of X is decreasing in $[0, x_0)$. This follows from the fact that $h(x) = \frac{f(x)}{S(x)}$ is decreasing in $[0, x_0)$ and $\frac{1}{S(x)}$ is increasing in x.

5.6 OTHER FORMS OF NON-MONOTONIC HAZARD RATES

Non-monotonic hazard rate shapes are not confined to the BT and UBT classes alone. Empirical versions of hazard rate for real data has confirmed that there can be a variety of forms that are not monotone or BT or UBT. We make a brief survey of some of these cases in what follows.

Periodic Hazard Rate

Definition 5.1. We say that a discrete random lifetime X has *periodic hazard rate* function if and only if for $c > 1$ and integer,

$$h(x+c) = h(x), \quad x = 0, 1, 2, \dots . \tag{5.33}$$

The constant c is called the *period* of $h(x)$.

The existence and utility of periodic hazard rates can be justified on the following grounds:

(a) Castillo and Sieworek (1981) considered reliability of computing systems and presented several data sets showing that hard disk failure seems to follow the work load. The influence of this work load can be taken into account quite satisfactorily by the addition of a periodic hazard rate;

(b) Lifetimes of various kind of components operating in environments for which the work load may induce failures occur in a periodic manner (Tanguy, 2009);

(c) Labour market participation, consumer behaviour, and many other phenomena exhibit strong periodic patterns that result from cyclic behaviour constraint of the timing of events or seasonal variations. While estimating the inter-purchasing times of ice cream packages, the data consisted of the day of purchase of 8400 households over a period of three years, the hazard rate was found to be periodic. The discrete daily hazard rate oscillated with maxima at 7, 14, 21 days, and so on.
The estimated hazard rates of job durations in Germany for 1975–1990, covering 400,000 job spells, showed large annual peaks, somewhat smaller quarterly peaks and also monthly peaks suggesting periodicity (Potter et al., 2002);

(d) Periodic random environments and mechanisms of their effect on embedded random variables represent either waiting time until some event occurs or the number of events in a given time interval. Dimitrov et al. (1997) considered the probability distributions of such random variable having periodic residual life functions and periodic hazard rates;

(e) A probabilistic interpretation of periodic rates is possible in terms of the almost lack of memory property which is related to the usual lack of memory property. Thus, we can have a meaning that is relevant to ageing.

Example 5.11. The geometric distribution $f(x) = q^x p$ has hazard rate $h(x) = p$ which trivially satisfies $h(x + c) = h(x)$, and hence periodic. The role of geometric distribution is important in the construction of distributions with periodic hazard rate.

Example 5.12. The distribution specified by

$$S(x) = q^{x+\frac{a}{c}\sin cx}, \quad x = 0, 1, 2, \ldots, \ a, c > 0, \tag{5.34}$$

has probability mass function

$$f(x) = q^{x+\frac{a}{c}\sin cx-(x+1)-\frac{a}{c}\sin c(x+1)},$$

and so

$$\begin{aligned} h(x) &= 1 - q^{1+\frac{a}{c}[\sin c(x+1)-\sin cx]}, \\ h\left(x + \frac{2\pi}{c}\right) &= 1 - q^{1+\frac{a}{c}[\sin c(x+1+\frac{2\pi}{c})-\sin c(x+\frac{2\pi}{c}]}, \\ &= 1 - q^{1+\frac{a}{c}[\sin c(x+1)-\sin cx]} = h(x). \end{aligned}$$

Thus, the distribution has periodic hazard rate with period $\frac{2\pi}{c}$.

Some properties of periodic hazard rate relevant to life time data analysis are as follows:

(i) If $h(x)$ is periodic, $h_1(x)$, $H(x)$ and $H_1(x)$ are also periodic with the same period;

(ii) A random variable X has periodic hazard rate if and only if its survival function is of the form

$$S_X(x) = q^{[\frac{x}{c}]}[q + (1-q)S_Y(x - [\frac{x}{c}]^c)], \quad 0 < q < 1, \tag{5.35}$$

where $S_Y(\cdot)$ is the survival function with support in $[a, 1, 2, \ldots c-1]$ and $[t]$ stands for the integer part of t (Chukova et al., 1993b). The probability mass function of X is

$$f_X(x) = q^{[\frac{x}{c}]}(1-q)f_Y\left(x - \left[\frac{x}{c}\right]c\right), \quad x = 0, 1, 2, \ldots. \tag{5.36}$$

From (5.35) and (5.36), we have

$$h_X(x) = \frac{(1-q)f_Y(x - [\frac{x}{c}]c)}{q + (1-q)S_Y(x - [\frac{x}{c}]c)}. \tag{5.37}$$

Alternatively,

$$f_Y(x) = \frac{h_X(x)}{1 - \prod_{t=0}^{c-1}(1 - h_X(t))} \prod_{t=0}^{x-1}(1 - h(t)) \tag{5.38}$$

(Dimitrov et al., 1997);

(iii) The distribution of residual life, $X_t = (X - t|X > t)$, has hazard rate $h(x + t + 1)$. Hence, if $h(x)$ is periodic, X_t also has a periodic hazard rate with period c;

(iv) If Y_n is the equilibrium random variable of order n, then Y_{n-1} has periodic mean residual life iff Y_n has periodic hazard rate;

(v) If X_1 and X_2 are independent random lives that follow a series system with periodic hazard rates, then the system lifetime is also periodic with the same period as that of the components;

(vi) When a parallel system is formed with independent components each of which has periodic hazard rates, the system lifetime need not have a periodic hazard rate;

(vii) The convolution of two independent random variables each with periodic hazard rate need not have a hazard rate with periodicity. For example, let X_1 and X_2 be two independent and identically distributed geometric random variables with parameter p. Then, their convolution Y has probability mass function

$$f_Y(x) = (x+1)p^2q^x,$$

survival function

$$S_Y(x) = (1 + px)q^x,$$

so that

$$h_Y(x) = \frac{(x+1)p^2}{1+px}$$

for which $h_Y(x + c) = h_Y(x)$ is not satisfied for any integer $c > 1$.

The interpretation of periodic hazard rates, discussion of further properties and construction of models become considerably simplified if we consider equivalent notion of *almost lack of memory property* (ALMP). Recall that the lack of memory (LMP) characterizes the geometric distribution and also constancy of the hazard rate and mean residual function indicating no-ageing. Notice that in (5.35), the survival function of the periodic hazard rate function also contains a geometric component and this raises several questions regarding the relationships between LMP, ALMP and periodicity of $h(x)$. We briefly discuss some of these aspects and suggest the works of Chukova and Dimitrov (1992),

Chukova et al. (1993a, 1993b), Dimitrov et al. (1995), Sandhya and Rajasekharan (2012a), and Buiculescu et al. (2011) for further reading.

Definition 5.2. If there exists a sequence of constants $\langle b_n \rangle$, $n = 1, 2, \ldots$ such that

$$P(X \geq x + c | X \geq c) = P(X \geq x) \tag{5.39}$$

holds for any integer $c = b_n$, $n = 1, 2, \ldots$ and $x \geq 0$, we say that X has almost lack of memory property (ALMP).

From Marsaglia and Tubilla (1975), a random variable X possesses ALMP only at points b_n which form a lattice. Thus, for integer-valued random variables, $a_n = k_n c$, where $\langle k_n \rangle$ is a sequence of positive integers. It follows that X has ALMP iff (5.39) holds for all x and infinitely many distinct values of c.

Consider the *k-geometric distribution* defined by (Sandhya and Rajasekharan, 2012a)

$$P(X = kx) = pq^x, \quad x = 0, 1, 2, \ldots.$$

When Y is geometric, kY has k-geometric distribution. The k-geometric has ALMP over $(0, k, 2k, \ldots)$. Further, X has periodic hazard rate of period $c > 1$ if and only if it can be represented as a sum $X = Y + cZ$ of two independent random variables Y and Z, where Y has support $(0, 1, \ldots, c-1)$ and Z is geometric with $P(Z = x) = q^x p$. When X is geometric, cX is c-geometric, and the above representation takes the form

$$X = Y + G, \tag{5.40}$$

where Y and G are independent and G is c-geometric. Representation in (5.40) provides an alternative methodology for constructing models with periodic hazard rates in terms of generating function $P(t)$ as

$$\begin{aligned} P_X(t) &= P_Y(t) P_G(t) \\ &= P_Y(t) \frac{1-q}{1-qt^c}. \end{aligned} \tag{5.41}$$

Example 5.13. When $P_Y(t) = 1$, the c-geometric has periodic $h(x)$. When Y is binomial $(c-1, \alpha)$, we have

$$P_X(t) = (1 - \alpha + \alpha t)^{c-1} \frac{(1-q)}{1-qt^c}.$$

Remark 5.7. The above result also answers the following question: When can the convolution of two discrete random variables have periodic hazard rate. It should be a sum with one component with geometric distribution independently of the other, which should be a finite range distribution. Notice that Y need not have periodic hazard rate, as the binomial case in Example 5.13.

The nature of the mean residual life function when X has periodic hazard rate is discussed in the next theorem.

Theorem 5.6. *When $h(x)$ is periodic, the mean residual life $m(x)$ is also periodic.*

Proof. When $h(x)$ is periodic, (5.39) holds and so

$$\begin{aligned} & S(x+c)=S(x)S(c) \\ & \Rightarrow \sum_{x+1}^{\infty} S(t+c)=S(c)\sum_{x+1}^{\infty} S(t) \\ & \Rightarrow \sum_{x+t+1}^{\infty} S(t)=S(c)\sum_{x+1}^{\infty} S(t) \\ & \Rightarrow S(x+t+1)m(x+c)=S(c)S(x+1)m(x) \\ & \Rightarrow m(x+c)=m(x). \end{aligned}$$

Thus, $m(x)$ is periodic with the same periodicity as $h(x)$. □

Other properties of periodic hazard rates such as invariance with respect to relevation transforms, preservation of service time distribution on unreliable servers and applications to queues, environmental studies have also been pointed out in Dimitrov et al. (1997).

In the literature, there are some other shapes of the hazard function such as roller-coaster, modified bathtub, polynomial type, etc. that have been discussed. In the absence of many data sets exhibiting such shapes or models for them, the details of such hazard functions have not been seriously pursued. Wong (1991) has stated that generation of the roller-coaster shape starts with the basic failure mechanism which leads to a decreasing hazard rate. The bumps observed later in life could be caused by changing hazard conditions, wear out failure, distribution of flawed items, distribution of flow sizes, or residue small size flaws left in the equipment because of test and inspection limitations are the physical bases for the formation of roller-coaster shape. For details, one may refer to Wong (1988, 1991) and Jensen (1989).

Chapter 6

Multivariate Reliability Concepts

6.1 INTRODUCTION

In Chapter 2, we have discussed at some length various reliability functions such as hazard rate, mean residual life, etc., and various functions in reversed time. These correspond to the univariate random lifetime of a device or component. However, there are many complex devices and systems whose functioning depends on several components that may be independent or dependent. In such a scenario, the lifetime of the system is composed of the lifetimes of components that may depend on different physical properties. If the system consists of p components with lifetimes $X_1, \ldots, X_p$, the study of system reliability can be facilitated only through the joint distribution of the vector $(X_1, \ldots, X_p)$ and its reliability functions. Thus, there is a need to extend the univariate reliability concepts to higher dimensions. Accordingly, we discuss various multivariate reliability functions in this chapter. A fascinating aspect, when one tries to generalize from the univariate to the multivariate context is that several alternative definitions could emerge depending on the particular definition or property that is employed in each situation. Consequently, the resulting models will also change. It is difficult to evolve any set of criteria based upon which generalization of univariate formulations have to be chosen in evolving multivariate concepts. However, it may be reasonable to have a multivariate definition if (i) it coincides with the existing definition for a single variable when appropriately reduced to one dimension and (ii) the implications and chain of relationships between multivariate concepts should follow the patterns in the univariate case. In spite of such and similar desirable proposals, a unique formulation of the definition of reliability functions, ageing classes etc., in the multivariate setup is still elusive. In this chapter, we discuss different types of definitions of important reliability functions and the properties each definition implies. We will consider the p-dimensional cases only when it is not a direct extension of the bivariate case; otherwise, for the sake of brevity, only the bivariate results will be presented.

Reliability Modelling and Analysis in Discrete Time. http://dx.doi.org/10.1016/B978-0-12-801913-9.00006-3

6.2 MULTIVARIATE HAZARD RATE

6.2.1 Scalar Hazard Rate

Following the notation in Section 1.8, let $\mathbf{X} = (X_1, \dots, X_p)$ be a random vector in N^p, $N = (0, 1, 2, \dots)$, with joint probability mass function $f(\mathbf{x})$ and survival function $S(\mathbf{x})$. We define the multivariate scalar hazard rate of $\mathbf{X}$ as

$$a_p(\mathbf{x}) = \frac{f(\mathbf{x})}{S(\mathbf{x})} \tag{6.1}$$

at those points of $\mathbf{x}$ for which $S(\mathbf{x}) > 0$.

The definition in (6.1) implies that

$$a_p(\mathbf{x}) = P(X_1 = x_1, \dots, X_p = x_p | X_1 \geq x_1, X_2 \geq x_2, \dots, X_p \geq x_p)$$

and so it is the conditional probability that, in a p-component device, the ith component fails at age x_i, given that it has survived age x_i, $i = 1, 2, \dots, p$. Also, $0 \leq a(\mathbf{x}) \leq 1$. Thus, the definition in (6.1) is in tune with the concept of univariate hazard rate. However, it cannot provide the marginal hazard rate except in the case of independence of $X_1, \dots, X_p$.

Example 6.1. Let (X_1, X_2) follow the bivariate geometric law

$$\begin{aligned} f(x_1, x_2) = q_1^{x_1} q_2^{x_2} \theta^{x_1 - x_2 - 1}[(1 - q_1\theta^{x_2+1})(1 - q_2\theta^{x_1+1} + \theta - 1)], \\ 0 < q_1, q_2 < 1,\ 0 \leq \theta \leq 1,\ 1 - \theta \leq (1 - q_1\theta)(1 - q_2\theta), \end{aligned} \tag{6.2}$$

and survival function

$$S(x_1, x_2) = q_1^{x_1} q_2^{x_2} \theta^{x_1 x_2}.$$

Then,

$$a_2(x_1, x_2) = [(1 - q_1\theta^{x_2+1})(1 - q_2\theta^{x_1+1}) + \theta - 1]\theta^{-1}.$$

The variables X_1 and X_2 are independent when $\theta = 1$ in which case X_1 and X_2 are geometric with parameters p_1 and p_2, respectively. In this case,

$$a_2(x_1, x_2) = (1 - q_1)(1 - q_2),$$

the product of the hazard rates of X_1 and X_2.

Some properties of $a_p(\mathbf{x})$ are as follows:

1. The scalar hazard rate $a_p(\mathbf{x})$ does not determine the distribution of $\mathbf{X}$ uniquely.

Example 6.2. Let $S_1(x_1, x_2) = (0.3)^{x_1}(0.7)^{x_2}$, $x_1, x_2 = 0, 1, 2, \ldots$, the product of two geometric survival functions, and

$$S_2(x_1, x_2) = \frac{1}{2}((0.3)^{x_1}(0.7)^{x_2} + (0.7)^{x_1}(0.3)^{x_2}),$$

a mixture of two geometric distributions. Both have the same hazard rate $a(x_1, x_2) = 0.21$;

2. In view of the above property, it is of interest to know with what additional condition can $a(\mathbf{x})$ determine a life distribution. This is stated in the next theorem;

Theorem 6.1. *The scalar hazard rates* $a_r(\mathbf{x}_r)$, $r = 1, \ldots, p$, *where* $\mathbf{x}_r = (x_1, \ldots x_r)$ *determine* $S(\mathbf{x})$.

Proof. From (6.1) we can write

$$a(\mathbf{x})S(\mathbf{x}) = f(x) = P(X_1 = x_1, \ldots, X_p = x_p).$$

The distribution of X_1 is known from the relationship

$$S(\mathbf{x}_1) = \prod_{0}^{x_1-1}(1 - a_1(t)),$$

and

$$\begin{aligned} a(\mathbf{x}_2)S(\mathbf{x}_2) &= P(X_1 = x_1, X_2 = x_2) \\ &= P(X_1 = x_1, X_2 \geq x_2) - P(X_1 = x_1, X_2 \geq x_2 + 1). \end{aligned}$$

Summation from x_1 to ∞ yields

$$P(X_1 \geq x_1, X_2 \geq x_2 + 1) = S(\mathbf{x}_2) - \sum_{t=x_1}^{\infty} a_2(t, x_2)S(t, x_2). \tag{6.3}$$

Setting $x_2 = 0$, we get

$$P(X_1 \geq x_1, X_2 \geq 1) = S_1(x_1) - \sum_{t=x_1}^{\infty} a_2(t, 0)S_1(t), \tag{6.4}$$

where $S_1(x)$ is the survival function of X_1 which is known. Since $a_2(\mathbf{x})$ is also given, (6.4) is completely known. Upon applying the recurrence relation (6.3) with (6.4) as the starting value, $P(X_1 \geq x_1, X_2 \geq x_2)$ is determined. Following the same kind of steps in arriving at (6.3) using $S(x_1, x_2)$, we can obtain

$$P(X_1 \geq x_1, X_2 \geq x_2, X_3 \geq x_3 + 1)$$
$$= S(\mathbf{x}_2) - \sum_{t_1=x_1}^{\infty} \sum_{t_2=x_2}^{\infty} a_3(t_1, t_2, x_3) S(t_1, t_2). \quad (6.5)$$

Setting $x_3 = 0$ and using this result in (6.5), recursively we find $S(\mathbf{x}_3)$. Proceeding like this, $S(\mathbf{x})$ can be reached for

$$P(X_1 \geq x_1 \ldots X_{p-1} \geq x_{p-1}, X_p \geq x_p + 1)$$
$$= S(\mathbf{x}_{p-1}) - \sum_{t=x_1}^{\infty} \cdots \sum_{t=x_{p-1}}^{\infty} a_{p-1}(\mathbf{t}_{p-1}, x_p) S(\mathbf{t}_{p-1}, x_p). \quad (6.6)$$

□

Remark 6.1. One marginal distribution of any dimension is enough for the conclusion of Theorem 6.1 to hold. Instead of the marginal distribution, one can also assume the functional form of a marginal hazard rate.

Example 6.3. Let X have a scalar hazard rate of the form

$$a_2(\mathbf{x}_2) = \frac{n(n+1)}{(m+n+x_1+x_2)(m+n+x_1+x_2+1)}, \quad \mathbf{x} \in N^2,\ m, n > 0,$$

and marginal survival function

$$S_1(x_1, 0) = \frac{(m)_{x_1}}{(m+n)_{x_1}}, \quad m, n > 0,\ x_1 \in N. \quad (6.7)$$

From (6.3) with $p = 2$, we have

$$S(x_1, 1) = \frac{(m)_{x_1}}{(m+n)_{x_1}} - \sum_{t=x_1}^{\infty} \frac{(m)_t}{(m+n)_t} \frac{n(n+1)}{(m+n+t)(m+n+t+1)},$$
$$= \frac{(m)_{x_1}}{(m+n)_{x_2}} - \sum_{t=x_1}^{\infty} \frac{(m)_t}{(m+n)_t} \left[1 - \frac{2(m+t)}{m+n+t} + \frac{m(t)(m+t+1)}{(m+n+t)(m+n+t)}\right],$$
$$= \frac{(m)_{x_1}}{(m+n)_{x_1}} - \sum_{t=x_1}^{\infty} \frac{(m)_t}{(m+n)_t} + 2\sum_{t=x_1}^{\infty} \frac{(m)_{t+1}}{(m+n)_{t+1}} - \sum \frac{m(t)_{t+2}}{(m+n)_{t+2}},$$
$$= \frac{(m)_{x_1+1}}{(m+n)_{x_1+1}}, \quad \text{using Waring expansion.} \quad (6.8)$$

Now assume that

$$S(x_1, x_2 - 1) = \frac{(m)_{x_1+x_2-1}}{(m+n)_{x_1+x_2-1}}.$$

Using (6.4), we get

$$S(x_1,x_2) = \frac{(m)_{x_1+x_2-1}}{(m+n)_{x_1+x_2-1}} - \sum_{t=x_1}^{\infty} \frac{(m)_{t+x_2-1}}{(m+n)_{t+x_2-1}}\left[1+\frac{2(m+t+x_2-1)}{(m+n+t+x_2-1)}\right.$$

$$\left. + \left[\frac{m(m+t+x_2-1)(m+t+x_2)}{(m+n+t+x_2-1)(m+n+t+x_2)}\right]\right],$$

$$= \frac{(m)_{x_1+x_2-1}}{(m+n)_{x_1+x_2-1}} - \sum \frac{(m)_{t+x_2-1}}{(m+n)_{t+x_2-1}}$$

$$+2\sum \frac{(m)_{t+x_2}}{(m+n)_{t+x_2}} - \sum \frac{(m)_{t+x_2+1}}{(m+n)_{t+x_2+1}},$$

$$= \frac{(m)_{x_1+x_2}}{(m+n)_{x_1+x_2}}, \quad (x_1,x_2) \in N^2,\ m,n > 0. \tag{6.9}$$

From (6.7), (6.8) and (6.9), the result is true for all integers $x_2 = 2, 3, \ldots$ by induction. Note that (6.9) is a *bivariate Waring distribution.*

Example 6.4. When

$$a_2(\mathbf{x}) = \frac{k(k-1)}{(k+n-x_1-x_2)(k+n-x_1-x_2-1)},$$

$k, n > 0$, $x_1, x_2 = 0, 1, \ldots n$, $x_1 + x_2 \leq n$, and

$$S_1(x) = \binom{k+n-x_1}{n-x_1} \Big/ \binom{k+n}{n}, \quad k, n > 0, x_1 = 0, 1, \ldots n, \tag{6.10}$$

the negative hyper-geometric survival function, the distribution of (X_1, X_2) is specified by

$$S(x_1,x_2) = \binom{k+n-x_1-x_2}{n-x_1-x_2} \Big/ \binom{k+n}{n}, \tag{6.11}$$

$x_1, x_2 = 0, 1, \ldots n$, $x_1 - x_2 \leq n$, $k, n > 0$, which is *bivariate negative hyper-geometric distribution*;

3. The marginal hazard rates $h_i(x_i)$ of X_i, $i = 1, \ldots, p$, can not be directly expressed in terms of $a(\mathbf{x})$, except in the case of independence.

Theorem 6.2. *If the random variables $X_1, \ldots, X_p$ are independent, then*

$$a_p(\mathbf{x}) = h_1(x_1)\ldots h_p(x_p), \tag{6.12}$$

and conversely if (6.10) *holds, $X_1, \ldots, X_p$ are independent.*

Proof. The proof of the first part is easily obtained from (6.1). To prove the converse, assume (6.12) for $p = 2$, that is,

$$a_2(x_1, x_2) = h_1(x_1)h_2(x_2).$$

Then, from (6.4), we have

$$\begin{aligned} S(x_1, 1) &= S_1(x_1) - \sum_{t=x_1}^{\infty} h_1(t)h_2(0)S_1(t) \\ &= S_1(x_1) - \sum_{t=x_1}^{\infty} f_1(t)h_2(0) \\ &= S_1(x_1)(1 - h_2(0)) = S_1(x_1)S_2(1). \end{aligned}$$

Using this, (6.4) recursively gives

$$S(x_1, x_2) = S_1(x_1)S_2(x_2)$$

so that X_1 and X_2 are independent. Now, this fact is employed in (6.5) and we have the independence of X_1, X_2 and X_3, and finally the result in the theorem with the aid of (6.6); □

4. In modelling data using bivariate distributions, the dependence relation between X_1 and X_2 play a significant role. When the functional form of $a_2(x_1, x_2)$ is known, it is possible to study the nature of dependence between X_1 and X_2.

Theorem 6.3. *When the random vector (X_1, X_2) is such that $E(X_1, X_2) < \infty$,*

$$Cov\,(X_1, X_2) = E\left[\frac{1}{a_2(X_1, X_2)}\right] - E\left[\frac{1}{h_1(X_1)}\right]E\left[\frac{1}{h_2(X_2)}\right].$$

Proof. We have

$$\begin{aligned} E\left[\frac{1}{a_2(X_1, X_2)}\right] &= \sum_{x_1=0}^{\infty}\sum_{x_2=0}^{\infty} S(x_1, x_2) \\ &= S(0,0) + \sum_{x_1=1}^{\infty} S_1(x_1) + \sum_{x_2=1}^{\infty} S_2(x_2) + \sum_{x_1=1}^{\infty}\sum_{x_2=1}^{\infty} S(x_1, x_2) \\ &= 1 + \mu_{X_1} + \mu_{X_2} + E(X_1, X_2), \end{aligned}$$

where μ_{X_1} and μ_{X_2} are the means of X_1 and X_2. Hence,

$$\begin{aligned} \text{Cov}\,(X_1, X_2) &= E(X_1, X_2) - \mu_{X_1}\mu_{X_2} \\ &= E\left[\frac{1}{a_2(X_1, X_2)}\right] - 1 - \mu_{X_1} - \mu_{X_2} - \mu_{X_1}\mu_{X_2} \end{aligned}$$

$$= E\left[\frac{1}{a_2(X_1, X_2)} - (1 + \mu_{X_1})(1 + \mu_{X_2})\right]$$

$$= E\left[\frac{1}{a_2(X_1, X_2)}\right] - E\left[\frac{1}{h_1(X_1)}\right] E\left[\frac{1}{h_2(X_2)}\right] \text{ from (2.31).}$$

□

More properties of the scalar hazard rate, like its relationship with other reliability functions, characterizations, etc., will be considered in the subsequent sections. Properties 1–4 above have been established in the bivariate case by Nair and Sankaran (2015a).

6.2.2 Conditional Hazard Rate

Shaked et al. (1995) have proposed another definition of bivariate and multivariate hazard rates for a random vector **X**. For the random vector (X_1, X_2), where $X_1, X_2 = 1, 2, \ldots,$ the conditional hazard rate is defined by

$$\begin{aligned}
b_1(x) &= P(X_1 = x, X_2 > x | X_1 \geq x, X_2 \geq x), \ x = 1, 2, 3, \ldots, \\
b_2(x) &= P(X_2 = x, X_1 > x | X_1 \geq x, X_2 \geq x), \\
b_{12}(x) &= P(X_1 = x, X_2 = x | X_1 \geq x, X_2 \geq x), \\
b_1(x|x_2) &= P(X_1 = x | X_1 \geq x, X_2 = x_2), \ x > x_2, \ x_2 = 1, 2, \ldots, \\
b_2(x|x_1) &= P(X_2 = x | X_1 = x_1, X_2 \geq x), \ x > x_1, \ x_1 = 1, 2, \ldots,
\end{aligned}$$

provided the conditioned events in the conditional probabilities have positive probabilities. In the above format, only one component 1 (2) fails with probability $b_1(x)$ $(b_2(x))$ and both components fail with probability $b_{12}(x)$. Also, when one component failed at age x_1 (x_2) and the other is alive, then the conditional hazard rate of the live component at age $x > x_1$ $(x > x_2)$ is given by $b_2(x|x_1)$ $(b_1(x|x_2))$. The above set of hazard rates determine the joint distribution of $\mathbf{X}_2$ through the formula

$$f(x_1, x_2) = \begin{cases}
\left[\prod_{i=1}^{x_1-1}(1 - b^*(i))\right] b_1(x_1) \left[\prod_{i=x_1+1}^{x_2-1}(1 - b_2(i|x_2))\right] b_2(x_2|x_1), & x_1 < x_2, \\
\left[\prod_{i=1}^{x_2-1}(1 - b^*(i))\right] b_2(x_2) \left[\prod_{i=x_2+1}^{x_1-1}(1 - b_1(i|x_1))\right] b_1(x_1|x_2), & x_1 > x_2, \\
\left[\prod_{i=1}^{x-1}(1 - b^*(i))\right] \lambda_{12}(x), & x_1 = x_2 = x,
\end{cases}$$

where $b^*(i) = b_1(i) + b_{12}(i)$.

Theorem 6.4. *A necessary and sufficient condition that* $(b_1, b_2, b_{12}, b_1(x|x_2), b_2(x|x_1))$ *is the bivariate hazard rate of a random vector* (X_1, X_2), *where* $X_1, X_2 = 1, 2, \ldots,$ *is that*

$$\sum_{i=1}^{\infty}[b_1(i) + b_2(i) + b_{12}(i)] = \infty,$$

$$\sum_{x=x_2+1}^{\infty} b_1(x|x_2) = \infty \quad \textit{and} \quad \sum_{x=x_1+1}^{\infty} b_1(x|x_1) = \infty.$$

Example 6.5. Consider the bivariate geometric distribution in Example 6.1.

In this case, we have

$$\begin{aligned}
P(X_1 \geq x_1, X_2 \geq x_2) &= q_1^{x_1} q_2^{x_2} \theta^{x_1 x_2}, \\
P(X_1 = x_1, X_2 \geq x_2) &= q_1^{x_1} q_2^{x_2} \theta^{x_1 x_2} - q_1^{x_1+1} q_2^{x_2} \theta^{(x_1+1)x_2} \\
&= q^{x_1} q_2^{x_2} \theta^{x_1 x_2}(1 - q_1 \theta^{x_1}), \\
P(X_1 = x_1, X_2 = x_2) &= q_1^{x_1} q_2^{x_2} \theta^{x_1 x_2 - 1}[(1 - q_1 \theta^{x_2+1})(1 - q_2^{x_1+1}) + \theta - 1].
\end{aligned}$$

With the above results, we find

$$\begin{aligned}
b_1(x) &= q_2^x \theta^x (1 - q_1 \theta^x), \\
b_2(x) &= q_1^x \theta^x (1 - q_2 \theta^x), \\
b_{12}(x) &= [(1 - q_1 \theta^{x+1})(1 - q_2 \theta^{x+1}) + \theta - 1]\theta^{-1}, \\
b_1(x|x_2) &= \frac{(1 - q_1 \theta^{x+1})(1 - q_2 \theta^{x_2+1}) + \theta - 1}{\theta(1 - q_1 \theta^x)}, \quad x > x_2, \\
b_2(x|x_1) &= \frac{(1 - q_2 \theta^{x+1})(1 - q_1 \theta^{x_1+1}) + \theta - 1}{\theta(1 - q_2 \theta^x)}, \quad x > x_2.
\end{aligned}$$

The multivariate extension of the conditional hazard rate is accomplished in the following manner. Take $\mathbf{I} = (i_1, i_2, \ldots, i_k) \subset (1, 2, \ldots, p)$ and $\mathbf{t_I} = (t_{i_1} \ldots t_{i_k})$, I^c as the complement of I, and $\mathbf{e} = (1, 1, \ldots, 1)$. Then, the multivariate conditional hazard rate is defined as

$$b_{\mathbf{J}|\mathbf{I}}(x|\mathbf{x}_I) = P(\mathbf{X_J} = x\mathbf{e}, X_{\mathbf{I}^c - \mathbf{J}} > x\mathbf{e} | X_\mathbf{I} = x_\mathbf{I}, X_{\mathbf{I}^c} \geq x\mathbf{e})$$

for some $\mathbf{J} \subset I^c \subset (1, 2, \ldots, p)$ and $x_\mathbf{I} < x\mathbf{e}$.

Application of the above hazard rate in multivariate imperfect repair has also been pointed out by Shaked et al. (1995).

6.2.3 Vector Hazard Rate

In a variant approach analogous to the continuous case in Johnson and Kotz (1975), Nair and Nair (1990) defined the vector-valued bivariate hazard rate as

$$\mathbf{c}(\mathbf{x}_2) = (c_1(\mathbf{x}_2), c_2(\mathbf{x}_2)), \quad \mathbf{x}_2 = (x_1, x_2) \in N^2,$$

where

$$c_i(\mathbf{x}_2) = P(X_i = x_i | \mathbf{X}_2 \geq \mathbf{x}_2), \quad i = 1, 2. \tag{6.13}$$

We interpret $c_1(\mathbf{x}_2)$ as the conditional probability that component 1 fails at age x_i, given that components 1 and 2 survive age x_1 and x_2, respectively. A similar interpretation is given to $c_2(\mathbf{x}_2)$. Note that

$$1 - c_1(\mathbf{x}_2) = \frac{S(x_1 + 1, x_2)}{S(x_1, x_2)} \tag{6.14}$$

and

$$1 - c_2(\mathbf{x}_2) = \frac{S(x_1, x_2 + 1)}{S(x_1, x_2)}. \tag{6.15}$$

The vector failure rate for a p-dimensional vector $\mathbf{X} = (X_1, \ldots, X_p)$ is a direct extension of (6.13) in the form (Nair and Asha, 1997)

$$\mathbf{c}(\mathbf{x}) = (c_1(\mathbf{x}), \ldots, c_p(\mathbf{x})),$$

where

$$c_i(\mathbf{x}) = P[X_i = x_i | \mathbf{X} \geq \mathbf{x}] = 1 - \frac{S(\mathbf{x}_{(i)})}{S(\mathbf{x})} \tag{6.16}$$

with $\mathbf{x}_{(i)} = (x_1, x_2, \ldots, x_{i-1}, x_i + 1, x_{i+1} \ldots, x_p)$. The p-dimensional vector hazard rate determines the distribution of $\mathbf{X}$ uniquely through the formula

$$S(\mathbf{x}) = \prod_{r=0}^{x_1-1} [1 - h_1(x_1 - r - 1, \mathbf{x}_1^*)] \times \prod_{r=0}^{x_2-1} [1 - h_2(0, x_2 - r - 1, \mathbf{x}_2^*)]$$
$$\cdots \times \prod_{r=0}^{x_p-1} [1 - h_p(\mathbf{O}_{p-1}, x_p - r - 1)], \tag{6.17}$$

where $\mathbf{x}_i^* = (x_{i+1}, \ldots, x_p)$.

Remark 6.2. The result in (6.17) is a consequence of the multiplicative rule

$$P(\mathbf{X} > \mathbf{x}) = P(X_p \geq x_p) P(X_{p-1} \geq x_{p-1} | X_p \geq x_p),$$
$$\ldots P(X_1 \geq x_1 | \mathbf{X}_1^* > \mathbf{x}_1^*).$$

Since the left hand side can be expressed as a product of expressions in $X_{i_1}, X_{i_2}, \ldots, X_{i_p}$, where $(i_1, \ldots, i_p)$ are permutations of $(1, \ldots, p)$, there are p expressions of the form (6.17) depending on which of the variables we begins with. Although all p forms give the same $S(\mathbf{x})$, they may be useful in characterization problems.

Remark 6.3. The component vectors in $\mathbf{c}(\mathbf{x})$ are not independent of one another. Certain conditions have to be satisfied among them; for example, in the bivariate case, we need to have

$$[1 - c_1(x_1, x_2 + 1)][1 - c_2(x_1, x_2)] = [1 - c_2(x_1 + 1, x_2)][1 - c_1(x_1, x_2)].$$

Example 6.6. For the same bivariate geometric distribution in Example 6.5,

$$\begin{aligned} c_1(\mathbf{x}_2) &= 1 - \frac{S(x_1+1, x_2)}{S(x_1, x_2)} \\ &= 1 - \frac{q_1^{x_1+1} q_2^{x_2} \theta^{(x_1+1)x_2}}{q_1^{x_1} q_2^{x_2} \theta^{x_1 x_2}} \\ &= 1 - q_1 \theta^{x_2}, \end{aligned} \tag{6.18}$$

and similarly

$$c_2(\mathbf{x}_2) = (1 - q_1 \theta^{x_1}). \tag{6.19}$$

Notice that $c_1(\mathbf{x}_2)$ is independent of x_1 and $c_2(\mathbf{x}_2)$ is independent of x_2. Thus, the vector hazard rate can be written in the form $c(\mathbf{x}) = (\alpha_1(x_2), \alpha_2(x_1))$, where $\alpha_1(x_2)$ and $\alpha_2(x_1)$ are given by (6.18) and (6.19), respectively.

Remark 6.4. The univariate hazard rates of $X_1, \ldots, X_p$ can be derived from (6.16) as

$$h_i(x_i) = C_i(0, \ldots, 0, x_i, 0, \ldots, 0), \quad i = 1, 2, \ldots, p.$$

Some properties of the vector hazard rates are as follows:

(i) Being conditional probabilities, $0 \le c_i(\mathbf{x}) \le 1$;
(ii) $c_i(\mathbf{x})$ is the one-dimensional hazard rate of the univariate conditional distribution of X_i, given $\mathbf{X} \ge \mathbf{x}$, and as such enjoys appropriate properties of $h_i(x_i)$;
(iii) When $X_1, \ldots, X_p$ are independent, $\mathbf{c}(\mathbf{x}) = \mathbf{k}$, $\mathbf{k} = (k_1, \ldots, k_p) > \mathbf{0}$ and conversely;
(iv) In the bivariate case, the scalar hazard rate and the vector hazard rate are connected through

$$a(\mathbf{x}_2) = c_1(x_1, x_2) - c_1(x_1, x_2 + 1) + c_1(x_1, x_2 + 1)c_2(x_1, x_2),$$

or

$$a(\mathbf{x}_2) = c_2(x_1, x_2) - c_2(x_1+1, x_2) + c_1(x_1, x_2)c_2(x_1+1, x_2);$$

(v) In the bivariate case, $(h_1(x_1), c_1(x_1, x_2))$ or $(c_1(x_1, x_2), h_2(x_2))$ is enough to evaluate $S(x_1, x_2)$. However, in the multivariate case, we would require one marginal failure rate and the functional forms of the remaining elements of $\mathbf{c}(\mathbf{x})$;

(vi) Characterization of multivariate distributions can be accomplished by special forms of $\mathbf{c}(\mathbf{x})$. This will be discussed later in Chapter 8.

6.2.4 Total Hazard Rate

The failure of a two-component parallel system can be considered to consist of two stages: first, when one of the components fail and then the remaining component fails. Thus, a bivariate hazard rate can be decomposed into (a) the hazard rate of the minimum lifetime of the two components and (b) the conditional hazard rate of one of the components, given that the other failed first. In this way, Sun and Basu (1995) defined the total hazard rate of (X_1, X_2) as the vector $(d(x), d_1(x_1|x_2), d_2(x_2|x_1))$, where

$$d(x) = \frac{P(\min(X_1, X_2) = x)}{S(x_1, x_2)}, \tag{6.20}$$

$$d_1(x_1|x_2) = \frac{f(x_1, x_2)}{\sum_{t=x_1}^{\infty} f(t, x_2)}, \quad x_1 > x_2, \tag{6.21}$$

$$d_2(x_2|x_1) = \frac{f(x_1, x_2)}{\sum_{t=x_2}^{\infty} f(x_1, t)}, \quad x_1 < x_2,\ x_1, x_2 = 0, 1, 2, \ldots \tag{6.22}$$

Sun and Basu (1995) have shown that

(1) If (X_1, X_2) has constant total hazard rate $(1-p, 1-p_1, 1-p_2)$, then the distribution of (X_1, X_2) is specified by the bivariate geometric distribution

$$S(x_1, x_2) = \begin{cases} q^{x_2} q_1^{x_1 - x_2}, & x_1 \geq x_2, \\ q^{x_1} q_2^{x_2 - x_1}, & x_1 \leq x_2, 0 < q < q_1 \\ & q_2 < 1, 1 + q \geq q_1 + q_2; \end{cases} \tag{6.23}$$

(2) (X_1, X_2) has bivariate geometric distribution in (6.23) if and only if (X_1, X_2) has geometric marginals and constant total failure rate.

A further investigation into the conditions for constancy of total hazard rate by Asha et al. (2003) revealed that geometric marginals as mentioned above is not an essential condition. The geometric marginals can be replaced by others and accordingly there exist many bivariate distributions for which $d(x)$,

$d_1(x_1|x_2)$ and $d_2(x_2|x_1)$ are constants. Main results in this connection are reviewed in Chapter 8.

Example 6.7. Consider the distribution in (6.23). The probability mass function is

$$f(x_1, x_2) = \begin{cases} (q_1 - q_0)(1-q)q^{x_2}q_1^{x_1-x_2-1}, & x_1 > x_2, \\ (q_2 - q)(1-q)q^{x_1}q_2^{x_1-x_2-1}, & x_1 < x_2, \\ (1-p)q^{x_2}, \quad x_1 = x_2, \end{cases}$$

$$d(x) = \frac{P(\min(X_1, X_2) = x)}{S(x_1, x_2)} = P(Z = x | Z \geq x) = 1 - q,$$

where $Z = \min(X_1, X_2)$. Also,

$$\sum_{t=x_1}^{\infty} f(t, x_2) = \frac{(q_1 - q)(1-q)}{1-q_1} q^{x_2} q_1^{x_1-x_2-1},$$

so that

$$d_1(x_1|x_2) = 1 - q_1.$$

By symmetry,

$$d_2(x_2|x_1) = 1 - q_2.$$

6.2.5 Alternative Conditional Hazard Rate

A fifth bivariate hazard rate based on conditional distributions, proposed by Kotz and Johnson (1991), is defined by the vector

$$\mathbf{e}(x_1, x_2) = (e_1(x_1, x_2), e_2(x_1, x_2)),$$

where

$$e_i(\mathbf{x}_2) = P(X_i = x_i | X_i \geq x_i, X_j = x_j), \ i, j = 1, 2, \ i \neq j. \tag{6.24}$$

Notice that

$$1 - e_1(\mathbf{x}_2) = \frac{S(x_1+1, x_2) - S(x_1+1, x_2+1)}{S(x_1, x_2) - S(x_1, x_2+1)}$$

and

$$1 - e_2(\mathbf{x}_2) = \frac{S(x_1, x_2+1) - S(x_1+1, x_2+1)}{S(x_1, x_2) - S(x_1+1, x_2)}.$$

Asha and Nair (1997) showed that subject to the compatibility conditions on the conditional probability mass functions $f_{12}(x_1|x_2)$ and $f_{21}(x_2|x_1)$, $\mathbf{e}(\mathbf{x}_2)$ determines the distribution of $\mathbf{X}$ uniquely through the identities

$$f_{12}(x_1|x_2) = \prod_{t=0}^{x_1-1} [1 - e_1(t, x_2)] e_1(\mathbf{x}_2)$$

and

$$f_{21}(x_2|x_1) = \prod_{t=0}^{x_2-1} [1 - e_2(x_1, t)] e_2(\mathbf{x}_2).$$

Example 6.8. The bivariate geometric conditional distribution is represented by the probability mass function

$$f(\mathbf{x}) = \alpha q_1^{x_1} q_2^{x_2} \theta^{x_1 x_2}, \quad 0 < q_1, q_2 < 1,\ 0 \le \theta \le 1,$$

and

$$\alpha = \sum_{t=0}^{\infty} \frac{q_1^t}{(1 - p_1\theta^t)} = \sum_{t=0}^{\infty} \frac{q_2^t}{(1 - p_2\theta^t)}.$$

From (6.24),

$$\begin{aligned} e_1(\mathbf{x}) &= \frac{P(X_1 = x_1, X_2 = x_2)}{P(X_1 \ge x_1, X_2 = x_2)} = \frac{\alpha q_1^{x_1} q_2^{x_2} \theta^{x_1 x_2}}{\alpha q_2^{x_2} \sum_{t=x_1}^{\infty} (q_1 \theta^{x_2})^t} \\ &= 1 - q_1 \theta^{x_2}. \end{aligned}$$

Similarly,

$$e_2(\mathbf{x}) = 1 - q_2 \theta^{x_1}.$$

Note that $e_1(\mathbf{x})$ is independent of x_1 and $e_2(\mathbf{x})$ is independent of x_2.

6.2.6 Constancy of Hazard Rates

In the univariate case, we have seen that a constant hazard rate implies that the corresponding component does not age, and as such is of importance in deriving ageing concepts. Moreover, in that case, the distribution of the lifetime is geometric. In this case, we shall investigate the interpretation and distributions when the various definitions provide constant multivariate hazard rates. The multivariate geometric distribution with independent marginals given by

$$S(\mathbf{x}) = q_1^{x_1} \cdots q_p^{x_p}, \quad 0 < q_i < 1,\ i = 1, 2, \ldots, p, \tag{6.25}$$

plays an important role in this connection. First, we consider the conditions under which scalar hazard rate is a constant.

Theorem 6.5. *The scalar hazard rate is of the form*

$$a_r(\mathbf{x}_r) = k_1 \dots k_r \text{ for each } r = 1, 2, \dots p \text{ and all } \mathbf{x}, \tag{6.26}$$

with k_i, $i = 1, \dots, r$, *being constants in* $(0, 1)$, *if and only if the distribution of* $\mathbf{X}$ *is of the form* (6.25).

Proof. By assumption, $a(x_1, x_2) = k_1 k_2$. Using (6.4), we have

$$\begin{aligned} S(x_1, 1) &= (1-k_1)^{x_1} - \sum_{t=x_1}^{\infty} k_1 k_2 (1-k_1)^t \\ &= (1-k_1)^{x_1}(1-k_2). \end{aligned}$$

By induction, from (6.3),

$$S(x_1, x_2) = (1-k_1)^{x_1}(1-k_2)^{x_2}.$$

Thus, the result is true for $p = 2$. Now, assume the result for $p = s$ so that

$$S(x_1, x_2 \dots x_s) = (1-k_1)^{x_1}(1-k_2)^{x_2} \dots (1-k_r)^{x_s}.$$

Using (6.6), we have

$$\begin{aligned} S(x_1, x_2, \dots x_s, 1) &= (1-k_1)^{x_1} \dots (1-k_r)^{x_s} - \sum_{t=x_1}^{\infty} \cdots \sum_{t=x_r}^{\infty} k_1 k_2 \\ &\quad \dots k_s k_{s+1} \prod_{i=1}^{s} (1-k_i)^{x_i} \\ &= (1-k_1)^{x_1} \dots (1-k_s)^{x_s}(1-k_{s+1}). \end{aligned}$$

Again, from (6.6) by induction

$$S(x_1 \dots x_{s+1}) = \prod_{i=1}^{s+1} (1-k_i)^{x_i}.$$

Thus, for any finite p, the distribution is of the form (6.25) with $q_i = 1 - k_i$. The converse is easy to see from (6.25) and

$$f(\mathbf{x}) = \prod_{i=1}^{p} (1-q_i) q_i^{x_i}$$

with $q_i = 1 - k_i$. □

Remark 6.5. The relationship $a(\mathbf{x}_r) = k^p$ holds for all $\mathbf{x}$ if the X_i's are independent and identically distributed as geometric with parameter $(1-k)$, provided X_1 is geometric $(1-k)$. Without the condition on all $\mathbf{X}_r$, $\mathbf{X}$ is neither geometric nor contains independent components. See Example 6.1.

Theorem 6.6. *A necessary and sufficient condition that* $\mathbf{X}$*, defined on* N^p*, has a scalar failure rate of the form*

$$a_r(\mathbf{x}_r) = K_{r \cdot i_1 i_2 \ldots i_r}$$

for every $\mathbf{x}_r$ *and* $r = 1, \ldots, p$ *satisfying* $x_{i_1} > \cdots > x_{i_k} = x_{i_{k+1}} = \cdots = x_{i,k+j} > \cdots > x_{i_r}$, $j = 0, 1, 2, \ldots, r-k$*, is that the survival function of* $\mathbf{X}$ *has the form*

$$S(\mathbf{x}) = q_{i_1}^{x_{i_1}} \left(\frac{q_{i_1 i_2}}{q_{i_1}}\right)^{x_{i_2}} \left(\frac{q_{i_1,i_2,i_3}}{q_{i_1,i_2}}\right)^{x_{i_3}} \cdots \left(\frac{q_{i_1 \ldots i_p}}{q_{i_1 \ldots i_{p-1}}}\right)^{x_{i_p}} \tag{6.27}$$

for $x_{i_1} \geq x_{i_2} \geq \cdots \geq x_{i_p}$.

The parameters of the model in (6.27) *are such that*

$$0 \leq q_{i_1}, \cdots \leq q_{i_p} \leq \cdots \leq q_{i_1,i_2} \leq \cdots \leq q_1 q_2 \ldots q_p < 1,$$
$$q_{i_1,i_2 \ldots i_K} = q_{123 \ldots k} \text{ for } k = 2, 3, \ldots, p,$$

and

$$1 - \sum_{k=1}^{p} q_k - \sum_{k<l} q_{kl} + \cdots + (-1)^{p-1} q_{123 \ldots p} \geq 0.$$

The proof has been given by Asha (1995). The above result is more visible in the bivariate case in which the distribution.

$$S(x_1, x_2) = \begin{cases} q^{x_2} q_1^{x_1 - x_2}, & x_1 \geq x_2, \\ q^{x_1} q_2^{x_2 - x_1}, & x_2 \geq x_1, \end{cases}$$

$x_1, x_2 = 0, 1, 2, \ldots, 1 + q \geq q_1 + q_2, 0 < q < q_1, q_2 < 1$, is characterized by

$$a(\mathbf{x}) = \begin{cases} k_1, & x_1 > x_2, \\ k_2, & x_1 < x_2, \\ k_3, & x_1 = x_2, \end{cases}$$

where k_1, k_2 and k_3 are constants, $0 < k_1, k_2, k_3 < 1$, and the marginal distribution of X_i is geometric with parameter q_i, $i = 1, 2$.

Now, we prove a similar result for the vector-valued hazard rate.

Theorem 6.7. *A necessary and sufficient condition for the vector-valued hazard rate*

$$\mathbf{c}(\mathbf{x}) = \mathbf{k},\ k = (k_1, k_2, \dots k_p),$$

is that the distribution of $\mathbf{X}$ *is* (6.25).

Proof. Since $\mathbf{c}(\mathbf{x}) = \mathbf{k}$,

$$\mathbf{c}(\mathbf{O}_{i-1}, x_i, \mathbf{O}_{p-i}) = h_i(x_i) = k_i.$$

Substituting in (6.17), we have (6.25) with $q_i = 1 - k_i$. The converse is obvious. □

Theorem 6.8. *For all* $\mathbf{x}$ *and* $\mathbf{t} = (t, t, \dots, t)$,

$$c_i(x_1 + t, \dots x_r + t) = c_i(x_1, x_2, \dots x_r), \quad r = 1, 2, \dots, p,$$

holds if and only if $\mathbf{X}$ *has distribution in* (6.27).

The conditions for the constancy of the total hazard rate ($d(x)$, $d_1(x_1|x_2)$, $d_2(x_2|x_1)$) have been investigated by Asha et al. (2003). They have found that there many distributions possessing the property.

Theorem 6.9. *The random vector* (X_1, X_2) *has bivariate distribution in* N^2 *with survival function*

$$S(x_1, x_2) = \begin{cases} q^{x_2}[(1-\alpha)q^{x_1-x_2} + \alpha q_1^{x_1-x_2}], & x_1 \geq x_2, \\ q^{x_1}[(1-\alpha)q^{x_1-x_2} + \alpha q_2^{x_2-x_1}], & x_2 \geq x_1, \end{cases}$$

$0 < q < q_1, q_2 < 1$, $q \neq q_1 q_2$, $1 - q + \alpha(2q - q_1 - q_2) > 0$, $0 < d < 1$, *if and only if* $\mathbf{X}_2$ *has constant total hazard rate* (c_1, c_2, c_3) *and the marginals are*

$$S_i(x_i) = \alpha q_i^{x_i} + (1-\alpha)q^{x_i},\ x = 0, 1, 2, \dots,\ q = S(1, 1),$$

and

$$c_1 = 1 - q,\ c_2 = 1 - q_1 \text{ and } c_3 = 1 - q_2.$$

Finally, the alternative conditional hazard rate $\mathbf{e}(x_1, x_2) = (c_1, c_2)$ if and only if X_1 and X_2 are independent geometric with parameter $1 - c_1$ and $1 - c_2$, respectively. This follows from the representation of $f_{12}(x_1|x_2)$ and $f_{21}(x_2|x_1)$ in terms of e_1 and e_2 in Section 6.2.5 and the distribution in (6.25).

The functional forms of various types of hazard rates for some bivariate distributions are given in Tables 6.1, 6.2, 6.3 and 6.4. Note that the same distributions labelled as $A, B, C, \dots$ in Table 6.1 are carried forward in other tables to avoid repetitions of expressions of their survival functions.

TABLE 6.1 Bivariate scalar hazard rates

Distribution	Survival function	$a(x_1, x_2)$
A	$q_1^{x_1} q_2^{x_2}$	$(1-q_1)(1-q_2)$
B	$q_1^{x_1} q_2^{x_2} \theta^{x_1 x_2}$	$[(1-q_1\theta^{x_2+1})(1-q_2\theta^{x_1+1})+\theta-1]\theta^{-1}$
C	$\frac{(1-p)q_1^{x_1+x_2}}{1-pq_1^{x_1+x_2}}$	$1-\frac{2q_1(1-pq_1^{x_1+x_2})}{1-pq_1^{x_1+x_2+1}}+\frac{q_1^2(1-pq_1^{x_1+x_2})}{1-pq_1^{x_1+x_2+2}}$ $0<p<1, 0<q_1<1$
D	$q^{(x_1+x_2)^\beta}$	$1-2q^{(x_1+x_2+1)^\beta-(x_1+x_2)^\beta}+q^{(x_1+x_2+2)^\beta-(x_1+x_2)^\beta}$ $0<q<1, \beta>0$
E	$\frac{2}{1+e^{\frac{x_1+x_2}{\sigma}}}$	$1-\frac{2(1+e^{\frac{x_1+x_2}{\sigma}})}{1+e^{\frac{x_1+x_2+1}{\sigma}}}+\frac{1+e^{\frac{x_1+x_2}{\sigma}}}{1+e^{\frac{x_1+x_2+2}{\sigma}}}$
F	$\frac{e^{-\theta(x_1+x_2)}(1+\theta+\theta(x_1+x_2))}{1+\theta}$	$\frac{1+\theta(x_1+x_2)-2(1+\theta(x_1+x_2+1)e^{-\theta})+[1+\theta(x_1+x_2+2)e^{-2\theta}]}{1+\theta(x_1+x_2)}$
G	$\prod_{i=1}^{x_1+x_2-1}(1-p+p\pi^i)$	$1-\{(1-p)+p\pi^{x_1+x_2}\}\{1+p-p\pi^{x_1+x_2+1}\}$
H	$\frac{(m)_{x_1+x_2}}{(m+n)_{x_1+x_2}}$	$\frac{n(n+1)}{(m+n+x_1+x_2)(m+n+x_1+x_2+1)}, m, n>0$
I	$\frac{\binom{k+n-x_1-x_2}{n-x_1-x_2}}{\binom{k+n}{n}}$	$\frac{k(k-1)}{(k+n-x_1-x_2)(k+n-x_1-x_2-1)}$ $x_1, x_2=0, 1, \ldots, n, k, n>0$
J	$\frac{2}{1+(2q^{-x_1}-1)(2q^{-x_2}-1)}$	$2\Big[1-1+(2q^{-x_1}-1)(2q^{-x_2}-1)$ $\Big\{\frac{1}{1+(2q^{-x_1-1}-1)(2q^{-x_2}-1)}+\frac{1}{1+(2q^{-x_1}-1)(2q^{-x_2-1}-1)}-\frac{1}{1+(2q^{-x_1-1}-1)(2q^{-x_2-1}-1)}\Big\}\Big]$
K	$p^{x_2}p_1^{x_1-x_2}, x_1 \geq x_2$ $p^{x_1}p_2^{x_2-x_1}, x_1 \leq x_2$	$\frac{(p_1-p)(1-p_1)}{p_1}, x_1>x_2$ $\frac{(p_2-p)(1-p_2)}{p_2}, x_1<x_2$ $1+p-p_1-p_2, x_1=x+2$

TABLE 6.2 Bivariate vector hazard rates

Distribution	hazard rate $c_1(\mathbf{x}), c_2(\mathbf{x})$
A	$(1-q_1, 1-q_2)$
B	$(1-q_1\theta^{x_2}, 1-q_2\theta^{x_1})$
C	$\left(\frac{1-q_1}{1-pq_1^{x_1+x_2+1}}, \frac{1-q_1}{1-pq_1^{x_1+x_2+1}}\right)$
D	$(1-q_1^{(x_1+x_2+1)^\beta-(x_1+x_2)^\beta}, 1-q^{(x_1+x_2+1)-(x_1+x_2)^\beta})$
E	$\left(\frac{e^{\frac{x_1+x_2}{\sigma}}(e^{\frac{1}{\sigma}}-1)}{1+e^{\frac{x_1+x_2+1}{\sigma}}}, \frac{e^{\frac{x_1+x_2}{\sigma}}(e^{\frac{1}{\sigma}}-1)}{1+e^{\frac{x_1+x_2+1}{\sigma}}}\right)$
F	$\left(1-\frac{(1+\theta(x_1+x_2+2)e^{-\theta})}{1+\theta(x_1+x_2+1)}, 1-\frac{1+\theta(x_1+x_2+2)e^{-\theta}}{1+\theta(x_1+x_2+1)}\right)$
G	$(p(1-\pi^{x_1+x_2}), p(1-\pi^{x_1+x_2}))$
H	$\left(\frac{n}{m+n+x_1+x_2}, \frac{n}{m+n+x_1+x_2}\right)$
I	$\left(\frac{k}{k+n-x_1-x_2}, \frac{k}{k+n-x_1-x_2}\right)$
J	$\left(1-\frac{1+(2q^{-x_1}-1)(2q^{-x_2}-1)}{1+(2q^{-x_1+1}-1)(2q^{-x_2+1}-1)}, 1-\frac{1+(2q^{-x_1}-1)(2q^{-x_2}-1)}{1+(2q^{-x_1+1}-1)(2q^{-x_2+1}-1)}\right)$
K	$c_1(\mathbf{x}) = \begin{cases} (1-p_1, 1-pp_1^{-1}) & x_1 > x_2,\ c_2(\mathbf{x}) = c_1(\mathbf{x}) \\ (1-pp_2^{-1}, 1-p_2) & x_1 < x_2, \\ (1-p_1, 1-p_2) & x_1 = x_2. \end{cases}$

TABLE 6.3 Distributions with constant total hazard rates

	Survival function	Total hazard rates $(d(x), d(x_1\|x_2), d(x_2\|x_1))$
1	$\begin{cases} p^{x_2}p_1^{x_1-x_2}, & x_1 \geq x_2 \\ p^{x_1}p_1^{x_2-x_1}, & x_1 \leq x_2 \end{cases}$	$(1-p, 1-p_1, 1-p_2)$
2	$\begin{cases} p^{x_2}[\alpha p^{x_1-x_2}+(1-\alpha)p_1^{x_1-x_2}], & x_1 \geq x_2 \\ p^{x_1}[\alpha p^{x_2-x_1}+(1-\alpha)p_2^{x_1-x_2}], & x_1 \leq x_2 \end{cases}$ $0 \leq p, p_1, p_2 \leq 1$	$(1-p, 1-p_1, 1-p_2)$
3	$p^{x_2}\frac{(1-p)(1-p_1)}{1-p_1p_2}\left[1-\frac{p_1p_2-p}{p_1-p}\frac{p_1^{x_1-x_2}}{1-p_1}+\frac{p_1p_2-p}{p_1-p}\frac{p_1^{x_1-x_2}}{1-p}\right], x_1 \geq x_2$ $p^{x_1}\frac{(1-p)(1-p_2)}{1-p_1p_2}\left[1-\frac{p_1p_2-p}{p_2-p}\frac{p_1^{x_2-x_1}}{1-p_2}+\frac{p_1p_2-p}{p_2-p}\frac{p_2^{x_2-x_1}}{1-p}\right], x_1 \leq x_2$	$(1-p, 1-p_1, 1-p_2)$

6.3 MULTIVARIATE MEAN RESIDUAL LIFE

As in the case of the multivariate hazard rates, several definitions have been offered for the mean residual life functions in higher dimensions. These depend on

TABLE 6.4 Conditional hazard rates

Distribution	Conditional hazard rates $(e_1(\mathbf{x}), e_2(\mathbf{x}))$
A	$(1-q_1, 1-q_2)$
B	$\left(1-q_1\theta^{x_2}+\frac{q_1q_2\theta^{x_1+x_2}(\theta-1)}{1-q_2\theta^{x_1}}, 1-q_2\theta^{x_1}+\frac{(\theta-1)q_1q_2\theta^{x_1+x_2}}{1-q_1\theta^{x_2}}\right)$
C	$\left(1-\frac{q^{(x_1+x_2-1)^\beta}-q^{(x_1+x_2+2)^\beta}}{q^{(x_1+x_2)^\beta}-q^{(x_1+x_2+1)^\beta}}, 1-\frac{q^{(x_1+x_2-1)^\beta}-q^{(x_1+x_2+2)^\beta}}{q^{(x_1+x_2)^\beta}-q^{(x_1+x_2+1)^\beta}}\right)$
G	$\left(1-\frac{p(1-\pi)\pi^{x_1+x_2}}{p(1-\pi^{x_1+x_2})}, 1-\frac{p(1-\pi)\pi^{x_1+x_2}}{p(1-\pi^{x_1+x_2})}\right)$
H	$\left(\frac{n+1}{m+n+x_1+x_2+1}, \frac{n+1}{m+n+x_1+x_2+1}\right)$
K	$\begin{cases}(1-p_1, 1-pp_1^{-1}), & x_1 > x_2\\ (1-pp_2^{-1}, 1-p_2), & x_1 < x_2\\ \left(\frac{1+p-p_1-p_2}{1-p_2}, \frac{1-p-p_1-p_2}{1-p_1}\right), & x_1 = x_2\end{cases}$

the manner in which the extension of the univariate concept is carried over in the multivariate case. In all cases, the interest is on the lifetime of a p-component system after the ith component has attained age x_i, for $i = 1, \ldots, p$. If $\mathbf{X}$ represents the lifelength as mentioned in Section 6.2, the random variable

$$\mathbf{X_x} = (\mathbf{X} - \mathbf{x}|\mathbf{X} > \mathbf{x}) \tag{6.28}$$

is the residual life of the system after it has survived age $\mathbf{x}$. The survival function of $\mathbf{X_x}$ is

$$S_{\mathbf{x}}(\mathbf{t}) = \frac{S(\mathbf{x}+\mathbf{te})}{S(\mathbf{x}+\mathbf{e})}, \tag{6.29}$$

where $\mathbf{t} = (t_1, \ldots, t_p)$ and $\mathbf{e} = (1, \ldots, 1)$ and $t_i = 0, 1, \ldots$ for all i.

The first definition of the multivariate mean residual life function conceives it as the mean of the distribution in (6.29). In other words, the multivariate mean residual life function is defined as the vector

$$\mathbf{m}(\mathbf{x}) = (m_1(\mathbf{x}), \ldots, m_p(\mathbf{x}))$$

where

$$m_i(\mathbf{x}) = E(X_i - x_i|\mathbf{X} > \mathbf{x}), \quad i = 1, \ldots, p, \ x_i = -1, 0, 1, \ldots \tag{6.30}$$

From (6.30), we have

$$m_i(\mathbf{x}) = \frac{1}{S(\mathbf{x}+\mathbf{e})} \sum_{t_1=x_1+1}^{\infty} \cdots \sum_{t_p=x_p+1}^{\infty} (t_i - x_i) f(\mathbf{x})$$

$$= \frac{1}{S(\mathbf{x}+\mathbf{e})} \sum_{t_i=x_i+1}^{\infty} S(x_1+1, \ldots, x_{i-1}+1, t, x_{i+1}+1, \ldots, x_p+1).$$

Example 6.9. When X is bivariate geometric with

$$S(\mathbf{x}) = q_1^{x_1} q_2^{x_2} \theta^{x_1 x_2}$$

as in Example 6.1, we have

$$m_1(\mathbf{x}_2) = \frac{1}{S(\mathbf{x}_2+\mathbf{e}_2)} \sum_{t_1=x_1+1}^{\infty} q_1^{t_1} q_2^{x_2+1} \theta^{(x_2+1)t}$$

$$= \frac{1}{S(\mathbf{x}_2+\mathbf{e}_2)} q_2^{x_2+1} \sum_{t_1=x_1+1}^{\infty} (q_1\theta^{x_2+1})^{t_1}$$

$$= \frac{1}{1-q_1\theta^{x_2+1}}.$$

Likewise, $m_2(\mathbf{x}) = (1-q_2\theta^{x_1+1})^{-1}$. Notice that $m_1(\mathbf{x})$ is independent of x_1 and $m_2(\mathbf{x})$ is independent of x_2, and in the bivariate case, we have the formula

$$m_1(x_1, x_2) = \frac{1}{S(x_1+1, x_2+1)} \sum_{t=x_1+1}^{\infty} S(t, x_2+1) \tag{6.31}$$

and

$$m_2(x_1, x_2) = \frac{1}{S(x_1+1, x_2+1)} \sum_{t=x_2+1}^{\infty} S(x_1+1, t). \tag{6.32}$$

The bivariate mean residual life function satisfies the following properties:

(i) $m_1(-1,-1) = 1+\mu_1$ and $m_2(-1,-1) = 1+\mu_2$, where $\mu_i = E(X_i)$, $i = 1, 2$;

(ii) The mean residual life functions of the marginal distributions of X_1 and X_2 are derived as

$$m_{X_1}(x_1) = E(X_1 - x_1 | X_1 > x_1) = m_1(x_1, -1)$$

and

$$m_{X_2}(x_2) = E(X_2 - x_2 | X_2 > x_2) = m_2(-1, x_2).$$

Further, if X_1 and X_2 are independent, then $m_1(x_1, x_2) = m_1(x_1, -1)$ and $m_2(x_1, x_2) = m_2(-1, x_2)$;

(iii) Since the univariate mean residual life function determines the corresponding distribution uniquely, it is of interest to know whether the same is true for the bivariate extension.

Theorem 6.10. *Given* $\mathbf{m}(\mathbf{x})$, *the distribution of* $\mathbf{X}$ *is determined as*

$$S(\mathbf{x}) = \prod_{r=0}^{x_1-1} \frac{m_1(r-1, x_2) - 1}{m_1(r, x_2)} \prod_{r=0}^{x_2-1} \frac{m_2(0, r-1) - 1}{m_2(0, r)}.$$

Proof. From (6.31),

$$m_1(x_1, x_2) S(x_1 + 1, x_2 + 1) = \sum_{t=x_1+1}^{\infty} S(t, x_2 + 1)$$

and so

$$m_1(x_1 - 1, x_2) S(x_1, x_2 + 1) - m_1(x_1, x_2) S(x_1 + 1, x_2 + 1) = S(x_1, x_2 + 1).$$

This gives a recurrence relation in $S(x_1, x_2)$ as

$$S(x_1 + 1, x_2 + 1) = \frac{[m_1(x_1 - 1, x_2) - 1]}{m(x_1, x_2)} S(x_1, x_2 + 1).$$

Reducing x_2 successively, we get

$$S(x_1, x_2) = \prod_{r=0}^{x_1-1} \frac{m_1(r-1, x_2) - 1}{m(r, x_2)} S(0, x_2). \tag{6.33}$$

Similarly, working with (6.32), we have

$$S(x_1, x_2) = \prod_{r=0}^{x_2-1} \frac{m_2(x_1, r-1) - 1}{m_2(x_1, r)} S(x_1, 0). \tag{6.34}$$

Setting $x_1 = 0$ in (6.34) we get $S(0, x_2)$ and substituting it in (6.33), we get the required result. □

Remark 6.6. We can start with (6.32) and then use (6.31) to arrive at an equivalent alternative expression

$$S(x_1, x_2) = \prod_{r=0}^{x_1-1} \frac{m_1(r_1 - 1, 0) - 1}{m_1(0, r)} \prod_{r=0}^{x_2-1} \frac{m_2(x_1, r-1) - 1}{m_2(x_1, r)}; \tag{6.35}$$

(iv) The bivariate hazard rates and mean residual life functions are related to one another:

(a)

$$a(x_1+1,x_2+1)=1-\frac{m_1(\mathbf{x}_2)-1}{m_1(x_1+1,x_2)}-\frac{m_2(\mathbf{x}_2)-1}{m_2(x_1,x_2+1)}-\frac{[m_2(x_1+1,x_2)-1][m_1(\mathbf{x})-1]}{m_2(x_1+1,x_2+1)-1]m_1(x_1+1,x_2)},$$

(b)

$$c_1(x_1+1,x_2+1)=1-\frac{m_1(\mathbf{x}_2)-1}{m_1(x_1+1,x_2)}$$

and

$$c_2(x_1+1,x_2+1)=1-\frac{m_2(\mathbf{x}_2)-1}{m_2(x_1,x_2+1)},$$

(c)

$$e_1(x_1+1,x_2+1)=1-\frac{m_1(\mathbf{x}_2-1)\left[1-\frac{m_2(x_1+1,x_2)-1}{m_2(x_1+1,x_2+1)}\right]}{m_1(x_1+1,x_2)\left[1-\frac{m_2(\mathbf{x})-1}{m_2(x_1,x_2+1)}\right]}$$

and

$$e_2(x_1+1,x_2+1)=1-\frac{m_2(\mathbf{x}_2)-1\left[1-\frac{m_1(x_1,x_2+1)-1}{m_1(x_1+1,x_2+1)}\right]}{m_2(x_1,x_2+1)\left[1-\frac{m_1(\mathbf{x})-1}{m_1(x_1+1,x_2)}\right]};$$

(v) The mean residual life function is of the form

$$\mathbf{m}(\mathbf{x}_2)=(k_1,k_2),\quad k_1,k_2>1,$$

where k_1 and k_2 are constants independent of x_1 and x_2 if and only if X_1 and X_2 are independent and geometrically distributed with parameters $\frac{k_1-1}{k_1}$ and $\frac{k_2-1}{k_2}$ (Nair and Nair, 1988a). The proof is obtained by an application of Theorem 6.10.

Expressions of mean residual life functions of some distributions are given in Table 6.5.

6.3.1 Conditional Mean Residual Life

From Asha et al. (2004) with a slight change in definition, we write

TABLE 6.5 Bivariate mean residual life functions

Distribution	$(m_1(\mathbf{x}), m_2(\mathbf{x}))$
A	$((1-q_1)^{-1}, (1-q_2)^{-1})$
B	$((1-q_1\theta^{x_2+1})^{-1}, (1-q_2\theta^{x_1+1})^{-1})$
C	$(1-pq_1^{x_1+x_2})\,(\sum_{t=x_1+1}^{\infty}(1-pq_1^{x_2+1})q_1^t, \sum_{t=x_1+1}^{\infty}(1-pq_1^{x_1+1})q_2^t)$
F	$\left(\frac{1+\theta(x_1+x_2+x_3)}{1-e^{-\theta}}+\frac{\theta e^{-\theta}}{(1-e^{-\theta})^2}, \frac{1+\theta(x_1+x_2+x_3)}{1-e^{-\theta}}+\frac{\theta e^{-\theta}}{(1-e^{-\theta})^2}\right)$
H	$\left(\frac{m+n+x_1+x_2+1}{n-1}, \frac{m+n+x_1+x_2+1}{n-1}\right)$
I	$\left(\frac{k+n-x_1-x_2-1}{k+1}, \frac{k+n-x_1-x_2-1}{k+1}\right)$

$$l_1(x_1,x_2) = E(X_1 - x_1 | X_1 > x_1,\ X_2 = x_2),\ x_1 = -1, 0, 1, \ldots,\quad x_2 = 0, 1, 2, \ldots,$$

and

$$l_2(x_1,x_2) = E(X_2 - x_2 | X_1 = x_1,\ X_2 > x_2),\ x_1 = 0, 1, 2 \ldots,\quad x_2 = -1, 0, 1, \ldots,$$

and then call the vector

$$\mathbf{l}(x_1,x_2) = (l_1(x_1,x_2), l_2(x_1,x_2))$$

as the *conditional mean residual life function* of (X_1, X_2). The function $l_1(\mathbf{x})$, $i = 1, 2$, can be interpreted as the expected remaining life of component 1 when it has survived age x_1 given the lifetime of the other component is x_2.

Notice that

$$l_1(x_1,x_2) = \sum_{t=x_1+1}^{\infty} \frac{P(X_1 \geq t_1, X_2 = x_2)}{P(X_1 \geq x_1, X_2 = x_2)}. \tag{6.36}$$

Following Asha et al. (2004), we have the following results:

(a) The alternative conditional hazard rate in (6.24) is related to $\mathbf{l}(x_1,x_2)$ through the identities

$$e_1(x_1,x_2) = 1 - \frac{l_1(x_1-1,x_2)-1}{l_1(x_1,x_2)}$$

and

$$e_2(x_1,x_2) = 1 - \frac{l_2(x_1,x_2-1)-1}{l_1(x_1,x_2)},\ x_1, x_2 = 0, 1, 2, \ldots; \tag{6.37}$$

(b) By virtue of (6.37), we can write from the results in Section 6.2.5 that

$$f_{12}(x_1|x_2) = \left[1 - \frac{l_1(x_1-1,x_2)-1}{l_1(x_1,x_2)}\right] \prod_{t=0}^{x-1} \frac{l_1(t_1-1,x_2)-1}{l_1(t_1,x_2)} \tag{6.38}$$

and

$$f_{21}(x_2|x_1) = \left[1 - \frac{l_2(x_1,x_2-1)-1}{l_2(x_1,x_2)}\right] \prod_{t=0}^{x_2-1} \frac{l_2(x_1,t-1)-1}{l_1(x_1,t)}. \tag{6.39}$$

When the conditional probability mass functions in (6.38) and (6.39) satisfy the compatibility conditions, the vector $\mathbf{l}(x_1,x_2)$ determines the distribution of (X_1,X_2) uniquely;

(c) When X_1 and X_2 are independent, $l_i(x) = m_{X_i}(x)$, $i = 1,2$;

(d) The components of $l_i(x_1,x_2)$ are linear in both x_1 and x_2 of the form

$$l_i(x_1,x_2) = a_i x_i + b_j x_j + c_i, \quad i \neq j,\ i,j = 1,2,$$

if and only if the components of $\mathbf{m}(x_1,x_2)$ is linear in both x_1 and x_2;

(e) The conditional mean residual life function is locally constant, that is, $\mathbf{l}(x_1,x_2) = (\alpha_1(x_2),\alpha_2(x_1))$ if and only if the distribution of (X_1,X_2) is

$$f(x_1,x_2) = \alpha q_1^{x_1} q_2^{x_2} \theta^{x_1,x_2}, \quad 0 < q_i < 1,\ 0 < \theta \le 1,\ x_1,x_2 = 0,1,2,\ldots,$$

with

$$\alpha^{-1} = \sum_{r=0}^{\infty} \frac{q_1^r}{1-q_2\theta^r} = \sum_{r=0}^{\infty} \frac{q_2^r}{1-q_1\theta^r}.$$

Further, if $\mathbf{l}(x_1,x_2) = (k_1,k_2)$, where k_1 and k_2 are constants, then X_1 and X_2 are independent and each is geometric;

(f) Let $\mathbf{X}$ and $\mathbf{Y}$ be two bivariate lifetimes. Then, $\mathbf{X}$ is less than $\mathbf{Y}$ in alternative conditional rate, denoted by $\mathbf{X} \le_{\text{chr}} \mathbf{Y}$, if $e_{\mathbf{X}}(x_1,x_2) \ge e_{\mathbf{Y}}(x_1,x_2)$ for all x_1,x_2 in N. Similarly, $\mathbf{X}$ is less than $\mathbf{Y}$ in conditional mean residual life, $\mathbf{X} \le_{\text{cmrl}} \mathbf{Y}$, if $l_{\mathbf{X}}(x_1,x_2) \le l_{\mathbf{Y}}(x_1,x_2)$. Further,

$$\mathbf{X} \le_{\text{chr}} \mathbf{Y} \Rightarrow \mathbf{X} \le_{\text{cmrl}} \mathbf{Y}.$$

Shaked and Shanthikumar (1991) defined another type of conditional mean residual life function in the continuous case, called the *dynamic mean residual life*. Its analogue in the discrete case is the vector $(n_1(x), n_2(x), n_1(x|x_2), n_2(x|x_1))$ in which

$$n_i(x) = E(X_i - x | X_1 > x, X_2 > x), \quad i = 1,2,\ x \ge 0,1,2,\ldots,$$

$$n_1(x|x_2) = E(X_1 - x|X_1 > x, X_2 = x_2), \quad x \geq x_2 = 0, 1, 2, \ldots,$$

and

$$n_2(x|x_1) = E(X_2 - x|X_1 = x_1, X_2 > x), \quad x \geq x_1 = 0, 1, 2, \ldots.$$

The interpretation to the above notion is that as time progresses failures occur one by one and the functions given above predict the remaining lives of the surviving components by conditioning them on the observed past up to time x.

6.4 VARIANCE RESIDUAL LIFE FUNCTIONS

In the bivariate case, generally the variations in X_1 and X_2 are measured by the variance-covariance matrix

$$\Sigma = \begin{pmatrix} \sigma_1^2 & \sigma_{12} \\ \sigma_{12} & \sigma_2^2 \end{pmatrix},$$

where σ_i^2 is the variance of X_i and σ_{12} is the covariance between X_1 and X_2 given by

$$\sigma_{12} = \text{Cov}(X_1, X_2) = E(X_1 X_2) - E(X_1)E(X_2).$$

In an analogous manner, we have to consider the variance-covariance matrix of the residual life distribution

$$\Sigma_{x_1 x_2} = \begin{pmatrix} \sigma_1^2(\mathbf{x}) & \sigma_{12}(\mathbf{x}) \\ \sigma_{12}(\mathbf{x}) & \sigma_2^2(\mathbf{x}) \end{pmatrix}.$$

The elements of $\Sigma_{x_1 x_2}$ can be expressed as

$$\sigma_1^2(\mathbf{x}) = V(X_1 - x_1|X_1 > x_1, X_2 > x_2), \tag{6.40}$$

$$\sigma_2^2(\mathbf{x}) = V(X_2 - x_2|X_1 > x_1, X_2 > x_2), \tag{6.41}$$

and

$$\sigma_{12}(\mathbf{x}) = E[(X_1 - x_1)(X_2 - x_2)|\mathbf{X} > \mathbf{x}] - m_1(\mathbf{x})m_2(\mathbf{x}). \tag{6.42}$$

We call $\sigma_1^2(\mathbf{x})$ and $\sigma_2^2(\mathbf{x})$ the variance residual life functions of X_1 and X_2, respectively and $\sigma_{12}(\mathbf{x})$ as the covariance of residual life. Also,

$$M(\mathbf{x}) = E[(X_1 - x_1)(X_2 - x_2)|\mathbf{X} > \mathbf{x}]$$

will be referred to as the product moment of residual life. Apart from providing a measure of variation in residual life of a two-component device, both $\sigma_1^2(\mathbf{x})$ and

$\sigma_2^2(\mathbf{x})$ can be treated as reliability functions in the same manner as the hazard rate or mean residual life. At the same time, $\sigma_{12}(\mathbf{x})$ measures the joint variation in the residual lives of the two components and gives an account of the nature of dependence between them.

6.4.1 Bivariate Variance Residual Life

From (6.39), the variance residual life function simplifies to

$$\sigma_1^2(\mathbf{x}) = E[(X - x_1)^2|\mathbf{X} > \mathbf{x}] - m_1^2(\mathbf{x}), \quad x_1, x_2 = -1, 0, 1, \ldots. \tag{6.43}$$

For practical purposes, evaluation of (6.43) can be accomplished in terms of the survival function using the expression provided in the next theorem.

Theorem 6.11.

$$\sigma_1^2(\mathbf{x}) = \frac{2}{S(x_1+1, x_2+1)} \sum_{t_1=x_1+1}^{\infty} \sum_{u=t_1+1}^{\infty} S(u, x_2+1) - m_1(\mathbf{x})(m_1(\mathbf{x}) - 1). \tag{6.44}$$

Proof. We have

$$E[(X - x_1)^2|\mathbf{X} > \mathbf{x}] = \frac{1}{S(x_1+1, x_2+1)} \sum_{t_1=x_1+1}^{\infty} \sum_{t_2=x_2+1}^{\infty} (t_1 - x_1)^2 f(t_1, t_2)$$

$$= \frac{1}{S(x_1+1, x_2+1)} \sum_{t_1=x_1+1}^{\infty} (t_1 - x_1)^2 P(X_1 = t_1, X_2 \geq x_2 + 1)$$

$$= \frac{1}{S(x_1+1, x_2+1)} \sum_{t_1=x_1+1}^{\infty} (t_1 - x_1)^2 [S(t_1, x_2+1) - S(t_1+1, x_2+1)]$$

$$= \frac{1}{S(x_1+1, x_2+1)} \sum_{t_1=x_1+1}^{\infty} (2(t_1 - x_1) - 1) S(t_1, x_2+1)$$

$$= 2\frac{\sum_{t_1=x_1+1}^{\infty} (t_1 - x_1) S(t_1, x_2+1)}{S(x_1+1, x_2+1)} + m_1(x_1, x_2)$$

$$= \frac{2}{S(x_1+1, x_2+1)} \sum_{t_1=x_1+1}^{\infty} \sum_{u=t_1+1}^{\infty} S(u, x_2+1) + m_1(x_1, x_2).$$

Now using (6.43), we obtain the formula in (6.44). □

Remark 6.7. Similarly,

$$\sigma_2^2(\mathbf{x}) = \frac{2}{S(x_1+1, x_2+1)} \sum_{t_1=x_1+1}^{\infty} \sum_{u=t_1+1}^{\infty} S(x_1+1, u) - m_2(\mathbf{x})[m_2(\mathbf{x}) - 1].$$

Example 6.10. Applying formula (6.44) for the bivariate geometric distribution

$$S(x_1, x_2) = q_1^{x_1} q_2^{x_2} \theta^{x_1 x_2},$$

whose mean residual life function is

$$(m_1(\mathbf{x}) = (1 - q_1\theta^{x_2+1})^{-1},\ m_2(\mathbf{x}) = (1 - q_2\theta^{x_1+1})^{-1}),$$

we have

$$\begin{aligned}
&E[(X_1 - x_1)^2 | \mathbf{X} > \mathbf{x}] \\
&= \frac{2}{q_1^{x_1+1} q_2^{x_2+1} \theta^{(x_1+1)(x_2+1)}} \sum_{t_1=x_1}^{\infty} (1 + q_1\theta^{x_2+1})^{-1} q_1^{t_1+1} q_2^{x_2+1} \theta^{(t_1+1)(x_2+1)} \\
&\quad - \frac{1}{1 - q_1\theta^{x_2+1}} \\
&= \frac{2(1 + q_1\theta^{x_2+1})^{-1}}{q_1^{x_1+1}\theta^{x_1+1}} \sum_{t=x_1}^{\infty} (q_1\theta^{x_2+1})^t - \frac{1}{1 - q_1\theta^{x_2+1}} \\
&= \frac{2}{(1 - q_1\theta^{x_2+1})^2} - \frac{1}{1 - q\theta^{x_2+1}} \\
&= \frac{1 + q_1\theta^{x_2+1}}{(1 - q_1\theta^{x_2+1})^2}.
\end{aligned}$$

Hence,

$$\begin{aligned}
\sigma_1^2(\mathbf{x}) &= \frac{1 + q_1\theta^{x_2+1}}{(1 - q_1\theta^{x_2+1})^2} - \frac{1}{(1 - q_1\theta^{x_2+1})^2} \\
&= \frac{q_1\theta^{x_2+1}}{(1 - q_1\theta^{x_2+1})^2}.
\end{aligned} \tag{6.45}$$

Similarly,

$$\sigma_2^2(\mathbf{x}) = \frac{q_2\theta^{x_1+1}}{(1 - q_2\theta^{x_1+1})^2} \tag{6.46}$$

As in the case of $m_1(\mathbf{x})$ and $m_2(\mathbf{x})$, $\sigma_1^2(\mathbf{x})$ is independent of x_1 and $\sigma_2^2(\mathbf{x})$ is independent of x_2.

The bivariate variance residual life function possesses the following properties:

1. $\sigma_1(-1,-1)=\sigma_1^2$ and $\sigma_2(-1,-1)=\sigma_2^2$;
2. The variance residual life function can be expressed in terms of the mean residual life function.

Theorem 6.12.

$$\sigma_1^2(x_1+1,x_2)=\frac{\sigma_1^2(\mathbf{x})-(m_1(\mathbf{x})-1)c_1(x_1+1,x_2+1)m_1(x_1+1,x_2)}{1-c_1(x_1+1,x_2+1)} \tag{6.47}$$

and

$$\sigma_2^2(x_1,x_2+1)=\frac{\sigma_2^2(\mathbf{x})-(m_2(\mathbf{x})-1)c_2(x_1+1,x_2+1)m_2(x_1,x_2+1)}{1-c_2(x_1+1,x_2+1)}, \tag{6.48}$$

where c_1 and c_2 are the components of the vector hazard rate (6.13), *which can be written as*

$$c_1(x_1+1,x_2+1)=1-\frac{m_1(\mathbf{x})-1}{m_1(x_1+1,x_2)}$$

and

$$c_2(x_1+1,x_2+1)=1-\frac{m_2(\mathbf{x})-1}{m_2(x_1,x_2+1)}.$$

Proof. From (6.44), we have

$$\sigma_1^2(\mathbf{x})+m_1(\mathbf{x})(1+m_1(\mathbf{x}))S(x_1+1,x_2+1)=2\sum_{t=x_1}^{\infty}m_1(t,x_2)S(t+1,x_2+1). \tag{6.49}$$

Now, changing x_1 to x_1+1 and subtracting from (6.49), we get

$$\begin{aligned}2m_1(\mathbf{x})S(x_1+1,x_2+1)&=[\sigma_1^2(\mathbf{x})+m_1(\mathbf{x})(1+m_1(\mathbf{x}))S(x_1+1,x_2+1)]\\&-[\sigma_1^2(x_1+1,x_2)+m_1(x_1+1,x_2)(1+m_1(x_1+1,x_2))]S(x_1+2,x_2+1).\end{aligned}$$

Dividing by $S(x_1+1,x_2+1)$ and using (6.16), we get

$$\begin{aligned}2m_1(\mathbf{x})&=\sigma_1^2(\mathbf{x})+m_1(\mathbf{x})(1+m_1(\mathbf{x}))\\&+[\sigma_1^2(x_1+1,x_2)+m_1(x_1+1,x_2)(1+m_1(x_1+1,x_2))](1-c_1(x_1+1,x_2+1).\end{aligned}$$

Upon simplifying using (6.16), we get (6.47). The proof of (6.48) is quite similar. □

Remark 6.8. With the application of (6.16) in (6.47) and (6.48), we deduce the alternate forms

$$\sigma_1^2(x_1+1,x_2)-\sigma_1^2(\mathbf{x})=c_1[\sigma_1^2(x_1+1,x_2) - (m_1(\mathbf{x})-1)(m_1(x_1+1,x_2))] \tag{6.50}$$

and

$$\sigma_2^2(x_1,x_2+1)-\sigma_2^2(\mathbf{x})=c_2[\sigma_2^2(x_1,x_2+1) - (m_2(\mathbf{x})-1)(m_2(x_1,x_2+1))]. \tag{6.51}$$

These formulas are generalizations of the univariate identity in (2.43). When the mean residual life vector is known, (6.50) and (6.51) serve as recurrence relations for computing $\sigma_1^2(\mathbf{x})$ and $\sigma_2^2(\mathbf{x})$ for given x_2 and x_1, respectively, with the use of $\sigma_1(-1,-1)=\sigma_1^2$ and $\sigma_2(-1,-1)=\sigma_2^2$.

Theorem 6.13.

$$(\sigma_1^2(\mathbf{x}),\sigma_2^2(\mathbf{x}))=(k_1,k_2)\ \textit{for all}\ \mathbf{x} \tag{6.52}$$

if and only of X_1 and X_2 are independent geometric random variables.

Proof. When the condition (6.52) is true, we have

$$k_1=(m_1(\mathbf{x})-1)(m_1(x_1+1,x_2). \tag{6.53}$$

From (6.16), we have

$$1-c_1(\mathbf{x})=\frac{m_1(\mathbf{x})-1}{m_1(x_1+1,x_2)}$$

so that (6.53) is equivalent to the statement that $c_1(\mathbf{x})=$ a constant. Similarly, $c_2(\mathbf{x})$ is a constant on considering $\sigma_2^2(\mathbf{x})$.

The theorem follows from Property (iii) in Section 6.2.3 of vector hazard rates. □

6.4.2 Covariance Residual Life

The covariance residual life of $\mathbf{X}=(X_1,X_2)$ is defined as

$$\sigma_{12}(\mathbf{x})=M(\mathbf{x})-m_1(\mathbf{x})m_2(\mathbf{x}),\quad x_1,x_2=-1,0,1,\ldots \tag{6.54}$$

where

$$M(\mathbf{x}) = E[(X_1 - x_1)(X_2 - x_2)|X_1 > x_1, X_2 > x_2] \tag{6.55}$$

is the product moment of residual life. We note that

$$\begin{aligned} M(\mathbf{x}) &= \frac{1}{S(x_1+1, x_2+1)} \sum_{t_1=x_1+1}^{\infty} \sum_{t_2=x_2+1}^{\infty} (t_1 - x_1)(t_2 - x_2) f(t_1, t_2) \\ &= \frac{1}{S(x_1+1, x_2+1)} \sum_{x_1+1}^{\infty} \sum_{x_2+1}^{\infty} S(t_1, t_2). \end{aligned} \tag{6.56}$$

Example 6.11. Let $\mathbf{X}$ be distributed as Waring with

$$S(\mathbf{x}) = \frac{(m)_{x_1+x_2}}{(m+n)_{x_1+x_2}}, \quad x_1, x_2 = 0, 1, 2, \ldots.$$

Then,

$$\begin{aligned} M(\mathbf{x}) &= \frac{(m+n)_{x_1+x_2+2}}{(m)_{x_1+x_2+2}} \sum_{t_1=x_1+1}^{\infty} \left[\frac{(m)_{t_1+x_2+1}}{(m+n)_{t_1+x_2+1}} + \frac{(m)_{t_1+x_2+2}}{(m+n)_{t_1+x_1+2}} + \ldots \right] \\ &= \frac{(m+n)_{x_1+x_2+2}}{(m)_{x_1+x_2+2}} \sum_{t_1=x_1+1}^{\infty} \frac{(m)_{t_1+x_2+1}}{(m+n)_{t_1+x_2+1}} + \frac{m+n+t_1+x_2}{n-1} \\ &= \frac{1}{n-1} \frac{(m+n)_{x_1+x_2+2}}{(m)_{x_1+x_2+2}} \sum_{t=x_1+1}^{\infty} \frac{(m)_{t_1+x_2+1}}{(m+n)_{t_1+x_2}} \\ &= \frac{(m+n+x_1+x_2+1)(m+n+x_1+x_2)}{(n-1)^2}. \end{aligned}$$

From Table 6.5, we have

$$m_1(\mathbf{x}) = m_2(\mathbf{x}) = \frac{m+n+x_1+x_2+1}{n-1}.$$

Substituting in (6.55), we get

$$\sigma_{12}(\mathbf{x}) = -\frac{m+n+x_1+x_2+1}{(n-1)^2}.$$

In the above calculations, the Waring expansion in (2.6) has been utilized in finding the sums.

The properties of $\sigma_{12}(\mathbf{x})$ consist of the following:

(i) $\sigma_{12}(-1,-1) = \text{Cov}\,(X_1, X_2)$;

(ii) Sometimes, it is easier for calculation to consider a reduced version of (6.54) *viz.*,

$$\sigma_{12}(\mathbf{x}) = E(X_1 X_2 | X_1 > x_1, X_2 > x_2) \\ - E(X_1 | X_1 > x_1, X_2 > x_2) E(X_2 | X_1 > x_1, X_2 > x_3);$$

(iii) There exists relationships between $M(\mathbf{x})$ and other reliability functions. The first pair is

$$M(x_1+1, x_2) - M(x_1, x_2) \\ = c_1(x_1+1, x_2+1) M(x_1+1, x_2) - m_2(x_1, x_2) \quad (6.57)$$

and

$$M(x_1, x_2+1) - M(x_1, x_2) \\ = c_2(x_1+1, x_2+1) M(x_1, x_2+1) - m_1(x_1, x_2). \quad (6.58)$$

From the above, we can eliminate c_1 and c_2 using (iv) (b) in Section 6.3 to reach identities connecting the product moment and mean residual life functions as

$$m_1(x_1+1, x_2) M(\mathbf{x}) - (m_1(\mathbf{x}) - 1) M(x_1+1, x_2) \\ = m_2(\mathbf{x}) m_1(x_1+1, x_2) \quad (6.59)$$

and

$$m_2(x_1, x_2+1) M(\mathbf{x}) - (m_2(\mathbf{x}) - 1) M(x_1, x_2+1) \\ = m_1(\mathbf{x}) m_2(x_1, x_2+1). \quad (6.60)$$

These two relationships show that a necessary condition for a function $M(x_1, x_2)$ to be a product moment residual life is that

$$m_2(\mathbf{x}) + \frac{m_1(\mathbf{x}) - 1}{m_1(x_1+1) x_2} M(x_1+1, x_2) \\ = m_1(\mathbf{x}) + \frac{m_2(\mathbf{x}) - 1}{m_2(x_1, x_2+1)} M(x_1, x_2+1),$$

or

$$m_2(\mathbf{x}) + (1 - c_1(x_1+1, x_2+1)) M(x_1+1, x_2) \\ = m_1(\mathbf{x}) + (1 - c_2(x_1+1, x_2+1)) M(x_1, x_2+1);$$

(iv) A characterization of independence of X_1 and X_2 in terms of σ_{12} is spelled out in the next theorem.

Theorem 6.14. *The covariance residual life* $\sigma_{12}(\mathbf{x}) = 0$ *for all* x_1, x_2 *if and only if* X_1 *and* X_2 *are independent.*

Proof. When X_1 and X_2 are independent,

$$\begin{aligned}\sigma_{12}(\mathbf{x}) &= E(X_1 - x_1)(X_2 - x_2|\mathbf{X} > \mathbf{x}) - m_1(\mathbf{x})m_2(\mathbf{x}) \\ &= E(X_1 - x_1|X_1 > x_1)E(X_2 - x_2|X_2 > x_2) - m_1(x_1)m_2(x_2) = 0,\end{aligned}$$

where m_1 and m_2 are the mean residual lives of X_1 and X_2. Conversely, $\sigma_{12}(\mathbf{x}) = 0$ implies $M(x_1, x_2) = m_1(x_1, x_2)m_2(x_1, x_2)$.

Using this in (6.59), we have

$$m_1(x_1, x_2)m_2(x_1, x_2) - m_1(x_1, x_2)m_2(x_1 + 1, x_2) - m_2(x_1, x_2) = 0,$$

or

$$[m_1(x_1, x_2) - 1][m_2(x_1, x_2) - m_2(x_1 + 1, x_2)] = 0.$$

Since the first factor on the left hand side is greater than unity,

$$m_2(x_1 + 1, x_2) = m_2(x_1, x_2) = m_2(-1, x_2).$$

Similarly, considering (6.60), we get

$$m_1(x_1, x_2) = m_1(x_1, -1).$$

These yields

$$M(x_1, x_2) = m_1(x_1, -1)m_2(-1, x_2),$$

and therefore X_1 and X_2 are independent. □

Remark 6.9. Generally, $\text{Cov}(X_1, X_2) = 0$ does not guarantee independence of X_1 and X_2, but in the case of $\sigma_{12}(x_1, x_2) = 0$, it is a necessary and sufficient condition for independence;

(v) In general, $M(x_1, x_2)$ and $\sigma_{12}(x_1, x_2)$ do not determine the distribution uniquely. This can be seen from

$$S_1(x_1, x_2) = p_1^{x_1} p_2^{x_2},\ 0 < p_1,\ p_2 < 1,\ x_1, x_2 = 0, 1, 2, \ldots,$$

and

$$S_2(x_1, x_2) = p_2^{x_1} p_1^{x_2},$$

in which $M(x_1, x_2) = (1 - p_1)^{-1}(1 - p_2)^{-1}$ in both cases, where as

$$f_1(2,1) \neq f_2(2,1), \text{ when } p_1 = \frac{1}{4}, p_2 = \frac{3}{4};$$

(vi) $a(x_1, x_2) = k \Leftrightarrow M(x_1, x_2) = \frac{1}{k}$ for all x_1, x_2.

For an application of $M(x_1, x_2)$ in formulating a measure of association, see Section 6.8.

6.5 MULTIVARIATE REVERSED HAZARD RATE

All the reliability functions in this and subsequent sections are extensions of the univariate concepts discussed in Sections 2.6–2.8.

The definition and properties regarding them exhibit close similarity with the reliability functions discussed in the earlier sections due to the change in replacing $\mathbf{X} > \mathbf{x}$ by $\mathbf{X} \leq \mathbf{x}$ or $S(\mathbf{x})$ by $F(\mathbf{x})$. Various results in this section are due to Sankaran et al. (2016b).

6.5.1 Scalar Reversed Hazard Rate

The scalar reversed hazard rate of $\mathbf{X} = (X_1, \ldots, X_p)$ is defined by

$$\lambda(\mathbf{x}) = P(\mathbf{X} = \mathbf{x} | \mathbf{X} \leq x) = \frac{f(\mathbf{x})}{F(\mathbf{x})},$$

where $F(x) = P(\mathbf{X} \leq \mathbf{x})$ is the distribution function of $\mathbf{X}$. Obviously, $0 \leq \lambda(\mathbf{x}) \leq 1$. It is interpreted as the conditional probability that a p-component device fails at $\mathbf{x}$, given that it has a maximum lifetime $\mathbf{x}$.

Example 6.12. Let (X_1, X_2) be distributed as bivariate uniform with

$$f(x_1, x_2) = \frac{1}{b_1 b_2}\left[1 + \theta\left(1 - \frac{2x_1 - 1}{b_1}\right)\left(1 - \frac{2x_2 - 1}{b_2}\right)\right],$$

$x_i = 1, 2, \ldots, b_i$, $i = 1, 2$.

Then, the distribution function is

$$\begin{aligned} F(x_1, x_2) &= \sum_{t_1=1}^{x_1}\sum_{t_2=1}^{x_2} f(t_1, t_2) \\ &= \frac{x_1 x_2}{b_1 b_2}\left[1 + \theta\left(1 - \frac{x_1}{b_1}\right)\left(1 - \frac{x_2}{b_2}\right)\right] \end{aligned}$$

so that

$$\lambda(x_1, x_2) = \frac{1 + \theta(1 - \frac{2x_1 - 1}{b_1})(1 - \frac{2x_2 - 1}{b_2})}{x_1 x_2[1 + \theta(1 - \frac{x_1}{b_1})(1 - \frac{x_2}{b_2})]}.$$

We now give some important results concerning multivariate scalar reversed hazard rates:

(i) The distribution of $\mathbf{X}$ is not uniquely determined by $\lambda(\mathbf{x})$. This can be verified in the case of bivariate distribution functions

$$F_1(x_1, x_2) = \frac{x_1(x+1)}{b_1(b_1+1)} \frac{x_2(x_2+1)}{b_2(b_2+1)}, \quad x_i = 0, 1, \ldots, b_i,$$

and

$$F_2(x_1, x_2) = \frac{1}{2}\left[\frac{x_1x_2(x_1+1)(x_2+1)}{b_1(b_1+1)b_2(b_2+1)} + \frac{x_1x_2(x_1+1)(x_2+1)}{c_1(c_1+1)b_2(b_2+1)}\right], \quad c_1 \neq b_1,$$

where both have the same $\lambda(\mathbf{x}) = \frac{4}{(x_1+1)(x_2-1)}$;

(ii) Consider the case when X_1 and X_2 are independent random variables following (2.57). Then,

$$F(x_1, x_2) = (1+c_1)^{x_1-b}(1+c_2)^{x_2-b}, \; x_1x_2 = 0, 1, 2, \ldots,$$

and the corresponding probability mass function is

$$f(x_1, x_2) = \begin{cases} c_1c_2(1+c_1)^{x_1-b_1-1}(1+c_2)^{x_2-b_2-1}, & x_1, x_2 = 1, 2, \ldots, b, \\ c_1(1+c_1)^{x_1-b_1-1}(1+c_2)^{-b_2}, & x_1 = 1, 2, \ldots, b, x_1 = 0, \\ c_2(1+c_2)^{x_2-b_2-1}(1+c_1)^{-b_1}, & x_1 = 0, x_2 = 1, 2, \ldots, b, \\ (1+c_1)^{-b_1}(1+c_2)^{-b_2}, & x_1 = 0, x_2 = 0. \end{cases}$$

Then, the bivariate reversed hazard rate is constant of the form

$$\lambda(x_1, x_2) = \begin{cases} c_1c_2(1+c_1)^{-1}(1+c_2)^{-1}, & x_1, x_2 = 1, 2, \ldots, b, \\ c_1(1+c_1)^{-1}, & x_1 = 1, 2, \ldots, b, x_2 = 0, \\ c_2(1+c_2)^{-1}, & x_1 = 0, x_2 = 1, 2, \ldots, b, \\ 1, & x_1 = 0, x_2 = 0; \end{cases}$$

(iii) At $\mathbf{O}_p$, $\lambda(\mathbf{O}_p) = 1$, irrespective of the distribution and $0 \leq \lambda(\mathbf{x}) \leq 1$;

(iv) If $X_1, \ldots X_p$ are independent random variables, then

$$\lambda(\mathbf{x}) = \prod_{i=1}^{p} \lambda_i(x_i),$$

where $\lambda_i(x_i)$ is the univariate reversed hazard rate of X_i.

There are several open problems relating to $\lambda(\mathbf{x})$. The conditions under which $\lambda(\mathbf{x})$ determines $F(\mathbf{x})$, characterization of life distributions based on the functional forms of $\lambda(\mathbf{x})$ and the distribution(s) associated with constant $\lambda(\mathbf{x})$ are some of them.

6.5.2 Vector Reversed Hazard Rate

The multivariate reversed hazard rate can also be viewed as a vector quantity. It is defined as

$$\delta(\mathbf{x}) = (\delta_1(\mathbf{x}), \ldots, \delta_p(\mathbf{x})),$$

where

$$\delta_i(x) = P(X_i = x_i | \mathbf{X} \leq \mathbf{x}). \tag{6.61}$$

Eq. (6.61) shows that the ith component of the multivariate reversed hazard rate is the conditional probability that component i of a p-component device fails at age x_i when it is known that the device fails at age $\mathbf{x}$. Since

$$P(X_i = x_i) = F(\mathbf{x}) - F(x_1, x_2, \ldots, x_{i-1}, x_i - 1, x_{i+1}, \ldots, x_p),$$

we write (6.61) as

$$\delta_i(x) = 1 - \frac{F(\mathbf{x}_{i-1}, x_i - 1, \mathbf{x}_i^*)}{F(\mathbf{x})}, \tag{6.62}$$

where as noted earlier $\mathbf{x}_i = (x_1, \ldots, x_i)$ and $\mathbf{x}_i^* = (x_{i+1}, \ldots, x_p)$, $i = 1, 2, \ldots, p-1$. In the bivariate case,

$$\delta(x_1, x_2) = (\delta_1(x_1, x_2), \delta_2(x_1, x_2))$$

with

$$\delta_1(x_1, x_2) = 1 - \frac{F(x_1 - 1, x_2)}{F(x_1, x_2)} \tag{6.63}$$

and

$$\delta_1(x_1, x_2) = 1 - \frac{F(x_1, x_2 - 1)}{F(x_1, x_2)}. \tag{6.64}$$

Then, we have from (6.62) that

$$\begin{aligned} F(\mathbf{x}_{i-1}, x_i - 1, \mathbf{x}_i^*) &= (1 - \delta_i(\mathbf{x})\bar{F}(\mathbf{x})) \\ &= (1 - \delta_i(\mathbf{x}))(1 - \delta_i(\mathbf{x}_{i-1}, x_i + 1, \mathbf{x}_i^*))F(\mathbf{x}_{i-1}, x_i + 1, \mathbf{x}_i^*) \\ &= (1 - \delta_i(\mathbf{x}))\ldots(1 - \delta_i(\mathbf{x}_i, \infty, \mathbf{x}_i^*))F(\mathbf{x}_{i-1}, \mathbf{x}_i^*). \end{aligned}$$

The distribution function on the right hand side is $(p-1)$-dimensional. Successive reduction can be achieved in the same manner, reaching finally at a one-dimensional marginal distribution function. Thus, when the limit of $\delta_i(x)$

as each of the arguments tends to infinity is finite, we can arrive at an expression for $F(\mathbf{x})$ in terms of $\delta(\mathbf{x})$ as

$$F(\mathbf{x}) = \prod_{t_1=x_1+1}^{\infty} (1-\delta_1(t_1, x_2, \ldots, x_p)) \prod_{t_2=x_2+1}^{\infty} (1-\delta_2(t_2, x_3, \ldots, x_p)) \times \cdots \prod_{t_p=x_p+1}^{\infty} (1-\delta_1, x_2, \ldots t_p)) \tag{6.65}$$

In view of (6.65), we have the following theorem.

Theorem 6.15. *The distribution of* $\mathbf{X}$ *is uniquely determined by* $\delta(\mathbf{x})$.

Remark 6.10. Because we can begin with any one of the $\delta_i(\mathbf{x})$ in the reduction process indicated above, there are p different forms in which $F(\mathbf{x})$ can be written in terms of $\delta(\mathbf{x})$.

Remark 6.11. The bivariate form of (6.65) that is quite useful in theoretical work is

$$\begin{aligned} F(x_1, x_2) &= \prod_{t_1=x_1+1}^{\infty} \prod_{t_2=x_2+1}^{\infty} (1-\delta_1(t_1, x_2))(1-\delta_2(\infty, t_2)), \\ &\qquad x_1, x_2 = 0, 1, 2, \ldots \\ &= \prod_{t_1=x_1+1}^{\infty} \prod_{t_2=x_2+1}^{\infty} (1-\delta_2(x_1, t_2))(1-\delta_1(t_1, \infty)). \end{aligned} \tag{6.66}$$

Example 6.13. Consider the bivariate uniform distribution of Example 6.12 with distribution function

$$F(x_1, x_2) = \frac{x_1 x_2}{b_1 b_2}\left[1+\theta\left(1-\frac{x_1}{b_1}\right)\left(1-\frac{x_2}{b_2}\right)\right],$$
$$x_i = 1, 2, \ldots, b_i,\ i = 1, 2,$$

we have

$$\begin{aligned} \delta_1(x_1, x_2) &= 1 - \frac{(x_1-1)x_2[1+\theta(1-\frac{x_1-1}{b_1})(1-\frac{x_2}{b_2})]}{x_1 x_2[1+\theta(1-\frac{x_1}{b_1})(1-\frac{x_2}{b_2})]} \\ &= \frac{1+\theta(1-\frac{x_2}{b_2})(1-\frac{2x_1-1}{b_1})]}{x_1[1+\theta(1-\frac{x_1}{b_1})(1-\frac{x_2}{b_2})]}. \end{aligned}$$

Calculating $\delta_2(x_1, x_2)$ similarly, we have

$$\delta_2(x_1,x_2) = \frac{1+\theta(1-\frac{x_1}{b_1})(1-\frac{2x_2-1}{b_1})]}{x_1[1+\theta(1-\frac{x_1}{b_1})(1-\frac{x_2}{b_2})]}.$$

When prescribing models with specified functional forms of δ_1 and δ_2, it may noted that they cannot be chosen arbitrarily. There exists a consistency condition for two hypothesized bivariate reversed hazard rates. This can be seen as follows:

$$\begin{aligned}\lambda(x_1,x_2) &= \frac{f(x_1,x_2)}{F(x_1,x_2)}\\ &= \frac{F(x_1,x_2)-F(x_1,x_2-1)-F(x_1-1,x_2)+F(x_1-1,x_2-1)}{F(x_1,x_2)}.\end{aligned} \tag{6.67}$$

We may express

$$\begin{aligned}\frac{F(x_1-1,x_2-1)}{F(x_1,x_2)} &= \frac{F(x_1-1,x_2-1)}{F(x_1,x_2-1)}\frac{F(x_1,x_2-1)}{F(x_1,x_2-1)}\\ &= (1-\delta_1(x_1,x_2-1))(1-\delta_2(x_1,x_2)).\end{aligned}$$

Substituting this in (6.67) and using (6.63) and (6.64), we get

$$\lambda(x_1,x_2) = \delta_1(x_1,x_2) - \delta_1(x_1,x_2-1) + \delta_1(x_1,x_2-1)\delta_2(x_1,x_2). \tag{6.68}$$

We can also write

$$\begin{aligned}\frac{F(x_1-1,x_2-1)}{F(x_1,x_2)} &= \frac{F(x_1-1,x_2-1)}{F(x_1-1,x_2)}\frac{F(x_1-1,x_2)}{F(x_1,x_2)}\\ &= (1-\delta_2(x_1-1,x_2))(1-\delta_1(x_1,x_2)).\end{aligned}$$

Upon substituting this in (6.67), we get

$$\lambda(x_1,x_2) = \delta_2(x_1,x_2) - \delta_2(x_1-1,x_2) + \delta_1(x_1,x_2)\delta_2(x_1-1,x_2). \tag{6.69}$$

Thus,

$$\begin{aligned}&\delta_1(x_1,x_2) - \delta_1(x_1,x_2-1) + \delta_1(x_1,x_2-1)\delta_2(x_1,x_2)\\ &\quad = \delta_2(x_1,x_2) - \delta_2(x_1-1,x_2) + \delta_1(x_1,x_2)\delta_2(x_1-1,x_2).\end{aligned}$$

Notice that (6.68) or (6.69) is an identity connecting scalar and vector reversed hazard rates.

Unlike scalar reversed hazard rate, the vector reversed hazard rate can provide univariate reversed hazard rates as particular cases as

$$\lambda_i(x_i) = \delta_i(\infty,\ldots,\infty,x_i,\infty,\ldots,\infty), \quad i=1,\ldots,p.$$

Being conditional probabilities, $0 \le \delta_i(\mathbf{x}) \le 1$ and hence unlike the multivariate hazard functions in the continuous time, the range of $\delta_i(\mathbf{x})$ is limited to unit interval.

Theorem 6.16. *The multivariate reversed hazard rate is of the form*

$$\delta(\mathbf{x}) = \left(\frac{c_1}{1+c_1}, \frac{c_2}{1+c_2}, \ldots, \frac{c_p}{1+c_p}\right), \quad c_i > 0, i = 1, 2, \ldots, p,$$

if and only if the distribution of $\mathbf{X}$ *is multivariate reversed geometric with*

$$F(\mathbf{x}) = \prod_{i=1}^{p} (1 - c_i)^{x_i - b_i}, \quad x_i = 0, 1, \ldots b_i. \tag{6.70}$$

Proof. When the distribution is (6.70), from (6.62), we have

$$\delta_i(\mathbf{x}) = 1 - \frac{1}{1+c_i} = \frac{c_i}{1+c_i}$$

so that $\delta(\mathbf{x})$ is as required. Conversely, assuming $\delta(\mathbf{x})$ to be of the given form, formula (6.65) leads to (6.70). □

Remark 6.12. The distribution of $\mathbf{X}$ is the product of independent univariate reversed geometric distributions, where X_i has parameter c_i.

A more general result in this connection is presented in the next theorem, which is a direct consequence of (6.62).

Theorem 6.17. *The random variables* $X_1, \ldots, X_p$ *are independent if and only if*

$$\delta(\mathbf{x}) = (\lambda_1(x_1), \ldots, \lambda_p(x_p)).$$

A bivariate distribution of interest can be obtained if we assume that the reversed hazard rate is locally constant.

Theorem 6.18. *When the bivariate reversed hazard rate is of the form* $\delta(\mathbf{x}) = (\alpha_1(x_2), \alpha_2(x_1))$*, the only distribution with finite support that satisfies this condition is given by*

$$F(x_1, x_2) = (1 - \alpha_1)^{b_1 - x_1} (1 - \alpha_2)^{b_2 - x_2} k_1^{(b_1 - x_1)(b_2 - x_2)}, \quad x_i = 0, 1, 2, \ldots, b_i, \tag{6.71}$$

where

$$\alpha_1 = \alpha_1(\infty),\ \alpha_2 = \alpha_2(\infty) \text{ and } [1 - (1 - \alpha_1)k_1^{b_1}][1 - (1 - \alpha_2)k_1^{b_2}] + k_1 \ge 0,$$
$$0 < k_1 < 1,\ 0 < \alpha_1, \alpha_2 < 1.$$

Proof. Under the given condition, Eq. (6.66) leads to

$$F(x_1, x_2) = (1 - \alpha_1(x_2))^{b_1 - x_1}(1 - \alpha_2)^{b_2 - x_2}$$
$$= (1 - \alpha_1)^{b_1 - x_1}(1 - \alpha_2(x_1))^{b_2 - x_2}. \tag{6.72}$$

This implies

$$\left(\frac{1 - \alpha_1(x_2)}{1 - \alpha_1}\right)^{\frac{1}{b_2 - x_2}} = \left(\frac{1 - \alpha_2(x_1)}{1 - \alpha_1}\right)^{\frac{1}{b_1 - x_1}} \quad \text{for all } x_1, x_2. \tag{6.73}$$

Eq. (6.73) means that each of the expressions in (6.72) is a constant, say k_1. Hence,

$$(1 - \alpha_1(x_2)) = (1 - \alpha_1)k_1^{b_2 - x_2}.$$

Inserting this in (6.72), we obtain (6.71). Conversely, assuming (6.69),

$$\delta_1(x_1, x_2) = 1 - \frac{F(x_1 - 1, x_2)}{F(x_1, x_2)} = 1 - (1 - \alpha_1)k_1^{b_2 - x_2}$$

and similarly

$$\delta_2(x_1, x_2) = 1 - (1 - \alpha_2)k_1^{b_1 - x_1}.$$

Thus, $\delta(x_1, x_2)$ is of the given form, which completes the proof. □

The alternative reversed hazard rate defined in (2.61) allows extension to the multivariate case. Defining analogously, we have in the bivariate case

$$\rho(\mathbf{x}) = (\rho_1(\mathbf{x}), \rho_2(\mathbf{x})),$$

where

$$\rho_1(x_1, x_2) = \log \frac{F(x_1, x_2)}{F(x_1 - 1, x_2)} \tag{6.74}$$

and

$$\rho_2(x_1, x_2) = \log \frac{F(x_1, x_2)}{F(x_1, x_2 - 1)}. \tag{6.75}$$

The components of this alternative reversed hazard rate is related to (δ_1, δ_2) through

$$\delta_1 = 1 - e^{-\rho_1} \text{ and } \delta_2 = 1 - e^{-\rho_2}.$$

Denoting the marginal distribution function of X_i by $F_i(x_i)$ and using

$$F(x_1, x_2) = F(x_1 + 1, x_2)\exp[-\rho_1(x_1 + 1, x_2)]$$

derived from (6.73), by successive iteration, we obtain

$$F(x_1, x_2) = \exp\left[-\sum_{t=x_1+1}^{\infty} \rho_1(t, x_2)\right] F_2(x_2).$$

When $x_2 \to \infty$ in (6.74), we get

$$\rho_1(x_1, \infty) = \log \frac{F(x_1)}{F(x_1 - 1)} = \lambda_1(x_1)$$

and similarly

$$\rho_2(\infty, x_2) = \lambda_2(x_2),$$

where $\lambda_i(x_i)$ is the alternative reversed hazard rate of X_i. Thus,

$$F(x_1, x_2) = \exp\left[-\sum_{t=x_1+1}^{\infty} \rho_1(t, x_2) - \sum_{t=x_2+1}^{\infty} \rho_2(\infty, x_2)\right]. \tag{6.76}$$

Similarly,

$$F(x_1, x_2) = \exp\left[-\sum_{t=x_2+1}^{\infty} \rho_2(x_1, t) - \sum_{t=x_1+1}^{\infty} \rho_1(x_1, \infty)\right]. \tag{6.77}$$

The last two equations show that (ρ_1, ρ_2) determines $F(x_1, x_2)$ uniquely through one of them.

Several new bivariate distributions can be derived from (6.76) or (6.77) using appropriate functional forms of ρ_1 and ρ_2. One such class of distributions is presented in the next theorem.

Theorem 6.19. *The alternative reversed hazard rate is of the form*

$$(\rho_1, \rho_2) = (\lambda_1(x_1)C_1(x_2), \lambda_2(x_2)C_2(x_1)),$$

where C_1 and C_2 are non-negative functions such that C_i depends only on x_{3-i}, $i = 1, 2$, if and only if

$$F(x_1, x_2) = F_1(x_1)F_2(x_2)[F_2(x_2)]^{k \log F_1(x_1)},$$

$$0 \le k \le 1, \; x_1, x_2 = 0, 1, 2, \ldots \tag{6.78}$$

Proof. The alternative reversed hazard rate of (6.78) is obtained by employing (6.73) to be

$$\rho_1(x_1, x_2) = \lambda_1(x_1)[1 + k \log F_2]$$

and

$$\rho_2(x_1, x_2) = \lambda_2(x_2)[1 + k \log F_1],$$

which are of the required form in the theorem. On the other hand, when ρ_1 and ρ_2 are of the stated forms, we use (6.76) and (6.77) to arrive at the functional equation

$$[F_1(x_1)]^{c_1(x_2)-1} = [F_2(x_2)]^{c_2(x_1)-1},$$

or

$$[F_1(x_1)]^{\frac{1}{c_2(x_1)-1}} = [F_2(x_2)]^{\frac{1}{c_1(x_2)-1}}. \tag{6.79}$$

The solution of (6.79) is

$$[F_i(x_i)]^{\frac{1}{c_{3-i}(x_i)-1}} = C, \text{ a constant, } i = 1, 2,$$

or

$$c_{3-i}(x_i) = 1 + \frac{\log F_i(x_i)}{K}, \quad K = \log C.$$

This leads to

$$\begin{aligned} F(x_1, x_2) &== [F_1(x_1)]^{C_1(x_2)} F_2(x_2) \\ &= F_1(x_1) F_2(x_2) F_1(x_1)^{K \log F_2(x_2)}, \end{aligned}$$

which is the same as (6.78). □

Remark 6.13. Eq. (6.78) defines a family of bivariate distributions with marginals F_1 and F_2. Thus, substituting appropriate marginal distributions, we can realize the bivariate model. Alternatively, if one knows the form of the marginal alternative reverse hazard rates λ_1 and λ_2, $F(x_1, x_2)$ is determined. In modelling problems, it is not difficult to capture the marginal distributions or the reversed hazard rates which does make the identification of the bivariate model easier.

6.5.3 Conditional Reversed Hazard Rate

A third definition of bivariate reversed hazard rate is based on the conditional distributions. It is defined as the vector

$$\xi(\mathbf{x}) = (\xi_1(\mathbf{x}), \xi_2(\mathbf{x})), \quad \mathbf{x} = (x_1, x_2),$$

where

$$\xi_1(\mathbf{x}) = \frac{f_{12}(x_1|X_2 = x_2)}{F_{12}(x_1|X_2 = x_2)} \text{ and } \xi_2(\mathbf{x}) = \frac{f_{21}(x_2|X_1 = x_1)}{F_{21}(x_2|X_1 = x_1)}.$$

Its interpretation is similar to the univariate hazard rate with the change that it is conditioned here on the event $X_2 = x_2 (X_1 = x_1)$ in the cases of ξ_1 and ξ_2, respectively.

Example 6.14. Consider the bivariate uniform distribution

$$f(x_1, x_2) = \frac{1}{b_1 b_2}\left[1 + \theta\left(1 - \frac{2x_1 - 1}{b_1}\right)\left(1 - \frac{2x_2 - 1}{b_2}\right)\right],$$

$x_i = 1, 2, \ldots b_i, i = 1, 2.$

The marginal probability mass functions are

$$f_i(x_i) = \frac{1}{b_i}, \quad i = 1, 2.$$

Also, the joint distribution function

$$F(x_1, x_2) = \frac{x_1 x_2}{b_1 b_2}\left[1 + \theta\left(1 - \frac{x_1}{b_1}\right)\left(1 - \frac{x_2}{b_2}\right)\right]$$

gives the marginals as

$$F_i(x_i) = \frac{x_i}{b_i}, \quad i = 1, 2.$$

Thus, the conditional hazard rate is

$$(\xi_1, \xi_2) = \frac{\left[1 + \theta\left(1 - \frac{2x_1 - 1}{b_1}\right)\left(1 - \frac{2x_2 - 1}{b_2}\right)\right]}{1 + \theta\left(1 - \frac{x_1}{b_1}\right)\left(1 - \frac{x_2}{b_2}\right)}\left(\frac{1}{x_1}, \frac{1}{x_2}\right)$$

Working as in the univariate case, we have

$$F_{12}(x_1|x_2) = \prod_{t=x_1+1}^{\infty} (1 - \xi_1(t, x_2))$$

and

$$F_{21}(x_2|x_1) = \prod_{t=x_2+1}^{\infty} (1 - \xi_2(x_1, t)).$$

The last two equations determine the joint distribution function of (X_1, X_2), provided that

$$\frac{F_{12}(x_1|x_2)}{F_{21}(x_2|x_1)} = \frac{A_1(x_1)}{A_2(x_2)},$$

where A_1 and A_2 are distribution functions. Notice that when X_1 and X_2 are independent, $\xi_i(x_1, x_2) = \lambda_i(x_i)$, $i = 1, 2$.

6.6 REVERSED MEAN RESIDUAL LIFE FUNCTION

The concept of reversed mean residual life (mean inactivity time) introduced earlier in Section 2.7 can be extended to the case of p random variables specified by the vector $\mathbf{X} = (X_1, \ldots, X_p)$.

Definition 6.1. The multivariate reversed mean residual life function of $\mathbf{X}$ is defined as the vector

$$\begin{aligned} \mathbf{r}(\mathbf{x}) &= E(\mathbf{x} - \mathbf{X} | \mathbf{X} < \mathbf{x}) \\ &= (r_1(\mathbf{x}), \ldots, r_p(\mathbf{x})), \end{aligned} \tag{6.80}$$

where

$$\begin{aligned} r_i(\mathbf{x} - \mathbf{e}) &= E(x_i - X_i | \mathbf{X} < \mathbf{x}) \\ &= \frac{1}{F(\mathbf{x})} \sum_{t_1=0}^{x_1-1} \sum_{t_{i-1}=0}^{x_{i-1}-1} \sum_{t_i=0}^{x_i-1} \cdots \sum_{t_p=0}^{x_p-1} (x_i - t_i) f(\mathbf{t}) \end{aligned} \tag{6.81}$$

with $\mathbf{t} = (t_1, \ldots, t_p)$.

$$r_i(\mathbf{x}) = \frac{1}{F(\mathbf{x} - \mathbf{e})} \sum_{t_i=1}^{x_i} F(x_1 - 1, \ldots, x_{i-1} - 1, t_i - 1, x_{i+1} - 1, \ldots, x_p - 1).$$

In the bivariate case, the two components of $\mathbf{r}(x_1, x_2)$ are

$$r_1(x_1, x_2) = \frac{1}{F(x_1 - 1, x_2 - 1)} \sum_{t_1=1}^{x_1} F(t_1 - 1, x_2 - 1) \tag{6.82}$$

and

$$r_2(x_1, x_2) = \frac{1}{F(x_1 - 1, x_2 - 1)} \sum_{t_2=1}^{x_2} F(x_1 - 1, t_2 - 1). \tag{6.83}$$

Example 6.15. Let (X_1, X_2) have the uniform distribution with

$$F(x_1, x_2) = \frac{x_1 x_2}{b_1 b_2} \left[1 + \theta \left(1 - \frac{x_1}{b_1} \right) \left(1 - \frac{x_2}{b_2} \right) \right].$$

Then,

$$\sum_{t_1=1}^{x_1} F(t_1-1,x_2-1)$$
$$= \frac{x_2-1}{b_1b_2}\sum_{t_1=1}^{x_1}\left\{(t_1-1)+\theta\left(1-\frac{x_2-1}{b_2}\right)\left(t_1-1-\frac{(t_1-1)^2}{b_1}\right)^2\right\}$$
$$= \frac{(x_2-1)x_1(x_1-1)}{2b_1b_2}\left\{1+\theta\left(1-\frac{x_1-1}{3b_1}\right)\left(1-\frac{x_2-1}{b_2}\right)\right\}.$$

Hence,

$$r_1(x_1,x_2)=\frac{x_1}{2}\frac{1+\theta\left(1-\frac{2x_1-1}{3b_1}\right)\left(1-\frac{x_2-1}{b_2}\right)}{1+\theta\left(1-\frac{x_1-1}{b_1}\right)\left(1-\frac{x_2-1}{b_2}\right)},$$

and similarly

$$r_2(x_1,x_2)=\frac{x_2}{2}\frac{1+\theta\left(1-\frac{x_1-1}{b_1}\right)\left(1-\frac{2x_2-1}{3b_2}\right)}{1+\theta\left(1-\frac{x_1-1}{b_1}\right)\left(1-\frac{x_2-1}{b_2}\right)}.$$

The properties of the p-variate reversed mean residual life are transparent from those of the bivariate case. These are as follows:

(i) $0 \le r_1(x_1,x_2) \le x_i$;

(ii)

$$\delta_1(x_1,x_2-1)=\frac{r_1(x_1,x_2)-r_1(x_1+1,x_2)+1}{r_1(x_1,x_2)},\quad x_1,x_2=1,2,\ldots, \tag{6.84}$$

$$\delta_2(x_1-1,x_2)=\frac{r_2(x_1,x_2)-r_2(x_1,x_2+1)+1}{r_2(x_1,x_2)}. \tag{6.85}$$

Proof. We have

$$r_1(x_1,x_2)F(x_1-1,x_2-1)=\sum_{t=1}^{x_1}F(t-1,x_2-1),\ \text{on using (6.85)}.$$

Thus,

$$r_1(x_1+1,x_2)F(x_1,x_2-1)-r_1(x_1,x_2)F(x_1-1,x_2-1)=F(x_1,x_2-1),$$
$$r_1(x_1+1,x_2)=1+r_1(x_1,x_2)\frac{F(x_1-1,x_2-1)}{F(x_1,x_2-1)},$$
$$=1+r_1(x_1,x_2)(1-\delta_1(x_1,x_2-1)),$$

which is the first identity in (ii) above. Similarly, we arrive at the second identity;

(iii) $x_i - r_i(x_1, x_2)$ is an increasing function of x_i.

Proof. For $i = 1$,

$$x_i + 1 - r_1(x_1 + 1, x_2) - [x_1 - r_1(x_1, x_2)] = 1 - r(x_1 + 1, x_2) + r(x_1, x_2)$$
$$= r_1(x_1, x_2)\delta(x_1, x_2 - 1) \geq 0 \text{ from (6.84).}$$

Similarly, for $i = 2$, from (6.85);

(iv) Since $\delta_1(x_1, x_2) = 1 - e^{-\rho_1(x_1, x_2)}$, using (6.84) and (6.85), we have

$$\rho_1(x_1, x_2 - 1) = \log \frac{r_1(x_1, x_2)}{r_1(x_1 + 1, x_2) - 1}$$

and

$$\rho_2(x_1 - 1, x_2) = \log \frac{r_2(x_1, x_2)}{r_2(x_1, x_2 + 1) - 1};$$

(v) When X_1 and X_2 are independent,

$$r_1(x_1, x_2) = r_{X_1}(x_1) \text{ and } r_2(x_1, x_2) = r_{X_2}(x_2),$$

where $r_{X_i}(x_i)$ is the univariate reversed mean residual life function of X_i. Further, if the support of X_i is finite, say, $(0, 1, \ldots, b_i)$, and

$$(r_1, r_2) = \left(\frac{1 + c_1}{c_1}, \frac{1 + c_2}{c_2}\right),$$

where c_1 and c_2 are positive constants, then

$$F(x_1, x_2) = F_1(x_1)F_2(x_2)$$

with

$$F_i(x_i) = \begin{cases} c_i^{-1}(1 + c_i)^{1-b_i}, & x_i = 0, \\ (1 + c_i)^{x_i - b_i}, & x_i = 1, 2, \ldots, b_i; \end{cases}$$

(vi) If $(r_1, r_2) = (c_1(x_2), c_2(x_1))$, then $\delta_i = \frac{1}{r_i(x_1, x_2)}$, but the converse is not true.

6.7 BIVARIATE REVERSED VARIANCE RESIDUAL LIFE

As the developments in this section are quite similar to those in Section 6.4, only the main results are presented. The quantities of interest are the elements of the matrix

$$V_{x_1,x_2} = \begin{pmatrix} v_1^2(\mathbf{x}) & v_{12}(\mathbf{x}) \\ v_{12}(\mathbf{x}) & v_2^2(\mathbf{x}) \end{pmatrix}, \mathbf{x} = (x_1, x_2),$$

where

$$v_1^2(\mathbf{x}) = V(x_1 - X_1|\mathbf{X} < \mathbf{x})$$

and

$$v_2^2(\mathbf{x}) = V(x_2 - X_2|\mathbf{X} < \mathbf{x})$$

are the reversed variance residual life functions (variance inactivity time) and

$$v_{12}(\mathbf{x}) = \mathrm{Cov}\,[(x_1 - X_1)(x_1 - X_2)|\mathbf{X} < \mathbf{x}]$$

is the corresponding covariance function. We have

$$v_1^2(\mathbf{x}) = E[(x_1 - X_1)^2|\mathbf{X} < \mathbf{x}] - r_1^2(\mathbf{x}),\ x_1, x_2 = 0, 1, 2, \ldots \tag{6.86}$$

Further,

$$\begin{aligned} &E[(x_1 - X_1)^2|\mathbf{X} < \mathbf{x}] \\ &= \frac{2}{F(x_1 - 1, x_2 - 1)} \sum_{t_1=1}^{x_1} (x_1 - t_1) F(t_1 - 1, x_1 - 1) - r_1(\mathbf{x}) \\ &= \frac{2}{F(x_1 - 1, x_2 - 1)} \sum_{t_1+1=1}^{x_1} \sum_{u=1}^{t_1} F(u - 1, x_2 - 1) - r_1(\mathbf{x}). \end{aligned}$$

Substituting it in (6.86), we get

$$v_1^2(\mathbf{x}) = \frac{2}{F(x_1 - 1, x_2 - 1)} \times \sum_{t_1=1}^{x} \sum_{u=1}^{t_1} F(u - 1, x_2 - 1) - r_1(\mathbf{x})(r_1(\mathbf{x}) + 1). \tag{6.87}$$

Similar calculations also yield

$$v_1^2(\mathbf{x}) = \frac{2}{F(x_1 - 1, x_2 - 1)} \times \sum_{t_2=1}^{x} \sum_{u=1}^{t_2} F(x_1 - 1, u) - r_2(\mathbf{x})(r_2(\mathbf{x}) + 1). \tag{6.88}$$

□

Regarding the covariance residual life, we have

$$v_{12}(\mathbf{x}) = E((x_1 - X_1)(x_2 - X_2)|\mathbf{X}_2 < \mathbf{x}) - r_1(\mathbf{x})r_2(\mathbf{x}). \tag{6.89}$$

The first term on the right hand side is the product moment of reversed residual life of (X_1, X_2) and will be of special interest. We will denote it by $r_{12}(\mathbf{x})$. Then, we have

$$r_{12}(\mathbf{x}) = \frac{1}{F(x_1 - 1, x_2 - 1)} \sum_{t_1=0}^{x_1-1} \sum_{t_2=0}^{x_2-1} (x_1 - t_1)(x_2 - t_2) f(t_1, t_2)$$

which simplifies to

$$r_{12}(\mathbf{x}) = \frac{1}{F_1(x_1 - 1, x_2 - 1)} \sum_{t_1=1}^{x_1} \sum_{t_2=1}^{x_2} F(t_1 - 1, t_2 - 1).$$

Thus,

$$v_{12}(\mathbf{x}) = r_{12}(\mathbf{x}) - r_1(\mathbf{x})r_2(\mathbf{x}). \tag{6.90}$$

Remark 6.14. In the multivariate case, when $\mathbf{X}$ has p elements, we have to consider the variance-covariance matrix

$$V(\mathbf{x}) = (v_{ij}(\mathbf{x})), \quad i, j = 1, 2, \ldots, p,$$

where

$$v_{ij}(\mathbf{x}) = E((x_i - X_i)(x_j - X_j)|\mathbf{X} < \mathbf{x}), \ \mathbf{x} = (x_1, x_2, \ldots, x_p).$$

All the elements can be calculated using formulas similar to (6.87)–(6.89).

The roles of v_1^2, v_2^2 and v_{12} in the analysis of data can be two-fold. The first one is the identification of the model, using either the functional form of v_i or its behaviour. Secondly, they are needed in discussing the sampling behaviour of r_1 or r_2, where v_1^2 and v_2^2 play the role of the usual variance in connection with inference on the mean.

6.8 RELIABILITY FUNCTIONS AND TIME DEPENDENT MEASURES OF ASSOCIATION

As noted earlier, while modelling any kind of multivariate data, the nature of dependence between the constituent variables have a vital role to play. So is the case with reliability function. Instead of considering each aspect separately, the matching of observed facts with the population characteristics in respect of association and behaviour of reliability functions can be simultaneously verified if we can find some relationships between the two. The objective of this section is to explore such relationships.

Recall that the Clayton (1978) measure of association from (1.71) is

$$\theta(x_1,x_2)=\frac{f(x_1,x_2)S(x_1,x_2)}{[S(x_1,x_2)-S(x_1+1,x_2)][S(x_1,x_2)-S(x_1,x_2+1)]}.$$

Dividing by $[S(x_1,x_2)]^2$, we get

$$\theta(x_1,x_2)=\frac{a(x_1,x_2)}{c_1(x_1,x_2)c_2(x_1,x_2)}$$

which expresses $\theta(x_1,x_2)$ in terms of the scalar failure rate and the two components of the vector hazard rate. Making use of (iv) in Section 6.2.3, we have

$$\theta(x_1,x_2)=\frac{c_1(x_1,x_2+1)c_2(x_1,x_2)+[c_1(x_1,x_2)-c_1(x_1,x_2+1)]}{c_1(x_1,x_2)c_2(x_1,x_2)}. \quad (6.91)$$

From (6.91), $\theta > (< 1)$ according as

$$a(x_1,x_2) > (<) c_1(x_1,x_2)c_2(x_1,x_2), \quad (6.92)$$

or

$$[1-c_2(x_1,x_2)][c_1(x_1,x_2+1)-c_1(x_1,x_2)] < (>) 0. \quad (6.93)$$

Since $0 \le c_1 \le 1$, (6.93) means that $\theta(x_1,x_2) > (< 1)$ according as

$$c_1(x_1,x_2+1) < (>) c_1(x_1,x_2). \quad (6.94)$$

An alternative interpretation of $\theta(x,y)$ in terms of hazard rates is also possible. With the help of the conditional probability formula

$$P(AB)=P(A)P(B|A),$$

we can write

$$\theta(x_1,x_2)=\frac{P[X_1=x_1,X_2=x_2]P[X_1\ge x_1,X_2\ge x_2]}{P[X_1=x_1,X_2\ge x_2]P[X_1\ge x_1,X_2=x_2]}$$

as

$$\theta(x_1,x_2)=\frac{P[X_1=x_1|X_2=x_2]}{P[X_1\ge x_1|X_2=x_2]}\frac{P[X_1\ge x_1|X_2\ge x_2]}{P[X_1=x_1|X_2\ge x_2]}$$
$$=\frac{h(x_1|X_2=x_2)}{h(x_1|X_2\ge x_2)}, \quad (6.95)$$

where $h(\cdot)$ is the univariate hazard rate of X_1. Then, $\theta(x_1, x_2) > (<)1$ if and only if $h(x_1|X_2 = x_2) > (<)h(x_1|X_2 \geq x_2)$. Further, condition (6.94) implies that $\theta(x_1, x_2) > 1$ if and only if

$$\frac{P[X_1 > x_1 + 1, X_2 > x_2]}{P[X_1 > x_1, X_2 > x_2]} > \frac{P[X_1 > x_1 + 1, X_2 > x_2 + 1]}{P[X_1 > x_1, X_2 > x_2 + 1]}.$$

The last inequality shows that (X_1, X_2) is RCSI. We have thus established the following theorem.

Theorem 6.20. *The following statements are equivalent:*

(i) *X_1 and X_2 are positively associated in the sense of Clayton measure if $\theta(x_1, x_2) > 1$;*
(ii) *$c_1(x_1, x_2 + 1) < c_2(x_1, x_2)$ for all x_1, x_2;*
(iii) *$a(x_1, x_2) > c_1(x_1, x_2)c_2(x_1, x_2)$;*
(iv) *(X_1, X_2) is RCSI;*
(v) *$h(x_1|X_2 = x_2) > h(x_1|X_2 \geq x_2)$.*

Remark 6.15. By reversing the inequalities in (ii), (iii) and (v), we have $\theta(x_1, x_2) < 1$ or X_1 and X_2 are negatively associated, and (X_1, X_2) is RCSD.

In Theorem 6.20 and in the above remark, we have characterized positive and negative dependence through various properties. A similar result exists for the case of independence.

Theorem 6.21. *A necessary and sufficient condition that X_1 and X_2 are independent is that any one of the following equivalent conditions hold:*

(i) *$\theta(x_1, x_2) = 1$;*
(ii) *$c_1(x_1, x_2 + 1) = c_1(x_1, x_2)$ for all x_1, x_2;*
(iii) *$c_2(x_1, x_2) = c_2(x_1 + 1, x_2)$;*
(iv) *$h(x_1|X_2 = x_2) = h(x_1|X_2 > x_2)$.*

Example 6.16. Consider the bivariate geometric distribution (Example 6.1) with

$$S(x_1, x_2) = q_1^{x_1} q_2^{x_2} \theta^{x_1 x_2},$$
$$c_1(x_1, x_2) = 1 - q_1\theta^{x_2},$$

giving

$$c_1(x_1, x_2 + 1) = (1 - q_1\theta^{x_2+1}) < 1 - q_1\theta^{x_2} = c_1(x_1, x_2).$$

Thus, (X_1, X_2) is negatively associated. For this distribution, the conditional survival functions of X_1, given $X_2 = x_2$, is

$$S(x_1|X_2 = x_2) = q_1^{x_1}[1 - q_2\theta^{x_1}]\theta^{x_1 x_2}$$

and the corresponding probability mass function is

$$f(x_1|X_2 = x_2) = q_1^{x_1}\theta^{x_1x_2}[1 - q_2\theta^{x_1} - q_1\theta^{x_2} + q_1q_2\theta^{x_1+x_2+1}].$$

Hence,

$$h(x_1|X_2 = x_2) = \frac{1 - q_2\theta^{x_1} - q_1\theta^{x_2} + q_1q_2\theta^{x_1+x_2+1}}{1 - q_2\theta^{x_1}}.$$

Also, the survival function and the probability mass function of $(X_1|X_2 \geq x_2)$ are

$$S(x_1|X_2 \geq x_2) = q^{x_1}\theta^{x_1x_2}$$

and

$$f(x_1|X_2 \geq x_2) = q^{x_1}(1 - q_1\theta^{x_2})\theta^{x_1x_2},$$

giving

$$h(x_1|X_2 \geq x_2) = 1 - q_1\theta^{x_2}. \tag{6.96}$$

Finally, from (1.71) and (6.96), we have

$$\theta(x_1, x_2) = \frac{1 - q_1\theta^{x_2} - q_2\theta^{x_1} + q_1q_2\theta^{x_1+x_2+1}}{(1 - q_1\theta^{x_2})(1 - q_2\theta^{x_1})}.$$

Since $0 < q_1, q_2 < 1$ and $0 \leq \theta \leq 1$, we have $\theta(x_1, x_2) < 1$, showing that X_1 and X_2 are negatively associated. We can also arrive at the above expression for $\theta(x_1, x_2)$ by substituting $a(x_1, x_2)$ in Example 6.1 and c_1 and c_2 from (6.18) and (6.19).

The Anderson, Louis, Holm and Harwald measure in terms of the mean residual life is

$$\phi(x_1, x_2) = \frac{m_1(x_1, x_2)}{m_1(x_1, -1)} = \frac{m_1(x_1, x_2)}{m(x_1)}, \tag{6.97}$$

where $m(x_1)$ is the mean residual life of the marginal distribution of X_1. Gupta (2003) has pointed out that the numerator is the life expectancy of X_1 with the information that $X_1 > X_2$, while the denominator is the life expectancy of X_1 without the information on X_2. Thus, values of ϕ are much different from unity indicate strong influence of X_2 on X_1. The measure has potential on twin research since $\phi_1(x, x)$ measures the percentage of additional expected life for twin 1 due to twin 2 being alive at x. In terms of survival functions, (6.97) can be written as

$$\phi(x_1, x_2) = \frac{S_1(x_1 + 1)\sum_{t=x_1+1}^{\infty} S(t, x_2 + 1)}{S(x_1 + 1, x_2 + 1)\sum_{t=x_1+1}^{\infty} S_1(t)},$$

where $S_1(\cdot)$ is the marginal survival function of X_1.

Theorem 6.22.

$$\theta(x_1, x_2) > (<)\phi(x_1, x_2) > (<)1.$$

Proof. To prove this result, we appeal to Eq. (6.94) which states that

$$\begin{aligned}\theta(x_1, x_2) > 1 &\Rightarrow c_1(x_1, x_2 + 1) < c_1(x_1, x_2) \text{ for all } x_1\\ &\Rightarrow c_1(x_1, x_2) < c_1(x_1, 0)\\ &\Rightarrow m_1(x_1, x_2 + 1) > m_1(x_1, -1)\\ &\Rightarrow \phi(x_1, x_2) > 1.\end{aligned}$$

□

Example 6.17. The bivariate geometric distribution in Example 6.17 gives

$$m_1(x_1, x_2) = (1 - q_1\theta^{x_2+1})^{-1} \text{ and } m_1(x_1, -1) = (1 - q_1)^{-1}$$

and therefore

$$\phi(x_1, x_2) = \frac{1 - q_1)}{1 - q_1\theta^{x_2+1}} < 1.$$

Azadian et al. (2009) defined the concept of *hazard positive dependence* (HPD) (*hazard negative dependence* (HND)) for absolutely continuous random variables by the relation that Basu's hazard rate is greater (less) than or equal to the product of the components of the vector hazard rate. In the discrete case, we have

$$a(x_1, x_2) \geq (\leq)c_1(x_1, x_2)c_2(x_1, x_2).$$

Obviously, this is the same as the condition for positive (negative) dependence using the Clayton measure. It becomes clear from the earlier discussions that the properties they have mentioned can be written as

$$\begin{aligned}&\text{RCSD}(X_1, X_2) \Leftrightarrow S(x_1, x_2) \text{ is } \text{TP}_2 \Rightarrow \theta(x_1, x_2) < 1,\\ &\text{LCSI}(X_1, X_2) \Leftrightarrow F(x_1, x_2) \text{ is } \text{RR}_2 \Rightarrow \theta(x_1, x_2) > 1.\end{aligned}$$

Anderson et al.'s second measure is (see (1.73))

$$\psi(x_1, x_2) = \frac{S(x_1, x_2)}{S_1(x_1)S_2(x_2)}.$$

We have the following properties of ψ stated in the next theorem.

Theorem 6.23. ***(i)*** *If $\psi(x_1, x_2) > (< 1)$, then (X_1, X_2) is PQD (NQD);*
(ii) *$\psi(x_1, x_2) = 1$ if and only if X_1 and X_2 are independent;*
(iii) *Since RCSI $\Rightarrow$ PQD, $\theta(x_1, x_2) > 1 \Rightarrow \psi(x_1, x_2) > 1$.*

Finally, Nair and Sankaran's measure (1.77) can be written in terms of reliability functions as

$$\alpha(x_1, x_2) = \frac{M(x_1, x_2)}{m_1(x_1, x_2)m_2(x_1, x_2)}, \tag{6.98}$$

where $M(x_1, x_2)$ is the product moment of residual life. The measure in (6.98) is based on the residual lives of X_1 and X_2 and hence takes into consideration the entire life history of the device they represent beyond (x_1, x_2). This is in sharp contrast to the Clayton and other measures that represent the lifetime at $X_1 = x_2$, $X_2 = x_2$. Moreover, the involvement of the product moment in the definition brings in the effect of covariance in the concept. Further, (6.98) is a symmetric measure and satisfies the properties mentioned below.

Theorem 6.24. *The variables X_1 and X_2 are independent if and only if $\alpha(x_1, x_2) = 1$.*

Proof. When X_1 and X_2 are independent, we have

$$M_1(x_1, x_2) = E[(X_1 - x_1)(X_2 - x_2)|X_1 > x_1, X_2 > x_2]$$

which can be written as

$$M_1(x_1, x_2) = m_1(x_1, -1)m_2(-1, x_2),$$

$m_1(x_1, x_2) = m_1(x_1, -1)$, and $m_2(x_1, x_2) = m_2(-1, x_2)$ arising from

$$S(x_1, x_2) = S_1(x_1)S_2(x_2)$$

and the definitions of M, m_1 and m_2. Thus, $\alpha(x_1, x_2) = 1$. To prove the converse, we write $\alpha(x_1, x_2) = 1$ as

$$M_1(x_1, x_2) = m_1(x_1, x_2)m_2(x_1, x_2),$$

or

$$\sum_{t_1=x_1+1}^{\infty} \sum_{t_2=x_2}^{\infty} S(t_1, t_2) = S(x_1 + 1, x_2 + 1). \tag{6.99}$$

Changing x_2 to $x_2 + 1$ and substituting from (6.99), we have

$$\sum_{t_1=x_1+1}^{\infty} S(t_1, x_2 + 1) = S(x_1 + 1, x_2 + 1)m_1(x_1, x_2)m_2(x_1, x_2)$$
$$- S(x_1 + 1, x_2 + 2)m_1(x_1, x_2 + 1)m_2(x_1, x_2 + 1).$$

The last equation can be written as

$$\begin{aligned}&S(x_1+1,x_2+1)m_1(x_1,x_2)\\&\quad=S(x_1+1,x_2+1)m_1(x_1,x_2)[m_2(x_1,x_2)-m_2(x_1,x_2+1)]\\&\qquad+S(x_1+1,x_2+1)(m_1(x_1,x_2)-m_1(x_1,x_2+1)m_2(x_1,x_2+1))\\&\qquad+m_1(x_1,x_2+1)m_2(x_1,x_2+1)(S(x_1+1,x_2+1)-S(x_1+1,x_2+2)).\end{aligned}$$

Dividing now by $S(x_1+1,x_2+1)$ and then using (6.15) and Part (iv)(b) in Section 6.3 results in, after some algebra,

$$[1-m_2(x_1,x_2)][m_1(x_1,x_2)][m_1(x_1,x_2)-m_1(x_1,x_2+1)]=0.$$

Since $m_2(x_1,x_2)>1$,

$$m_1(x_1,x_2+1)=m_1(x_1,x_2)$$

and accordingly $m_1(x_1,x_2)=m_1(x_1,-1)$. Similarly, changing x_1 to x_1+1 in (6.99) and working as above, we get $m_2(x_1,x_2)=m_2(1,x_2)$. Thus, X_1 and X_2 are independent as seen from result (ii) in Section 6.3. □

Theorem 6.25. *(i)* *Cov* $(X_1,X_2)\ge 0 \Leftrightarrow \alpha(-1,-1)\ge 1$;
(ii) $\alpha(x_1,x_2)\ge 1$ *for all* $x_1,x_2 \Rightarrow$ *Cov* $(X_1,X_2)\ge 0$.

Proof. We have

$$\alpha(-1,-1)=\frac{E(X_1X_2)}{E(X_1)E(X_2)}=\frac{\mathrm{Cov}(X_1,X_2)+E(X_1)E(X_2)}{E(X_1)E(X_2)}$$

gives (i). Also,

$$\alpha(x_1,x_2)\ge 1 \text{ for all } x_1,x_2 \Rightarrow \alpha(-1,-1)\ge 1 \Leftrightarrow \mathrm{Cov}\ (X_1,X_2)\ge 0. \quad □$$

Remark 6.16. Of the six basic concepts of association in Fig. 1.1, Cov(X_1, X_2) ≥ 0 is the weakest. The α-measure is stronger than Cov$(X_1,X_2)\ge 0$ and $\alpha(x_1,x_2)\ge 1$ is a sufficient condition for positive association through covariance.

Theorem 6.26. (X_1,X_2) *is WPQD* $\Rightarrow \alpha(x_1,x_2)\ge S(x_1+1,x_2+1)$.

Proof. We have, as a result of WPQD,

$$S(x_1+1,x_2+1)M(x_1,x_2)\ge S-1(x_1+1)m_1(x_1,-1)S_2(x_2+2)m_2(-1,x_2). \tag{6.100}$$

Since

$$S(x_1+1,x_2+1)\le S_1(x_1+1),$$

we can write

$$\sum_{t_1=x_1+1}^{\infty} S(t_1, x_2+1) \le \sum_{t_1=x_1+1}^{\infty} S_1(t_1),$$

or

$$S(x_1+1, x_2+1)m_1(x_1, x_2) \le S_1(x_1+1)m_1(x_1, -1). \tag{6.101}$$

Likewise, we have

$$S(x_1+1, x_2+1)m_2(x_1, x_2) \le S_2(x_2+2)m_2(-1, x_2). \tag{6.102}$$

With the use of (6.101) and (6.102), the inequality in (6.100) takes on the form

$$S(x_1+1, x_2+1)M(x_1, x_2) \ge S(x_1+1, x_2+1)m_1(x_1, x_2) \\ S(x_1+1, x_2+1)m_2(x_1, x_2)$$

which means that

$$(X_1, X_2)\text{is WPQD } \Rightarrow \alpha(x_1, x_2) \ge S(x_1, x_2).$$ □

Remark 6.17. In view of Remark 6.16 and Theorem 6.26, the nature of dependence using $\alpha(x_1, x_2)$ has no implication with TP_2, RCSI, SI, RTI, RCSI, Association PQD and WPQD, except that it is weaker than positive (negative) covariance.

Theorem 6.27. *Let (Y_1, Y_2) be a discrete random vector with probability mass function*

$$p(x_1, x_2) = P(Y_1 = x_1, Y_2 = x_2) = \frac{S(x_1+1, x_2+1)}{E(Y_1 Y_2)}, \\ x_1, x_2 = 0, 1, 2, \ldots \tag{6.103}$$

such that $E(Y_1 Y_2) < \infty$. Then, the Clayton measure of (Y_1, Y_2) is the same as $\alpha(x_1, x_2)$ of (X_1, X_2) with survival function $S(x_1, x_2)$, that is,

$$\theta_{Y_1,Y_2}(x_1, x_2) = \alpha(x_1, x_2).$$

Proof.

$$\theta_{Y_1,Y_2}(x_1, x_2) = \frac{S(x_1+1, x_2)\sum_{t_1=x_1+1}^{\infty}\sum_{t_2=x_2+1}^{\infty} S(t_1, t_2)}{\sum_{t_2=x_2+1}^{\infty} S(x_1+1, t_2)\sum_{t_1=x_1+1}^{\infty} S(t_1, x_2+1)} \\ = \frac{M(x_1, x_2)}{m_1(x_1, x_2)m_2(x_1, x_2)} = \alpha(x_1, x_2).$$ □

Remark 6.18. From Definition 1.1, (6.103) is recognized as the bivariate equilibrium of (X_1, X_2). See also the discussion in Section 6.6 on the role of this distribution in reliability analysis.

Remark 6.19. The distribution of (Y_1, Y_2) can be easily deduced from that of (X_1, X_2). Hence, various properties of $\theta_{Y_1,Y_2}(x_1, x_2)$ hold good for $\alpha(x_1, x_2)$ as well.

Example 6.18. Consider the bivariate Waring distribution with

$$S(x_1, x_2) = \frac{(m)_{x_1+x_2}}{(m+n)_{x_1+x_2}}, \qquad x_1, x_2 = 0, 1, 2, \ldots .$$

Then, we have from Example 6.11,

$$M(x_1, x_2) = \frac{(m+n+x_1+x_2)(m+n+x_1+x_2+1)}{(n-1)^2},$$

$$m_1(x_1, x_2) = m_2(x_1, x_2) = \frac{m+n+x_1+x_2+1}{(n-1)},$$

so that

$$\alpha(x_1, x_2) = \frac{m+n+x_1+x_2}{m+n+x_1+x_2+1}.$$

We see that $\alpha(x_1, x_2) < 1$ for all $m, n > 0$, and accordingly X_1 and X_2 and negatively associated.

Following the literature on dependence measures and concepts discussed so far, there have been some parallel development to introduce association measures using concepts in reversed time. Sankaran and Gleeja (2008) and Domma (2011) have studied the properties of

$$\lambda(x_1, x_2) = \frac{F(x_1, x_2)\frac{\partial^2 F(x_1,x_2)}{\partial x_1 \partial x_2}}{\frac{\partial F(x_1,x_2)}{\partial x_1}\frac{\partial F(x_1,x_2)}{\partial x_2}} = \frac{F(x_1, x_2) f(x_1, x_2)}{\frac{\partial F(x_1,x_2)}{\partial x_1}\frac{\partial F(x_1,x_2)}{\partial x_2}},$$

where F is the distribution of a continuous random vector (X_1, X_2). Since the survival function $\bar{F}(x_1, x_2)$ in the Clayton measure for continuous random variables is replaced by the distribution function, the properties and interpretations are translated in terms of the reversed hazard rates. Since the discrete version of the above measure follows the same pattern as in Section 6.8 by replacing usual reliability functions with the corresponding reversed functions, this is not further discussed here.

6.9 RELIABILITY FUNCTIONS OF EQUILIBRIUM DISTRIBUTIONS

Recalling various definitions of bivariate equilibrium distributions mentioned in Section 1.9, we discuss their properties that have direct relevance to reliability modelling.

Model 1

In this case, a random vector (Y_1, Y_2) has equilibrium corresponding to (X_1, X_2) with survival function $S(x_1, x_2)$ if its probability mass function is specified by

$$f_E(x_1, x_2) = \frac{S(x_1+1, x_2+1)}{E(X_1 X_2)}, \quad x_1, x_2 = 0, 1, 2, \ldots. \tag{6.104}$$

Since

$$E(X_1 X_2) = \sum_{t_1=0}^{\infty} \sum_{t_2=0}^{\infty} S(t_1+1, t_2+1) = \sum_{t_1=1}^{\infty} \sum_{t_2=1}^{\infty} S(t_1, t_2),$$

f_E satisfies the requirements of a bivariate probability mass function.

Example 6.19. The equilibrium distribution of the bivariate geometric law

$$S(x_1, x_2) = q_1^{x_1} q_1^{x_2} \theta^{x_1 x_2}$$

is represented by

$$f_E(x_1, x_2) = C q_1^{x_1+1} q_1^{x_2+1} \theta^{(x_1+1)(x_2+1)},$$
$$x_1, x_2 = 01, 2, \ldots, \; 0 < q_1, q_2 < 1, \; 0 \le \theta \le 1,$$

where

$$C^{-1} = \sum_{r=0}^{\infty} \frac{\theta^{r_1+1} q_1^{r+2}}{1 - q_2 \theta^{r+1}} = \sum_{r=0}^{\infty} \frac{\theta^{r_2+1} q_2^{r_2+2}}{1 - q_1^r \theta}.$$

Some useful aspects of Model 1 in (6.104) are as follows:

(1) The survival function of (Y_1, Y_2) is

$$\begin{aligned} S_E(x_1, x_2) &= \sum_{t_1=x_1}^{\infty} \sum_{t_2=x_2}^{\infty} S(t_1+1, t_2+1) \Big/ E(X_1 X_2) \\ &= \sum_{t_1=x_1}^{\infty} \sum_{t_2=x_2}^{\infty} S(t_1, t_2) \Big/ E(X_1 X_2) \\ &= \frac{S(x_1+1, x_2+1) M(x_1, x_2)}{E(X_1 X_2)} \end{aligned}$$

From (6.104) and (6.100), the scalar hazard rate of (Y_1, Y_2) becomes

$$a_E(x_1, x_2) = \frac{1}{M(x_1, x_2)},$$

the reciprocal of the product moment of residual life;

(2) If X_1 and X_2 are independent random variables,

$$f_E(x_1, x_2) = \frac{S_1(x_1+1)}{\mu_1}\frac{S_2(x_2+1)}{\mu_2}, \quad \mu_i = E(X_i), i = 1, 2.$$

Observe from Eq. (1.23) that $\frac{S_i(x_i+1)}{\mu_i}$ is the probability mass function of the equilibrium distribution of X_i. Thus, when X_1 and X_2 are independent, Y_1 and Y_2 are also independent. The marginal distributions of the bivariate equilibrium distribution of (X_1, X_2) are univariate equilibrium distributions of the component variables X_1 and X_2. However, in the non-independent case, the marginals are not of equilibrium form;

(3) The bivariate weighted distribution of (X_1, X_2) can be written as (see Eq. (1.54))

$$f_W(x_1, x_2) = \frac{W(x_1, x_2) f(x_1, x_2)}{EW(X_1, X_2)}. \tag{6.105}$$

Choosing

$$W(x_1, x_2) = \frac{S(x_1+1, x_2+1)}{f(x_1, x_2)}, \tag{6.106}$$

Eq. (6.105) shows that $f_W(x_1, x_2) = f_E(x_1, x_2)$. Thus, the bivariate equilibrium model (6.104) is a weighted distribution with weight function as in (6.106). In the continuous case, the weight function is the reciprocal of the scalar failure rate. This is not the case in the discrete case. We can express (6.106) in terms of the hazard rates as

$$W(x_1, x_2) = \frac{1 + a(x_1, x_2) - c_1(x_1, x_2) - c_2(x_1, x_2)}{a(x_1, x_2)};$$

(4) The vector valued hazard rate of (Y_1, Y_2) has components

$$c_{1,E}(x_1, x_2) = 1 - \frac{S_E(x_1+1, x_2)}{S_E(x_1, x_2)} \tag{6.107}$$

and

$$c_{2,E}(x_1, x_2) = 1 - \frac{S_E(x_1, x_2+1)}{S_E(x_1, x_2)}. \tag{6.108}$$

Further simplifying the expression, we get

$$
\begin{aligned}
& c_{1,E}(x_1, x_2) \\
&= 1 - \frac{S(x_1+2, x_2+1)M(x_1+1, x_2))}{S(x_1+1, x_2+1)M(x_1, x_2)} \\
&= 1 - \frac{M(x_1+1, x_2)}{M(x_1, x_2)}[1 - c_1(x_1+1, x_2+1)] \\
&= 1 - \frac{M(x_1+1, x_2)}{M(x_1, x_2)}\left[\frac{m_1(x_1, x_2) - 1}{m_1(x_1+1, x_2)}\right] \\
&= \frac{M(x_1, x_2)m_1(x_1+1, x_2) - M(x_1+1, x_2)(m_1(x_1, x_2) - 1)}{M(x_1, x_2)m_1(x_1+1, x_2)} \\
&= \frac{m_2(x_1, x_2)m_1(x_1+1, x_2)}{M(x_1, x_2)m_1(x_1+1, x_2)}, \text{ on using (6.59)} \\
&= \frac{m_2(x_1, x_2)}{M(x_1, x_2)} \\
&= \frac{E(X_2 - x_2|X_1 > x_1, X_2 > x_2)}{E((X_1 - x_1)(X_2 - x_2)|X_1 > x_1, X_2 > x_2)}.
\end{aligned}
\tag{6.109}
$$

Similarly,

$$C_{2,E}(x_1, x_2) = \frac{m_1(x_1, x_2)}{M(x_1, x_2)}. \tag{6.110}$$

The measure of association is

$$\alpha(x_1, x_2) = \frac{a_E(x_1, x_2)}{c_{1E}(x_1, x_2)c_{2E}(x_1, x_2)}.$$

Model 2

This model is analogous to the continuous case discussed in Gupta and Sankaran (1998). It is defined as the joint distribution of the vector (W_1, W_2) whose conditional probability mass functions of $(W_1|W_1 > x_2)$ and $(W_2|W_1 > x_1)$ are defined by

$$g_1(x_1|W_2 > x_2) = \frac{P[X_1 > x_1|X_2 > x_2]}{E(X_1|X_2 > x_2)}, \quad x_2 = -1, 0, 1, \ldots, \tag{6.111}$$

and

$$g_2(x_2|W_1 > x_1) = \frac{P[X_2 > x_2|X_1 > x_1]}{E(X_2|X_1 > x_1)}, \quad x_1 = -1, 0, 1, \ldots \tag{6.112}$$

subject to the compatibility criterion described below. Two families of conditional survival functions $P[X_1 > x_1|X_2 > x_2] = a(x_1, x_2)$ and $P[X_2 > x_2|X_1 >$

$x_1] = b(x_1, x_2)$ are compatible (that is, will generate a valid joint survival function) if there exist functions $u(x_1)$ and $v(x_2)$ such that

$$\frac{a(x_1, x_2)}{b(x_1, x_2)} = \frac{u(x_1)}{v(x_2)},$$

where $u(x_1)$ is a one-dimensional survival function.

Eqs (6.111) and (6.112) can be employed to find the joint survival function of (W_1, W_2) by adopting the following procedure. That is, (a) compute $P(W_1 > x_1)$ from (6.111) by summation of x_1 over $[x_1 + 1, \infty)$ after setting $X_2 = -1$, (b) Find $P[W_2 > x_1 | W_1 > x_1]$ by summation of x_2 over $[x_2 + 1, \infty]$, and (c) taking the product

$$P(W_2 > x_2 | W_1 > x_1) P(W_1 > x_1)$$

to derive $P(W_1 > x_1, W_2 > x_2)$, and then to write $S_{W_1,W_2}(x_1, x_2) = P(W_1 \geq x_1, W_2 \geq x_2)$. We give an illustration of this procedure in the example below.

Example 6.20. Let (X_1, X_2) be distributed as

$$S(x_1, x_2) = q_1^{x_1} q_2^{x_2} \theta^{x_1 x_2}, \quad x_1, x_2 = 0, 1, 2, \ldots.$$

Then,

$$P(X_1 > x_1, X_2 > x_2) = q_1^{x_1+1} q_2^{x_2+1} \theta^{(x_1+1)(x_2+1)},$$
$$P(X_1 > x_1) = q_1^{x_1+1} \text{ and } P(X_2 > x_2) = q_2^{x_2+1},$$

by taking $x_1 = -1$ and $x_2 = -1$, respectively. Thus,

$$P(X_1 > x_1 | X_2 > x_2) = (q_1 \theta^{x_2+1})^{x_1+1}, \quad x_1 = -1, 0, 1, \ldots,$$

and

$$P(X_2 > x_2 | X_1 > x_1) = (q_2 \theta^{x_1+1})^{x_2+1}, \quad x_2 = -1, 0, 1, \ldots.$$

Now,

$$E(X_1 | X_2 > x_2) = \sum_{t_1=0}^{\infty} (q_1 \theta^{x_2+1})^{t_1+1} = \frac{q_1 \theta^{x_1+1}}{1 - q_1 \theta^{x_2+1}}$$

and likewise

$$E(X_2 | X_1 > x_1) = \frac{q_2 \theta^{x_1+1}}{1 - q_2 \theta^{x_2+1}}.$$

From (6.111), we have

$$g_1(x_1|W_2 > x_2) = (q_1\theta^{x_2+1})^{x_1}(1 - q_1\theta^{x_2+1}),$$

or

$$g_1(x_1) = (1 - q_1)q_1^{x_1}.$$

Thus,

$$P(W_1 > x_1) = \sum_{t_1=x_1+1}^{\infty} (1 - q_1)q_1^{x_1} = q_1^{x_1+1}.$$

Using (6.112), we have

$$g_2(x_2|W_1 > x_1) = (q_2\theta^{x_1+1})^{x_2}(1 - q_2\theta^{x_1+1})$$

and so by (b) above,

$$P(W_2 > x_2) = (q_2\theta^{x_1+1})^{x_2+1}.$$

The joint probability

$$P[W_1 > x_1, W_2 > x_2] = q_1^{x_1+1} q_2^{x_2+1} \theta^{(x_1+1)(x_2+1)},$$

or

$$S_{W_1,W_2} = q_1^{x_1} q_2^{x_2} \theta^{x_1 x_2},$$

which is identical with the baseline survival function $S(x_1, x_2)$.

From the definition, it is clear that the distributions of $(W_1|W_2 > x_2)$ and $(W_2|W_1 > x_1)$ are the equilibrium versions of $(X_1|X_2 > x_2)$ and $(X_2|X_1 > x_1)$. So are the marginal distributions of $(W_1 W_2)$ with respect to the marginals of (X_1, X_2). We also note that

$$c_{i,W_i}(x_1, x_2) = \frac{1}{m_i(x_1, x_2)}, \quad i = 1, 2;$$

that is, the vector hazard rate components of (W_1, W_2) are the reciprocal of mean residual life components of (X_1, X_2).

Remark. Note that

$$c_1(x_1, x_2) = 1 - \frac{S(x_1 + 1, x_2)}{S(x_1, x_2)} = 1 - \frac{S_1(x_1 + 1|X_2 > x_2)}{S_1(x_1|X_2 > x_2)},$$

where S_1 is the survival function of $(X_1|X_2 > x_2)$. Similarly, $c_2(x_1, x_2)$. Likewise,

$$m_1(x_1, x_2) = m(x_1|X_2 > x_2),$$

where m is the univariate mean residual life of $(X_1|X_2 > x_2)$.

Some relationships exist between the reliability functions of Models 1 and 2. Starting with Model 1 in (6.104), the marginal distributions of $f_E(x_1, x_2)$ are

$$f_{1,E}(x_1) = K \sum_{x_2=0}^{\infty} S(x_1+1, x_2+1), \quad \text{with } K = \frac{1}{E(X_1X_2)}$$

$$= K P(X_1 > x_1) \sum_{x_2=0}^{\infty} P(X_2 > x_2|X_1 > x_1)$$

$$= K P(X_1 > x_1) E(X_2|X_1 > x_1)$$

and

$$f_{2,E}(x_2) = K P(X_2 > x_2) E(X_1|x_2 > x_2).$$

Hence,

$$f_{2,E}(x_1|x_2) = \frac{S(x_1+1, x_2+1)}{S(x_2+1)E(X_1|X_2 > x_2)} = \frac{P(X_1 > x_1|X_2 > x_2)}{E(X_1|X_2 > x_2)}$$

$$= g_1(x_1|W_2 > x_2).$$

The hazard rates of Y_1 and W_1 satisfy

$$h(x_1|X_2 = x_2) = C_{1,W_1}(x_1, x_2) = \frac{1}{m_1(x_1, x_2)}.$$

Similarly,

$$h(x_2|X_1 = x_1) = C_{2,W_2}(x_1, x_2) = \frac{1}{m_2(x_1, x_2)}.$$

Combining these with Theorem 6.20, we have the following result.

Theorem 6.28. *If (X_1, X_2) is a discrete random vector defined on N^2, then the following statements are equivalent:*

(i) (Y_1, Y_2) *is RCSI;*
(ii) $h_1(x_1|X_2 = x_2) \leq \frac{1}{m_1(x_1,x_2)}$ *for all* x_1, x_2*;*
(iii) $\theta(x_1, x_2) > 1$*;*
(iv) $h_1(x_1|Y > x_2) \leq h_1(x_1|W_2 > x_2)$ *for all* x_1, x_2*;*
(v) $(Y_1|Y_2 > x_2) \geq_{hr} (W_1|W_2 > x_2)$*, where* $\geq_{hr}$ *is the hazard rate order.*

Since all conditional concepts utilized above are univariate in nature, these can have relationships with the corresponding conditional score functions similar to those described in Section 5.6.

Model 3

The third version of bivariate equilibrium distribution is defined in (1.61) and (1.62). It is defined through the conditional probability mass functions of (Z_1, Z_2), where

$$g_{12}(z_1|Z_2=z_2)=\frac{P(X_1>z_1|X_2=z_2)}{E(X_1|X_2=z_2)} \tag{6.113}$$

and

$$g_{21}(z_2|Z_1=z_1)=\frac{P(X_2>z_2|X_1=z_2)}{E(X_2|X_1=z_1)}. \tag{6.114}$$

Navarro and Sarabia (2010) have pointed out a real situation wherein such a model might be useful when X_1 and X_2 are continuous lifetimes. The conditional probability mass functions in (6.113) and (6.114) will provide a joint distribution for (Z_1, Z_2) if

$$\frac{\sum_{t=z_1+1}^{\infty} f(t,z_2)}{\sum_{t=z_2+1}^{\infty} f(z_1,t)}=\frac{a(z_1)}{f_1(z_1)E(X_2|X_2|X_1=z_1)}\frac{b(z_2)}{f_2(z_2)E(X_1|X_1|X_2=z_2)} \tag{6.115}$$

$$=u(z_1)v(z_2),$$

where u and v represent mass functions. The joint distribution of (Z_1, Z_2) is then

$$g(z_1,z_2)=Ku(z_1)\sum_{t=x_2+1}^{\infty} f(z_1,t).$$

Example. Consider

$$f(x_1,x_2)=\alpha q_1^{x_1}q_2^{x_2}\theta^{x_1x_2}$$

discussed earlier in Example 6.8. We have the marginal probability mass functions

$$f_1(x_1)=\frac{\alpha q_1^{x_1}}{1-q_2\theta^{x_1}} \text{ and } f_2(x_2)=\frac{\alpha q_2^{x_2}}{1-q_1\theta^{x_2}}.$$

Hence,

$$f_{12}(x_1|x_2)=(1-q_1\theta^{x_2})(q_1\theta^{x_2})^{x_1}$$

and

$$f_{21}(x_2|x_1)=(1-q_2\theta^{x_1})(q_2\theta^{x_1})^{x_2},$$

which are geometric. Hence,

$$E(X_1|X_2 = x_2) = \frac{q_1\theta^{x_2}}{1 - q_1\theta^{x_2}} \tag{6.116}$$

and

$$E(X_2|X_1 = x_1) = \frac{q_2\theta^{x_1}}{1 - q_2\theta^{x_1}}. \tag{6.117}$$

Thus,

$$g_{12}(z_1|z_2) = (q_1\theta^{x_2})^{x_1}(1 - q_1\theta^{x_2}),$$
$$g_{21}(z_2|z_2) = (q_2\theta^{x_1})^{x_2}(1 - q_2\theta^{x_1}),$$

which are the same as $f_1(x_1|x_2)$ and $f_2(x_2|x_1)$. Hence, for the bivariate geometric conditional distribution, the equilibrium distribution is identical with the baseline distribution.

Notice that in (6.115),

$$\frac{\sum_{t=z_1+1}^{\infty} f(t, z_2)}{\sum_{t=z_2+1}^{\infty} f(z_1, t)} = \frac{(1 - q_2\theta^{z_1})q_1\theta^{z_2}}{(1 - q_1\theta^{z_2})q_2\theta^{z_1}},$$

which is of the form $u(z_1)v(z_2)$. The alternative conditional hazard rate vector of (Z_1, Z_2), $(e_1(z_1, z_2), e_2(z_1, z_2))$ and the conditional mean residual life, $(l_1(x_1, x_2), l_2(x_1, x_2))$ are related through the identity

$$e_i(x_, x_2) = [l_i(x_1, x_2)]^{-1}.$$

Multivariate weighted distributions in their general form in (1.57) can also be considered to develop identities connecting the reliability functions of $\mathbf{X}_W$. Except for some theoretical work in this direction, they do not appear to have been applied for modelling or analyzing real multivariate data even in the continuous domain. Hence, we do not enter into a detailed discussion of the topic other than remarking that the methodology in this connection follow the same logic as those of equilibrium distributions.

Chapter 7

Multivariate Ageing Concepts

7.1 INTRODUCTION

Different classes of life distributions based on the monotonicity of reliability functions, their properties and importance in the univariate case were discussed earlier in Chapter 4. We extend these classes to the multivariate case in this chapter. Instead of a single ageing class and its dual corresponding to a specific property of a reliability function when the lifetime is represented by a single variable, in the multi-component systems, several alternatives can be considered. For instance, in the univariate case, the IHR and DHR classes are uniquely defined depending on whether the hazard rate $h(x)$ is increasing or decreasing. Two types of complexities arise when such a concept is attempted in the multivariate case. First is that there are several ways in which multivariate hazard rate can be defined and therefore, there can be IHR and DHR classes existing with respect to each one of these definitions. Secondly, the monotonicity can be thought of from different angles at which the age is considered. Ageing can be assessed either when all the components are of the same age or when their ages are different. Also, the duration for which the behaviour of the reliability function is studied can be the same or different for the components. Thus, in a two component system, for example, the ageing is reckoned either at $(x+t, x+t)$, (x_1+t, x_2+t), $(x+t_1, x+t_2)$ or at (x_1+t_1, x_2+t_2), giving four different ageing criteria. We will look at all such criteria and the implications among them with respect to the multivariate reliability functions described in the last chapter. As in Chapter 4, there are criteria for positive ageing, negative ageing and no-ageing which are interpreted in the same way as in the univariate case.

7.2 MULTIVARIATE STOCHASTIC ORDERS

We consider random vectors $\mathbf{X}$ and $\mathbf{Y}$, taking values in N^p, with survival functions $S_{\mathbf{X}}(\mathbf{x})$ and $S_{\mathbf{Y}}(\mathbf{x})$ and probability mass function $f_{\mathbf{X}}(\mathbf{x})$ and $f_{\mathbf{Y}}(\mathbf{y})$. Following Shaked and Shanthikumar (2007), we define the stochastic orders given below.

Reliability Modelling and Analysis in Discrete Time. http://dx.doi.org/10.1016/B978-0-12-801913-9.00007-5

Usual Stochastic Order

Let **X** and **Y** be such that

$$P(\mathbf{X} \in U) \leq P(Y \in U),$$

where U is a subset of N^p such that $\mathbf{y} \geq \mathbf{x}$ whenever $\mathbf{x} \in Y$ for $\mathbf{y} \in \mathbf{Y}$. Then, we say that **X** is smaller **Y** in multivariate usual stochastic order and write $\mathbf{X} \leq_{\text{st}} Y$. This definition is equivalent to saying that

$$E[\phi(\mathbf{X})] \leq E[\phi(\mathbf{Y})]$$

for all increasing functions ϕ.

Hazard Rate Orders

Writing $x \vee y = \max(x, y)$ and $x \wedge y = \min(x, y)$, we define

$$\mathbf{x} \vee \mathbf{y} = (x_1 \vee y_1, x_2 \vee y_2, \ldots, x_p \vee y_p),$$
$$\mathbf{x} \wedge \mathbf{y} = (x_1 \wedge y_1, x_2 \wedge y_2, \ldots, x_p \wedge y_p).$$

Then, **X** is smaller than **Y** in multivariate hazard rate order, denoted by $\mathbf{X} \leq_{\text{hr}} \mathbf{Y}$, iff

$$S_{\mathbf{X}}(\mathbf{x}) S_{\mathbf{Y}}(\mathbf{x}) \leq S_{\mathbf{X}}(\mathbf{x} \wedge \mathbf{y}) S_{\mathbf{Y}}(\mathbf{x} \vee \mathbf{y}),$$

for every **x** and **y** in N^p.

Further **X** is smaller that **Y** in weak hazard rate order, $\mathbf{X} \leq_{\text{whr}} \mathbf{Y}$ iff,

$$\frac{S_{\mathbf{Y}}(\mathbf{x})}{S_{\mathbf{X}}(\mathbf{x})} \quad \text{is increasing for all } \mathbf{x},$$

for which the denominator is non-zero. We have the implication

$$\mathbf{X} \leq_{\text{hr}} \mathbf{Y} \Rightarrow \mathbf{X} \leq_{\text{whr}} \mathbf{Y}.$$

Theorem 7.1.

$$\mathbf{C}_{\mathbf{X}}(\mathbf{x}) \geq \mathbf{C}_{\mathbf{Y}}(\mathbf{x}) \Leftrightarrow \mathbf{X} \leq_{whr} \mathbf{Y},$$

where $\mathbf{C}_{\mathbf{X}}(\mathbf{x}) = (c_1(\mathbf{x}), \ldots, c_p(\mathbf{x}))$ *is the vector hazard rate of* **x**.

Proof. We have

$$\begin{aligned}\mathbf{C}_{\mathbf{X}}(\mathbf{x}) \geq \mathbf{C}_{\mathbf{Y}}(\mathbf{x}) &\Leftrightarrow c_{i,\mathbf{X}}(\mathbf{x}) \geq c_{i,\mathbf{Y}}(\mathbf{x}) \quad \text{for } i = 1, 2, \ldots, p \\ &\Leftrightarrow 1 - \frac{S_{\mathbf{X}}(x_1, \ldots, x_{i-1}, x_i + 1, x_{i+1}, \ldots, x_p)}{S_X(\mathbf{x})} \\ &\quad \geq 1 - \frac{S_{\mathbf{Y}}(x_1, \ldots, x_{i-1}, x_i + 1, x_{i+1}, \ldots, x_p)}{S_Y(\mathbf{x})}\end{aligned}$$

$$\Leftrightarrow \frac{S_{\mathbf{X}}(x_1, \ldots, x_{i-1}, x_i + 1, x_{i+1}, \ldots, x_p)}{S_{\mathbf{Y}}(x_1, \ldots, x_{i-1}, x_i + 1, x_{i+1}, \ldots, x_p)} \leq \frac{S_X(\mathbf{x})}{S_Y(\mathbf{x})},$$

$$i = 1, 2, \ldots, p$$

$$\Leftrightarrow \frac{S_{\mathbf{Y}}(\mathbf{x})}{S_{\mathbf{X}}(\mathbf{x})} \text{ is increasing for all } \mathbf{x} \Leftrightarrow \mathbf{X} \leq_{\text{whr}} \mathbf{Y}. \qquad \square$$

Likelihood Ratio Order

We say that $\mathbf{X}$ is smaller that $\mathbf{Y}$ in likelihood ratio order, denoted by $\mathbf{X} \leq_{\text{lr}} \mathbf{Y}$, if

$$f_{\mathbf{X}}(\mathbf{x}) f_{\mathbf{Y}}(x) \leq f_{\mathbf{X}}(\mathbf{x} \wedge \mathbf{y}) f_{\mathbf{Y}}(\mathbf{x} \vee \mathbf{y})$$

for all $\mathbf{x}$ and $\mathbf{y}$ in N^p. The order $\leq_{\text{lr}}$ satisfies

(a) $\mathbf{X} \leq_{\text{lr}} \mathbf{Y} \Rightarrow \mathbf{X} \leq_{\text{hr}} Y$,
(b) $\mathbf{X} \leq_{\text{lr}} \mathbf{Y} \Rightarrow \mathbf{X} \leq_{\text{st}} \mathbf{Y}$.

Orthant Orders

The upper and lower orthant orders are some sort of generalizations of the usual stochastic order in one dimension. If $S_{\mathbf{X}}(\mathbf{x}) \leq S_{\mathbf{Y}}(\mathbf{x})$ for all $\mathbf{x}$, then $\mathbf{X}$ is said to be smaller in upper orthant order than $\mathbf{Y}$, $\mathbf{X} \leq_{\text{uo}} \mathbf{Y}$. Similarly, if $F_{\mathbf{X}}(\mathbf{x}) \geq F_{\mathbf{Y}}(x)$ for all $\mathbf{x}$, we say that $\mathbf{X}$ is less than $\mathbf{Y}$ in lower orthant order, $\mathbf{X} \leq_{\text{lo}} \mathbf{Y}$, where $F(\cdot)$ stands for the distribution function. These orders are such that

(i) $\mathbf{X} \leq_{\text{st}} \mathbf{Y} \Rightarrow \mathbf{X} \leq_{\text{uo}} \mathbf{Y}$ and $\mathbf{X} \leq_{\text{lo}} \mathbf{Y}$,
(ii) $\mathbf{X} \leq_{\text{uo}} (\leq_{\text{lo}})\mathbf{Y} \Rightarrow E(\mathbf{X}) \leq E(\mathbf{Y})$,
(iii) $(\mathbf{X}|\mathbf{X} > \mathbf{x}) \leq_{\text{uo}} (\mathbf{Y}|\mathbf{Y} > \mathbf{y}) \Leftrightarrow \mathbf{X} \leq_{\text{whr}} \mathbf{Y}$,

$$\mathbf{X} \leq_{\text{whr}} \mathbf{Y} \Rightarrow \mathbf{X} \leq_{\text{uo}} Y.$$

Mean Residual Life Order

Let $\mathbf{X}$ and $\mathbf{Y}$ have respective mean residual life functions

$$m_{\mathbf{X}}(\mathbf{x}) = E(\mathbf{X} - \mathbf{x}|\mathbf{X} > \mathbf{x}) \text{ and } m_{\mathbf{Y}}(\mathbf{x}) = E(\mathbf{Y} - \mathbf{x}|\mathbf{Y} > \mathbf{x}).$$

Then, $\mathbf{X}$ is less than $\mathbf{Y}$ in multivariate mean residual life order, $\mathbf{X} \leq_{\text{mrl}} \mathbf{Y}$, if $m_{\mathbf{X}}(\mathbf{x}) \leq m_{\mathbf{Y}}(\mathbf{x})$ for all $\mathbf{x}$. From the implication (iii) on upper orthant order, we see that

$$\mathbf{x} \leq_{\text{whr}} \mathbf{Y} \Rightarrow \mathbf{X} \leq_{\text{mrl}} \mathbf{Y}.$$

7.3 MULTIVARIATE NO-AGEING

There are different versions of multivariate no-ageing property. In this section, we provide the definitions in each case and derive the distribution possessing such a property

Definition 7.1. Let $\mathbf{X} = (X_1, \ldots, X_p)$ be a p-dimensional random vector assuming values in N^p. Then, $\mathbf{X}$ possesses the *strong no-ageing property* (also called the *strong lack of memory property*) if

$$P(X_1 \geq x_1 + t_1, \ldots, X_p \geq x_p + t_p | X_1 \geq t_1, \ldots, X_p \geq t_p) \\ = P(X_1 \geq x_1, \ldots, X_p \geq x_p) \quad (7.1)$$

for all $\mathbf{x}$ and $\mathbf{t}$ in N^p.

Theorem 7.2. *The random vector* $\mathbf{X}$ *satisfies* (7.1) *if and only if* $\mathbf{X}$ *has survival function*

$$S(\mathbf{X}) = q_1^{x_1} \ldots q_p^{x_p}, \quad 0 < q_1, \ldots, q_p < 1, \quad (7.2)$$

Proof. It is easy to verify that property (7.1) holds when the survival function is as in (7.2). To prove the converse, we observe that (7.1) is equivalent to

$$S(x_1 + t_1, x_2 + t_2, \ldots, x_p + t_p) = S(x_1, x_2, \ldots, x_p) S(t_1, t_2, \ldots, t_p).$$

Setting $x_2 = \cdots = x_p = 0$ and $t_2 = t_3 = \cdots = t_p = 0$, we get

$$S_1(x_1 + t_1) = S_1(x_1) S_1(t_1).$$

Thus, the marginal distribution of X_1 and similarly for any X_i, $i = 1, \ldots, p$, satisfies the univariate no-ageing property, and so

$$S_i(x_i) = q_i^{x_i} \quad \text{for some } 0 < q_i < 1.$$

This shows that (7.2) is true for $p = 1$ or any $i = 1, 2, \ldots, p$. Now, assume that the result is true for any m variables which gives, for example,

$$S_m(x_1 + t_1, \ldots, x_m + t_m) = S(x_1 \ldots x_m) S(t_1, \ldots, t_p)$$

and

$$S(x_1, \ldots, x_m) = q_1^{x_1} q_2^{x_2} \ldots q_m^{x_m}.$$

Consider

$$S_{m+1}(x_1 + t_1, \ldots, x_{m+1} + t_{m+1}) = S(x_1, \ldots, x_{m+1}) S(t_1, \ldots, t_{m+1}).$$

Now, set $t_1 = 0$ and $x_2 = x_3 = \cdots = x_{m+1} = 0$. Then,

$$S_{m+1}(x_2, t_2, \ldots, t_{m+1}) = S_1(x_1) S_m(t_2, \ldots, t_{m+1}) \\ = q_1^{x_1} q_2^{t_2} \ldots q_{m+1}^{x_{m+1}},$$

showing that the result holds for $p = m + 1$. Thus, by induction, (7.2) holds for all positive integer values. □

Remark 7.1. Theorem 7.2 also means that when the strong no-ageing property is satisfied, the variables must be independent with each being geometrically distributed. The definition is not useful when there is dependency between the lifetimes of the system.

Remark 7.2. From the form of the distribution characterizing the above property, it becomes obvious that (7.1) is equivalent to

$$a(x_1, x_2, \dots, x_p) = h_1(x_1)\dots h_p(x_p),$$
$$c_i(\mathbf{x}) = h_i(x_i),$$
$$m_i(\mathbf{x}) = m_i(x_i),$$

where h_i and m_i are the univariate hazard rate and mean residual life functions of the component X_i.

In view of the restrictions imposed on the application of the strong version to independent lives only, other versions have been proposed.

Definition 7.2. The vector $\mathbf{X}$ is said to be locally no-ageing if it satisfies

$$\begin{aligned} &P[X_i \geq x_i + t_i | \mathbf{X} \geq \mathbf{t}] \\ &\quad = P[X_i \geq x_i | X_1 \geq t_1, X_{i-1} \geq t_{i-1}, X_{i+1} \geq t_{i+1} \dots X_p \geq t_p] \end{aligned} \tag{7.3}$$

for all $i = 1, 2, \dots, p$.

This definition means that the univariate no-ageing holds for the conditional distribution of X_i, given $X_1 \geq x_1, \dots, X_{i-1} \geq x_{i-1}, \dots, X_{i+1} \geq x_{i+1}, \dots, X_p \geq x_p$ for all i. Since the proof is similar to that of Theorem 7.2, we give the proof of the distribution having (7.3) only in the bivariate case.

Theorem 7.3. *A random vector* (X_1, X_2), *defied on* N^2, *satisfies the locally no-ageing property*

$$P[X_1 \geq x_1 + t_1 | X_1 \geq x_1, X_2 \geq x_2] = P[X_1 \geq t_1 | X_2 \geq x_2], \tag{7.4}$$
$$P[X_2 \geq x_2 + t_2 | X_1 \geq x_1, X_2 \geq x_2] = P[X_2 \geq t_2 | X_1 \geq x_1], \tag{7.5}$$

if and only if its survival function is the bivariate geometric distribution with

$$S(x_1, x_2) = q_1^{x_1} q_2^{x_2} \theta^{x_1 x_2}. \tag{7.6}$$

Proof. The conditions (7.4) and (7.5) can be easily verified when the distribution is (7.6). Eqs (7.4) and (7.5) can be rewritten as

$$S(0, x_2) S(x_1 + t_1, x_2) = S(x_1, x_2) S(t_1, x_1), \tag{7.7}$$

$$S(x_1, 0)S(x_1, x_2 + t_2) = S(x_1, x_2)S(x_1, t_2). \tag{7.8}$$

Eq. (7.7) can be written as

$$G_{x_2}(x_1 + t_1) = G_{x_2}(x_1)G_{x_2}(t_1),\ G_{x_2}(u) = \frac{S(u, x_2)}{S(0, x_2)}.$$

The non-trivial solution that gives a survival function is

$$G_{x_2}(x_1) = [q_1(x_2)]^{x_1}, \quad x_1 = 0, 1, 2, \ldots$$

for some $0 < p_1(x_2) < 1$. Thus,

$$S(x_1, x_2) = S(0, x_2)[q_1(x_2)]^{x_1}. \tag{7.9}$$

By a similar calculation, (7.8) leads to

$$S(x_1, x_2) = S(x_1, 0)[q_2(x_1)]^{x_2}, 0 < q_2(x_1) < 1. \tag{7.10}$$

Evaluating $S(0, x_2)$ from (7.10) and substituting in (7.9) and again $S(x_1, 0)$ from (7.9) and substituting in (7.10), we get

$$q_2(x_2)[q_1(x_2)]^{x_1} = q_1^{x_1}[q_2(x_1)]^{x_2},$$

where $q_i = q_1(0)$, $i = 1, 2$. Thus,

$$\left(\frac{q_1(x_2)}{q_1}\right)^{\frac{1}{x_2}} = \left(\frac{q_2(x_1)}{q_2}\right)^{\frac{1}{x_1}},$$

which holds for all x_1, x_2 if and only if each is a constant, say θ. Thus,

$$q_1(x_2) = q_1\theta^{x_2} \quad \text{and} \quad q_2(x_1) = q_2\theta^{x_1}.$$

Substituting now in (7.9) or (7.10), we get the bivariate geometric law. □

Remark 7.3. In the p-variate case, the geometric law characterized by (7.3) is specified by the survival function

$$S(\mathbf{x}) = \left(\prod_{i=1}^{p} q_i^{x_i}\right)\left(\prod_{i<j}^{p} q_{ij}^{x_i x_j}\right)\cdots\left(q_{12\ldots p}^{x_1 x_2 \ldots x_p}\right), \tag{7.11}$$

$0 < q_{0i}, q_{1,j}, \ldots q_{12\ldots p} < 1$, $1 - \sum q_i - \sum_{i<j} q_{ij} + \cdots + (-1)^p q_{12\ldots p} \geq 0$.

Definition 7.3. The vector **X** possesses the *weak no-ageing property* if it satisfies

$$P[\mathbf{X} \geq \mathbf{x} + \mathbf{t}] = P[\mathbf{X} \geq \mathbf{x}]P[\mathbf{X} \geq \mathbf{t}], \tag{7.12}$$

where $\mathbf{t} = (t, t, \ldots, t)$ for $\mathbf{x}$ in N^p and $t = 0, 1, 2, \ldots$.

Theorem 7.4. *The random vector X satisfies* (7.12) *if and only if its distribution is multivariate geometric distribution, for all $n \leq p$,*

$$S(\mathbf{x}) = q_{i_1}^{x_{ii}} \left(\frac{q_{i_1 i_2}}{q_{i_1}}\right)^{x_{i_2}} \left(\frac{q_{i_1 i_2 i_3}}{q_{i_1 i_2}}\right)^{x_{i_3}} \left(\frac{q_{i_1 \ldots i_n}}{q_{i_1} \cdots q_{i_{n-1}}}\right)^{x_{in}},$$
$$x_{i_1} \geq x_{i_2} \geq \cdots \geq x_{i_n}, \tag{7.13}$$

for each permutation $(i_1 i_2 \ldots i_n)$ of the integers $(1, 2, \ldots, n)$. The parameters satisfy

$$0 < q_{i_1 \ldots i_n} \leq q_{i_1 i_2} \leq q_1 q_2 \ldots q_n < 1$$
$$q_{i_1 \ldots i_j} = q_{12 \ldots j},\ j = 1, 2, \ldots, n$$
$$1 - \sum_{j=1}^{m} q_i - \sum_{j<k} q_{jk} + \cdots + (-1)^{n-1} q_{12 \ldots n} \geq 0.$$

Proof. We shall sketch the proof below. Assume that (7.12) is satisfied. For $n = 1$, the only solution is

$$S_1(x_j) = q_1^{x_j} \text{ for some } 0 < p_1 < 1,$$

and for $n = 2$,

$$S(x_1 + t, x_2 + t) = S(x_1, x_2) S(t, t).$$

Setting $x_2 = 0$,

$$S(x_1 + t, t) = q_1^{x_1} s(t, t). \tag{7.14}$$

Also,

$$S(x + t, x + t) = S(x, x) S(t, t)$$

gives $S(t, t) = q_{12}^t$, $0 < q_{12} < 1$. Thus, for $x_1 \geq x_2$, from (7.14), we have

$$S(x_1, x_2) = q_1^{x_1} \left(\frac{q_{12}}{q_1}\right)^{x_2}.$$

Now, let the no-ageing property hold for m variables in **X**. Then, from

$$S(x_1 + t, \ldots, x_m + t) = S(x_1, \ldots, x_m) S(t, \ldots, t),$$

we get

$$S(x_1+t,\dots,x_m+t,t)=S(x_1,\dots,x_m,0)S(t,\dots,t)$$
$$=q_1^{x_1}\left(\frac{q_{12}}{q_1}\right)^{x_2}\cdots\left(\frac{q_{12\dots m}}{q_{12\dots m-1}}\right)^{x_m}S(t,\dots,t). \quad (7.15)$$

Further,

$$S(x+t,\dots,x+t)=S(x,\dots,x)S(t,\dots,t)$$

yields

$$S(t,\dots,t)=q_{12\dots m+1}^{t},\ 0<q_{12\dots m+1}<1. \quad (7.16)$$

Combining (7.15) and (7.16), we get

$$S(x_1,\dots,x_{m+1})=q_1^{x_1}\left(\frac{q_{12}}{q_1}\right)^{x_2}\cdots\left(\frac{q_{1\dots m+1}}{q_{1\dots m}}\right)^{x_{m+1}}$$
$$x_1\ge x_2\ge\cdots\ge x_{m+1}.$$

The result in the theorem follows by induction for the permutation $(1, 2, \dots, n)$ and the other permutations are obtained by change of variables. Note that the conditions on the parameters are deduced from

$$S(x_1,\dots,x_n)\ge S(x_1,\dots,x_{n+1}),\ n=1,2,\dots,p,$$

and $f(0,0,\dots,0)\ge 0$. □

Definition 7.4. Let (X_1, X_2) be a random vector, defined on N^2, with marginal probability mass functions $f_i(x_i) > 0$, $i = 1, 2$, and conditionals

$$f_{ij}(x_i|x_j)=P[X_i=x_i|X_j=x_j],\quad i,j=1,2,\ i\ne j.$$

Then, (X_1, X_2) satisfies the conditional no-ageing property iff

$$P[X_i\ge x_i+t_i|X_i\ge x_i,X_j=x_j]=P[X\ge t_i|X_j=x_j] \quad (7.17)$$

for all x_1, x_2.

Theorem 7.5. *The vector (X_1, X_2) satisfies the conditional no-ageing property if and only if the probability mass function of (X_1, X_2) is*

$$f(x_1,x_2)=\phi q_1^{x_1}q_2^{x_2}\theta^{x_1x_2},\ 0<q_1,q_2<1,0<\theta\le 1. \quad (7.18)$$

Proof. Eq. (7.17) is the same as

$$S_1(x_1+t_1|X_2=x_2)=S_1(x_1|X_2=x_2)S_2(t_1|X_2=x_2)$$

for $i=1$, which is of the form

$$g_{x_2}(x_1+t_1)=g_{x_2}(x_1)g_{x_2}(t_1)$$

and hence its non-trivial solution is

$$g_{x_2}(x_1)=[q_1(x_2)]^{x_1}, \quad 0<q_1(x_2)<1,\ x_1=0,1,2,\ldots.$$

This leads to

$$S_1(x_1|X_2=x_2)=[q_1(x_2)]^{x_1} \text{ and } f_{12}(x_1|x_2)p_1(x_2)[q_1(x_2)]^{x_1},$$

where $p_1(x_2)=1-q_1(x_2)$. Also, when $i=2$ in (7.17), we have

$$f_{21}(x_2|x_1)=p_2(x_1)[q_2(x_1)]^{x_2}.$$

Thus,

$$f_1(x_1)p_2(x_1)[q_2(x_1)]^{x_2}=f_2(x_2)p_1(x_2)[q_1(x_2)]^{x_1}. \tag{7.19}$$

Changing x_1 to x_1+1 and dividing, we get

$$q_2(x_1)=\frac{f_2(x_2+1)p_1(x_2+1)q_1[(x_2+1)]^{x_1}}{f_2(x_2)p_1(x_2)q_1[(x_2)]^{x_1}}. \tag{7.20}$$

Now changing x_1 to x_1+1 in (7.20) and dividing this by (7.20), we get

$$\frac{q_2(x_1+1)}{q_2(x_1)}=\frac{q_1(x_2+1)}{q_1(x_2)}.$$

For the last equation, to hold for all x_1, x_2, each side must be a constant, say θ, and then

$$q_1(x_2)=\theta^{x_2}q_2 \text{ and } q_2(x_1)=\theta^{x_1}q_1,\ q_i=q_i(0),\ i=1,2.$$

Inserting the above in (7.19), we obtain

$$\frac{f_1(x_1)p_2(x_1)}{q_1^{x_1}}=\frac{f_2(x_2)p_2(x_2)}{q_2^{x_2}}=\phi, \text{ a constant.}$$

Hence,

$$f_1(x_1)=\phi(1-\theta^{x_1}q_2)^{-1}q_1^{x_1}.$$

This, together with (7.19), gives (7.18). When the distribution is (7.18), it is easy to verify that (7.17) holds. This completes the proof. □

Remark 7.4. It may be noted that when conditional no-ageing holds, the conditional distributions of $(X_i|X_j = x_j)$ are geometric as expected. The above definition and theorem are from Nair and Nair (1991).

In the case of bivariate distributions determined by the conditional mass functions, the form of the survival function is often not tractable. It will, therefore, be mathematically more convenient to deal with mass functions in defining no-ageing. The next theorem gives an alternative definition of conditional no-ageing as well as some other useful relationships.

Theorem 7.6. *The following statements are equivalent:*

(i) (X_1, X_2) *is conditionally no-ageing;*
(ii) $f(x_1, x_2) = \phi q_1^{x_1} q_2^{x_2} \theta^{x_1 x_2}$;
(iii) $S_i(x_i + t_i|X_j = x_j) = S_i(x_i|X_j = x_j)S_i(t_i|X_j = x_j)$;
(iv) $f_i(0|X_j = x_j)f_i(x_i + t_i|X_j = x_j) = f_i(x_i|X_j = x_j)f_i(t_i|X_j = x_j)$;
(v) $e_1(x_1 + t_1, x_2)e_1(0, x_2) = e_1(x_1, x_2)e(t_1, x_2)$,
$e_2(x_1, x_2 + t_2)e_1(x_1, 0) = e_2(x_1, x_2)e_2(x_1, t_2)$.

Proof. The implication in (i), (ii) and (iii) have been established in Theorem 7.5. Now, suppose that (iii) is true. Then, for $i = 1$,

$$S_1(x_1 + t_1 + 1|x_2) - S_1(x_1 + t_1|x_2) = S_1(t_1|x_2)[S_1(x_1 + 1|x_2) - S_1(x_1|x_2)],$$

or

$$f_1(x_1 + t_1|x_2) = f_1(x_1|x_2)S_1(t_1|x_2). \tag{7.21}$$

Setting $x_1 = 0$,

$$f_1(t_1|x_2) = f_1(0|x_2)S_1(t_1|x_2).$$

Substituting for $s_1(t_1|x_2)$ in (7.21), we have (iv) for $i = 1$ and similarly for $i = 2$. Conversely if (iv) holds, taking $i = 1$ and summing over $[x_1, \infty)$,

$$f_1(0|x_2)S_1(x_1 + t_1|x_2) = S_1(x_1|x_2)f_1(t_1|x_2)$$

and in particular,

$$f_1(0|x_2)S(t_1|x_2) = f_1(t_1|x_2).$$

From the last two equations, (iii) holds for $i = 1$ and similarly for $i = 2$.

Dividing (iv) by (iii), we get (v). Conversely, (v) can be written as

$$g_{x_2}(x_1 + t_1) = g_{x_2}(x_1)g_{x_1}(t_1), \tag{7.22}$$

where

$$g_{x_2}(x_1) = \frac{e_1(x_1, x_2)}{e_1(0, x_2)}.$$

The solutions of (7.22) are $g_{x_2}(x_1) = 0, 1$ and $[p_1(x_2)]^{x_1}$ for some $0 < p_1(x_2) < 1$. Of these, $g_{x_2}(x_1) = 0$ is trivial, $g_{x_2}(x_1) = p_1(x_2)^{x_1}$ gives

$$e_1(x_1, x_2) = e_1(0, x_2)[p_1(x_2)]^{x_1}.$$

This leads to $f_1(0|x_2) = 1$, again a trivial case. Thus, the only admissible solution is

$$g_2(x_1) = 1 \text{ or } e_1(x_1, x_2) = e_1(0, x_2)$$

and similarly

$$e_2(x_1, x_2) = e_2(x_1, 0).$$

These two expressions characterize (ii), and therefore the chain of implications in Theorem 7.6 is completed. □

We have seen that when X follows the univariate geometric distribution, the hazard rate is $h(x) = p$. In a Bayesian point of view, p cannot be specified and there is uncertainty about the value of p so that the constancy of the hazard rate is conditional on p being known. Thus, for a geometric random variable X, the probability mass function is written as a conditional one with

$$f(x|p) = q^x p \tag{7.23}$$

so that the hazard rate is also conditional on p

$$h(x|p) = p,$$

which is often referred to as the model hazard rate (as it can be used to generate the model in (7.23)). In the Bayesian approach, the predictive hazard rate is defined as the unconditional hazard rate obtained by averaging $h(x|p)$ using an appropriate distribution, which is called the prior distribution of p. Thus, when the life distribution is geometric, the predictive hazard rate takes the form

$$h^*(x) = \frac{\int_0^1 p(1-p)^x f(p)dp}{\int_0^1 (1-p)^x f(p)dp},$$

where $f(p)$ is the probability density function of p treated as a random variable. Rodrigues and Wechsler (1993) have shown that $h^*(x)$ is a decreasing function for any choice of the prior distribution $f(p)$. Thus, the predictive hazard rate of the geometric law is no longer constant in contrast to that in the conventional approach where X is no-ageing due to the constancy of hazard rate. However, in the multivariate case, there is a definition of non-ageing due to Barlow and Mendel (1992) which we shall refer to as the Bayesian no-ageing property.

Definition 7.5. The random vector $\mathbf{X}$ is said to have the *Bayesian no-ageing property* iff

$$P[X_i \geq x_i + t|\mathbf{X} \geq \mathbf{x}] = P[X_j \geq x_j + t|\mathbf{X} \geq \mathbf{x}], \quad i, j = 1, 2, \ldots p,\ i \neq j. \tag{7.24}$$

The above definition means that the residual lifetimes of the younger and older components with the same survival history have identical distributions. In a Bayesian viewpoint, regardless of the ages of the components one would bet the same amount on the next increment in the life of either component. Eq. (7.24) is satisfied if the survival function of $\mathbf{X}$ satisfies

$$S(\mathbf{x}) = S_1(x_1 + \cdots + x_p), \tag{7.25}$$

where $S_1(\cdot)$ is the survival function of a univariate distribution such that $S(\mathbf{x})$ is a p-dimensional survival function. Distributions of the form (7.25) are called *Schur-constant models*. A detailed study of all the distributions encountered in connection with various versions of no-ageing will be taken up in Chapter 8.

7.4 MULTIVARIATE IHR CLASSES

In this section, we extend the concept of univariate monotone hazard rate classes to the multivariate case. There are several cases of multivariate increasing (decreasing) hazard rate, MIHR (MDHR), distributions. We distinguish them by attaching subscripts $1, 2, \ldots$. The definitions are made from different approaches, for example, based on the monotonicity of hazard rate or of the survival function.

Definition 7.6 (Nair and Asha, 1997). A random vector $\mathbf{X}$, in the support of N^p or its subset, is said to have

(i) MIHR-1 (MDHR-1) if, for all $\mathbf{x}$ and $\mathbf{t} = (t_1, t_2, \ldots, t_p)$ in N^p,

$$\mathbf{c}(\mathbf{x} + \mathbf{t}) \geq (\leq) c(\mathbf{x});$$

(ii) MIHR-2 (MDHR-2) if, for all $\mathbf{x}$ in N^p and t in N,

$$c_i(x_1, \ldots, x_{i-1}, x_i + t, x_{i+1}, \ldots, x_p) \geq (\leq) c_i(\mathbf{x}),\ i = 1, 2, \ldots, p;$$

(iii) MIHR-3 (MDHR-3), if for all $m \leq p$, $\mathbf{x}$ in N^m and t in N,

$$c_i(x_1 + t, \ldots, x_m + t) \geq (\leq) c_i(x_1, \ldots x_m);$$

(iv) MIHR-4 (MDHR-4) if, for all $m \leq p$,

$$c_i(x + t, x + t, \ldots, x + t) \geq (\leq) c_i(x, x, \ldots, x).$$

The above classes of life distributions based on the vector hazard rates satisfy the following properties:

(a) The only distribution that is both MIHR-1 and DIHR-1 is given by

$$S(\mathbf{x}) = q_1^{x_1} q_2^{x_2} \ldots q_p^{x_p};$$

(b) The vector $\mathbf{X}$ is both MIHR-2 and MDHR-3 if and only if its distribution is (7.11), and it is both MIFR-3 and MDHR-3 if and only if the distribution is (7.13);

(c)

$$\begin{array}{ccccc} \text{MIHR-1} & \Rightarrow & \text{MIHR-3} & \Rightarrow & \text{MIHR-4} \\ \Downarrow & & & & \\ \text{MIHR-2} & & & & \end{array}$$

and

$$\begin{array}{ccccc} \text{MDHR-1} & \Rightarrow & \text{MDHR-3} & \Rightarrow & \text{MDHR-4} \\ \Downarrow & & & & \\ \text{MDHR-2} & & & & \end{array}$$

Nair and Asha (1997) have given counter examples to show that no other implications exist among the various classes;

(d) MIHR-2 (MDHR-2) is equivalent to saying that $c_i(\mathbf{x})$ is increasing (decreasing) in x_i for all x_j, $j = 1, 2, \ldots, p$. Accordingly, if $\mathbf{X}$ is MIHR-2 (MDHR-2), each of the one-dimensional marginal hazard rates are IHR (DHR);

(e) $\mathbf{X}$ is MIHR-2 (MDHR-2) if and only if

$$\frac{S(x_i + t, x^*_{(i)})}{S(x_i, x^*_{(i)})}$$

is decreasing in t for all $\mathbf{x} > \mathbf{0}$, where $x^*_{(i)} = \mathbf{x} - \{x_i\}$;

(f) $X_i \geq_{\text{hr}} (X_i - x_i | X_1 > x_1, X_2 > x_2)$, for $i = 1, 2$, if and only if $\mathbf{X}$ is MIHR-2.

Following the works of Harris (1970), Brindley and Thompson (1972) and Marshall (1975) in the continuous case, there are other ways of defining multivariate monotone hazard rates for a discrete random vector $\mathbf{X}$.

Definition 7.7.

(A) **X** is MIHR-5 iff

$$\frac{S(x+t,\ldots,x+t)}{S(x,\ldots,x)}$$

decreases in $x \geq 0$ for all t in N;

(B) **X** is MIHR-6 iff

$$\frac{S(x+t_1,\ldots,x+t_m)}{S(x,\ldots,x)}$$

decreases in x for all t_i in N and $m \leq p$; $i = 1, 2, \ldots, p$;

(C) **X** is MIHR-7 iff

$$\frac{S(x_1+t,\ldots,x_m+t)}{S(x_1,\ldots,x_m)}$$

decreases in **x** for all t in N and $m \leq p$;

(D) **X** is MIHR-8 iff

$$\frac{S(x_1+t_1,\ldots,x_p+t_p)}{S(x_1,\ldots,x_p)}$$

decreases in **x** for all **t** in N^p, $i = 1, 2, \ldots, p$.

In the univariate case, X is IHR iff $\frac{S(x+t+1)}{S(t+1)}$ is decreasing in x for all $t = 0, 1, 2, \ldots$. Accordingly, Definitions (A) and (D) can be seen as some sort of generalization of the definition in the single variable case. While the univariate definition is equivalent to monotone increase in $h(x)$, some such implications are possible with respect to a bivariate hazard rate in all the above cases. This fact has not been investigated either when the lifetimes are discrete or continuous. We investigate this aspect in one typical definition in the bivariate case.

Let **X** be bivariate increasing hazard rate-8 in which $\frac{S(x+t+1)}{S(t+1)}$ is decreasing in x_1, x_2 for all t_1, t_2. Thus,

$$\frac{S(x_1+t_1,x_2+t_2)}{S(x_1,x_2)} \geq \frac{S(x_1+t_1+1,x_2+t_2+1)}{S(x_1+1,x_2+1)} \text{ for all } t_1, t_2$$

$$\Rightarrow \frac{S(x_1+1,x_2+1)}{S(x_1,x_2)} \geq \frac{S(x_1+t_1+1,x_2+t_2+1)}{S(x_1+t_1,x_2+t_2)}. \tag{7.26}$$

Now

$$\frac{S(x_1+1,x_2+1)}{S(x_1,x_2)}$$
$$= \frac{S(x_1,x_2) - S(x_1+1,x_2) - S(x_1,x_2+1) + S(x_1+1,x_2+1)}{S(x_1,x_2)}$$

$$
\begin{aligned}
&- \frac{S(x_1,x_2)-S(x_1+1,x_2)-S(x_1,x_2+1)}{S(x_1,x_2)} \\
&= \frac{f(x_1,x_2)}{S(x_1,x_2)} - \left[1 - \frac{S(x_1+1,x_2)}{S(x_1,x_2)} - \frac{S(x_1,x_2+1)}{S(x_1,x_2)}\right] \\
&= a_2(x_1,x_2) - [1-(1-c_1(x_1,x_2)) - (1-c_2(x_1,x_2))].
\end{aligned}
$$

Using the identity connecting $a_2(x_1,x_2)$ and c_1 and c_2 in Section 6.2.3, we have

$$
\frac{S(x_1+1,x_2+1)}{S(x_1,x_2)} = (1-c_1(x_1+1,x_2))(1-c_2(x_1,x_2)). \tag{7.27}
$$

By virtue of (7.27), the inequality in (7.26) asserts that

$$
(1-c_1(x_1+1,x_2))(1-c_2(x_1,x_2))
$$

is decreasing in x_1 and x_2 for all t_1, t_2. Thus, when c_1 and c_2 are increasing in x_1, x_2, then (7.26) holds. Equivalently,

$$
\text{BIHR-1} \Rightarrow \text{BIHR-8}.
$$

Similar conclusions can be reached for other ageing concepts. Among the MIHR classes 5–8, it is easy to see that

$$
\text{MIHR-8} \Rightarrow \text{MIHR-7} \Rightarrow \text{MIHR-6} \Rightarrow \text{MIHR-5}.
$$

The corresponding implications among MDHR-5–MDHR-8 can be obtained by replacing 'decreases' in Definitions (A)–(D) by 'increases'.

Example 7.1. The bivariate hypergeometric distribution with

$$
S(x_1,x_2) = \frac{\binom{k+n-x_1-x_2}{n-x_1-x_2}}{\binom{k+n}{n}}, \; x_1+x_2 \le n, \; x_i = 0,1,2,\ldots,n, \; i=1,2,
$$

has

$$
c_1(x_1,x_2) = \frac{k}{k+n-x_1-x_2}, \qquad c_2(x_1,x_2) = \frac{k}{k+n-x_1-x_2}.
$$

It is easy to recognize that c_1 is decreasing in x_1 for all x_2, and c_2 is decreasing in x_2 for all x_1. Hence, (X_1, X_2) is MDHR-2. Also,

$$
\begin{aligned}
\frac{S(x_1+t_1,x_2+t_2)}{S(x_1,x_2)} &= \frac{(n-x_1-x_2)}{k+n-x_1-x_2} \\
&\times \frac{(n-x_1-x_2-1)\ldots(n-x_1-t_1-x_2-t_2+1)}{(k+n-x_1-x_2-1)\ldots(k+n-x_1-t_1-x_2-t_2+1)}
\end{aligned}
$$

is decreasing in x_1 and x_2 for all t_1, t_2, and hence (X_1, X_2) is MIHR-8.

Other definitions of hazard rate can also be employed to derive different classes of multivariate hazard rate models.

Definition 7.8. We say that **X** has increasing (decreasing) multivariate hazard rate-9, MIHR-9 (MDHR-9), if and only if $a_r(\mathbf{x}_r)$ is increasing (decreasing) in $x_1, x_2, \ldots x_r$ for $r = 1, 2, \ldots, p$ (Nair and Sankaran, 2015a).

Example 7.2. The bivariate Waring distribution with survival function

$$S(x_1, x_2) = \frac{(m)_{x_1+x_2}}{(m+n)_{x_1+x_2}}, \; m, n > 0, \; x_1, x_2 = 0, 1, 2, \ldots,$$

has

$$a_2(x_1, x_2) = \frac{n(n+1)}{(m+n)_{x_1+x_2}} \text{ and } a_1(x_1) = \frac{n}{(m+n)_{x_1}}$$

which are both decreasing in x_1 and x_2. Hence, (X_1, X_2) is MDHR-9.

For the negative hypergeometric law, we have

$$a_2(x_1, x_2) = \frac{k(k-1)}{(k+n-x_1-x_2)(k+n-x_1-x_2-1)},$$

$$a_1(x_1) = \frac{k}{k+n-x_1}.$$

Thus (X_1, X_2) is MIHR-9. On the other hand,

$$S(x_1, x_2) = \frac{(m)_{x_1}}{(m+n)_{x_1}} \frac{\binom{k+n-x_1}{n-x_1}}{\binom{k+n}{n}}, \quad x_1 = 0, 1, 2, \ldots, \; x_2 = 0, 1, 2, \ldots, n,$$

has hazard rates

$$a_2(x_1, x_2) = \frac{nk}{(m+n+x_1)(k+p-x_2)}, \quad a_1(x_1) = \frac{n}{(m+n)_{x_1}}.$$

In this case, $a_1(x_1)$ is decreasing in x_1, while $a_2(x_1, x_2)$ can be either decreasing or increasing by the choice of parameters. Hence, (X_1, X_2) that corresponds to this distribution is neither MIHR-9 nor MDHR-9.

While expressing positive (negative) ageing through MIHR-9 (MDHR-9) based on $a(\mathbf{x})$, it is of interest to know the conditions for no-ageing when $a(\mathbf{x})$ is utilized to distinguish that property. The lifetime **X** has multivariate no-ageing property if and only if it satisfies

$$P[\mathbf{X}_r = \mathbf{x}_r | \mathbf{X}_r \geq \mathbf{x}_r] = P[\mathbf{X}_r = \mathbf{O}_r | \mathbf{X}_r \geq \mathbf{O}_r] \tag{7.28}$$

for all $\mathbf{x}_r$ taking values in N^r and for $r = 0, 1, 2, \ldots, p$. This means that the conditional probability of failure of a p-component system at age $\mathbf{x}_r$,

$r = 1, 2, \ldots, p$ is the same as the probability of the failure at the starting point of its operation. Equivalently, in terms of the scalar failure rate, the no-ageing property is expressed as $a_r(\mathbf{x}_r) = k_r$, a constant for all $\mathbf{x}_r$, $r = 1, 2, \ldots, p$. To see this, we note that (7.28) can be written as $a_r(\mathbf{x}_r) = f_r(0, 0, \ldots, 0)$. Recall that the strong no-ageing property in Definition 7.1, in the bivariate case, becomes

$$S_2(x_1 + t_1, x_2 + t_2) = S(x_1, x_2)S(t_1, t_2) \tag{7.29}$$

and so

$$\begin{aligned} S(x_1 + t_1, x_2 + t_2) &- S(x_1 + t_1 + 1, x_2 + t_2) - S(x_1 + t_1, x_2 + t_2 + 1) \\ &+ S(x_1 + t_1 + 1, x_2 + t_2 + 1) \\ &= S(t_1, t_2)[S(x_1, x_2) - S(x_1 + 1, x_2) - S(x_1, x_2 + 1) \\ &\quad + S(x_1 + 1, x_2 + 1)]. \end{aligned}$$

Thus,

$$f(x_1 + t_1, x_2 + t_2) = f(x_1, x_2)S(t_1, t_2).$$

Setting $x_1 = 0$, $x_2 = 0$, we get

$$f(t_1, t_2) = f(0, 0)S(t_1, t_2).$$

Hence, we restate (7.29) as

$$f(x_1 + t_1, x_2 + t_2)f(0, 0) = f(x_1, x_2)f(t_1, t_2). \tag{7.30}$$

From (7.29) and (7.30), we have

$$a_2(x_1 + t_1, x_2 + t_2)a_2(0, 0) = a_2(x_1, x_2)a_2(t_1, t_2)$$

as a condition for no-ageing. Finally, (7.28) is satisfied if and only if $\mathbf{X}$ has strong no-ageing property and the distribution is (7.2).

We can have yet another bivariate increasing hazard rate class by recalling the definitions of the alternative conditional hazard rate $(e_1(x_1, x_2), e_2(x_1, x_2))$.

Definition 7.9. The discrete vector (X_1, X_2), defined on N^2, has the bivariate increasing hazard rate-10 (BIHR-10) if $e_1(x_1, x_2)$ is increasing in x_1 for all x_2 and $e_2(x_1, x_2)$ is increasing in x_2 for all x_1. By replacing 'increasing' by 'decreasing', we have the bivariate decreasing hazard rate-10 (BDHR-10).

Remark 7.5. The only distribution for which (X_1, X_2) is both increasing and decreasing in conditional hazard rate is (7.18).

Example 7.3. Consider the bivariate survival function

$$S(x_1, x_2) = \frac{2}{x_1 + x_2}, \qquad x_1, x_2 = 1, 2, 3, \ldots.$$

Then,

$$e_1(x_1, x_2) = 1 - [1 + \frac{2}{x_1 + x_2}]^{-1}$$

and

$$e_2(x_1, x_2) = e_1(x_1, x_2)$$

which are decreasing in x_i for all x_j, $j \neq i$, $j = 1, 2$. Thus, (X_1, X_2) is BDHR-10.

Example 7.4. Let

$$S(x_1, x_2) = \frac{\binom{k-n-x_1-x_2}{n-x_1-x_2}}{\binom{k+n}{n}}, \qquad x_1, x_2 = 0, 1, \ldots, n,\ x_1 + x_2 \leq n,\ k > 0.$$

Direct calculations yield

$$e_i(x_1, x_2) = \frac{k-1}{k+n-x_1-x_2-1}, i = 1, 2.$$

It is easy to see that (X_1, X_2) is BIHR-10.

7.5 BAYESIAN APPROACH TO AGEING CLASSES

The Bayesian approach to multivariate ageing classes is substantially different from the conventional notions so far discussed in the present chapter. Motivation and development of the Bayesian ideas in defining ageing came in a series of works notably from Proschan (1975), Barlow and Mendel (1992), Barlow and Spizzichino (1993), Bassan and Spizzichino (1999), Spizzichino (2001), and Bassan et al. (2002). The discrete analogues of some important results and definitions are be presented in this section.

In the Bayesian paradigm, probabilities are conceived to be personal as the degree of belief in the truth of an event and is indexed by the person who evaluates it. Hence, probability statements are not universal or objective as in the classical sense, but are subject to change with individuals as well as with the amount of information one possesses. Stochastic events may exhibit interdependence even if they are physically independent since the knowledge of the values taken by some events modifies the state of information due to the influence of certain unobserved events. Thus, in consideration of system life, physical independence of the components as assumed in the conventional theory needs modification when it is the case that some unobserved facts may induce dependence among the components. In such a scenario, to elicit appropriate subjective probability the notion of exchangeability of events is introduced. This leads to *exchangeable random variables* and corresponding distributions. Let $(i_1, i_2, \ldots, i_p)$ be any permutation of the integers $1, 2, \ldots, p$. Then, $X_1, X_2 \ldots X_p$ are exchangeable if they have the same joint distribution as $X_{i_1}, X_{i_2}, \ldots, X_{i_p}$. In other words

$$S(x_1, x_2, \ldots, x_p) = S(x_{i_1}, x_{i_2}, \ldots, x_{i_p}), \tag{7.31}$$

so that $S(\cdot)$ is a symmetric function of the arguments $x_1, x_2, \ldots, x_p$. Exchangeability is in some sense analogous to the concept of random sample in the conventional analysis in which case $X_1, X_2, \ldots X_p$ are independent and identically distributed. There is no loss of generality in assuming exchangeability instead of independent and identical distribution for $X_1, X_2, \ldots, X_p$, because the former is weaker than the latter. When (7.31) holds, it is not difficult to see that the joint probability mass function $f(x_1, x_2, \ldots, x_p)$ is also symmetric.

Let $X_1, X_2, \ldots$ be discrete lifetimes satisfying

$$\sum_{i=1}^{p} x_i = \sum_{i=1}^{p} y_i \Rightarrow f(x_1, x_2, \ldots, x_p) = f(y_1, y_2, \ldots, y_p). \tag{7.32}$$

Then there exist non-negative functions $g_2, g_3, \ldots$ such that

$$f(x_1, x_2, \ldots, x_p) = g_p\left(\sum_{i=1}^{p} x_i\right), \quad p = 2, 3, \ldots.$$

A distribution satisfying (7.32) is called *Schur-constant distribution*. Such a distribution possess the no-ageing property. In the case of exchangeable units ageing relative to lifetime, Schur-concavity can be used to define increasing hazard rate notion. The survival function is Schur-concave if and only if

$$P[X_i \geq x_i + t | \mathbf{X} > \mathbf{x}] \geq P[X_j \geq x_j + t | \mathbf{X} \geq \mathbf{x}], \quad x_i < x_j. \tag{7.33}$$

This means that if p-components with the same life history are considered, and the component lifetimes are exchangeable, the younger is the best when the survival function is Schur-concave. Further, the Schur-concavity of the conditional survival function in (7.33) is equivalent to the Schur-concavity of the unconditional survival probabilities.

Definition 7.10. The vector $\mathbf{X}$ is said to be multivariate increasing hazard rate-11 (MIHR-11) if $S(\mathbf{x})$ is Schur-concave, or $(x_1 - x_2)[S(x_1 + 1, x_2) - S(x_1, x_2 + 1)] \leq 0$.

Example 7.5. Consider the exchangeable uniform distribution with

$$S(x_1, x_2) = \left(1 - \frac{x_1}{b}\right)\left(1 - \frac{x_2}{b}\right)\left[1 + \frac{\theta x_1 x_2}{b^2}\right],$$
$$x_1, x_2 = 0, 1, 2, \ldots, b-1, \ -1 \le \theta \le 1, \ b > 1.$$

Then,

$$S(x_1+1, x_2) - S(x_1, x_2+1) = \frac{(x_2 - x_1)}{b^2}$$
$$\times \left[\theta - \frac{\theta}{b} + 1 - \frac{\theta}{b}(x_1 + x_2) + \frac{\theta}{b^2}(x_1 + x_2) + \frac{2\theta x_1 x_2}{b^2}\right].$$

Since the term inside the square brackets is positive as

$$\theta\left(1 - \frac{x_1 + x_2}{b} + \frac{x_1 x_2}{b^2}\right) \ge 0,$$

we get

$$(x_1 - x_2)[S(x_1+1, x_2) - S(x_1, x_2+1)] \le 0.$$

Thus, $S(x_1, x_2)$ is Schur-concave and therefore, (X_1, X_2) is BIHR-11.

The BIHR-11 concept can also be expressed in terms of a stochastic order in the residual life. An exchangeable random vector (X_1, X_2) is BIHR-11 if and only if

$$(X_1 - t_1 | X_1 > t_1, X_2 > t_2) \ge_{\text{st}} (X_2 - t_2 | X_1 > t_1, X_2 > t_2), \quad t_1 \le t_2. \quad (7.34)$$

This is interpreted by saying that a component with smaller age is more probable to have a larger residual life than an older component when (X_1, X_2) is IHR. Thus, the BIHR-11 definition is a natural extension of the property of IHR life in the univariate case. Rewriting (7.34) in terms of the survival function as

$$S(x_1 + t_1, x_2) - S(X_1, x_2 + t_2) \ge 0, \quad t_1 \le t_2,$$

we see that $S(x_1, x_2)$ is Schur-concave.

A second definition in Bassan and Spizzichino (1999) uses the hazard rate order. Its discrete analogue is given in the next definition.

Definition 7.11. The discrete exchangeable lifetime vector (X_1, X_2) is BIHR-11 if

$$(X_1 - t_1 | X_1 > t_1, X_2 > t_2) \ge_{\text{hr}} (X_2 - t_2 | X_1 > t_1, X_2 > t_2) \quad (7.35)$$

for $t_1 \le t_2$.

We can express (7.35) also as

$$c_1(x_1+t_1,x_2) \le c_1(x_2+t_2,x_1), \quad t_1 \le t_2 \text{ and } x_1 \le x_2$$

or

$$c_1(x_1+t,x_2) \text{ is increasing in } t \text{ for } x_1 \le x_2.$$

A further simplification is

$$\frac{S(x_1+t,x_2)}{S(x_2+t,x_1)} \text{ is increasing in } t \text{ for } x_1 \le x_2.$$

Unless X_1 and X_2 are independent, BIHR-11 is not equivalent to BIHR-12, but

$$\text{BIHR-12} \Rightarrow \text{BIHR-11}.$$

It has been observed in Bassan and Spizzichino (1999) that marginals of a BIFR random vector need not have increasing hazard rates unless the component variables are independent. A sufficient condition for the marginals also to be IHR is that X_1 is RTD in X_2. In the above discussions, we have the dual ageing class MDHR when the inequalities defining them are reversed or 'increasing' is replaced by 'decreasing'.

7.6 MULTIVARIATE INCREASING HAZARD RATE AVERAGE

The notion of multivariate increasing (decreasing) hazard rate average, MIHRA (MDHRA), has been formulated by several researchers in the continuous case. Some of them are direct extensions of the univariate case while others are based on those reliability properties which explains the importance of the univariate IHRA concept. Though not all of them are adaptable to discrete lifetimes directly, we state them for a possible future work to modify them appropriately. A continuous non-negative random vector $\mathbf{I} = (T_1, T_2)$ is said to be bivariate IHRA if

(i)

$$E^{\alpha}[g(T_1,T_2)] \le E[\alpha^{-1} g^{\alpha}(T_1,T_2)]$$

for all continuous non-negative increasing functions g and $0 < \alpha < 1$ (Block and Savits, 1976);

(ii) there exist independent random variables $X_1, \dots, X_k$ such that

$$T_i = \sum_{j \in t_i} X_j,\ I_k = (1,2,\dots k),\ i = 1,2 \quad \text{(Block and Savits, 1976);}$$

(iii) $\overline{F}^{\alpha}(t_1,t_2) \le \overline{F}(\alpha t_1, \alpha t_2)$, $0 \le \alpha \le 1$, for all t_1, t_2, where $\overline{F}$ is the survival function of **I**;

(iv) each monotone system from **I** is univariate IHRA;

(v) there exist independent IHRA random variables $Y_1, Y_2, \dots Y_k$ and coherent life functions g_i such that $T_i = g_i(Y_1, Y_2, \dots Y_k)$, $i = 1, 2$;

(vi) $T_i = \min_{j \in I} Y_j$, $i = 1, 2$;

(vii) $\min(T_1, T_2)$ is IHRA;

(viii) $\min(\alpha_1 T_1, \alpha_2 T_2)$ is IHRA, $\alpha_1, \alpha_2 \ge 0$.

The conditions (iii)–(viii) were given by Esary and Marshall (1979).

A different approach to obtain MIHRA motions is to consider the average behaviour of the multivariate hazard rate. Since there are various hazard rate forms, each one can be utilized to derive the corresponding IHRA concept. The method is illustrated in some cases below.

Definition 7.12. A discrete random vector **X**, defined on N^P is multivariate increasing (decreasing) hazard rate average 1, MIHRA-1 (MDHRA-1), if and only if

$$H_r(\mathbf{x}_r) = \frac{1}{x_1 x_2 \dots x_r} \sum_{t_1=0}^{x_1-1} \dots \sum_{t_r=0}^{x_r-1} a_r(\mathbf{t}_r)$$

is increasing (decreasing) or decreasing in $x_1, x_2, \dots, x_r$, for $r = 1, 2, \dots, p$.

Here, $\mathbf{x}_r = (x_1, x_2, \dots x_r)$ and $\mathbf{t}_r = (t_1, t_2, \dots, t_r)$; see Nair and Sankaran (2015a).

Theorem 7.7. *MIHR-9 (MDHR-9)⇒ MIHRA-1 (MDHRA-1).*

Proof. The proof of the p-variate case is apparent from that of the bivariate case. Since (X_1, X_2) is BIHR-9, both $a_1(x_1)$ and $a_2(x_1, x_2)$ are increasing functions of their arguments. Now,

$$\begin{aligned} H_1(x_1+1) - H_1(x_1) &= \frac{1}{x_1+1} \sum_{t_1=0}^{x_1} a_1(t) - \frac{1}{x_1} \sum_{t_1=0}^{x_1-1} a_1(t) \\ &= \frac{1}{x_1(x_1+1)} \sum_{t_1=0}^{x_1-1} [a_1(x_1) - a_1(t)]. \end{aligned}$$

Since $a_1(x_1) \ge a_1(t)$ for all t in $(0, 1, \dots, x_1 - 1)$ we see that $H_1(x_1)$ is increasing. Also,

$$
\begin{aligned}
&H_1(x_1+1,x_2+1)-H_1(x_1,x_2)\\
&\quad=\frac{1}{(x_1+1)(x_2+1)}\sum_{t_1=0}^{x_1}\sum_{t_2=0}^{x_2}a(t_1,t_2)-\frac{1}{x_1x_2}\sum_{t_1=0}^{x_1-1}\sum_{t_2=0}^{x_2-1}a(t_1,t_2)\\
&\quad=[x_1x_2(x_1+1)(x_2+1)]^{-1}\Bigg[x_1x_2\Bigg\{\sum_{t_1=0}^{x_1-1}\sum_{t_2=0}^{x_2-1}a(t_1,t_2)\\
&\qquad+\sum_{t_1=0}^{x_1-1}a(t_1,x_2)+\sum_{t_2=0}^{x_2-1}a(x_1,t_2)+a(x_1,x_2)\Bigg\}\\
&\qquad-(x_1+1)(x_2+1)\sum_{t_1=0}^{x_1-1}\sum_{t_2=0}^{x_2-1}a(t_1,t_2)\Bigg]\\
&\quad=[x_1x_2(x_1+1)(x_2+1)]^{-1}\Bigg[x_2\Bigg\{\sum_{t_1=0}^{x_1-1}(a(t_1,x_2)-a(t_1,t_2))\Bigg\}\\
&\qquad+x_1\Bigg\{\sum_{t_2=0}^{x_2-1}(a(x_1,t_2)-a(t_1,t_2))\Bigg\}\\
&\qquad+x_1x_2\left(a(x_1x_2)-\sum_{t_1=0}^{x_1-1}\sum_{t_2=0}^{x_2-1}a(t_1,t_2)\right)\Bigg]\geq 0,
\end{aligned}
$$

since $a(x_1,t_2)>a(t_1,t_2)$ for all t_1 in $(0,1,\ldots x_1-1)$ and $a(t_1,x_2)>a(t_1,t_2)$ for all t_2 in $(0,1,2,\ldots,x_2-1)$. Thus, by definition, (X_1,X_2) is BIHRA. The proof of BDHRA is obtained by replacing 'increasing' with 'decreasing'. □

If we replace $a(\mathbf{x})$ by the vector hazard rate $\mathbf{c}(\mathbf{x})$, another definition of IHRA is obtained.

Definition 7.13. The discrete random vector $\mathbf{X}$ belongs to the MIHRA-2 (MDHRA-2) class if

$$
\frac{1}{x_i}\sum_{t=0}^{x_i-1}c_i(x_1\ldots,x_{i-1},t,x_{i+1},\ldots,x_p)
$$

is increasing (decreasing) in x_i, for $i=1,2,\ldots,p$, for all x_j, $j\neq i$.

Theorem 7.8.

$$
\begin{aligned}
&\mathit{MIHR\text{-}2}\Rightarrow \mathit{MIHRA\text{-}2},\\
&\mathit{MDHR\text{-}2}\Rightarrow \mathit{MDHRA\text{-}2}.
\end{aligned}
$$

Proof. Since $\mathbf{X}$ is MIHR-2,

$$c_i(x_1, \ldots x_{i-1}, x_i + t, x_{i+1}, \ldots, x_p) \geq c_i(\boldsymbol{\alpha}) \text{ for } t_1 = 0, 1, 2, \ldots. \tag{7.36}$$

Now,

$$\begin{aligned}
&\frac{1}{x_1+1}\sum_{t=0}^{x_i} c_i(x_1, \ldots, x_{i-1}, t, x_{i+1}, \ldots, x_p) \\
&\quad - \frac{1}{x_i}\sum_{t=0}^{x_i-1} c_i(x_1 \ldots x_{i-1}, t, x_{i+1}, \ldots, x_p) \\
&= \frac{1}{x_i(x_i-1)}[x_i c_i(x_1, \ldots, x_{i-1}, x_i+1, x_{i+1} \ldots x_p) \\
&\qquad - \sum_{t=0}^{x_i-1} c_i(x_1 \ldots x_{i-1}, t, x_{i+1}, \ldots, x_p) \\
&\geq 0 \quad \text{by (7.36).}
\end{aligned}$$

Thus, $\mathbf{X}$ is MIHRA-2. Proof of MDHRA-2 is obtained by reversing the inequalities. □

Remark 7.6. The only distribution which is both MIHRA-2 and MDHRA-2 is (7.11)

Definition 7.14. We say that (X_1, X_2) is MIHRA-3 (NDHRA-3) if

$$\frac{1}{x_1}\sum_{t=0}^{x_1-1} e_1(t, x_2) \text{ and } \frac{1}{x_2}\sum_{t=0}^{x_2-1} e_2(x_1, t)$$

are increasing in x_i for all $x_j, i, j = 1, 2, i \neq j$.

Using the same method as in Theorem 7.8, we can show that

$$\text{BIHR-10 (BDHR-10)} \Rightarrow \text{MIHRA-3 (MDHRA-3)}$$

and also that (X_1, X_2) is MIHRA-3 and MDHRA-3 if and only if the distribution is (7.18). In all the above statements, it has been assumed that (e_1, e_2) is compatible.

7.7 NEW BETTER THAN USED IN HAZARD RATE

Recall from Section 4.3.4, that a random lifetime X is said to be new better (worse) than used in hazard rate if $h(x) \geq (\leq) h(0)$. The definitions of various versions of multivariate hazard rate allow us to extend this concept to the multivariate setup.

Definition 7.15. The random vector **X** is multivariate new better (worse) than used in hazard rate, MNBUHR-1 (MNWUHR-1), if

$$a_r(\mathbf{x}_r) \geq (\leq) a_r(\mathbf{0})$$

for all $\mathbf{x}_r$ in N^r and $r = 1, 2, \ldots, p$.

Using this definition, the MNBUHR-1 class of multivariate model possesses the following properties:

(a) MIHR-9 (MDHR-9) $\Rightarrow$ MNBUHR-1 (MNWUHR-1);
(b) **X** is both MNBUHR-1 and MNWUHR-1 if and only if the distribution is (7.2);
(c) If **X** is MNBUHR-1, all its subsets are also NBUHR-1.

Definition 7.16. We say that **X** is MNBUHR-2 (MNWUHR-2) if

$$c_i(\mathbf{x}) \geq (\leq) c_i(x_1, \ldots, x_{i-1}, 0, x_{i+1}, \ldots, x_p)$$

for all x_i and $i = 1, 2, \ldots, p$.

From the above definition, it follows that

(1) MIHR-2$\Rightarrow$ MNBUHR-2 and MDHR-2 $\Rightarrow$ MNWUHR-2;
(2) the only distribution that is both MNBUHR-2 and MNWUHR-2 is (7.11);
(3) all the univariate marginals are NBUHR-2 if **X** is MNBUHR-2;
(4) The joint distribution of the random variables $(X_1 - x_1|X_1 > x_1)$ and X_2 is specified by the survival function

$$S_{x_1}(t_1, x_2) = \frac{S(x_1 + t_1, x_1)}{S(x_1, x_2)} = \frac{S(x_1 + t_1|x_2)}{S(x_1|x_2)},$$

where $S(x_1|x_2) = P(X_1 > x_1|X_2 > x_2)$. Similarly, with respect to $(X_2 - x_2|X_2 > x_1)$ and X_1, we have

$$S_{x_2}(x_1, t_2) = \frac{S(x_2 + t_2|x_2)}{S(x_1|x_2)}.$$

In this notation, we can write

$$S(x_1 + 1|x_2) \leq S(x_1|x_2)S(1|x_2) \tag{7.37}$$

and

$$S(x_1, x_2 + 1) \leq S(x_2|x_1)S(1|x_1) \tag{7.38}$$

so, (7.37) and (7.38) $\Rightarrow$ **X** is MNBUHR-2;

(5) In terms of the initial hazard rate order defined in (vi) of Section 4.3.4, we find from the above discussion that

$$(X_i - t_i | X_j > t_j) \leq_{h(0)} X_i, \quad i, j = 1, 2,\ i \neq j,$$

if and only if (X_1, X_2) is MNBUHR-2.

Definition 7.17. The random vector (X_1, X_2) belongs to the bivariate NBUHR-3 (NWUHR-3) if $e_i(x_1, x_2) \geq e_1(0, x_2)$ for all x_2 and $x_1 = 0, 1, 2, \ldots$ and $e_2(x_1, x_2) \geq e_2(x_1, 0)$ for all x_1 and $x_2 = 0, 1, 2, \ldots$.

Among the properties of BNBUHR-3 class, we have the following:

(a) (X_1, X_2) is both BNBUHR-3 and BNWOHR-3 if and only if (X_1, X_2) has distribution (7.18);

(b) BIHR-10 (BDHR-10) $\Rightarrow$ BNBUHR-3 (BNWUHR-3);

(c)

$$S_{12}(x_1 + 1 | x_2) \leq S_{12}(x_1 | x_2) S_{12}(1 | x_2)$$

and

$$S_{21}(x_2 + 1 | x_1) \leq S_{21}(x_2 | x_1) S_{21}(1 | x_1)$$

is necessary and sufficient for (X_1, X_2) to be BNBUHR-3.

Remark 7.7. By a simple averaging process in the definitions of MNBUHR definitions, weaker classes of multivariate life distributions can be deduced. Accordingly we have the following:

(a) $\mathbf{X}$ is MNBUHRA-1 (MNWUHRA-1) if

$$\frac{1}{x_1 x_2 \ldots x_r} \sum_{t_1=0}^{x_1-1} \cdots \sum_{t_r=0}^{x_r-1} a_r(\mathbf{x}_r) \geq (\leq) a_r(\mathbf{0})$$

for all $\mathbf{x}_r$ in N^r and each $r = 1, 2, \ldots, p$;

(b) $\mathbf{X}$ is MNBUHRA-2 (MNWUHRA-2) if

$$\frac{1}{x_i} \sum_{t_i=0}^{x_i-1} c_i(x_1, \ldots, x_{i-1}, t_i, x_{i+1}, \ldots, x_p) \geq (\leq) c_i(x_1, \ldots, x_{i-1}, 0, x_{i+1}, \ldots, x_p)$$

for all x_j and each $i = 1, 2, \ldots, p$, $j \neq i$;

(c) (X_1, X_2) is BNBUHRA-3 (BNWUHRA-3) if

$$\frac{1}{x_i} \sum_{t_i=0}^{x_i-1} e_i(t_i, x_j) \geq e_i(0, x_j) \text{ for all } x_j.$$

Obviously,
MNBUHR-1(2)$\Rightarrow$ MNBUHRA-1(2),
BNBUHR-3 $\Rightarrow$ BNWUHRA-3.

7.8 MULTIVARIATE DECREASING MEAN RESIDUAL LIFE

In the univariate case, the monotonicity of the mean residual life function reflects the nature of ageing of a device. This has been successfully employed in replacement problems, comparison of life distributions, selection of models for data analysis, and so on. This section is devoted to the study of extension of these notions to the multivariate framework. Buchanan and Singpurwalla (1977) and Zahedi (1985) have proposed various approaches to the generalization of multivariate monotone mean residual life function when the lifetimes are continuous. The discrete version of the Buchanan-Singpurwalla definitions are as follows

(i)

$$\sum_{x=0}^{\infty} \frac{S(x+t, x+t)}{S(x, x)} \text{ decreases in } x \text{ for all } t \text{ in } N; \tag{7.39}$$

(ii)

$$\sum_{x=0}^{\infty} \frac{S(x+t_1, x+t_2)}{S(x, x)} \text{ decreases in } x \text{ for all } t_1, t_2 \text{ in } N; \tag{7.40}$$

(iii)

$$\sum_{x_1=0}^{\infty}\sum_{x_2=0}^{\infty} \frac{S(x_1+t, x_2+t)}{S(x_1, x_2)} \text{ decreases in } x_1 \text{ and } x_2 \text{ for all } t; \tag{7.41}$$

(iv)

$$\sum_{x_1=0}^{\infty}\sum_{x_2=0}^{\infty} \frac{S(x_1+t_1, x_2+t_2)}{S(x_1, x_2)} \text{ decreases in } x_1 \text{ and } x_2 \text{ for all } t_1, t_2. \tag{7.42}$$

The behaviour of the survival function is used as the basis for all definitions and the conditional probability of survival of the components with same or different ages are compared with the corresponding probabilities of two new components with same or different ages. Zahedi (1985) remarked that the arguments which prompted the above definitions are not necessarily the natural extensions of the univariate case as they do not possess the intuitive physical interpretation of the univariate definitions.

Retaining the notation and results in Section 6.3, the versions of multivariate decreasing (increasing) mean residual life, MDMRL (MIMRL), classes are considered below.

Definition 7.18. A p-dimensional discrete random vector $\mathbf{X}$, defined on N^P, is

(i) MDMRL-1 (MIMRL-1) if

$$m_i(\mathbf{x}+\mathbf{t}) \leq (\geq) m_i(\mathbf{x}), i = 1, 2, \dots, p,$$

for all $\mathbf{x}$ and $\mathbf{t}$ in N^p;

(ii) MDMRL-2 (MIMRL-2) if

$$m_i(x_1, \dots, x_{i-1}, x_i + t, x_{i+1} \dots, x_p) \leq (\geq) m_i(\mathbf{x}),$$

for $i = 1, 2, \dots, p$ and t in N;

(iii) MDMRL-3 (MIMRL-3) if

$$m_i(x_1 + t, \dots, x_n + t) \leq (\geq) m_i(x_1, \dots, x_n),$$

for all $n \leq p$ and t in N;

(iv) MDMRL-4 (MIMRL-4) if, for all $n \leq p$,

$$m_i(x+t, \dots, x+t) \leq (\geq) m_i(x, \dots, x), \text{ and } t \text{ in } N.$$

The interpretation of (i) is that the mean residual life of a p-component device, where the components are of different ages, decrease (increase) as the components age with different intensities. In (ii) also, the mean residual life decreases when a component is replaced by a younger one with the same life distribution. Case (iii) results when the time moves at the same rate for all components while initially the components are of different ages. Lastly, when the components are of the same age as well as time moves at the same rate for all of them, the mean residual life is decreasing (increasing) gives rise to version (iv).

We discuss the properties of the classes in some detail. An extension of Theorem 6.10 is obtained by adopting the same method of proof.

Theorem 7.9. *Given the multivariate mean residual life function* $\mathbf{m}(\mathbf{x})$*, the survival function is uniquely determined as*

$$S(\mathbf{x}) = \prod_{r=0}^{x_1-1} \frac{m_1(r-1, x_2, \dots, x_p) - 1}{m_1(r, x_2, \dots, x_p)} \prod_{r=0}^{x_2-1} \frac{m_2(0, r-1, x_3, \dots, x_p) - 1}{m_2(0, r, \dots, x_p)}$$
$$\times \dots \prod_{r=0}^{x_p-1} \frac{m_p(0, 0, \dots, 0, r-1) - 1}{m_p(0, 0, \dots, r)}. \tag{7.43}$$

Theorem 7.10. *The random vector* $\mathbf{X}$ *is both MDMRL-1 and MIMRL-1 if and only if the distribution of* $\mathbf{X}$ *is multivariate geometric with independent geometric marginals, viz.,*

$$S(\mathbf{x}) = q_1^{x_1} q_2^{x_2} \dots q_p^{x_p}.$$

Proof. When the distribution is as above, $m_i(\mathbf{x}) = \frac{1}{1-q_i}$ so that $m_i(\mathbf{x}+\mathbf{t}) = m_i(\mathbf{x})$, or $\mathbf{X}$ is MDMRL-1 and MIMRL-2. Conversely, the given property means that $m_i(\mathbf{x}) = k_i$, a constant greater than unity for all $\mathbf{x}$ and $i = 1, 2, \ldots, p$. Taking $\frac{k_i-1}{k_i} = q_i$, the stated distribution results on using (7.43). □

Remark 7.8. Eq. (7.43) is only one of several ways in which $S(\mathbf{x})$ can be written in terms of $m_i(\mathbf{x})$ expressions. In (7.43), we have started with the definition of $m_1(\mathbf{x})$ in terms of $S(\mathbf{x})$. On the other hand, if begin with $m_2(\mathbf{x})$,

$$S(\mathbf{x}) = \prod_{r=0}^{x_2-1} \frac{m_2(x_1, r-1, x_3, \ldots, x_p) - 1}{m_2(x_1, r, x_3, \ldots, x_p)} \times \prod_{r=0}^{x_2-1} \frac{m_3(x_1, 0, r-1, x_4, \ldots, x_p) - 1}{m_3(x_1, 0, r, \ldots, x_p)} \cdots \prod_{r=0}^{x_1-1} \frac{m_1(1, 0, \ldots, 0) - 1}{m_1(r, 0, \ldots, 0)}. \tag{7.44}$$

Thus, on the whole, there can be p expressions which are all equal to each other.

Theorem 7.11. *The only discrete distribution in N^p which belongs to both MDMRL-2 and MIMRL-3 class is the multivariate geometric law in (7.11).*

Proof. Assume that $\mathbf{X}$ is MDMRL-2 and MIMRL-3. Then,

$$m_i(x_1, \ldots, x_{i-1}, x_i + t, x_{i+1}, \ldots, x_p) = m_i(\mathbf{x}), \quad i = 1, 2, \ldots, p,$$

so that

$$m_i(\mathbf{x}) = g_i(x_1, \ldots, x_{i-1}, x_{i+1}, \ldots, x_p),$$

a function independent of x_i. Using this, (7.43) gives

$$S(\mathbf{x}) = [g_1(x_2, \ldots, x_p)]^{x_1} [g_2(0, x_3, \ldots, x_p)]^{x_2} \ldots [g_p(0, \ldots, 0)]^{x_p}.$$

Thus, log $S(\mathbf{x})$ is a linear function of x_1. Similarly, from (7.44), log $S(\mathbf{x})$ is linear in x_2, and so on. Equating p expressions, we see that the only general solution that satisfies all the p equations is

$$\prod_{i=1}^{p} q_i^{x_i} \prod_{i<j=1}^{p} q_{ij}^{x_i x_j} \cdots (q_{12\ldots p})^{x_1, x_2, \ldots, x_p}.$$

Thus, $S(\mathbf{x})$ is of the form (7.11). The conditions on the parameters are obtained from $S(x_1, \ldots x_{r+i}) \leq S(x_1, \ldots, x_r)$, $r = 1, 2, \ldots, n$, $f(\mathbf{x})$ is probability mass function and $f(\mathbf{0}) > 0$. □

Theorem 7.12. *The random vector* **X** *is MDMRL-3 and MIMRL-3 if and only if* **x** *is multivariate geometric with survival function (7.13).*

Proof. For simplicity, we prove the bivariate case, which easily extends to the multivariate vector. When $\mathbf{X} = (X_1, X_2)$, (7.13) becomes

$$S(x_1, x_2) = \begin{cases} p^{x_2} p_1^{x_1 - x_2}, & x_1 \geq x_2, \\ p^{x_1} p_2^{x_2 - x_1}, & x_1 \geq x_2. \end{cases} \tag{7.45}$$

Assuming (7.45), we have

$$m_1(x_1, x_2) = m_2(x_1, x_2) = \begin{cases} \left(\frac{1}{1-p_1}, \frac{1}{1-pp_1^{-1}}\right), & x_1 > x_2, \\ \left(\frac{1}{1-pp_1^{-1}}, \frac{1}{1-p_2}\right), & x_1 < x_2, \\ \left(\frac{1}{1-p_1}, \frac{1}{1-p_2}\right) & x_1 = x_2. \end{cases}$$

Thus,

$$m_i(x_1 + t, x_2 + t) = m_i(x_1, x_2) \tag{7.46}$$

for all x_1, x_2 and t taking values in N, showing that (X_1, X_2) is MDMRL-3 as well as MIMRL-3.

Conversely, if (7.46) holds, using the relation

$$c_1(x_1 + 1, x_2 + 1) = 1 - \frac{m_1(x_1, x_2) - 1}{m_1(x_1 + 1, x_2)}$$

between vector hazard rate and mean residual life in (iv) (b) in Section 6.4, we have

$$\begin{aligned} c_1(x_1 + t + 1, x_2 + t = 1) &= 1 - \frac{m_1(x_1 + t, x_2 + t) - 1}{m_1(x_1 + t + 1, x_2 + t)} \\ &= 1 - \frac{m_1(x_1, x_2) - 1}{m_1(x_1 + 1, x_2)} \quad \text{by (7.46)} \\ &= c_1(x_1 + 1, x_2 + 1). \end{aligned}$$

By invoking Theorem 6.8 with $p = 2$, we arrive at (7.45) and this completes the proof. □

Some other interesting aspects of the above classes are mentioned below:

(a)

MDMRL-1 (MIMRL-1)	⇒	MDMRL-3 (MIMRL-3)	⇒	MDMRL-4 (MIMRL-4)
⇓				
MDMRL-2 (MIMRL-2)				

(b) If **X** is MDMRL-l (MIMRL-l), $l = 1, 2, 3, 4$, each non-empty subset **X** has the corresponding property;
(c) If **X** is MDMRL-l (MIMRL-l) and **Y** is also likewise, then (**X**, **Y**) is also MDMRL-l (MIMRL-l) provided **X** and **Y** are independent;
(d) If **X** is MDMRL-l (MIMRL-l) $\Rightarrow a\mathbf{X}$ is MDMRL-l (MIMRL), $a > 0$;
(e) If **X** is MDMRL-l (MIMRL-l), then $(X_{i_1} \dots, X_{i_p})$ is also MDMRL-l (MIMRL);
(f) **X** is MIFR-l (MDFR-l) $\Rightarrow$ **X** is MDMRL-l (MIMRL-l).

Proof. Since the proof is similar for $l = 1, 2, 3, 4$, we prove it when $l = 1$ in the bivariate case. (X_1, X_2) is BIHR-1 if

$$c_i(x_1 + t_1, x_2 + t_2) \geq c_i(x_1, x_2) \text{ for all } x_1, x_2, t_1, t_2 \text{ and } i = 1, 2.$$

When $i = 1$, this implies

$$1 - \frac{S(x_1 + t_1 + 1, x_2 + t_2)}{S(x_1 + t_1, x_2 + t_2)} \geq 1 - \frac{S(x_1 + 1, x_2)}{S(x_1, x_2)},$$

or

$$\frac{S(x_1 + t_1 + 1, x_2 + t_2)}{S(x_1 + t_1, x_2 + t_2)} \leq \frac{S(x_1 + 1, x_2)}{S(x_1, x_2)} \text{ for all } x_1, x_2. \tag{7.47}$$

Now,

$$\begin{aligned}
m_1(x_1, x_2) &= \frac{1}{S(x_1 + 1, x_2 + 1)} \sum_{r=x_1+1}^{\infty} S(r, x_2 + 1) \\
&= 1 + \frac{S(x_1 + 2, x_2 + 1)}{S(x_1 + 1, x_2 + 1)} + \frac{S(x_1 + 3, x_2 + 1)}{S(x_1 + 2, x_2 + 1)} \\
&\quad \frac{S(x_1 + 2, x_2 + 1)}{S(x_1 + 1, x_2 + 1)} + \dots \\
&\geq 1 + \frac{S(x_1 + t_1 + 2, x_2 + t_2 + 1)}{S(x_1 + t_1 + 1, x_2 + t_2 + 1)} \\
&\quad + \frac{S(x_1 + t_1 + 3, x_2 + t_2 + 1)}{S(x_1 + t_1 + 2, x_2 + t_2 + 1)} \frac{S(x_1 + t_1 + 2, x_2 + t_2 + 1)}{S(x_1 + t_1 + 1, x_2 + t_2 + 1)} \\
&\quad + \dots = m_i(x_1 + t_1, x_2 + t_2) \text{ for all } x_1, x_2, t_1, t_2.
\end{aligned}$$

Similarly, when $i = 2$, $m_2(x_1, x_2) \geq m_2(x_1 + t_1, t_2)$, which proves the result. □

(g) Let (W_1, W_2) be the equilibrium random variable corresponding to (X_1, X_2) defined in (6.111) and (6.112). Then, (X_1, X_2) is BDRL-l if and only if (W_1, W_2) is BIFR-l; see Section 6.9;

(h) $\mathbf{X}$ is MDMRL-1 (MIMRL-1) if and only if $S_{\mathbf{x}}(\mathbf{t})$ is decreasing $\mathbf{x}$ for all $\mathbf{t}$, where $S_{\mathbf{x}}(\mathbf{t})$ is the residual life distribution of $\mathbf{X}$. Taking $x_i = x$ and/or $t_i = t$, $i = 1, 2, \ldots, p$, we have a similar characterization of the other MDMRL classes;

(i) Since

$$c_1(x_1+1, x_2+1) = 1 - \frac{m_1(x_1, x_2) - 1}{m_1(x_1+1, x_2)}$$

and

$$c_2(x_1+1, x_2+1) = 1 - \frac{m_2(x_1, x_2) - 1}{m_2(x_1, x_2+1)},$$

(X_1, X_2) is BDMRL-2 (BIMRL-2) if and only if

$$c_1(x_1+1, x_2+1)m_1(x_1+1, x_2) \text{ and } c_2(x_1+1, x_2+1)m_2(x_1, x_2+1)$$

are less than unity;

(j) $W_{\mathbf{X}} \leq_{\text{whr}} W_{\mathbf{Y}} \Rightarrow \mathbf{X} \leq_{\text{mrl}} \mathbf{Y}$, when $\mathbf{X}$ and $\mathbf{Y}$ are bivariate random vectors.

7.9 DECREASING CONDITIONAL MEAN RESIDUAL LIFE

In the case of life distributions mentioned in this section, we consider the monotonicity of the conditional mean residual life functions proposed in Section 6.3.1.

Definition 7.19. We say that (X_1, X_2) is bivariate decreasing (increasing) mean residual life-5, BDMRL-5 (BIMRL-5), if $l_1(x_1, x_2)$ is decreasing (increasing) in x_1 for all x_2 and $l_2(x_1, x_2)$ is decreasing in x_2 for all x_1. This will mean that

$$l_1(x_1+t, x_2) \leq (\geq) l_1(x_1, x_2) \quad \text{and} \quad l_2(x_1, x_2+t) \leq (\geq) l_2(x_1, x_2)$$

for all t in N.

Based on the above definition, we observe the following:

(i) (X_1, X_2) is BDMRL-5 (BIMRL-5) if and only if the vector (Z_1, Z_2), with equilibrium distributions (6.113) and (6.114), is BIHR-10 (BDHR-10), by virtue of the relationship

$$e_{i,Z_1,Z_2}(x_1, x_2) = [l_i(x_1, x_2)]^{-1};$$

(ii) From the last identity and discussions in Section 6.9, that (X_1, X_2) is both BDMRL-5 and BIMRL-5 if and only if its distribution is

$$f(x_1, x_2) = \alpha q_1^{x_1} q_2^{x_2} \theta^{x_1 x_2}$$

presented in Example 6.8;

(iii) (X_1, X_2) is BIHR-10 $\Rightarrow$ (x_1, x_2) is BDMRL-5.

7.10 BAYESIAN DEFINITION OF DMRL DISTRIBUTIONS

There are different definitions for BDMRL in the Bayesian approach also. We follow the discussions in Bassan et al. (2002) in providing the discrete analogues.

Definition 7.20. An exchangeable random vector (X_1, X_2) is BDMRL-6 distribution if, for $x_1 < x_2$,

$$E(X_1 - x_1|X_1 > x_1, X_2 > x_2) \geq E(X_2 - x_2|X_1 > x_1, X_2 > x_2).$$

The above definition is satisfied if and only if

(a) $\sum_{t=x_1}^{\infty} S(t, x_2) \geq \sum_{t=x_2}^{\infty} S(x_1, t)$ for $x_1 < x_2$,

(b) $m_1(x_1, x_2) \geq m_2(x_1, x_2)$, $x_1 < x_2$.

Also, note that BIFR-11 $\Rightarrow$ BDMRL-6.

Definition 7.21. An exchangeable random vector (X_1, X_2) is BDMRL-7 if, for all $x_1 < x_2$,

$$(X_1 - x_1|X_1 > x_1, X_2 > x_2) \geq_{\text{mrl}} (X_2 - x_2|X_1 > x_1, X_2 > x_2).$$

This is equivalent to saying that

$$\frac{\sum_{t=x+t_1}^{\infty} S(t, x_2)}{\sum_{t=x+t_2}^{\infty} S(t, x_1)} \text{ is increasing in } x.$$

Some properties related to the above two definitions are as follows:

a. A sufficient condition for (X_1, X_2) to be BDMRL-6 is that
(i) $m_1(x, y)$ be increasing in y for every $x > 0$, and
(ii) $m_1(x, y)$ be decreasing in x for every $y > 0$;

b. If (X_1, X_2) is BDMRL-6 and $E(X_1|X_2 > x_2) \leq E(X_1)$ for $x_2 > 0$, then X_1 is NBUE.

Proof. Since (X_1, X_2) is BDMRL-6,

$$m_1(x_1, x_2) \geq m_2(x_1, x_2), \quad x_1 < x_2.$$

When $x_1 = -1$,

$$m_1(x_2) \leq m_1(0, x_2) = E(X_1|X_2 > x_2) \leq E(X_1),$$

and therefore X_1 is NBUE; □

c. If (X_1, X_2) is DMRL-7 and $m_1(x, y) \leq m_1(x)$ for $x, y > 0$, then X_1 is DMRL.

Proof. (X_1, X_2) is DMRL-7 $\Leftrightarrow$

$$E(X_1 \leq t - x_2)|X_1 > x + x_2, X_2 > x_1) \leq E[X_1 - x - x_1|X_1 > x + X_1, X_2 > -x_2].$$

As before, taking $x_1 = -1$

$$E(X_1 - t - x_2|X_2 > 0) \leq E(X_1 - x|X_2 > x_2), \quad x_1 < x_2,$$
$$m_1(x + x_2) \leq m_1(x),$$

and hence X_1 is DMRL. □

7.11 MULTIVARIATE NEW BETTER THAN USED CLASS

Different versions of multivariate new better than used concept have been proposed when lifetime is continuous by authors like Buchanan and Singpurwalla (1977), Marshall and Shaked (1982, 1986), El-Neweihi (1981), and Ghosh and Ebrahimi (1981). It appears that the only work that deal with the discrete case is that of Klefsjo (1980) who took the bivariate survival function as $P(X_1 > x_1, X_2 > x_2)$ in proposing the bivariate definitions of various forms of the new better than used classes. In the spirit of Klefsjo (1980), we have the following definitions in the p-variate case using our definitions of the survival function.

Definition 7.22. A discrete random vector $\mathbf{X} = (X_1, \ldots, X_p)$, taking values in N^p, is said to have multivariate

(i) new better than used-1 (MNBU-1) if

$$S(x_1 + t_1, x_2 + t_2, \ldots, x_p + t_p) \leq S(x_1, x_2, \ldots, x_p)S(t_1, t_2, \ldots, t_p)$$

for all $\mathbf{x} \geq \mathbf{0}$ and $\mathbf{t} \geq \mathbf{0}$;

(ii) MNBU-2 if

$$S(x_1 + t, x_2 + t, \ldots, x_p + t) \leq S(x_1, \ldots, x_p)S(t, t, \ldots, t), \mathbf{x} > 0,\ t > 0;$$

(iii) MNBU-3 if

$$S(x + t_1, x + t_2, \ldots, x + t_p) \leq S(x, x \ldots, x_p)S(t_1, t_2, \ldots, t_p);$$

(iv) MNBU-4 if

$$S(x + t, \ldots, x + t) \leq S(x, x, \ldots, x)S(t, t, \ldots, t), \quad x, t \geq 0;$$

(v) MNBU-5 if

$$S(x_1 + t_1, x_2 + t_2, \ldots, x_p + t_p) \leq S(x_1, \ldots, x_p)S(t_1, \ldots, t_p)$$

for every $\mathbf{x} \geq \mathbf{0}$ and $\mathbf{t} \geq \mathbf{0}$ for which $(x_i - x_j)(t_i - t_j) \geq 0$, $i, j = 1, 2, \ldots, p$. When the vectors satisfy the condition $(x_i - x_j)(t_i - t_j) \geq 0$, they are said to be similarly ordered. By reversing the inequality in (i)–(iv), we have the new worse than used forms MNWU-1, MNWU-2, MNWU-3 and MNWU-4, respectively.

Among these versions, it may be noted that

$$\text{MNBU-1} \Rightarrow \text{MNBU-3} \Rightarrow \text{MNBU-4} \Rightarrow \text{MNBU-5}$$

and also MNBU-1 $\Rightarrow$ MNBU-2.

We also see that

$$\text{MIHR-8} \Rightarrow \text{MNBU-1 and MIHR-5} \Rightarrow \text{MNBU-4}.$$

Theorem 7.13. *If* $\mathbf{X}$ *is MNBU-4, then* $X_{(1)}$ *is NBU, where* $X_{(1)}$ *is the smallest order statistic.*

Proof. The survival function of $X_{(1)}$ is $S_{X_1}(\mathbf{x}) = S(x, \ldots, x)$. So,

$$\begin{aligned} S_{X_{(1)}}(x+t) &= S(x+t, x+t, \ldots, x+t) \\ &\leq S(x, x, \ldots, x) S(t, t, \ldots, t) \text{ by MNBU-4 property} \\ &= S_{X_1}(x) S_{X_1}(t) \end{aligned}$$

showing that $X_{(1)}$ is NBU. □

Marshall and Shaked (1982) called $\mathbf{X}$ to be multivariate NBU if $P(X_i \leq 0) = 1$, $i = 1, 2, \ldots, p$, and if

$$P(\mathbf{X} \in (\alpha + \beta)A) \leq P(\mathbf{X} \in \alpha A) P(\mathbf{x} \in \beta A)$$

for every $\alpha, \beta \geq 0$ and every open upper set contained in N^P. Notice that A is an upper set if $\mathbf{x} \in A$ and $x_i \leq y_i$, $i = 1, 2, \ldots, p$, implies $\mathbf{y} \in A$.

7.12 MULTIVARIATE NEW BETTER THAN USED IN EXPECTATION

Like the MNBU concept, Klefsjo (1980) had prescribed four versions of bivariate new better than used classes which can be extended to our definition of $S(\mathbf{x})$ in the multivariate case.

Definition 7.23. The random vector is said to be

(a) multivariate new better than used in expectation-1 (MNBUE-1) if

$$\sum_{t_1=1}^{\infty}\cdots\sum_{t_p=1}^{\infty} S(x_1+t_1, x_2+t_2, \ldots, x_p+t_p)$$
$$\leq S(x_1+1, \ldots, x_p+1)\sum_{t_1=1}^{\infty}\cdots\sum_{t_p=1}^{\infty} S(t_1, \ldots, t_p)$$

for every $x_i = 0, 1, 2, \ldots, i = 1, 2, \ldots, p$;

(b) MNBUE-2 if

$$\sum_{t=1}^{\infty} S(x_1+t, \ldots, x_p+t) \leq S(x+1, \ldots, x_p+1)\sum_{t_1}^{\infty} S(t, \ldots, t) \text{ for every } x_i;$$

(c) MNBUE-3 if

$$\sum_{t=1}^{\infty}\cdots\sum_{t_p=1}^{\infty} S(x+t_1, \ldots, x+t_p) \leq S(x_1+1, \ldots, x+1)\sum_{t_1=1}^{\infty}\cdots\sum_{t_p=1}^{\infty} S(t_1, \ldots, t_p)$$

for every x;

(d) MNBUE-4 if

$$\sum_{t=1}^{\infty} S(x+t, \ldots, x+t) \leq S(x, \ldots, x)\sum_{t=1}^{\infty} S(t, \ldots, t)$$

for every x.

Other than visualizing (a)–(d) as generalizations of the univariate NBUE, Klefsjo (1980) has not given any interpretations to these concepts. Some remarks seem to be in order in this connection. In the univariate case, the NBUE concept means that (Definition 4.21) the mean residual life of a device at age x is smaller than the corresponding quality at age zero, thereby giving a clear indication that the device is ageing positively in the sense of the mean residual life. By comparison, the definitions in (a) and (b) do not involve the mean residual life of a multi-component device nor the mean itself. In some way, the summations are more in relationship with the covariance functions of the residual life in the bivariate case.

The idea of defining MNBUE in terms of the multivariate mean residual life will be a more realistic generalization of the univariate notion. In this regard, taking the ith component of the multivariate mean residual life in (6.30), viz.

$$m_i(\mathbf{x}) = E(X_i - x_i|\mathbf{X} > \mathbf{x}),$$

we have the following definitions (see Basu and Ebrahimi, 1986).

Definition 7.24. We say that $\mathbf{X}$ is

(a) MNBUE-5 if $m_i(\mathbf{x}) \leq m_i(x_1, \ldots, x_{i-1}, -1, x_{i+1}, \ldots, x_p)$ for all $\mathbf{x}$;
(b) MNBUE-6 if $m_i(x, \ldots, x, \ldots, x) \leq m_i(x, \ldots, x, -1, x, \ldots, x)$ for all x and $i = 1, 2, \ldots, p$;
(c) MNBUE-7 if $m_i(\mathbf{x}) \leq E(X_i)$, $i = 1, 2, \ldots, p$;
(d) MNBUE-8 if $m_i(x, \ldots, x) \leq E((X_i)$, $i = 1, 2, \ldots, p$, and similar inequalities in (a) and (b) for all subsets of $\mathbf{X}$.

It is not difficult to see that

MNBUE-1 ⇒ MNBUE-2, MNBUE-3 ⇒ MNBUE-4,
MNBUE-5 ⇒ MNBUE-6, MNBUIE-7 ⇒ MNBUE-8,
MDMRL-1 ⇒ MNBUE-5 and MDMRL-4 ⇒ MNBUE-6.

The dual classes, multivariate new worse than used in expectation, MNWUE-i, $i = 1, 2, \ldots, 8$, can be obtained readily by reversing the inequalities in MNBUE-i.

7.13 MULTIVARIATE HARMONIC NEW BETTER THAN USED IN EXPECTATION

In the univariate case, the HNBUE class was obtained by comparing the harmonic mean residual life of a given lifetime distribution with that of the geometric law. Klefsjo (1980) extended the same logic in the case of a bivariate distribution to arrive at bivariate HNBUE classes of life distributions. In the process, he had used the bivariate geometric law

$$G(x_1, x_2) = \begin{cases} p_{11}^{x_1}(p_{01}^{x_1} + p_{11})^{x_2 - x_1} & 0 \leq x_1 \leq x_2,\ x_1 = x_2 = 0, 1, 2, \ldots, \\ p_{11}^{x_2}(p_{10} + p_{11})^{x_1 - x_2} & 0 \leq x_2 \leq x_1, \end{cases} \tag{7.48}$$

for comparison.

Definition 7.25. The random vector $\mathbf{X}$ is said to be

(a) multivariate HNBUE-1 (MHNBUE-1) if

$$\sum_{t_1=1}^{\infty} \sum_{t_2=1}^{\infty} S(x_1+t_1, x_2+t_2) \leq \sum_{t_1=1}^{\infty} \sum_{t_2=1}^{\infty} G(x_1+t_1, x_2+t_2) \text{ for all } x_1, x_2 \text{ in } N;$$

(b) MHNBUE-2 if

$$\sum_{t=1}^{\infty} S(x_1 + t, x_2 + t) \leq \sum_{t=1}^{\infty} G(x_1 + t, x_2 + t) \text{ for all } x_1, x_2;$$

(c) MHNBUE-3 if

$$\sum_{t_1=1}^{\infty}\sum_{t_2=1}^{\infty} S(x+t_1, x+t_2) \le \sum_{t_1=1}^{\infty}\sum_{t_2=1}^{\infty} G(x+t_1, x+t_2) \text{ for all } x \text{ in } N;$$

(d) MHNBUE-4 if

$$\sum_{t=1}^{\infty} S(x+t, x+t) \le \sum_{t=1}^{\infty} G(x+t, x+t) \text{ for all } x \text{ in } N,$$

$$\sum_{t=1}^{\infty} S(x_1, x_2+t) \le \sum_{t=1}^{\infty} G(x_1, x_2+t) \text{ for } 0 \le x_1 \le x_2,$$

and

$$\sum_{t=1}^{\infty} S(x_1+t, x_2) \le \sum_{t=1}^{\infty} G(x_1+t, x_2) \text{ for } 0 \le x_2 < x_1.$$

In all cases, G is the survival function of the bivariate geometric distribution that satisfies

$$\sum_{x_1=1}^{\infty} G(x_1, 0) = \sum_{x_1=1}^{\infty} S(x_1, 0), \quad \sum_{x_2=1}^{\infty} G(0, x_2) = \sum_{x_2=1}^{\infty} S(0, x_2)$$

and

$$\sum_{x_1=1}^{\infty}\sum_{x_2=1}^{\infty} G(x_1, x_2) = \sum_{x_1=1}^{\infty}\sum_{x_1=1}^{\infty} S(x_1, x_2).$$

In order that these conditions are satisfied, the parametrics of the model in (7.48) should satisfy

$$p_{10} = \frac{r_1 + r_2 - 1}{r_{12}} - \frac{1}{r_1}, \quad p_{11} = 1 - \frac{r_1 + r_2 - 1}{r_{12}},$$

where

$$r_1 = \sum_{x_1=1}^{\infty} S(x_1, 0),\ r_2 = \sum_{x_2=1}^{\infty} S(0, x_2) \quad \text{and} \quad r_{12} = \sum_{x_1=1}^{\infty}\sum_{x_2=1}^{\infty} S(x_1, x_2).$$

A second set of MHNBUE classes can be conceived along the lines of Basu et al. (1983). For this, we define

$$M_{1,2,\ldots p}(\mathbf{x}) = E\left[(X_1 - x_1)\ldots(X_p - x_p)|\mathbf{X} > \mathbf{x})\right]$$

as the joint product moment of the residual life of **X** and

$$N(x,\ldots,x)=\sum_{t=x}^{\infty}\frac{S(t,t,\ldots,t)}{S(x,x,\ldots,x)}.$$

Definition 7.26. We say that **X** is

(a) MHNBUE-5 if

$$\frac{x_1\ldots x_p}{\sum_{\mathbf{t}=\mathbf{e}} M^{-1}_{1,2,\ldots p}(\mathbf{t})}\le\sum_{\mathbf{t}=\mathbf{e}}^{\infty}S(t)\text{ for all }\mathbf{x}\text{ in }N^p,$$

where $\sum_{\mathbf{t}=\mathbf{e}}^{x}$ is used to denote the p-fold sum $\sum_{t_1=1}^{x_1}\cdots\sum_{t_p=1}^{x_p}$;

(b) MHNBUE-6 if

$$\frac{x^p}{\sum_{t=\mathbf{e}}^{\mathbf{x}} M^{-1}_{1,2,\ldots p}(\mathbf{t})}\le\sum_{\mathbf{t}=\mathbf{e}}^{\infty}S(t),\quad \mathbf{x}=(x,\ldots x),$$

for all $x=0,1,2\ldots$;

(c) MHNBUE-7 if

$$\frac{x}{\sum_{t=1}^{x} M^{-1}_{1,2,\ldots p}(x,\ldots,x)}\le\sum_{t=\mathbf{e}}^{\infty}S(t),\quad\text{for }x>0,$$

and similar inequalities hold in all the cases for every subset of **X**;

(d) MHNBUE-8 if

$$\frac{x}{\sum_{t=1}^{x} N^{-1}(t,\ldots t)}\le\sum_{t=1}^{\infty}S(t,t,\ldots,t)$$

for $(t,\ldots,t)>N(x,x,\ldots,x)$, and similar inequalities for all subsets of **X**.

The inequalities in (a)–(c) can be interpreted as the inequalities between harmonic mean of the product moment of residual life and the product moment of lifetime of a unit with new components. At the same time, (d) has a different meaning in that the time to failure of a series system is HNBUE if and only if the time to failure of the components are HNBUE (Basu and Ebrahimi, 1986). Reversing the inequalities, we have the dual MHNWUE classes.

We have MHNBUE-5 $\Rightarrow$ MHNBUE-6.

7.14 INCREASING PRODUCT MOMENT RESIDUAL LIFE

It was observed in the earlier discussions that the product moment of residual life (PMRL) was the basis for introducing the MHNBUE concept and also the properties of some other concepts that are closely related to it.

In this section we present classes of life distributions based on the monotonicity of the PMRL.

Definition 7.27. The vector of lifetimes (X_1, X_2) belongs to the class of

(a) bivariate decreasing PMRL-1 (BDPMRL-1) if

$$M(x_1+t_1, x_2+t_2) \le M(x_1, x_2) \text{ for all } x_1, x_2 \text{ and } t_1, t_2 \text{ in } N;$$

(b) bivariate decreasing PMRL-2 (BDPRML-2) if

$$M(x_1+t, x_2+t) \le M(x_1, x_2) \text{ for all } x_1, x_2 \text{ and } t \text{ in } N.$$

When the inequalities are reserved, we have the corresponding definitions of BIPMRL-1 and BIPMRL-2.

Example 7.6. Let

$$S(x_1, x_2) = \frac{(m)x_1 + x_2}{(m+n)_{x_1+x_2}}, \qquad x_1, x_2 = 0, 1, 2, \ldots,$$

as in Example 6.11. Then,

$$M(x_1, x_2) = \frac{(m+n+x_1+x_2+1)(m+n+x_1+x_2)}{(n-1)^2}.$$

It is easy to see that (X_1, X_2) is BIPMRL-1.

Example 7.7. Let

$$S(x_1, x_2) = \frac{\binom{k+n-x_1-x_2}{n-x_1-x_2}}{\binom{k+n}{n}},$$

$$x_1, x_2 = 0, 1, \ldots, n,\ x_1+x_2 \le n,\ k, n > 0.$$

By direct calculation, we find

$$M(x_1, x_2) = \frac{(k+n-x_1-x_2)(k+n-x_1-x_2-1)}{k(k+1)},$$

so that (X_1, X_2) is BDPMRL-1.

Among the two PMRL classes, we have the following implications:

$$\text{BDPMRL-1 (BIPMRL-1)} \Rightarrow \text{BDPMRL-2 (BIPMRL -2)},$$
$$\text{BDPMRL-1 (BIPMRL-1)} \Rightarrow \text{MNNBUE-5 (MHNWUE-5)}.$$

Also, if $a(x_1, x_2)$ is the bivariate scalar hazard rate of the equilibrium distribution in (6.104) representing (Y_1, Y_2),

$$a_{Y_1,Y_2}(x_1, x_2) \text{ is increasing } \Rightarrow (X_1, X_2) \text{ is BDPMRL-1,}$$

or

$$(Y_1, Y_2) \text{ is BIHR-9 } \Rightarrow (X_1, X_2) \text{ is BPPMRL-1.}$$

We have presented several ageing concepts in this chapter based on joint survival functions as well as by employing the properties of multivariate reliability functions. Even in one concept of ageing, there exists several versions imparting flexibility in their use. Employing a particular version depends on the physical conditions that generate the corresponding data. Due to paucity of data on discrete systems and analysis, it will be premature to suggest the superiority of one definition over the other. The properties of various classes, as has been done in the univariate case, still remain open problems. Some discrete multivariate distributions and their classifications according to the ageing concepts described above are discussed in the next chapter.

Chapter 8

Multivariate Lifetime Models

8.1 INTRODUCTION

Apart from aiding in analyzing the patterns in which failures occur in a multi-component device, the reliability functions introduced in the last chapter can serve as tools in identifying appropriate models to failure time data. Since the functional forms of many of those concepts uniquely determine the life distribution, characterization of distributions based on such forms is one way of choosing a suitable model. The essential steps involved in such a procedure is to acquire first-hand knowledge of the mathematical form of the reliability functions either from prior assessment based on the physical conditions under which the device works or from some definite indication from a preliminary evaluation of the data set. There are many situations, especially in the case of multivariate data, in which neither of the above are feasible. We may have to choose one model that might be appropriate from among a class of alternatives. A somewhat similar situation emerges even if we bank on the notions of ageing in model selection. We need to have a fairly good number of models that belong to the positive and negative ageing classes for which the necessary criteria of ageing can be compiled. All these point out to the real need for developing enough multivariate discrete distributions with the primary purpose of being used in a reliability context. Other than in exemplifying a few theoretical properties or by way of characterizations through simple functional forms, not many discrete models are present in the in literature. Discussions on models out of real-data situations also appears to be very rare. In this scenario, the objective of the present chapter is to propose a few discrete models as plausible candidates for the analysis of discrete multivariate life data. Johnson et al. (1997) have provided an exhaustive survey of multivariate discrete models like multinomial, negative multinomial, Poisson, power series, hyper-geometric, Pólya-Eggenberger, Ewens and other related distributions. Reliability functions associated with many of these models do not have analytically tractable forms for theoretical manipulations. Computer-based techniques are required to analyze their reliability characteristics. For this reason, in this chapter, we present some simple multivariate distributions whose reliability functions are amenable to algebraic manipulations and at the same time serve as meaningful models for real situations. All the distributions considered here are extensions of the uni-

Reliability Modelling and Analysis in Discrete Time. http://dx.doi.org/10.1016/B978-0-12-801913-9.00008-7

variate laws that were discussed at length earlier in Chapter 3 to multivariate case.

8.2 MULTIVARIATE GEOMETRIC DISTRIBUTIONS

Geometric distributions play a pivotal role in lifetime data analysis as models that satisfy the no-ageing property, in some reliability sense. We present such models in this section as well as some others which can be used in a more general context.

8.2.1 Geometric Distribution-1

A random vector **X** is distributed as multivariate geometric-1 if it has a survival function of the form

$$S(\mathbf{x}) = \prod_{i=1}^{p} q_i^{x_i} \prod_{\substack{i,j=1\\ i<j}}^{p} q_{ij}^{x_i x_j} \cdots q_{12\ldots p}^{x_1 x_2 \ldots x_p}, \ \mathbf{x} \in N^p, \ 0 < q_i, q_{1j}, \ldots, q_{12\ldots p} < 1,$$

$$1 - \sum q_i - \sum_{\substack{i,j\\ i<j}} q_{ij} + \cdots + (-1)^p q_{12\ldots n} \geq 0, \quad (8.1)$$

and was proposed by Nair and Asha (1997). The bivariate form of (8.1) is

$$S(x_1, x_2) = q_1^{x_1} q_2^{x_2} \theta^{x_1 x_2}, \ x_1, x_2 = 0, 1, 2, \ldots, \ 0 < q_i < 1,$$
$$0 \leq \theta \leq 1 \text{ and } 1 - \theta \leq (1 - q_1\theta)(1 - q_2\theta). \quad (8.2)$$

The joint probability mass function of (X_1, X_2) is

$$f(x_1, x_2) = q_1^{x_1} q_2^{x_2} \theta^{x_1 x_2 - 1}[(1 - q_1\theta^{x_2+1})(1 - q_2\theta^{x_1+1}) + \theta - 1]. \quad (8.3)$$

Thus, the marginal distribution of X_i is

$$f_i(x_i) = (1 - q_i)q_i^{x_i}, \quad i = 1, 2,$$

and the conditional distribution of X_i, given X_j, is given by

$$f(x_i|X_j = x_j) = (1 - q_j)^{-1} q_i^{x_i} \theta^{x_i x_j - 1}[(1 - q_i\theta^{x_j+1})(1 - q\theta^{x_i+1}) + \theta - 1] \quad (8.4)$$

with $i, j = 1, 2, i \neq j$. Further, the survival function of X_i, given $X_j = x_j$, has the form

$$S(x_i|X_j = x_j) = \frac{q_i^{x_i}\theta^{x_i x_j}}{(1 - q_j)}[(1 - q_j\theta^{x_i})], \quad x_i = 0, 1, 2, \ldots, \ i = 1, 2. \quad (8.5)$$

From the above, we readily find

$$E(X_i) = \frac{q_i}{1-q_i}, \; i = 1, 2, \quad V(X_i) = \frac{q_i}{(1-q_i)^2}$$

and

$$E(X_1 X_2) = \sum_{x_2=1}^{\infty} \frac{q_1 (q_1 \theta)^{x_2}}{(1 - q_1 \theta^{x_2})} = A(\theta; q_1, q_2), \text{ say.}$$

Thus, the coefficient of correlation is

$$r = \frac{A(\theta; q_1, q_2) - \frac{q_1 q_2}{(1-q_1)(1-q_2)}}{\left\{ \frac{q_1 q_2}{(1-q_1)^2 (1-q_2^2)} \right\}^{\frac{1}{2}}}$$
$$= [(1-q_1)(1-q_2) A(\theta; q_1, q_2) - q_1 q_2](q_1 q_2)^{-\frac{1}{2}}.$$

For fixed q_1 and q_2, $A(\theta, q_1, q_2)$, and hence r, is a non-decreasing function of θ. Hence, the maximum and minimum of r is attained at $\theta = 1$ and $\theta = 0$, respectively. Since $A(1, q_1, q_2) = \frac{q_1 q_2}{(1-q_1)(1-q_2)}$ and $A(0, q_1, q_2) = 0$, we see that the correlation coefficient ranges from 0 to $-(q_1 q_2)^{\frac{1}{2}}$. The value zero is attained when $\theta = 1$, in which case X_1 and X_2 are independent. Thus, for the bivariate geometric distribution-1 (BVGD-1), the correlation is always negative meaning negative dependence between X_1 and X_2.

The conditional survival function of X_i, given $X_j \geq x_j$, is

$$S(x_i | X_i \geq x_j) = (q_i \theta^{x_j})^{x_i}, \quad x_i = 0, 1, 2, \ldots,$$

with probability mass function

$$f(x_i | X_i \geq x_j) = (1 - q_i \theta^{x_j})(q_i \theta^{x_j})^{x_i}$$

which is geometric with parameter $q_i \theta^{x_j}$. The product moments can be expressed only in the form of an infinite series as

$$E(X_1^r X_2^s) = \sum_{x_1=0}^{\infty} x_1^r (Q(x_1, s) - Q(x_1 + 1, s))$$

where

$$Q(x_1, s) = q_1^{x_1} (1 - q_2 \theta^{x_1}) \sum_{x_2=0}^{\infty} x_2^s (q_2 \theta^{x_1})^{x_2}.$$

We now give some features of this distribution pertaining to reliability studies:

1. The scalar hazard rate is

$$a(x_1, x_2) = \theta^{-1}[(1 - q_1\theta^{x_2+1})(1 - q_2\theta^{x_1+1}) + \theta - 1].$$

Since $0 \leq \theta \leq 1$,

$$1 - q_1\theta^{x_1+1} \geq 1 - q_1\theta^{x_1}$$

for all x_1, and similarly

$$1 - q_2\theta^{x_2+1} \geq 1 - q_2\theta^{x_2}.$$

Accordingly, $a(x_1, x_2)$ is increasing in x_1 and x_2, and (X_1, X_2) is BIHR-9;

2. The vector valued hazard rate

$$(c_1(x_1, x_2), c_2(x_1, x_2)) = (1 - q_1\theta^{x_2}, 1 - q_2\theta^{x_1})$$

has components which are locally constant. In fact,

$$(c_1, c_2) = (a_1(x_2), a_2(x_1))$$

for all x_1, x_2 in N, characterizing the bivariate geometric law in (8.2) (Nair and Nair, 1990);

3. Conditional hazard rate $(b_1(x), b_2(x), b_{12}(x), b_1(x|x_2), b_2(x|x_1))$ is given in Example 6.5. On the other hand, the alternative conditional hazard rate is

$$(e_1(x_1, x_2), e_2(x_1, x_2)) = \left(1 - \frac{q_1\theta^{x_2}(1 - q_2\theta^{x_1+1})}{1 - q_2\theta^{x_1}}, 1 - \frac{q_2\theta^{x_1}(1 - q_1\theta^{x_2+1})}{1 - q_1\theta^{x_2}}\right).$$

Thus,

$$e_1(x_1 + 1, x_2) - e_1(x_1, x_2) = \frac{-q_1q_2\theta^{x_1+x_2}(1 - \theta)^2}{(1 - q_1\theta^{x_1})(1 - q_2\theta^{x_1+1})} \leq 0$$

for all x_2, and similarly

$$e_2(x_1, x_2 + 1) - e_2(x_1, x_2) \leq 0 \text{ for all } x_1.$$

Thus, (X_1, X_2) is BDHR-10 by Definition 7.9;

4. Since

$$\frac{S(x_1 + t_1, x_2 + t_2)}{S(x_1, x_2)} = q_1^{t_1}q_2^{t_2}\theta^{x_1t_2+x_2t_1+t_1t_2}$$

decreases in x_1, x_2 for all t_1, t_2, we see that (X_1, X_2) is BIHR-8, and hence is BIHR-7, BIHR-6 and BIHR-5;

5.

$$\frac{S(x_1+t, x_2)}{S(x_2+t, x_1)} = \theta^{t(x_2-x_1)}$$

which is decreasing in t for $x_1 \le x_2$, and so by Definition 7.11, (X_1, X_2) is BDHR-12 and hence BDHR-11;

6. The mean residual life is given by

$$(m_1(x_1,x_2)m_2(x_1,x_2)) = \left(\frac{1}{1-q_1\theta^{x_2+1}}, \frac{1}{1-q_2\theta^{x_1+1}}\right).$$

Nair and Nair (1988a) established that the mean residual life is of the form $(\alpha_1(x_2), \alpha_2(x_1))$, where $\alpha_1(\cdot)$ and $\alpha_2(\cdot)$ are non-negative functions if and only if (X_1, X_2) has BVGD-1. Accordingly the distribution is both BDMRL-2 and BIMRL-2;

7. The conditional mean residual life function is obtained from (8.5) to be

$$\begin{aligned} l_1(x_1,x_2) &= \frac{1}{S(x_1+1|X_2=x_2)} \sum_{t=x_1+1}^{\infty} S(t|X_2=x_2) \\ &= \frac{1}{(1-q_1\theta^{x_2})(1-q_2\theta^{x_1+1})} - \frac{q_2\theta^{x_1+1}}{(1-q_1\theta^{x_2+1})(1-q_2\theta^{x_1+1})}. \end{aligned}$$

Upon performing some algebra, we find that

$$\begin{aligned} &l_1(x_1+1,x_2) - l_1(x_1,x_2) \\ &\quad = \frac{q_2\theta^{x_1+1}(1-\theta)^2}{(1-q_1\theta^{x_2})(1-q_1\theta^{x_2+1})(1-q_2\theta^{x_1+1})(1-q_2\theta^{x_1+2})} \end{aligned}$$

which is non-negative for all x_1 and any x_2. Thus, (X_1, X_2) is BIMRL-5;

8. From Nair and Nair (1988b)

$$E[(X_i - x_i)^k | X_1 > x_1, X_2 > x_2] = a_k^{(i)}(x_{3-i}), \quad i = 1, 2, \tag{8.6}$$

for all positive integer k and all x_1, x_2 in N, and $a_k^{(i)}(\cdot)$ is independent of x_i if and only if (X_1, X_2) follows BVGD-1. In particular, for $k = 1$ and $k = 2$, we have

$$E[(X_i - x_i)^2 | X_1 > x_1, X_2 > x_2] = a_2^{(i)}(x_{3-i})$$

and

$$E[(X_i - x_i) | X_1 > x_1, X_2 > x_2] = a_1^{(i)}(x_{3-i}).$$

Thus, the bivariate variance residual life components are

$$\sigma_1^2(x_1,x_2) = a_2^{(1)}(x_2) - [a_1^{(1)}(x_2)]^2,$$

$$\sigma_2^2(x_1,x_2)=a_2^{(2)}(x_1)-[a_2^{(1)}(x_1)]^2,$$

so that σ_1^2 is a function of x_2 only and σ_2^2 is a function of x_1 only. This property is thus a characterization of the considered bivariate geometric law. The exact expressions for $\sigma_1^2(x_1,x_2)$ and $\sigma_2^2(x_1,x_2)$ can be seen in Example 6.10;

9. The product moment residual life has the expression

$$M(x_1,x_2)=(q_1^{x_1+1}\theta^{(x_1+1)(x_2+1)})^{-1}\sum_{t_1=x_1+1}^{\infty}\frac{(q_1\theta^{(x_2+1)})^t}{1-q_1\theta^{t_1}};$$

10. There exists an interesting characterization that relate the BVGD-1 and Gumbel's bivariate exponential distribution. If (X,Y) is a non-negative random vector admitting absolutely continuous distribution with survival function $\overline{F}(x,y)=P(X>x,Y>y)$ and (X_r,Y_s) is a discrete random vector in N^2 satisfying

$$P[X_r>u,Y_s>v]=P[X>ur,Y>vs],$$

then (X,Y) is bivariate exponential with

$$\overline{F}(x,y)=\exp[-\lambda_1 x-\lambda_2 y-\theta xy],\ x,y>0,\ \lambda_1,\lambda_2>0,\ 0\le\theta\le\lambda_1\lambda_2,$$

if and only if

$$P[X_r>u,Y_s>v]=q_1^u q_2^v\theta^{uv}\ \text{(Nair, 1992)};$$

11. A non-negative random vector (X_1,X_2), in the support of N^2, follow BVGD-1 if and only if

$$\sigma_i^2(x_1,x_2)=V(X_i|X_1>x_1,X_2>x_2)=V_i(x_j),\quad i,j=1,2,\ i\neq j,$$

with $V_i(-1)=q_i(1-q_i)^{-2}$;

12. Some useful identities can be established among various reliability functions that characterize the distributions as seen in the following theorem.

Theorem 8.1. *The relationships*

$$\begin{aligned}c_1(x_1+1,x_2+1)m_1(x_1+1,x_2)&=1,\\ c_2(x_1+1,x_2+1)m_2(x_1,x_2+1)&=1\end{aligned}\tag{8.7}$$

hold for all (x_1,x_2) if and only if (X_1,X_2) follows BVGD-1.

Proof. The expressions of $c_i(x_1, x_2)$ and $m_i(x_1, x_2)$ mentioned above validate (8.7).

Conversely, if (8.7) is true, from the identities

$$c_1(x_1+1, x_2+1) = 1 - \frac{m_1(x_1, x_2) - 1}{m_1(x_1+1, x_2)},$$
$$c_2(x_1+1, x_2+1) = 1 - \frac{m_2(x_1, x_2) - 1}{m_2(x_1, x_2+1)}$$

and (8.7), the conclusions

$$m_1(x_1+1, x_2) = m_1(x_1, x_2) \text{ and } m_2(x_1, x_2+1) = m_2(x_1, x_2)$$

follow. This means that m_1 is a function of x_2 only, and m_2 is a function of x_1 only and therefore (X_1, X_2) follows BVGD-1 from Property 2 above. □

Remark 8.1. The identities in (8.7) characterize the bivariate ageing property.

Theorem 8.2. *The variance residual life functions satisfy the identities*

$$\sigma_1^2(x_1+1, x_2) = m_1(x_1+1, x_2)(m_1(x_1, x_2) - 1),$$
$$\sigma_2^2(x_1, x_2+1) = m_2(x_1, x_2+1)(m_2(x_1, x_2) - 1)$$

if and only if (X_1, X_2) *follows BVGD-1.*

Proof. One can verify the 'if' part from the expressions of σ_1^2 and σ_2^2 in Example 6.10 and those of m_1 and m_2 in Property 6 above. Conversely, using (6.51),

$$\sigma_1^2(x_1+1, x_2) = \sigma_1^2(x_1, x_2)$$

which is equivalent to saying that σ_1^2 is a function of x_2 only, and similarly we find σ_2^2 to be a function of x_1 only. Using Property 8 above, the proof of the theorem is completed. □

When we look at the time-dependent measures of association, for the BVGD-1 from Example 6.16, the Clayton measure is less than unity and so X_1 and X_2 are negatively associated. The same conclusion can be arrived from the Anderson measures in (1.72) and (1.73) that

$$\phi(x_1, x_2) = \frac{1 - q_1}{1 - q_1\theta^{x_2}}$$

and

$$\psi(x_1, x_2) = \theta^{x_1 x_2},$$

which are both less than 1. Regarding the dependence concepts, we note that

$$P(X_1 > x_1, X_2 > x_2) - P(X_1 > x_1)P(X_2 > x_2) = q_1^{x_1+1} q_2^{x_2+1} (\theta^{(x_1+1)(x_2+1)} - 1) \le 0,$$

so that (X_1, X_2) is negatively quadrant dependent (NQD), which is the most often used concept. However, the stronger concept LCSD also holds for (X_1, X_2), but (X_1, X_2) is not RR_2. It is easy to verify that though (X_1, X_2) is neither RR_2 nor TP_2, the survival function is RR_2.

Judging from the properties so far obtained, the BVGD-1 is appropriate in situations where (i) the component lives exhibit univariate no-ageing property (the marginals being geometric) and (ii) the joint survival function satisfies

$$S(x_1 + t_1, x_2) = S(x_1, x_2) S(t_1, x_2)$$

and

$$S(x_1, x_2 + t_2) = S(x_1, x_2) S(x_1, t_2)$$

for all (x_1, x_2) and (t_1, t_2) in N^2. Further, (X_1, X_2) is negatively dependent.

8.2.2 Multivariate Geometric Distribution-2

The survival function of this distribution is as in (6.27) with

$$S(\mathbf{x}) = q_{i_1}^{x_{i_1}} \left(\frac{q_{i_1 i_2}}{q_{i_1}}\right)^{x_{i_2}} \cdots \left(\frac{q_{i_1 i_2 \ldots i_p}}{q_{i_1 \ldots i_{p-1}}}\right)^{x_{i_p}},$$

with

$$x_{i_1} \ge x_{i_2} \ge \ldots \ge x_{i_p}, \qquad x_{i_1}, \ldots, x_{i_p} \in N,$$
$$0 \le q_{i_1} \le \ldots \le q_{i_p} \le \ldots \le q_{i_1 \ldots i_p} < 1,$$
$$1 - \sum_{j=1}^{p} q_j - \sum_{j<k} q_{jk} - \ldots - (-1)^{p-1} q_{1,2\ldots p} \ge 0.$$

It could be seen that the p-variate BVGD-2 given above is a direct extension of the bivariate form

$$S(x_1, x_2) = \begin{cases} q^{x_2} q_1^{x_1 - x_2}, & x_1 \ge x_2, \\ q^{x_1} q_2^{x_2 - x_1}, & x_1 \le x_2, \end{cases} \tag{8.8}$$

$0 < q < q_j < 1$, $1 + q \ge q_1 + q_2$, $j = 1, 2$.

Some reliability properties of this model have been discussed in the preceding chapters. In this section, the distributional aspects and more features from the reliability context will be presented. There are different ways by which (8.8) can be derived. Let A_1 and A_2 be two events with joint probability

	A_1	A_1^c	
A_1	p_{11}	p_{01}	P_2
A_2^c	P_{10}	p_{00}	Q_2
	p_1	Q_1	1

In a sequence of independent trials, let X_i be the number of trials completed up to the first occurrence of the event A_i. Then, the probability generating function is

$$P(t_1,t_2)=t_1t_2\left\{p_{11}+\frac{p_{10}P_2t_2}{1-Q_2t_2}+\frac{p_{10}P_1t_1}{1-Q_1t_1}+p_{00}P(t_1,t_2)\right\},$$

and hence $S(x_1,x_2)$ will be of the form (8.8) with $p_{00}=q$, $q_1=p_{10}+p_{00}$ and $q_2=p_{01}+p_{00}$. The above model translates into a shock model as soon as we regard each trial as a shock, while the event A_i as a fatal shock to the ith component; see Hawkes (1972) for details. An alternative way of interpreting BVGD-2 is by the concept of no-ageing. Assume that (i) the probability of survival of a two-component system with components of ages x_1 and x_2 is the same as that of a new system, that is,

$$\begin{aligned}P[(X_1,X_2)\geq(x_1+t,x_2+t)|(X_1,X_2)\geq(t,t)]\\=P[(X_1,X_2)>(x_1,x_2)]\end{aligned}$$

for all x_1,x_2 and t, and (ii) each of the components taken individually does not age, that is,

$$P[X_i\geq x_i+t|X_i\geq t]=P[X_i\geq x_i],\quad i=1,2.$$

Then, the distribution is (8.8). The method of trivariate reduction can also be employed to construct (8.8) by making use of three independent geometric variates Y_0,Y_1 and Y_2 with parameters α_0,α_1 and α_2, respectively, and defining $X_1=\min(Y_0,Y_1)$ and $X_2=\min(Y_0,Y_2)$. Then, evidently,

$$S(x_1,x_2)=\begin{cases}\alpha_0^{x_1}\alpha_1^{x_1}\alpha_2^{x_2}, & x_1\geq x_2,\\ \alpha_0^{x_2}\alpha_1^{x_2}\alpha_2^{x_2}, & x_1\geq x_2.\end{cases}$$

Reparametrizing by $q=\alpha_0\alpha_1\alpha_2$, $q_1=\alpha_0\alpha_1$ and $q_2=\alpha_0\alpha_2$, the form in (8.8) is recovered.

The joint probability mass function is

$$f(x_1,x_2)=\begin{cases}(q_1-q)(1-q_1)q^{x_2}q_1^{x_1-x_2-1}, & x_1>x_2,\\ (q_2-q)(1-q_2)q^{x_1}q_2^{x_2-x_1-1}, & x_1<x_2,\\ (1+q-q_1-q_2)q^{x_1}, & x_1=x_2.\end{cases} \tag{8.9}$$

Marginal distributions are geometric with

$$f_i(x_i)(1-q_i)q_i^{x_i}$$

and the conditional distributions have probability mass functions

$$f_{ij}(x_i|x_j)=\begin{cases}q_i^{x_i-1}(\frac{q}{q_1q_2})^{x_j}\frac{(1-q_i)(q_i-q)}{1-q_j}, & x_i>x_j,\\ q^{x_i}(q_j)^{-(1+x_i)}(q_j-q), & x_i<x_j,\ i,j=1,2,\ i\neq j,\\ \frac{(1+q-q_1-q_2)}{1-q_j}(\frac{q}{q_j})^{x_i} & x_i=x_j,\ i,j=1,2.\end{cases}$$

The product moment is

$$E(X_1X_2)=\frac{1-q_1q_2}{(1-q_1)(1-q_2)(1-q)},$$

so that

$$\text{Cov}\,(X_1,X_2)=\frac{q-q_1q_2}{(1-q_1)(1-q_2)(1-q)}.$$

Since $V(X_i)=\frac{q_i}{(1-q_i)^2}$, we have the Pearson coefficient of correlation to be

$$r=\frac{q-q_1q_2}{(q_1q_2)^{\frac{1}{2}}(1-q)}. \tag{8.10}$$

It follows that when $q=q_1q_2$, X_1 and X_2 are uncorrelated and in this case, we deduce from (8.9) that the variables are independent. Notice that r is an increasing function of q and attain the maximum value unity when $q=(q_1q_2)^{1/2}$, the geometric mean between q_1 and q_2. This happens when $q=q_1=q_2$, in which case the two components survive or fail together. At the same time, r is minimum when $q=\max(0,q_1+q_2-1)$. When $q=0$, $r=-(q_1q_2)^{\frac{1}{2}}$ and when $q=q_1+q_2-1$,

$$r=\frac{(1-q_1)(1-q_2)}{(q_1q_2)^{\frac{1}{2}}(q_1+q_2-2)}$$

which is also negative. Thus,

$$\min\left((-q_1q_2)^{\frac{1}{2}}, \frac{(1-q_1)(1-q_2)}{(q_1q_2)^{\frac{1}{2}}(q_1+q_2-2)}\right) \le r \le 1$$

showing that r can be positive or negative.

The probability that one of the two components fail earlier than the other is given by

$$P(X_1 > X_2) = \frac{q_1 - q}{1-q}$$

and

$$P(X_1 < X_2) = \frac{q_2 - q}{1-q}.$$

Also, the simultaneous failure of the two components has probability

$$P(X_1 = X_2) = \frac{1+q-q_1-q_2}{1-q}.$$

The distributions of $Z = \min(X_1, X_2)$ and $W = \max(X_1, X_2)$ have the forms

$$P(Z \ge z) = q^z$$

and

$$P(W \ge w) = q_1^w + q_2^w - q^w.$$

Moreover, the event $Z = z$ is independent of the events $X_1 < X_2$, $X_2 < X_1$ and $X_1 = X_2$.

This can be seen from

$$\begin{aligned} P(Z = z|X_1 < X_2) &= P(X_1 = z, X_2 > x|X_1 < X_2), \\ &= \frac{1-q}{q_2-q} P[X_1 = z, X_2 > z]' \\ &= \frac{1-q}{q_2-q}[S(z, z+1) - S(z+1, z+1)], \\ &= (1-q)q^z = P(Z = z), \end{aligned}$$

and similar results on the other two events $X_2 < X_1$ and $X_1 = X_2$ can be shown. An interpretation of the result is that which of the components fail first does not depend on the relative magnitude of the component life lengths.

If (X_1, X_2) and (Y_1, Y_2) are independent random vectors having distribution (8.8) with $Z_1 = \min(X_1, X_2)$ and $Z_2 = \min(Y_1, Y_2)$, then

$$
\begin{aligned}
P(Z_1 \geq z_1, Z_2 \geq z_2) &= P(X_1 \geq z_1, X_2 \geq z_1, Y_1 > z_1, Y_2 > z_2) \\
&= P(X_1 \geq z_1, X_2 \geq z_1) P(Y_1 \geq z_2, Y_2 \geq z_2) \\
&= \begin{cases} (q^2)^{z_2} (q_1^2)^{z_1 - z_2}, & z_1 \geq z_2, \\ (q^2)^{z_1} (q_2^2)^{z_2 - z_1}, & z_1 \leq z_2, \end{cases}
\end{aligned}
$$

which satisfies $0 < q^2 < q_i^2 < 1$ and $1 + q^2 \geq q_1^2 + q_2^2$ since $q_1 q_2 > q$. Thus, for independent random vectors following (8.8), their component-wise minimum also follows the same distribution.

There are some characteristic properties of BVGD-2 based on distributional properties as detailed below.

Theorem 8.3. *(a) If (X_1, X_2) is a random vector, with support N^2, such that the conditional distribution $f_{ij}(x_i|x_j)$ is of the form given above, then X_i is geometric with parameter q_i if and only if X_j is geometric with parameter $q_j, i, j = 1, 2; i \neq j$;*

(b) *The vector (X_1, X_2) has bivariate geometric law in* (8.8) *with $q_1 q_2 < q$ if and only if there exist independent geometric variables Y_1, Y_2 and Y_3 such that $X_1 = \min(Y_1, Y_3)$ and $X_2 = \min(Y_2, Y_3)$;*

(c) *The vector (X_1, X_2) with geometric marginals has joint distribution in* (8.8) *if and only if*

  (i) *$Z = \min(X_1, X_2)$ is geometric, and*

  (ii) *Z is independent of $V = X_1 - X_2$;*

(d) *X_1 and X_2 are geometric variables and Y, in the support of non-negative integers, is independent of X_1 and X_2 such that*

$$
\begin{aligned}
&P[X_1 - y \geq x_1 + t_1, X_2 - y \geq X_2 + t] \\
&\qquad = P[X_1 - y \geq t, X_2 - y \geq t] P[X_1 \geq x_1, X_2 \geq x_2]
\end{aligned}
$$

for every x_1, x_2 and t in N, if and only (X_1, X_2) follows (8.8)*;*

(e) *Let $Z = \min(X_2, X_2)$. Then, $(X_1 - Z, X_2 - Z)$ is independent of Z if and only if (X_1, X_2) is bivariate geometric* (8.8)*;*

(f) *The distribution of a continuous non-negative vector (X, Y) follows bivariate exponential law of Marshall and Olkin (1967) with survival function*

$$P(X > x, Y > y) = \exp[-\lambda_1 x - \lambda_2 y - \lambda_{12} \max(x, y)]$$

if and only if, for every $r, s > 0$,

$$S_{r,s}(x_1, x_2) = P[X > x_1 r, Y > x_2 s],$$

where $S(\cdot, \cdot)$ is the survival function in (8.8).

Proofs of the above propositions can be seen in Nair and Asha (1994) and Asha (1995).

In addition to the above distributional properties, there are many reliability aspects of potential use in modelling. As observed in Section 6.2.6 the scalar failure rate is piecewise constant, given by

$$a(x_1,x_2)=\begin{cases}k_1, & x_1<x_2,\\ k_2, & x_1>x_2,\\ k_3, & x_1=x_2,\end{cases}$$

where $k_1=(q_1-q)(1-q_1)q_1^{-1}$, $k_2=(q_2-q)(1-q_2)q_2^{-1}$ and $k_3=(1+q-q_1-q_2)$. This is equivalent to the bivariate no-ageing property

$$P(X_1\geq x_1+t, X_2\geq x_2+t|X_1>x_1,X_2>x_2)=P[X_1>t,X_2>t]$$

for all x_1,x_2 and t in N. The equivalence is possible only when, with each property, the geometric marginals are assumed. The vector failure rate is also piece wise constant taking the form

$$(c_1(x_1,x_2),c_2(x_1,x_2))=\begin{cases}(k_1,k_2), & x_1>x_2,\\ (k_3,k_4), & x_1<x_2,\\ (k_1,k_4), & x_1=x_2,\end{cases} \tag{8.11}$$

with

$k_1=1-q_1, k_2=1-qq_1^{-1}, k_3=1-qq_2^{-1}, 1-q_2$ and $k_4=1-q_2$.

A similar property holds for the alternative conditional hazard rate with

$$(e_1(x_1,x_2),e_2(x_1,x_2))=\begin{cases}(c_{11},c_{12}), & x_1>x_2,\\ (c_{21},c_{22}), & x_1<x_2,\\ (c_{31},c_{32}), & x_1=x_2,\end{cases} \tag{8.12}$$

with

$$c_{11}=1-q_1,\ c_{12}=1-qq_1^{-1},\ c_{21}=1-qq_2^{-1},\ c_{22}=1-q_2,$$
$$c_{31}=\frac{1+q-q_1-q_2}{1-q_2}\quad\text{and}\quad c_{32}=\frac{1+q-q_1-q_2}{1-q_1}.$$

The total hazard rate is constant with

$$(d(x),d_1(x_1|x_2),d_2(x_2|x_1))=(1-q,1-q_1,1-q_2).$$

Regarding the monotonicity of the hazard rate, we see that (X_1,X_2) is both BIHR-2 and BDHR-2. The BVGD-2 does not exhibit any specific association as it is neither PQD nor NQD. If we look at the time-dependent measure, the picture is quite similar. For example, the Anderson measure in (1.73) is

$$\psi(x_1, x_2) = \begin{cases} (\frac{q}{q_1 q_2})^{x_2}, & x_1 \geq x_2, \\ (\frac{q}{q_1 q_2})^{x_1}, & x_1 \leq x_2, \end{cases}$$

whose magnitude depends on how large $q_1 q_2$ is compared to q and a general conclusion is not obvious. Basu and Dhar (1995) have considered a closely related bivariate geometric model with probability mass function

$$f(x_1, x_2) = \begin{cases} (p_1 p_{12})^{x_1 - 1} p_2^{x_2 - 1} q_2 (1 - p_1 p_{12}), & x_2 < x_1, \\ p_1^{x_1 - 1} (p_2 p_{12})^{x_2 - 1} q_1 (1 - p_2 p_{12}), & x_1 > x_2, \\ (p_1 p_2 p_{12})^{x_1 - 1} [1 - p_2 p_{12} - p_1 p_{12} + p_1 p_2 p_{12}], & x_1 = x_2, \end{cases}$$

$x_1, x_2 = 1, 2, 3, \ldots, 0 < p_1, p_2 < 1, 0 \leq p_{12} \leq 1$ and $q_i = 1 - p_i$, with similar properties.

Distributions with a similar structure, but under different assumptions, have been reported in Dhar and Balaji (2006) and Li and Dhar (2013). Let p_1 be the probability that component 1 survives unit time and p_3, the probability that component 1 survives for unit time, given that it has already survived for some time at the end of which component 2 has failed. Define p_2 and p_4 similarly with respect to component 2. Then, the lifetime (X_1, X_2) of components 1 and 2 is given by

$$P(X_1 = x_1, X_2 = x_2) = \begin{cases} \frac{q_1 q_4}{p_1 p_4} (\frac{p_1 p_2}{p_4})^{x_1} p_4^{x_2}, & x_1 < x_2, \\ \frac{q_1 q_2 q_{12}}{1 - p_1 p_2} p_{12}^{x_1 - 1}, & x_1 = x_2, \\ \frac{q_2 q_3}{p_2 p_3} (\frac{p_1 p_2}{p_3})^{x_2} p_3^{x_1}, & x_1 > x_2, \end{cases}$$

$x_1, x_2 = 1, 2, 3, \ldots, 0 < p_i < 1, \ p_1 p_2 < p_3, \ p_1 p_2 < p_4, \ p_i + q_i = 1, \ p_{12} + q_{12} = 1$. For such a model, the alternative conditional hazard rates are

$$e_1(x_1, x_2) = q_3, \ x_1 > x_2; \ e_1(x_1, x_1) = q_1, \ x_1 = 1, 2, \ldots;$$
$$e_2(x_1, x_2) = q_4, \ x_2 > x_1; \ e_2(x_2, x_2) = q_2 \ .x_2 = 1, 2, \ldots.$$

The distribution has no-ageing property when $p_{12} = p_1 p_2$. The marginal distributions are

$$f_1(x_1) = \frac{p_1 - p_1 p_2}{p_3 - p_1 p_2} (1 - p_3) p_3^{x_1 - 1} + \frac{p_3 - p_1}{p_3 - p_1 p_2} (1 - p_1 p_2)(p_1 p_2)^{x_1 - 1},$$

$x_1 = 1, 2, \ldots,$ and

$$f_2(x_2) = \frac{p_2 - p_1 p_2}{p_4 - p_1 p_2}(1 - p_4)p_4^{x_2 - 1} + \frac{p_4 - p_2}{p_4 - p_1 p_2}(1 - p_1 p_2)(p_1 p_2)^{x_2 - 1},$$

$x_2 = 1, 2, \ldots.$

8.2.3 Multivariate Geometric Distribution-3

A random vector $\mathbf{X} = (X_1, \ldots, X_p)$ is said to be distributed as multivariate geometric-3 (BVGD-3) if it has a probability mass function of the form

$$f(\mathbf{x}) = \frac{(\sum_{i=1}^{p} x_i)!}{\Pi_{i=1}^{p} x_i!} \Pi_{i=1}^{p} q_i^{x_i}(1 - q_1 - \cdots - q_p), \; \sum q_i < 1, \; x_i \in N. \quad (8.13)$$

This is a special case of the multivariate negative binomial law of Lukacs and Laha (1964). The bivariate version of (8.13) is

$$f(x_1, x_2) = \binom{x_1 + x_2}{x_1} q_1^{x_1} q_2^{x_2}(1 - q_1 - q_2). \quad (8.14)$$

Nagaraja (1983) has characterized (8.14) by the property

$$f(x_1, x_2) = c_1 f(x_1 - 1, x_2) + c_2 f(x_1, x_2 - 1),$$

where $c_1, c_2 > 0$ and $c_1 + c_2 < 1$. The marginal distributions are geometric with

$$f_1(x_1) = \frac{1 - q_2 - q_1}{1 - q_2}\left(\frac{q_1}{1 - q_2}\right)^{x_1}$$

and

$$f_2(x_2) = \frac{1 - q_1 - q_2}{1 - q_1}\left(\frac{q_2}{1 - q_1}\right)^{x_2}.$$

Accordingly

$$E(X_i) = \frac{q_i}{1 - q_1 - q_2}, \quad i = 1, 2,$$

with variance

$$V(X_i) = \frac{q_i(1 - q_j)}{(1 - q_1 - q_2)^2}, \quad i, j = 1, \; i \neq j.$$

The conditional distributions are

$$f_{12}(x_1|x_2) = \binom{x_1 + x_2}{x_1} q_1^{x_1}(1 - q_1)^{x_2 + 1} \quad (8.15)$$

and

$$f_{21}(x_2|x_1) = \binom{x_1+x_2}{x_1} q_2^{x_2}(1-q_2)^{x_1+1}. \tag{8.16}$$

Also,

$$\begin{aligned}P(X_1=x_1, X_2 \geq x_2) &= \sum_{t=x_2}^{\infty} \binom{x_1+t}{x_1} q_1^{x_1} q_2^t (1-q_1-q_2)\\ &= q_1^{x_1}(1-q_1-q_2)\binom{x_1+x_2}{x_1}\left[1+\frac{x_1+x_2+1}{x_2+1}q_2\right.\\ &\quad \left.+\frac{(x_1+x_2+1)(x_1+x_2+2)}{(x_2+1)x_2+2)}q_2^2+\dots\right]\\ &= q_1^{x_1}(1-q_1-q_2)\binom{x_1+x_2}{x_1}{}_2F_1(x_1+x_2+1, x_2+1, q_2),\end{aligned} \tag{8.17}$$

where

$${}_2F_1(a,b,c,z) = \sum_{n=0}^{\infty} \frac{(a)_n(b)_n}{(c)_n} z^n$$

and $(a)_n = a(a+1)\dots(a+n-1)$, $(a)_0 = 1$, is the Gauss hypergeometric function.

Similarly,

$$\begin{aligned}&P(X_1 \geq x_1, X_2 = x_2)\\ &\quad = q_2^{x_2}(1-q_1-q_2)\binom{x_1+x_2}{x_1}{}_2F_1(x_1+x_2+1, 1, x_1+1, q_1).\end{aligned} \tag{8.18}$$

From (8.15)–(8.18), the conditional hazard rate has its components as

$$e_1(x_1,x_2) = \frac{P(X_1=x_1, X_2=x_2)}{P(X_1 \geq x_1, X_2 = x_2)} = \frac{q_1^{x_1}}{{}_2F_1(x_1+x_2+1, 1, x_1+1, q_1)}$$

and

$$e_2(x_1,x_2) = \frac{q_2 x_2}{{}_2F_1(x_1+x_2+1, 1, x_2+1, q_2)}.$$

It is not difficult to see that $e_1(x_1,x_2)(e_2(x_1,x_2))$ is decreasing $x_1(x_2)$ for every $x_2(x_1)$, and as so (X_1, X_2) BDFR-10.

Since the survival function involves summation of hypergeometric functions, the other versions of hazard rates, mean residual life, etc., are not in closed form to analyse them algebraically. The sign of

$$f(x_1,y_1)f(x_2,y_2) - f(x_1,y_2)f(x_2,y_1)$$

depends on

$$(x_1 + y_1)!\,(x_2 + y_2)! - (x_1 + y_2)!\,(x_2 + y_1)!$$

which may be positive (e.g., $x_1 = y_1 = 1, x_2 = y_2 = 2$) or negative (e.g., $x_1 = 3$, $x_2 = 1,\ y_1 = 1,\ y_2 = 6$), and so $f(x_1, x_2)$ is neither TP_2 nor RR-2. There are regions of the sample space for which (X_1, X_2) is positively or negatively associated. Thus, the BVGD-3 may be considered as a model in which the individual components have no-ageing, but jointly their conditional hazard rate function is decreasing.

We now present three new bivariate geometric models arising as discrete analogues of the family of bivariate exponential distributions introduced by Nair and Sankaran (2014c).

8.2.4 Multivariate Geometric Distribution-4

The p-dimensional random vector $\mathbf{X}$ with survival function

$$S(\mathbf{x}) = q^{(x_1^\alpha + \ldots + x_p^\alpha)^{\frac{1}{\alpha}}}, \quad x_i = 0, 1, 2, \ldots, \ \alpha > 0,\ 0 < q < 1,$$

will be referred to as multivariate geometric distribution-4. Its bivariate version is

$$S(x_1, x_2) = q^{(x_1^\alpha + x_2^\alpha)^{\frac{1}{\alpha}}}, \tag{8.19}$$

which is a two-parametric bivariate geometric distribution (BVGD-4) with geometric marginals. The probability mass function is

$$f(x_1, x_2) = q^{(x_1^\alpha + x_2^\alpha)^{\frac{1}{\alpha}}} - q^{((x_1+1)^\alpha + x_2^\alpha)^{\frac{1}{\alpha}}} - q^{(x_1^\alpha + (x_2+1)^\alpha)^{\frac{1}{\alpha}}} + q^{((x_1+1)^\alpha + (x_2+1)^\alpha)^{\frac{1}{\alpha}}}.$$

Note that

$$S(x_1, 0) = q^{x_1} \quad \text{and} \quad S(0, x_2) = q^{x_2}$$

and that (X_1, X_2) is exchangeable. The means and variances of X_1 and X_2 are the same, being $\frac{q}{p}$ and $\frac{q}{p^2}$. Model in (8.19) is also a discrete version of the Gumbel's (1960) bivariate exponential distribution of Type III.

The different types of hazard rates have the following expressions:

(a) Scalar hazard rate:

$$a(x_1, x_2) = 1 - \frac{q^{((x_1+1)^\alpha + x_2^\alpha)^{\frac{1}{\alpha}}} + q^{(x_1^\alpha + (x_2+1)^\alpha)^{\frac{1}{\alpha}}} - q^{((x_1+1)^\alpha + (x_2+1)^\alpha)^{\frac{1}{\alpha}}}}{q^{(x_1^\alpha + x_2^\alpha)^{\frac{1}{\alpha}}}};$$

(b) Vector hazard rate:

$$c_1(x_1,x_2)=1-\frac{q^{((x_1+1)^\alpha+x_2^\alpha)^{\frac{1}{\alpha}}}}{q^{(x_1^\alpha+x_2^\alpha)^{\frac{1}{\alpha}}}},$$

$$c_2(x_1,x_2)=1-\frac{q^{((x_1^\alpha+(x_2+1)^\alpha)^{\frac{1}{\alpha}}}}{q^{(x_1^\alpha+x_2^\alpha)^{\frac{1}{\alpha}}}};$$

(c) Conditional hazard rate:

$$e_1(x_1,x_2)=1-\frac{q^{((x_1+1)^\alpha+x_2^\alpha)^{\frac{1}{\alpha}}}-q^{((x_1+1)^\alpha+(x_2+1)^\alpha)^{\frac{1}{\alpha}}}}{q^{(x_1^\alpha+x_2^\alpha)^{\frac{1}{\alpha}}}-q^{(x_1^\alpha+(x_2+1)^\alpha)^{\frac{1}{\alpha}}}},$$

$$e_2(x_1,x_2)=1-\frac{q^{(x_1^\alpha+(x_2+1)^\alpha)^{\frac{1}{\alpha}}}-q^{((x_1+1)^\alpha+(x_2+1)^\alpha)^{\frac{1}{\alpha}}}}{q^{(x_1^\alpha+x_2^\alpha)^{\frac{1}{\alpha}}}-q^{((x_1+1)^\alpha+x_2^\alpha)^{\frac{1}{\alpha}}}}.$$

We see that (X_1,X_2) is BIHR-3 for $\alpha>1$, BDHR-3 for $\alpha<1$ and for $\alpha=1$, X_1 and X_2 are independent geometric satisfying the strong no-ageing property

$$\begin{aligned}S(x_1+1,x_2)\geq S(x_1,x_2+1)&\Rightarrow q^{[(x_1+1)^\alpha+x_2^\alpha]^{\frac{1}{\alpha}}}\geq q^{[x_1^\alpha+(x_2+1)^\alpha]^{\frac{1}{\alpha}}}\\&\Rightarrow (x_1+1)^\alpha+x_2^\alpha\leq x_1^\alpha+(x_2+1)^\alpha.\end{aligned}$$

This means that

$$(x_1-x_2)[S(x_1+1,x_2)-S(x_1,x_2+1)]\geq 0$$

for all x_1 and x_2 and accordingly (X_1,X_2) is BIHR-11.

Regarding the dependent measures, it is easier to consider the Anderson et al. (1992) measure

$$\psi(x_1,x_2)=\frac{S(x_1,x_2)}{S_1(x_1)S_2(x_2)}=\frac{q^{(x_1^\alpha+x_2^\alpha)^{\frac{1}{\alpha}}}}{q^{x_1+x_2}}$$

which is >1 for $\alpha>1$, <1 for $\alpha<1$, and X_1 and X_2 are independent when $\alpha=1$. Thus, the variables have positive (negative) association according as $>(<)1$. Further, X_1 and X_2 are PQD for $\alpha>1$ and NQD for $\alpha<1$ and so positively (negatively) correlated for $\alpha>1$ $(\alpha<1)$.

The bivariate mean, variance and covariance of residual lives have no closed-form expressions. A possible physical situation in which the model is appropriate is when a two-component system has identical components with no-ageing property, the Bayesian or vector hazard rate has behaviour as mentioned above.

Another geometric model with similar structure is the bivariate geometric-5 (BVGD-5) with

$$S(x_1, x_2) = q^{(x_1^c + x_2^c + x_1^c x_2^c)^{\frac{1}{c}}}, \quad c > 0. \tag{8.20}$$

It is also exchangeable with identical geometric marginals. The analysis of this distribution, being very similar to BVGD-4, is not discussed further.

8.2.5 Bivariate Geometric Distribution-6

This is a three-parameter distribution specified by the survival function

$$S(x_1, x_2) = \frac{2q_1^{x_1} q_2^{x_2}}{2 - \theta(1 - q_1^{x_1})(1 - q_2^{x_2})}, \quad x_1, x_2 = 0, 1, 2, \ldots,$$

$$0 < q_1, q_2 < 1, \; 0 \geq \theta \geq \frac{2}{1 - q_1 - q_2}.$$

The probability mass function is

$$\begin{aligned} f(x_1, x_2) = 2q_1^{x_1} q_2^{x_2} \big[&(2 - \theta(1 - q_1^{x_1})(1 - q_1^{x_2})^{-1}) - 2q_1(2 - \theta(1 - q_1^{x_1+1}) \\ &(1 - q_2^{x_2})^{-1}) - 2q_2(2 - \theta(1 - q_1^{x_1})(1 - q_2^{x_2+1})^{-1}) \\ &- 2q_1 q_2(2 - \theta(1 - q_1^{x_1+1})(1 - q_2^{x_2+1})^{-1}) \big]. \end{aligned}$$

As is clearly seen, the marginals are geometric with parameters q_1 and q_2. When $\theta = 0$, X_1 and X_2 are independent.

The vector valued hazard rate has components

$$\begin{aligned} C_1(x_1, x_2) &= 1 - \frac{S(x_1 + 1, x_2)}{S(x_1, x_2)} \\ &= 1 - \frac{q_1[2 - \theta(1 - q_1^{x_1})(1 - q_2^{x_2})]}{2 - \theta(1 - q_1^{x_1+1})(1 - q_2^{x_2})} = \frac{(1 - q_1)[2 - \theta(1 - q_2^{x_2})]}{2 - \theta(1 - q_1^{x_1+1})(1 - q_2^{x_2})} \end{aligned}$$

and

$$C_2(x_1, x_2) = \frac{(1 - q_2)[2 - \theta(1 - q_1^{x_1})]}{2 - \theta(1 - q_2^{x_2+1})(1 - q_1^{x_1})}.$$

It is easy to see that (X_1, X_2) is BIHR-2. Further, it is also BIHR-8 implying BDHR-5, 6 and 7. The scalar hazard rate is

$$\begin{aligned} a(x_1, x_2) = 1 - \big\{ &[2 - \theta(1 - q_1^{x_1+1})(1 - q_2^{x_2})]^{-1} q_1 \\ &- [2 - \theta(1 - q_1^{x_1})(1 - q_2^{x_2+1})^{-1}] q_2 \\ &+ 2q_1 q_2 [2 - \theta(1 - q_1^{x_1+1})(1 - q_2^{x_2+1})^{-1}] \big\} [2 - \theta(1 - q_1^{x_1})(1 - q_2^{x_2})]. \end{aligned}$$

The time-dependent association measure

$$\psi(x_1, x_2) = \frac{2}{2 - \theta(1 - q_1^{x_1})(1 - q_2^{x_2})} > 1$$

so that the association is positive. Also, X_1 and X_2 are PQD.

8.2.6 Bivariate Geometric Distribution-7

The random vector (X_1, X_2) has bivariate geometric distribution-7 (BVGD-7) if it has a survival function of the form

$$S(x_1, x_2) = \begin{cases} \alpha q^{x_1+x_2} + (1-\alpha)q^{x_1}, & x_1 \geq x_2, \\ \alpha q^{x_1+x_2} + (1-\alpha)q^{x_2}, & x_1 \leq x_2,\ 0 \leq \alpha \leq 1. \end{cases} \tag{8.21}$$

Thus, (8.21) represents a two-parameter distribution with joint marginal probability mass function

$$f(x_1, x_2) = \begin{cases} \alpha(1-q)^2 q^{x_1+x_2}, & x_1 \geq x_2 + 1, \\ \alpha(1-q)^2 q^{x_1+x_2}, & x_2 \geq x_1 + 1, \\ \alpha(1-q)^2 q^{2x_1} + (1-\alpha)(1-q)q^{x_1}, & x_1 = x_2, \end{cases} \tag{8.22}$$

and marginals

$$f_i(x_i) = (1-q)q^{x_i},\ x_i = 0, 1, 2, \ldots,\ i = 1, 2.$$

Thus, the marginals are identical, but (X_1, X_2) is not exchangeable. When $\alpha = 0$ we have

$$S(x_1, x_2) = \begin{cases} q^{x_1}, & x_1 \geq x_2, \\ q^{x_2}, & x_1 \leq x_2, \end{cases} \tag{8.23}$$

and $\alpha = 1$ gives

$$S(x_1, x_2) = q^{x_1+x_2},$$

the case when X_1 and X_2 are independent. Regarding the distributional features, we have

$$E(X_i) = \frac{q}{1-q}, \qquad V(X_i) = \frac{q}{(1-q)^2}.$$

The hazard rate functions are

$$a(x_1,x_2)=\begin{cases}\frac{\alpha(1-q)^2q^{x_2}}{\alpha q^{x_2}+(1-\alpha)}, & x_1>x_2,\\ 1-q, & x_1=x_2,\\ \frac{\alpha(1-q)^2q^{x_1}}{\alpha q^{x_1}+(1-\alpha)}, & x_1>x_2,\end{cases}$$

$$c_1(x_1,x_2)=\begin{cases}1-q, & x_1>x_2 \text{ or } x_2>x_1,\\ 1-[\frac{\alpha q^{x_1+1}+(1-\alpha)]}{\alpha q^{x_1}+(1-\alpha)}, & x_1=x_2,\end{cases}$$

and

$$c_2(x_1,x_2)=\begin{cases}(1-q), & x_1>x_2 \text{ or } x_2>x_1,\\ 1-\frac{\alpha q^{x_1+1}+1-\alpha}{\alpha q^{x_1}+(1-\alpha)}, & x_1=x_2.\end{cases}$$

Since $a(x_1,x_2)$ is increasing in x_1,x_2, we conclude that (X_1,X_2) is BIHR-9. Further, $\mathbf{c}(x_1,x_2)$ is BIFR-3 and hence BIFR-4. The mean residual life function is

$$m_1(x_1,x_2)=\begin{cases}\frac{1}{1-q}, & x_1>x_2 \text{ or } x_2>x_1,\\ \frac{(1-q^2)^{-1}\alpha q^{x_1+1}+(1-q)^{-1}(1-\alpha)}{\alpha q^{x_1+1}+(1-\alpha),} & x_2=x_1,\end{cases}$$

and $m_2(x_1,x_2)=m_1(x_1,x_2)$. It can be seen that $[X_1,X_2)$ is DMRL-3. The dependence measure

$$\phi(x_1,x_2)=\begin{cases}\alpha+\frac{1-\alpha}{q^{x_2}}, & x_1\geq x_2,\\ \alpha+\frac{1-\alpha}{q^{x_1}}, & x_1\leq x_2,\end{cases}$$

is greater than unity for all (x_1,x_2). Hence, there is positive dependence between X_1 and X_2 which also implies that the correlation is positive. Further, X_1 and X_2 are PQD. There are two other bivariate geometric distributions discussed in the literature due to Paulson and Uppuluri (1972) and Longford (1989). The former is expressed in terms of characteristic functions and the latter is derived from Markov chains and their reliability aspects are difficult to enumerate.

8.3 SCHUR-CONSTANT FAMILY

Multivariate Schur-constant family of distributions play the same role in Bayesian reliability as multivariate geometric distributions in the conventional reliability theory. As mentioned in Definition 7.5, Schur-constant laws satisfy the Bayesian definition of no-ageing.

Let $\mathbf{X}=(X_1,\dots,X_p)$ be a random vector, with X_i taking values in N, and $S(x)$ be a univariate survival function of a random variable X with associated probability mass function $f(x)$. A real-valued function $g(\cdot)$, defined on N, is said to be n-monotonic if it satisfies the condition

$$(-1)^j \Delta^j f(x) \geq 0, \quad j = 0, 1, \ldots, n,$$

where as usual Δ is the forward difference operator, and Δ^j is its jth iteration.

Definition 8.1. A function $R(\mathbf{x})$, taking values in N^p, is the Schur-constant survival function of $\mathbf{X}$ if and only if

$$R(\mathbf{x}) = P[\mathbf{X} \geq \mathbf{x}] = S(x_1 + x_2 + \ldots + x_p)$$

where S is p-monotone on N.

Several distributional properties of the family have been studied by Castaner et al. (2015), and these include the following:

(i) A function $d(\mathbf{x})$ is the Schur-constant probability mass function of $\mathbf{x}$ if and only if it sums up to 1 and the mass function $f(x)$ is $(p-1)$-monotone on N;

(ii) For $1 \leq j \leq p$,

$$P(x_1 \leq X_1 < x_1 + h_1, \ldots, x_j \leq X_j < p_{ij}) = (-1)^j \Delta_{1,h_1} \ldots \Delta_{j,h_j} S(x_1 + \ldots + x_j),$$

where

$$\Delta_{i,h_i} g(x_1, \ldots, x_p) = g(x_1, \ldots, x_i + h_i, \ldots, x_p) - g(x_1, \ldots, x_i, \ldots, x_p);$$

(iii) $P(X_1 = x_1, \ldots, X_j = x_j) = (-1)^j \Delta^j S(x_1 + \ldots + x_j)$;

(iv) $P(X_1 = x_1, \ldots, X_{j-1} = x_{j-1}, X_j \geq x_j) = (-1)^{j-1} \Delta^{j-1} S(x_1 + \ldots + x_j)$;

(v) $T_j = X_1 + X_2 + \ldots + X_j$, $1 \leq j \leq n$,

$$P(T_j = t_j) = (-1)^j \Delta^j S(t_j) \binom{t_j + j - 1}{j - 1},$$

$$P(T_1 = t_1, \ldots, T_j = t_j) = (-1)^j \Delta^j S(t_j),$$

and

$$P(T_1 = t_1, \ldots, T_{j-1} = t_{j-1} | T_j = t_j) = \frac{1}{\binom{t_j + j - 1}{j - 1}};$$

(vi) For $1 \leq j \leq n$ and $x_1, x_2, \ldots, x_j, z_j \geq 0$ with $x_1 + \ldots + x_j \leq z$,

$$P(X_1 \geq x_1, \ldots X_j \geq x_j | T_n = z) = \binom{z - (x_1 + \ldots + x_j) + n - 1}{n - 1} \Big/ \binom{z + n - 1}{n - 1}$$

and

$$P(X_1 = x_1, \ldots X_j = x_j | T_n = z) = \binom{z - (x_1 + \ldots + x_j) + n - j - 1}{n - j - 1} \Big/ \binom{z + n - 1}{n - 1};$$

(vii)

$$S(x_1 + \ldots + x_n) = E\left[\binom{Z - (x_1 + \ldots + x_n) + n - 1}{n - 1} \Big/ \binom{Z + n - 1}{n - 1}\right]$$

where Z is distributed as T_n.

Specializing to the bivariate case, the joint survival function of a bivariate-Schur constant model (BSM) is of the form

$$R(x_1, x_2) = S(x_1 + x_2), \quad x_1, x_2 = 0, 1, 2 \ldots, \tag{8.24}$$

where $S(\cdot)$ is a convex survival function. The convexity of S implies that

$$S(x+2) - 2S(x+1) + S(x) \geq 0. \tag{8.25}$$

From (8.3), the probability mass function of BSM is

$$d(x_1, x_2) = f(x_1 + x_2) - f(x_1 + x_2 + 1). \tag{8.26}$$

Since $d(\cdot)$ must be non-negative, f has to be a decreasing function. Thus, in order to construct BSM, one should have a decreasing probability mass function or one has to restrict the parameter space such that the probability mass function is decreasing in that region. Some bivariate Schur-constant models are listed in Table 8.1.

In view of the simple forms for the survival and probability mass function, several interesting properties of BSM can be derived as described next.

8.3.1 Distributional Features

1. The marginal survival functions are

$$R(x_1, 0) = S(x_1) \text{ and } R(0, x_2) = S(x_2),$$

   and consequently the corresponding probability mass functions are $f(x_1)$ and $f(x_2)$, respectively;
2. The random variables X_1 and X_2, with joint distribution (8.24), are independent if and only if X_1 and X_2 are geometric random variables;

TABLE 8.1 Bivariate Schur-constant distributions

Baseline Distribution	Survival function $S(x)$	Schur-constant model $R(x_1, x_2)$
geometric	q^x	$q^{x_1+x_2}, 0<q<1$
Warning	$\frac{(m)_x}{(m+n)_x}$	$\frac{(m)_{x_1+x_2}}{(m+n)_{x_1+x_2}}$ $m,n>0$
Power	$1-(\frac{x}{\alpha})^\beta, \beta<0$	$1-(\frac{x_1+x_2}{\alpha})^\beta$, $k=1,2\ldots 0<\beta<1$
logarithmic	$\frac{-\theta^{x+1}}{(x+1)\log(1-\theta)}, 0<\theta<1,$	$\frac{-\theta^{x_1+x_2}}{(1+x_1+x_2)\log(1-\theta)}$ $0<\theta<1$
Pareto	$\frac{1}{(1+x)^{\theta+1}\xi(1+\theta)}, \theta>0$	$\frac{1}{(1+x_1+x_2)^{\theta+1}\xi(1+\theta)}, \theta>0$
Weibull	$q^{x^\beta}, \beta>0$	$q^{(x_1+x_2)^\beta}, 0<\beta\le 1$
Gompertz	$q^{-\theta(1-e^x)}\theta>0$	$q^{-\theta(1-e^{x_1+x_2})}, \theta>1$

3. Castaner et al. (2015) have calculated the means, variances and coefficient of correlation of BSM as

$$E(X_i)=\mu=\mu_{T_n}/n, \quad i=1,2,$$

$$V(X_i)=\sigma^2=\frac{2\sigma^2_{T_n}}{n(n+1)}+\frac{\mu^2_{T_n}(n-1)}{n^2(n+1)}+\frac{\mu_{T_n}(n-1)}{n(n+1)},$$

and

$$r_{X_1,X_2}=\frac{\sigma^2_X-\mu^2_X-\mu_X}{2\sigma^2_X} \tag{8.27}$$

in terms of μ and σ^2, and

$$r=\frac{n\sigma^2_{T_n}-\mu^2_{T_n}-n\mu_{T_n}}{2n\sigma^2_{T_n}+(n-1)\mu^2_{T_n}+n(n-1)\mu_{T_n}},$$

where $T_n=X_1+\ldots+X_n$. Alternative expressions for the above quantities can be obtained in terms of the baseline survival functions as

$$\mu=\sum_{x=0}^{\infty}S(x+1)=\mu_X,$$

$$\sigma^2=2\sum_{x=0}^{\infty}xS(x+1)-\mu^2+\mu=\sigma^2_X,$$

and r is computed from (8.27). Thus the correlation between X_1 and X_2 can be positive or negative. In fact, $-1<r<\frac{1}{2}$. Maximum correlation the BSM can attain is $\frac{1}{2}$. When the mean is fixed, r is an increasing function of σ_X^2. In terms of coefficient of variation $c = \frac{\sigma}{\mu}$, we can write

$$r = \frac{c^2 - c - 1}{2c^2};$$

4. The distribution of the total lifetime $T = X_1 + X_2$ is given by

$$f_T(t) = P(T = t) = (t+1)[f(t) - f(t+1)],$$

or

$$S_T(t) = P(T \geq t) = S(t) - t\Delta S(t);$$

5. Conditional distribution of $(X_1|X_1 + X_2)$ is uniform over $[0, t_2]$;
6. The joint distribution of $M = \max(X_1, X_2)$ and $Z = \min(X_1, X_2)$ has survival function

$$S_{M,Z}(m, z) = 2S(m+z) - S(2m).$$

Thus, the probability that at least one of the components of the device with lifetime represented by (X_1, X_2) survives time t is $2S(t) - S(2t)$, and both survives beyond t is $S(2t)$.

8.3.2 Reliability Aspects

The structure of BSM connecting a bivariate distribution through a univariate survival function provides some interesting reliability properties. We obtain the bivariate reliability functions in terms of the corresponding functions of the lifetime X.

First, we consider the scalar hazard rate of (X_1, X_2) given by

$$\begin{aligned} a(x_1, x_2) = \frac{d(x_1, x_2)}{R(x_1, x_2)} &= \frac{f(x_1 + x_2) - f(x_1 + x_2 + 1)}{S(x_1 + x_2)} \\ &= h(x_1 + x_2) - h(x_1 + x_2 + 1)(1 - h(x_1, x_2)). \end{aligned} \tag{8.28}$$

It is easy to see that when $h(\cdot)$, the hazard rate of X, is increasing, $a(x_1, x_2)$ is increasing. Thus,

X is IHR$_1$ $\Leftrightarrow$ X is IHR $\Leftrightarrow$ (X_1, X_2) is BIHR-9,

X is DHR$_1$ $\Leftrightarrow$ X is DHR $\Leftrightarrow$ (X_1, X_2) is DIHR-9.

Also when X is geometric with $h(x) = 1 - q$, $a(x_1, x_2) = (1 - q)^2$. The vector hazard rate has the form

$$(c_1(x_1, x_2), c_2(x_1, x_2)) = (h(x_1 + x_2), h(x_1 + x_2)) \tag{8.29}$$

with equal components. This gives the following:

X is $\text{IHR}_1 \Leftrightarrow X$ is IHR $\Leftrightarrow (X_1, X_2)$ is BIHR-2,

X is $\text{DHR}_1 \Leftrightarrow X$ is DHR $\Leftrightarrow (X_1, X_2)$ is DIHR-2.

Let (X_1, X_2) and (Y_1, Y_2) represent two random vectors with Schur-constant distributions associated with two random variables X and Y. Then, we have the following result.

Theorem 8.4.

$$X \leq_{hr} Y \Leftrightarrow (X_1, X_2) \leq_{whr} (Y_1, Y_2).$$

Proof. We have

$$\begin{aligned} X \leq_{\text{hr}} Y &\Leftrightarrow h_X(x) \geq h_Y(x) \Leftrightarrow h_X(x_1 + x_2) \geq h_Y(x_1 + x_2) \\ &\Leftrightarrow c_{i,(X_1,X_2)}(x_1, x_2) \geq c_{i,(Y_1,Y_2)}(x_1, x_2) \\ &\Leftrightarrow (X_1, X_2) \leq_{\text{whr}} (Y_1, Y_2). \end{aligned}$$ □

We have the bivariate mean residual life function of the BSM with components derived, from (6.31) and (6.32), in the form

$$m_1(x_1, x_2) = \frac{1}{S(x_1 + x_2 + 1)} \sum_{t=x_1+1}^{\infty} S(t + x_2 + 1) = m(x_1 + x_2 + 1)$$

and

$$m_2(x_1, x_2) = m(x_1 + x_2 + 1), \tag{8.30}$$

where $m(\cdot)$ is the mean residual life function of X. Thus,

$$X \text{ is DMRL } \Leftrightarrow (X_1, X_2) \text{ is BDMRL -2}$$

and

$$X \leq_{\text{hr}} Y \Rightarrow X \leq_{\text{mrl}} Y \Leftrightarrow (X_1, X_2) \leq_{\text{mrl}} (Y_1, Y_2).$$

The conditional means and variances can be expressed by means of reliability functions.

Theorem 8.5.

$$\begin{aligned} E(X_i|X_j = x_j) &= \frac{1 - h(x_2)}{h(x_2)}, \quad i, j = 1, 2, \, i \neq j, \\ V(X_i|X_j = x_j) &= \frac{1 - h(x_2)}{h^2(x_2)} \, (2h(x_2)m(x_2) - 1). \end{aligned}$$

Proof. We have

$$\begin{aligned} E(X_1|X_2=x_2) &= \sum_{x_1=0}^{\infty} x_1[f(x_1+x_2)-f(x_1+x_2+1)]/f(x_2) \\ &= \frac{S(x_2+1)}{f(x_2)} = \frac{S(x_2+1)}{S(x_2)}\frac{S(x_2)}{f(x_2)} \\ &= \frac{1-h(x_2)}{h(x_2)}. \end{aligned} \tag{8.31}$$

Also

$$\begin{aligned} E(X_1^2|X_2=x_2) &= \sum_{x_2=0}^{\infty} x_1^2[f(x_1+x_2)-f(x_1+x_2+1)]/f(x_2) \\ &= \frac{S(x_2+1)+2\sum_{t=x_2+2}^{\infty}S(t)}{f(x_2)} \\ &= \frac{S(x_2+1)}{f(x_2)} + \frac{2S(x_2+2)m(x_2+1)}{f(x_2)} \\ &= \frac{1-h(x_2)}{h(x_2)} + \frac{2(1-h(x_2))(1-h(x_2+1)m(x_2+1)}{h(x_2)} \\ &= \frac{1-h(x_2)}{h(x_2)}[2m(x_2)-1], \end{aligned}$$

on using the relationship in (2.21) between $m(x_2)$ and $h(x_2)$. So,

$$\begin{aligned} V(X_1|X_2=x_2) &= \frac{1-h(x_2)}{h(x_2)}[2m(x_2)-1] - \left[\frac{1-h(x_2)}{h(x_2)}\right]^2 \\ &= \frac{1-h(x_2)}{h^2(x_2)}(2m(x_2)h(x_2)-1). \end{aligned} \tag{8.32}$$

Similar calculations from (6.45) and (2.42) yield

$$\sigma_1^2(x_1,x_2) = \sigma^2(x_1+x_2) = \sigma_2^2(x_1,x_2),$$

where $\sigma^2(\cdot)$ stands for the variance residual life of X. □

8.3.3 Dependence and Ageing

We examine the relationship between some time-dependent measures of association and the ageing properties of the bivariate Schur-constant family. From (6.91) and (8.29), the expression for the Clayton measure becomes

$$\theta(x_1,x_2) = \frac{h(x_1+x_2+1)h(x_1+x_2)+h(x_1+x_2)-h(x_1+x_2+1)}{h^2(x_1+x_2)}.$$

Hence, $\theta > 1$ is equivalent to

$$[1-h(x_1+x_2)][h(x_1+x_2)-h(x_1+x_2+1)] > 0.$$

This means that X_1 and X_2 are positively associated if

$$h(x_1+x_2) > h(x_1+x_2+1)$$

and so X must be strictly DHR. Similarly,

(i) $\theta < 1 \Leftrightarrow X$ is strictly IHR,
(ii) $\theta = 1$ or X_1 and X_2 are independent iff $h(x)$ is a constant in which X is geometric.

Example 8.1. Let

$$R(x_1,x_2) = \frac{(m)_{(x_1+x_2)}}{(m+n)_{(x_1+x_2)}}.$$

Then the distribution of X has survival function

$$S(x) = \frac{(m)_x}{(m+n)_x}$$

with hazard rate $h(x) = \frac{n}{m+n+x}$.

Then,

$$\theta(x_1,x_2) = \frac{(n+1)(m+n+x_1+x_2)}{m(m+n+x_1+x_2+1)} > 1,$$

and accordingly X_1 and X_2 are positively associated. The same conclusion is reached if we notice that $h(x)$ is strictly decreasing.

By virtue of (8.31) and (8.32), the Bjerve and Doksum (1993) measures in (1.74) and (1.75) can be expressed in terms of hazard rate and the mean residual life as follows:

$$\rho(x_1) = \frac{\sigma_1 \Delta[E(X_2|X_1=x_1)]}{[(\sigma_1 \Delta E(X_2|X_1=x_1))^2+\sigma^2(X_1)]^{\frac{1}{2}}}$$
$$= \frac{\sigma_1(h(x_1)-h(x_1+1))}{h(x_1)h(x_1+1)[\sigma_1^2 \frac{h(x_1)-h(x_1+1)}{h(x_1)h(x_1+1)} + (2m(x_1)h(x_1)-1]^{\frac{1}{2}}}$$

and

$$\xi(x_1) = \frac{\sigma_1(h(x_1)-h(x_1+1))}{h(x_1)h(x_1+1)[\frac{1-h(x_1)}{h^2(x_1)}(2m(x_1))h(x_1)-1]^{\frac{1}{2}}}$$
$$= \frac{\sigma_1(h(x_1)-h(x_1+1))}{h(x_1+1)[(1-h(x_1))(2m(x_1)h(x_1)-1)]^{\frac{1}{2}}}.$$

Recalling that $\rho(x_1) > (<)0$ implies positive (negative) dependence, we conclude that when X is strictly DHR (IHR), (X_1, X_2) are positively (negatively) associated for all x_1 and x_2, and further $\xi(x_1) < (>)0 \Leftrightarrow X$ is DHR (IHR). Also,

$$\begin{aligned}\theta(x_1, x_2) > (< 1) \Leftrightarrow & \text{ is strictly DHR (IHR)} \\ & \Leftrightarrow \rho(x) < (>)0, \Leftrightarrow \xi(x_1) < (>)0\end{aligned}$$

for Schur-constant models.

The two measures of Anderson et al. (1992), in (1.72) and (1.73), simplify to

$$\phi(x_1, x_2) = \frac{m(x_1 + x_2 + 1)}{m(x_1)} \tag{8.33}$$

and

$$\psi(x_1, x_2) = \frac{S(x_1 + x_2)}{S(x_1)S(x_2)}. \tag{8.34}$$

In (8.33),

$$\phi(x_1, x_2) > 1 \Leftrightarrow m(x_1 + x_2 + 1) > m(x_1) \Leftrightarrow X_1 \text{ is strictly IMRL},$$

and from (8.34)

$$\begin{aligned}\log \psi(x_1, x_2 + 1) - \log \psi(x_1, x_2) &= \log \frac{S(x_1 + x_2 + 1)}{S(x_1 + x_2)} - \log \frac{S(x_1 + x_2)}{S(x_2)} \\ &= h(x_2) - h(x_1 + x_2). \end{aligned} \tag{8.35}$$

As x_2 increases, $\psi(x_1, x_2)$ also increases which means that X_1 and X_2 are positively associated. Now from (8.35),

$$\begin{aligned}X_1 \text{ is strictly DHR } & \Leftrightarrow h(x_2) - h(x_1 + x_2) > 0 \\ & \Leftrightarrow X_1 \text{ and } X_2 \text{ are positively associated.}\end{aligned}$$

Thus, we have the following results.

Theorem 8.6. *If (X_1, X_2) has bivariate Schur-constant distribution, then:*

(a) *X is DHR $\Leftrightarrow \theta(x_1, x_2) > 1 \Leftrightarrow \rho(x) > 0 \Leftrightarrow \xi(x) > 0 \Leftrightarrow \psi(x_1, x_2) > 1 \Rightarrow \phi(x_1, x_2) > 1$,*

(b) *X is IMRL $\Leftrightarrow \phi(x, y) > 1$.*

The case when X is IHR is obtained by reversing the inequalities in (a) and that of DMRL is obtained by reversing the inequality in (b).

Theorem 8.7. *If* (X_1, X_2) *has bivariate Schur-constant distribution,*

(a) $\theta(x_1, x_2) > 1 \Leftrightarrow \rho(x_1) > 0 \Leftrightarrow \xi(x_1) > 0 \Leftrightarrow (X_1, X_2)$ *is RCSI;*
(b) $\rho(x_1) \geq 0 \Leftrightarrow (X_1, X_2)$ *is SI* (X_1, X_2)*;*
(c) $\psi(x_1, x_2) > 1 \Leftrightarrow (X_1, X_2)$ *is PQD.*

Remark 8.2. Among bivariate distributions, in general, there is no direct relationship between SI and RCSI. Further, PQD implies neither RCSI nor SI.

Remark 8.3. From the above discussions, we could see that various measures provide the same result. Further, negative (positive) ageing is seen to be equivalent to positive (negative) association.

We have given only some important aspects of discrete BSM along the lines of the work done in continuous bivariate Schur-constant models by Nair and Sankaran (2014a). For more insight into the properties and applications, we refer to the works in the continuous case by Barlow and Mendel (1993), Caramellino and Spizzichino (1994, 1996), Bassan and Spizzichino (2005), Nelsen (2005), Chi et al. (2009), and Nair and Sankaran (2014b).

The multivariate models seen so far in this chapter are basic in the sense that they represent lifetimes of devices which exhibit no-ageing property at the component level or system level in some sense. In reality in multi-component devices, the components individually or jointly may have positive or negative ageing property and therefore more general distributions are required to model them. Some models of this category is taken up for analysis in the subsequent sections.

8.4 BIVARIATE WARING DISTRIBUTION-1

The bivariate Waring distribution-1 (BWD-1) appeared previously in some examples has survival function

$$s(x_1, x_2) = \frac{(m)_{(x_1+x_2)}}{(m+n)_{(x_1+x_2)}}, \quad x_1, x_2 = 0, 1, 2, \ldots, m, n > 0, \tag{8.36}$$

and probability mass function

$$f(x_1, x_2) = \frac{n(n+1)(m)_{x_1+x_2}}{(m+n)_{x_1+x_2+2}}. \tag{8.37}$$

The marginal distributions are univariate Waring with survival functions

$$S(x_1, 0) = \frac{(m)_{x_1}}{(m+n)_{x_1}} \text{ and } S(0, x_2) = \frac{(m)_{x_2}}{(m+n)_{x_2}}.$$

We have

$$E(X_i) = \frac{m}{n-1}, \quad n > 1, \text{ and } V(X_i) = \frac{mn(m+n-1)}{(n-1)^2(n-2)}, \; n > 2, \; i = 1, 2,$$

$$\text{Cov}(X_1, X_2) = -\frac{m+n-1}{(n-1)^2}$$

and the correlation coefficient is

$$r = \frac{n-2}{mn},$$

so that X_1 and X_2 are negatively correlated.

Various reliability functions have simple forms. The scalar hazard rate function is

$$a(x_1, x_2) = \frac{n(n+1)}{(m+n+x_1+x_2)(m+n+x_1+x_2+1)}.$$

Since the univariate hazard rate function is decreasing, we see that (X_1, X_2) is BDHR-9.

Also, we have a reciprocal quadratic expression in x_1 and x_2 for the hazard rate. At the same time, vector hazard rate has the expression

$$(c_1(x_1, x_2), c_2(x_1, x_2)) = \left(\frac{n}{m+n+x_1+x_2}, \frac{n}{m+n+x_1+x_2}\right)$$

with both components being reciprocal linear in x_1 and x_2. On the basis of this hazard rate, (X_1, X_2) is BIHR-2, BIHR-3 and hence BDHR-4. Also, the alternative conditional hazard rate function is found to be

$$(e_1(x_1, x_2), e_2(x_1, x_2)) = \left(\frac{n+1}{m+n+x_1+x_2+1}, \frac{n+1}{m+n+x_1+x_2+1}\right)$$

which is again reciprocal linear. By Definition 7.9, the vector (X_1, X_2) is BDHR-10.

We obtain the bivariate mean residual life function as

$$(m_1(x_1, x_2), m_2(x_1, x_2)) = \left(\frac{m+n+x_1+x_2+1}{n-1}, \frac{m+n+x_1+x_2+1}{n-1}\right)$$

which is linear in x_1 and x_2. Further,

$$m_i(x_1, x_2+1)c_i(x_1+1, x_2+1) = \frac{n}{n-1}, \qquad i = 1, 2,$$

which is a constant. It is easy to see that (X_1, X_2) is BIMRL-2.

From Example 6.11, the product moment residual life given by

$$M(x_1, x_2) = \frac{(m+n+x_1+x_2+1)(m+n+x_1+x_2)}{(n-1)^2}$$

is increasing in both x_1 and x_2, while the covariance residual life

$$\sigma_{12} = -\frac{m+n+x_1+x_2+1}{(n-1)^2}$$

is decreasing. Some calculations are required to obtain the variance residual life function. Employing (6.45),

$$\begin{aligned} \textit{we have } \sigma_1^2(x_1, x_2) &= \frac{2 \cdot (m+n)_{x_1+x_2+2}}{(m)_{x_1+x_2+2}} \\ &\times \sum_{t=x_1+1}^{\infty} \sum_{u=t+1}^{\infty} \frac{(m)_{u+x_2+1}}{(m+n)_{u+x_2+1}} - m_1(x_1, x_2)[m_1(x_1 x_2) - 1]. \end{aligned} \tag{8.38}$$

To evaluate the summation term, we write

$$\begin{aligned} \sum_{t=x_1+1}^{\infty} \sum_{u=t+1}^{\infty} \frac{(m)_{u+x_2+1}}{(m+n)_{u+x_2+1}} &= \sum_{t=x_1+1}^{\infty} \frac{(m)_{t+x_2+2}}{(m+n)_{t+x_2+1}} \left(\frac{1}{m+n+t+x_2+1} \right. \\ &\quad \left. + \frac{m+t+x_2+2}{(m+n+t+x_2+1)(m+n+t+x_2+2)} + \ldots \right) \\ &= \sum_{t=x_1+1}^{\infty} \frac{(m)_{t+x_2+2}}{(m+n)_{t+x_2+1}} \frac{m+t+x_2+2}{n-1}, \text{ by Warang expansion} \\ &= \frac{1}{n-1} \sum_{x_1+1} \frac{m(m+1)_{t+x_2+1}}{(m+n)_{t+x_2+1}} \\ &= \frac{m}{(n-1)(n-2)} \frac{(m+1)_{x_1+x_2+2}}{(m+n)_{x_1+x_2+1}}. \end{aligned}$$

Hence, the first term in (8.38) simplifies to

$$\begin{aligned} &\frac{2(m+n)_{x_1+x_2+2}}{(m)_{x_1+x_2+2}} \frac{m(m+1)_{x_1+x_2+2}}{(m+n)_{x_1+x_2+1}} \\ &\qquad = \frac{2(m+x_1+x_2+2)(m+n+x_1+x_2+1)}{(n-1)(n-2)}, \end{aligned} \tag{8.39}$$

and the second term reduces to

$$
\begin{aligned}
m_1(x_1, x_2)(m_1(x_1, x_2) - 1) \\
&= \frac{m+n+x_1+x_2+1}{n-1}\left(\frac{m+n+x_1+x_2+1}{n-1} - 1\right) \\
&= \frac{(m+x_1+x_2+2)(m+n+x_1+x_2+1)}{(n-1)^2}. \qquad (8.40)
\end{aligned}
$$

Upon using (8.39) and (8.40) in (8.38), we get

$$
\sigma_1^2(x_1, x_2) = \frac{n(m+x_1+x_2+2)(m+n+x_1+x_2+1)}{(n-1)^2(n-2)}.
$$

We then find that

$$
\frac{\sigma_1^2(x_1, x_2)}{m_1(x_1, x_2)(m_1(x_1, x_2) - 1)} = \frac{n}{n-2} > 1, \quad n > 2.
$$

It is also true that

$$
\sigma_1^2(x_1, x_2) = \sigma_2^2(x_1, x_2)
$$

and that both are increasing functions in x_1 and x_2. Since the distributions belong to the Schur-constant family, inference relating to the nature of association between X_1 and X_2 is evident from the deliberations in Section 8.8.

The multivariate version of (8.35) is straightforward. A discrete random vector $\mathbf{X} = (X_1, \ldots, X_p)$, with X_i taking values in N, is said to have multivariate Waring distribution-1, if it has survival function of the form

$$
S(\mathbf{x}) = \frac{(m)_{x_1+x_2+\ldots+x_p}}{(m+n)_{x_1+x_2+\ldots+x_p}}, \quad m, n > 0. \qquad (8.41)
$$

The reliability properties of $\mathbf{X}$ are transparent from the bivariate case. We can identify the model from the following characterization result.

Theorem 8.8. *A random vector* $\mathbf{X}$ *follows multivariate Waring distribution-1 if and only if its scalar hazard rate* $a_r(\mathbf{x}_r)$ *is the reciprocal of a polynomial of degree* r *in* $t = x_1 + \ldots + x_n$*, the total lifetime, for* $r = 1, 2, \ldots, p$ *and all* $\mathbf{x}$ *in* N^p.

The proof of the theorem is obtained by a direct application of Theorem 6.1. We can construct the empirical scalar hazard rate from the data using the empirical survival function. If the rth forward difference of the empirical hazard rate is approximately constant, the distribution in (8.41) is a plausible model for the data.

8.5 BIVARIATE WARING DISTRIBUTION-2

In the BWD-1, the assumption of exchangeability of the components is not always realized in practice and therefore, we present a more general model which possesses almost all features of BWD-1 with the exception of exchangeability. A random vector (X_1, X_2) taking values in N^2, is said to follow bivariate Waring distribution-2 (BWD-2) if it has survival function of the form

$$S(x_1, x_2) = \frac{(m)_{x_2}}{(m+n)_{x_2}} \frac{(a_1 + b_1 x_2)x_1}{(a_1 + b_1 x_2 + n)x_1}, \qquad m, a_1, b_1, n > 0. \tag{8.42}$$

From (8.42), the joint probability mass function of $(X_1, X_2]$ is derived as

$$f(x_1, x_2) = \frac{(m)_{x_2}}{(m+n)_{x_2}} \left[\frac{n(a_1 + b_1 x_2)_{x_1}}{(a_1 + b_1 x_2 + n)_{x_1}} - \frac{(a + b_1 + b_1 x_2)_{x_1}}{(a + b_1 + b_1 x_2 + n)_{x_1}} \right.$$
$$\left. \left(1 - \frac{(m + x_2)(a_1 + b_1 + b_1 x_2 + x_1)}{(m + n + x_2)(a_1 + b_1 + b_1 x_2 + x_1 + n)} \right) \right].$$

We have the marginals as

$$S(x_1, 0) = \frac{(a_1)_{x_1}}{(a_1 + n)_{x_1}} \quad \text{and} \quad S(0, x_2) = \frac{(m)_{x_2}}{(m+n)_{x_2}}.$$

The distribution in (8.42) is derived from the consideration that the mean residual life function is of the linear form

$$m_1(x_1, x_2) = p_1 x_1 + q_1 x_2 + r_1$$

and

$$m_2(x_1, x_2) = p_2 x_1 + p_1 x_2 + r_2.$$

Accordingly, the survival function can also be expressed as

$$S(x_1, x_2) = \frac{(a_1)_{x_1}}{(a_1 + n)_{x_1}} \frac{(m + b_2 x_1)_{x_2}}{(m + b_2 x_1 + n)_{x_2}}, \tag{8.43}$$

wherein the relationships between $m_1(x_1, x_2)$, $m_2(x_1, x_2)$ and $S(x_1, x_2)$ establish the equivalence of (8.42) and (8.43).

To enumerate the reliability aspects of BWD-2, we first note that the hazard rate vector has elements

$$c_1(x_1, x_2) = \frac{n}{a_1 + b_1 x_2 + x_1 + n} \tag{8.44}$$

and

$$c_2(x_1, x_2) = \frac{n}{m + b_2 x_1 + x_2 + n}, \tag{8.45}$$

which are reciprocal linear in x_1 and x_2. Hence, (X_1, X_2) Is BDHR-2. Further, the mean residual life function is $(m_1(x_1, x_2), m_2(x_1, x_2))$, where

$$m_1(x_1, x_2) = \frac{a_1 + b_1x_2 + x_1 + n + b_1}{n - 1} \tag{8.46}$$

and

$$m_2(x_1, x_2) = \frac{m + b_2x_1 + x_2 + n + b_2}{n - 1}. \tag{8.47}$$

Thus, m_1 and m_2 are linear in x_1 and x_2 as stated earlier and (X_1, X_2) is IMRL-2. From the above expressions, it is easy to see that

$$C_i(x_1, x_2 + 1)m_i(x_1, x_2) = \frac{n}{n - 1} > 1, \quad i = 1, 2. \tag{8.48}$$

Following Remark 6.4, the parameters in the two versions of $S(x_1, x_2)$ satisfy

$$\frac{a_1 + b_1}{a_1 + b_1 + n} \frac{m}{m + n} = \frac{m + b_2}{m + b_2 + n} \frac{a_1}{a_1 + n}.$$

Using (8.46), the variance residual life is obtained as

$$\sigma_1^2(x_1, x_2) = \frac{n(a_1 + b_1x_2 + x_1 + b_1 + n)(a + b_1x_2 + x_1 + b_1 + 1)}{(n - 1)^2(n - 2)}$$

and

$$\sigma_2^2(x_1, x_2) = \frac{n(m + x_1 + x_2 + b_2 + n)(m + b_2x_1 + x_2 + b_2 + 1)}{(n - 1)^2(n - 2)}.$$

Thus, we have

$$\frac{\sigma_i^2(x_1, x_2)}{m_i(x_1, x_2)[m_i(x_1, x_2) - 1]} = \frac{n}{n - 2} > 1.$$

The product moment of residual life has the expression

$$\begin{aligned} M(x_1, x_2) =& \frac{1}{n - 1} \frac{(a + b(x_2 + 1))_{x_1+1}}{(a + n + b(x_2 + 1))_{x_1+1}} \\ & \times \sum_{t_2=x_2+1}^{\infty} \frac{(a + b_1t_2)_{x_1+1}}{(a + n + b_1t_2)_{x_1+1}} \frac{(m + x_2 + 1)_{t_2-1}}{(m + n + x_2 + 1)_{t_2-1}}. \end{aligned}$$

Various descriptive measures of the BWD-2 are as follows:

$$E(X_1) = \frac{a_1}{n - 1}, \quad E(X_2) = \frac{m}{n - 1},$$

$$V(X_1) = \frac{m^n(m+n-1)}{(n-1)^2(n-2)} \quad \text{and} \quad V(X_2) = \frac{a_1 n(a_1+n-1)}{(n-1)^2(n-2)}.$$

The covariance and correlation coefficient have no simple closed-form expressions.

Identification of the model can be accomplished through any one of the characteristic properties given in the next theorem.

Theorem 8.9. *A discrete random vector (X_1, X_2), in the support of N^2, follows the BWD-2 if and only if any one of the following properties is satisfied:*

(a) $(m_1(x_1,x_2), m_2(x_1,x_2)) = (A_1 + B_1x_1 + C_1x_2, A_2 + B_2x_1 + B_1x_2)$;

(b) $(c_1(x_1,x_2), c_2(x_1,x_2)) = ((P_1+Q_1x_1+Q_2x_2)^{-1}, (P_2+Q_3x_1+Q_1x_2)^{-1})$;

(c) $c_i(x_1, x_2+1)m_i(x_1,x_2) = c_2(x_1+P, x_2)m_2(x_1,x_2) = \alpha$, *a constant greater than unity, for* $i = 1, 2$;

(d)

$$\frac{\sigma_i^2(x_1,x_2)}{m_i(x_1,x_2)[m_i(x_1,x_2)-1]} = k,$$

a constant greater than unity, for $i = 1, 2$.

We will now examine the nature of association in BWD-2. Among the time-dependent measures,

$$\phi(x_1,x_2) = \frac{a_1 + b_1x_2 + x_1 + b_1 + n}{a_1 - b_1 + x_1 + b_1 + n} > 1$$

implying that X_1 and X_2 are positively associated. Similarly,

$$\psi(x_1,x_2) = \frac{(a_1+b_1x_2)_{x_1}}{(a_1+b_1x_2+n)_{x_1}} \frac{(a_1+n)_{x_1}}{(a_1)_{x_1}}$$

assumes values larger than unity, and hence (X_1, X_2) are positively associated. This also indicates that (X_1, X_2) is PQD.

Some other bivariate Waring distributions have been discussed by Xekalaki (1984, 1985, 1986a, 1986b) in connection with accident theory. They are bivariate forms specified by the probability mass function

$$f_1(x_1,x_2) = \frac{(m)_2}{(m+n)_2} \frac{(n)_{x_1+x_2}}{(m+n+2)_{x_1+x_2}}, \qquad x_1, x_2 = 0, 1, 2, \ldots, \; m, n > 0,$$

$$f_2(x_1,x_2) = \frac{(m)_{l+t}}{(m+n)_{l+t}} \frac{(n)_{x_1+x_2}(l)_{x_1}(t)_{x_2}}{(m+n+l+t)_{x_1+x_2}},$$

$$x_1, x_2 = 0, 1, 2\ldots, \; m, n, l, t > 0,$$

and their extensions to the multivariate case. As a special case, we have the bivariate Yule distribution with probability mass function

$$f(x_1, x_2) = \frac{(m)_2(x_1 + x_2)!}{(m + 1)_{x+y+2}}.$$

8.6 BIVARIATE NEGATIVE HYPERGEOMETRIC DISTRIBUTION

The survival function of the negative hypergeometric model is given by

$$S(x_1, x_2) = \frac{\binom{\alpha+n-x_2}{\alpha-x_2}}{\binom{\alpha+n}{n}} \frac{\binom{\beta+\delta_1 x_2+n-x_1}{\beta+\delta_1 x_2-x_1}}{\binom{\beta+\delta_1 x_2+n}{\beta+\delta_1 x_2}}, \tag{8.49}$$

$x_1 = 0, 1, \ldots, n$, $x_2 = 0, 1, \ldots, n$, $\alpha, n, \beta > 0$.

We have the marginal distributions of X_1 and X_2 as negative hypergeometric with survival functions

$$S_1(x_1) = \frac{\binom{\beta-n-x_1}{b-x_1}}{\binom{\beta+n}{n}} \text{ and } S_2(x_2) = \frac{\binom{\alpha-n-x_1}{\alpha-x_1}}{\binom{\alpha+n}{n}}. \tag{8.50}$$

Further, the joint probability mass function is

$$f(x_1, x_2) = \frac{1}{\binom{\alpha+n}{n}}$$
$$\left\{ \frac{[\binom{\alpha+n-x_2}{\alpha-x_2}\binom{\beta+\delta_1 x_2+n-x_1-1}{\beta+\delta_1 x_2-x_1}]}{\binom{\beta+\delta_1 x_2+n}{\beta+\delta_1 x_2}} - \frac{\binom{\alpha+n-x_2-1}{\alpha-x_2}\binom{\beta+\delta x_2+\delta_1+n-x_1-1}{\beta+\delta x_2+\delta_1-x_1}}{\binom{\beta+\delta_1 x_2+\delta_1+n}{\beta+\delta_1 x_2+\delta}} \right\}.$$

There is an alternate expression for the survival function in (8.49) of the form

$$S(x_1, x_2) = \frac{\binom{\beta+n-x_1}{n-x_1}\binom{\alpha+\delta_2 x_1+n-x_2}{\alpha+\delta_2 x_1-x_2}}{\binom{\alpha+n}{n}\binom{\alpha+\delta_1 x_1+n}{\alpha+\delta_2+x_1}}. \tag{8.51}$$

From (8.50) and (8.51), we compute the vector hazard rate components to be

$$c_1(x_1, x_2) = 1 - \frac{S(x_1 + 1, x_2)}{S(x_1, x_2)} = \frac{n}{\beta + \delta_1 x_2 - x_1 + n} \tag{8.52}$$

and

$$c_2(x_1, x_2) = \frac{n}{\alpha + \delta_2 x_1 - x_2 + n}. \tag{8.53}$$

For all x_1 and x_2,

$$c_1(x_1 + 1, x_2) - c_1(x_1, x_2) = 1 = c_2(x_1, x_2 + 1) - c_2(x_1, x_2)$$

so that (X_1, X_2) is BIHR-2. The mean residual life has components

$$\begin{aligned} m_1(x_1,x_2) &= \frac{1}{S(x_1+1,x_2+1)} \sum_{t=x_1+1}^{n} S(t,x_2+1) \\ &= \binom{\beta+\delta_1 x_2+\delta_1-x_1+n}{\beta+\delta_1 x_2+\delta_1-x_1-1} \Big/ \binom{\beta+\delta_1 x_2+\delta_1+n-x_1-1}{\beta+\delta_1 x_2+\delta_1-x_1-1} \\ &= \frac{\beta+\delta_1 x_2+\delta_1+n-x_1}{n+1}, \end{aligned} \tag{8.54}$$

upon using the summation formula

$$\sum_{t=0}^{\infty} \binom{\alpha+\beta-t-1}{\alpha-t} = \binom{\alpha+\beta}{\alpha},$$

and

$$m_2(x_1,x_2) = \frac{\alpha+\delta_2 x_1+\delta_2+n-x_2}{(n+1)}. \tag{8.55}$$

From (8.54) and (8.55), it is easy to see that (X_1, X_2) is DMRL-2. Also,

$$c_1(x_1+1,x_2)m_1(x_1,x_2) = c_2(x_1,x_2+1)m_2(x_1,x_2) = \frac{n}{n+1},$$

a constant less than unity. We also have

$$\begin{aligned} &\frac{1}{S(x_1+1,x_2+1)} \sum_{t=x_1+1}^{n} \sum_{u=t_1+1}^{n} S(u,x_2+1) \\ &= \sum_{t=x_1+1}^{n} \sum_{u=t+1}^{n} \binom{\beta_r+\delta_1 x_2+\delta+n-u}{\beta+\delta_1 x+\delta_1-u} \Big/ \binom{\beta+\delta_1 x_2+\delta_1+n}{\beta+\delta_1 x_2+\delta_1} \\ &= \sum_{t=x_1+1}^{n} \binom{\beta+\delta_1 x_2+n-t_1+\delta}{\beta+\delta_1 x_2+\delta-t_1-1} \Big/ \binom{\beta+\delta_1 x_2+\delta_1+n}{\beta_1+\delta_1 x_2+\delta_1} \\ &= \binom{\beta+\delta_1 x_2+n-x_1}{\beta+\delta_1 x_2+\delta_1-x_1-2} \Big/ \binom{\beta+\delta_1 x_2+\delta_1+n}{\beta_1+\delta_1 x_2+\delta_1} \\ &= \frac{(\beta+\delta_1 x_2+\delta_1+n-x_1)(\beta+\delta_1 x_2+\delta_1-x_1-1)}{(n+1)(n+2)}. \end{aligned}$$

Hence, using formula (6.45), we find

$$\sigma_1^2(x_1,x_2) = \frac{n(\beta+\delta_1 x_2+\delta_1+n-x_1)(\beta+\delta_1 x_2+\delta_1-x_1-1)}{(n+1)^2(n+2)}$$

and similarly

$$\sigma_2^2(x_1, x_2) = \frac{n(\alpha + \delta_2 x_1 + \delta_2 + n - x_2)(\alpha + \delta_2 x_1 + \delta_2 - x_2 - 1)}{(n+1)^2(n+2)}.$$

Thus,

$$\frac{\sigma_i^2(x_1, x_2)}{m_i(x_1, x_2)[m_i(x_1, x_2) - 1]} = \frac{n}{n+2} < 1.$$

Moreover, $\sigma_i^2(x_1, x_2)$ is a decreasing function of x_i for all x_{3-i}. We now present a set of characterizations of the negative hypergeometric law in (8.49).

Theorem 8.10. *A discrete random vector* (X_1, X_2), *taking values in* N^2, *follows the negative hypergeometric law in* (8.49) *if and only if any one of the following properties is satisfied:*

(a) $(C_1(x_1, x_2), C_2(x_1, x_2)) = ((P_1 + Q_2 x_2 - Q_1 x_1)^{-1}, (p_2 + Q_3 x_1 - Q_1 x_2)^{-1})$;

(b) $(m_1(x_1, x_2), m_2(x_1, x_2)) = (A_1 - B_1 x_1 + C_1 x_2, A_2 B_2 x_1 - B_1 x_2)$;

(c) $e_1(x_1 + 1, x_2) m_1(x_1, x_2) = C_2(x_1, x_2 + 1) m_2(x_1, x_2) = p$, *a constant greater than unity;*

(d)

$$\frac{\sigma_i^2(x_1, x_2)}{m_i(x_1, x_2)(m_i(x_1, x_2) - 1)} = k,$$

$i = 1, 2$, *where k is a constant exceeding unity.*

As in the case of BWD-2, we can evaluate $\phi_1(x_1, x_2)$, ${}_2\phi(x_1, x_2)$ to verify that (X_1, X_2) is negatively associated, and also that it is NQD.

It is of interest to notice that when $\alpha = \beta$, $\delta_1 = 1$, the model in (8.49) reduces to the Schur-constant bivariate negative hypergeometric law with

$$S(x_1, x_2) = \frac{\binom{\alpha + n - x_1 - x_2}{n - x_1 - x_2}}{\binom{\alpha + n}{n}}, \quad \alpha > 1.$$

The properties of this distribution can be easily deduced from those of (8.49).

8.7 BIVARIATE WEIBULL DISTRIBUTION

All the bivariate geometric distributions presented at the beginning of this chapter can be generalized to provide corresponding versions of Weibull distribution. In this section, we illustrate the procedure with respect to the bivariate geometric distribution-1. We say that (X_1, X_2) is distributed as bivariate Weibull if its survival function can be written as

$$S(x_1, x_2) = q_1^{x_1^{\beta_1}} q_2^{x_2^{\beta_2}} \theta^{x_1^{\beta_1} x_2^{\beta_2}}. \tag{8.56}$$

It is a five-parameter distribution with probability mass function

$$f(x_1, x_2) = q_2^{x_2^{\beta_2}} \{[A(x_2)]^{x_1^{\beta_2}} - [A(x_2)^{(x_1+1)^{\beta_1}}]\} \\ - q_2^{(x_2+1)^{\beta}} \{[A((x_2+1)^{x_1^{\beta}}] - [A(x_2+1)]^{(x_1+1)^{\beta_1}}\}, \quad (8.57)$$

with $A(x_2) = q_1\theta^{x_2^{\beta_2}}$. The parameters satisfy the conditions

$$0 < q_1, q_2 < 1, \quad 0 \le \theta \le 1, \quad 1 - q_1\theta - q_2\theta + q_1q_2\theta \ge 0.$$

We have the scalar hazard rate as

$$a(x_1, x_2) = 1 - q_1^{\Delta x_1^{\beta_1}} \theta^{x_2^{\beta_2}} \Delta \mathbf{x}_1^{\beta_1} q_2^{\Delta x_2^{\beta}} \theta^{x_1^{\beta_1}} \Delta x_2^{\beta_2} \\ + q_1^{\Delta x_1^{\beta_1}} q_2^{\Delta x_2^{\beta_2}} \theta^{x_1^{\beta_1} x_2^{\beta_2} - (x_1+1)^{\beta_1}} - (x_2+1)^{\beta_2},$$

where Δ is the usual difference operator. On the other hand, the vector hazard rate has a simpler form

$$c_1(x_1, x_2) = 1 - q^{\Delta x_1^{\beta_1}} \theta^{x_2^{\beta_2} \Delta x_1^{\beta_2}}$$

and

$$c_2(x_1, x_2) = 1 - q^{\Delta x_2^{\beta_2}} \theta^{x_1^{\beta_1} \Delta x_2^{\beta_2}}.$$

It can be seen that $c_1(x_1, x_2)$ is increasing (decreasing) in x_1 for all x_2 when $\beta_1 > 1$ ($\beta_1 < 1$), and similarly $c_2(x_1, x_2)$ is increasing (decreasing) in x_2 for all x_2 when $\beta_2 > (<)1$. Thus, (X_1, X_2) is BIHR(2) (BDHR(2)) for $\beta_1, \beta_2 > (<)1$. When $\beta_1 = \beta_2 = 1$, we have the BGD-1 which has local no-ageing property. The other reliability functions like the mean and variance of residual life are not in closed-form. The marginal distributions are discrete Weibull (3.89) with parameters β_1, for X_1 and β_2 for X_2. Direct calculations show that (X_1, X_2) is NQD and also negatively time-dependent.

8.8 MULTIVARIATE ZIPF DISTRIBUTIONS

Yeh (2002) has introduced six types of multivariate Zipf distributions of which one typical form with survival function

$$S(\mathbf{x}) = \left\{1 + \sum_{i=1}^{p} \frac{x_i}{\sigma_i}\right\}^{-d}, \quad x_i = 0, 1, 2, \ldots, \alpha > 0, \sigma_i > 0,$$

will be considered here in detail. The other forms can be treated similarly and so are not discussed. Specializing to $i = 1, 2$, the bivariate case has

$$S(x_1, x_2) = \left(1 + \frac{x_1}{\sigma_1} + \frac{x_2}{\sigma_2}\right)^{-\alpha}$$

and

$$\begin{aligned} f(x_1, x_2) = {} & \left(1 + \frac{x_1}{\sigma_1} + \frac{x_2}{\sigma_2}\right)^{-\alpha} - \left(1 + \frac{x_1+1}{\sigma_1} + \frac{x_2}{\sigma_2}\right)^{-\alpha} \\ & - \left(1 + \frac{x_1}{\sigma_1} + \frac{x_2+1}{\sigma_2}\right)^{-\alpha} + \left(1 + \frac{x_1+1}{\sigma_1} + \frac{x_2+1}{\sigma_2}\right)^{-\alpha}. \end{aligned}$$

Consequently,

$$\begin{aligned} a(x_1, x_2) = 1 & - \left(\frac{1 + \frac{x_1}{\sigma_1} + \frac{x_2}{\sigma_2}}{1 + \frac{x_1+1}{\sigma_1} + \frac{x_2}{\sigma_2}}\right)^{\alpha} - \left(\frac{1 + \frac{x_1}{\sigma_1} + \frac{x_2}{\sigma_2}}{1 + \frac{x_1}{\sigma_1} + \frac{x_2+1}{\sigma_2}}\right)^{\alpha} \\ & + \left(\frac{1 + \frac{x_1}{\sigma_1} + \frac{x_2}{\sigma_2}}{1 + \frac{x_1+1}{\sigma_1} + \frac{x_2+1}{\sigma_2}}\right)^{\alpha}, \\ c(x_1, x_2) = 1 & - \left(\frac{1 + \frac{x_1}{\sigma_1} + \frac{x_2}{\sigma_2}}{1 + \frac{x_1+1}{\sigma_1} + \frac{x_2}{\sigma_2}}\right)^{\alpha} \end{aligned}$$

and

$$c_2(x_1, x_2) = 1 - \left(\frac{1 + \frac{x_1}{\sigma_1} + \frac{x_2}{\sigma_2}}{1 + \frac{x_1}{\sigma_1} + \frac{x_2+1}{\sigma_2}}\right)^{\alpha}.$$

The distribution belongs to the BDHR-2 class. Also since

$$\left(1 + \frac{x_1}{\sigma_1} + \frac{x_2}{\sigma_2}\right) \le \left(1 + \frac{x_1}{\sigma_1}\right)\left(1 + \frac{x_2}{\sigma_2}\right),$$

we see that (X_1, X_2) is PQD, and the marginals of X_1 and X_2 being

$$S_i(x_i) = \left(1 + \frac{x_i}{\sigma_i}\right)^{-\alpha}, \quad i = 1, 2.$$

Using $\psi(x_1, x_2)$, it is not difficult to see that the time-dependent association is also positive.

In this chapter, we have presented a few basic models that are capable of further generalizations to provide more flexibility while modelling. Also, there is scope for new models demanded by special features observed in multivariate discrete data, the analysis of which is still in its infancy stage in our opinion. Of course, the wide array of multivariate continuous lifetime distributions available in the literature (see, for example, Kotz et al., 2000) and their characteristics and properties may provide a good basis for developing work in this direction.

Chapter 9

Applications

9.1 INTRODUCTION

Reliability theory has grown in the last six decades into an independent discipline by drawing tools from several areas including mathematics, statistics probability theory, and actuarial science. During the course of solving its problems, reliability analysis have contributed several concepts and theorems to many other subjects. These topics have resulted in opening up new areas of investigation in several fields of scientific activity. This chapter is devoted to identifying some of the applications of the ideas, notions and results in reliability theory to other branches of knowledge. Since the application of reliability concepts to different contexts is too enormous to survey in a chapter like this, we present only some key disciplines where the interaction has evoked substantial interest. Thus, the areas of study considered here are survival analysis, social sciences, risk theory, information theory, mathematics and statistics. Even in this limited exposition here, the analysis is not exhaustive and is limited to a few illustrative cases.

9.2 SURVIVAL ANALYSIS

Survival analysis describes a body of statistical procedures for analyzing data on a random variable X that represents the waiting time until the occurrence of an event of interest. In the terminology of survival analysis, the concerned event is referred to as death and the waiting time as survival time. The basic concepts used here also are the survival function $S(x)$ and hazard rate function $h(x)$, which bear the same relationships and properties as in reliability theory. Methods of analyzing discrete time survival data have been considered by many authors including Allison (1982), Aranda-Ordaz (1983), Willet and Singer (1993), Singer and Willet (1993), and Barber et al. (2000). They have pointed out the advantages of this approach as

(1) suited to the analysis of longitudinal data,
(2) it can handle time-variant models and time-variant predictors,
(3) model violation can be easily checked and corrected,
(4) it is easier to handle censored observations.

Reliability Modelling and Analysis in Discrete Time. http://dx.doi.org/10.1016/B978-0-12-801913-9.00009-9

It is appropriate when duration time is grouped or banded into discrete intervals, in which case the data can be truly discrete.

When the time axis is divided into disjoint intervals $[a_{j-1}, a_j]$, $j = 1, 2, \dots, k$, $a_0 = 0$, $a_j = b \leq \infty$, the survival function at the end of the j-th interval is

$$P(X > a_j) = S(a_j)$$

and the hazard rate

$$h(a_j) = 1 - \frac{S(a_j)}{S(a_{j-1})} = P[a_{j-1} < X \leq a_j | X > a_j]$$

giving

$$S(j) = \prod_{x=1}^{j} (1 - h(j)),$$

when the intervals are of unit length (like a week or a month). From

$$\log S(j) = \sum_{x=1}^{j} \log(1 - h(x)),$$

for small $h(x)$, using first-order Taylor approximation, we have

$$\log S(j) \circeq -\sum_{x=1}^{j} h(x) = -H(j).$$

9.2.1 Proportional Hazards Models

The analysis of survival data has its major emphasis on the hazard of death at some time after an individual enters into an event of interest, usually in medical studies. In such studies, the presence of several explanatory variables or covariates that may affect the survival time is generally assumed as supplementary information. Demographic characteristics, physiological factors, life styles, etc., are usually considered for this purpose. The main objective of the analysis is to ascertain the extent to which the explanatory variables affect the risk of death. A popular model that explains the relationship between the survival of an individual in relation to the covariates is the *Cox proportional hazards model* (PHM). Let $\mathbf{Z} = (Z_1, \dots, Z_p)$ be a vector of explanatory variables. The discrete time (interval) hazard function, under the PHM assumption, is

$$h(a_j, \mathbf{Z}) = h(\mathbf{Z}) = 1 - \frac{S(a_j, \mathbf{Z})}{S(a_{j-1}, \mathbf{Z})}$$
$$= 1 - \exp[\lambda(H(j-1) - H(j)],$$

where $\lambda = \exp[\boldsymbol{\beta}'\mathbf{Z}]$ with $\beta = (\beta_1, \ldots, \beta_p)'$. Hence,

$$\log(-\log(1 - h_j(\mathbf{Z})) = \boldsymbol{\beta}'\mathbf{Z} + \log(H(j) - H(j-1)).$$

Similarly, the discrete time (interval) baseline hazard for $(a_{j-1}, a_j]$ is

$$1 - h_0(j) = \exp[H(j) - H(j-1)],$$

or

$$\log[-\log(1 - h_0(j))] = \log(H(j) - H(j-1)) = \alpha(j).$$

Thus,

$$h(a_j, \mathbf{Z}) = 1 - \exp[-\exp(\boldsymbol{\beta}'\mathbf{Z} + \alpha(j)].$$

When each interval is of unit length,

$$h(j, \mathbf{Z}) = 1 - \exp[-\exp(\boldsymbol{\beta}'\mathbf{Z} + \alpha(j))].$$

Taking the survival times to be discretely measured, we may use the model

$$\frac{h(j, \mathbf{Z})}{1 - h(j, \mathbf{Z})} = \frac{h_0(j)}{1 - h_0(j)} \exp[\boldsymbol{\beta}'\mathbf{Z}],$$

where $h(j, \mathbf{Z})$ is the discrete time hazard for j and $h_0(j)$ is the corresponding baseline hazard.

If we use the alternative hazard rate $h_1(x)$ instead of $h(x)$, it can be seen that the model can be written in the form

$$h_{1,i}(x) = h_{1,0}(x) \exp\left(\sum_{j=1}^{p} \beta_j z_{ji}\right),$$

where $h_{1,i}(x)$ is the hazard function of the ith individual, $h_{1,0}(x)$ is the baseline hazard, Z_{ji} is the value of the jth explanatory variable on the ith individual, and β_j are the regression coefficients. Since the second term does not contain x, we may write the model as

$$h_{1,i}(x) = C h_{1,0}(x),$$

where C is a constant with regard to x. From (2.15), this is, however, equivalent to

$$S_i(x) = [S_0(x)]^C \tag{9.1}$$

with $S_i(x)$ being the survival function of the ith individual and $S(x)$ is the baseline survival function. Models of type (9.1) have some important applications in reliability theory. Consider a system consisting of n components whose lifetimes are independent and identically distributed. We call it a series system if the system functions only when all the components function. When $X_1, \ldots X_n$ are the component lifetimes, this means that the system lifetime is $\min(X_1, \ldots X_n)$ whose distribution is $(S_0(x))^n$, where S_0 is the survival function of X_i. Thus, when C is a positive integer n, (9.1) represents the reliability function of a series system. The hazard rate of the system is

$$h_S(x) = nh_1(x), \tag{9.2}$$

where $h_1(x)$ is the alternative hazard rate of X_i and $h_S(x)$ is the alternative hazard rate of the system. From (9.1) and (9.2), the ageing properties of a series system can be inferred.

9.2.2 Additive Hazards Model

The PH models assume multiplicative relationship with the baseline hazard. There are situations in which an additive relationship is more meaningful in which case we have an additive hazards model. For example, in tumorogenicity experiments, when the dose effect on tumor risk is of interest, the excess risk becomes an important factor. Clinical trials that seek the effectiveness of treatments often experience lag times of treatment effectiveness after which treatment procedures will be in full effect; see Chen et al. (2002), Sun et al. (2006) and Nair and Sankaran (2010c) for more details and examples in the continuous case. If X is the lifetime under study and Y is another random variable, both discrete, from which information on X can be gathered, the additive hazards model assumes a structural relationship of the form

$$h(x|Y = y) = a(y) + h(x), \quad x, y = 0, 1, 2, \ldots. \tag{9.3}$$

In (9.3), $h(x)$ is the baseline alternative hazard function and Y represents the changes in the conditions that has an additive effect on X represented by a positive function $a(y)$. From Eq. (2.51), the relationship (9.3) can be translated into

$$\frac{S(x|Y = y)}{S(x + 1|Y = y)} = \frac{S(x)}{S(x + 1)} e^{-xa(y)},$$

or

$$S(x|Y = y) = S(x)e^{-xa(y)} = S(x)q^{xa(y)}, \ q = e^{-1}.$$

The marginal survival function of the random variable X^* satisfying (9.3) becomes

$$S^*(x) = S(x)S_G(x),\ x, y = 0, 1, 2, \ldots, \tag{9.4}$$

where

$$S_G(x) = \sum_{y=0}^{\infty} q^{x\,a(y)} f_Y(y).$$

Notice that S_G is the survival function of a mixture geometric distribution, and from (9.4)

$$S^*(x+1) = S(x+1)S_G(x+1). \tag{9.5}$$

Dividing (9.4) by (9.5) and taking logarithms, we get

$$h_1^*(x) = h_1(x) + h_{1,G}(x) \tag{9.6}$$

where $h_{1,G}(x)$ and $h_1^*(x)$ are the alternative hazard functions of the geometric mixture random variable and X^*; respectively. We also have an alternate formula for evaluating $h_{1,G}$ given by

$$\begin{aligned} h_G(x) &= \log \frac{S_G(x)}{S_G(x+1)} \\ &= \log \frac{\sum q^{xa(y)} f_Y(y)}{\sum q^{(x+1)a(y)} f_Y(y)} \\ &= \log \frac{Eq^{xa(Y)}}{Eq^{(x+1)a(Y)}}. \end{aligned}$$

Since geometric mixtures have decreasing failure rates, $h_{1,G}(x) \le h_{1,G}(0)$, and therefore

$$h_{1,G}(x) \le -\log E[q^{a(Y)}].$$

Thus, the unconditional hazard rates are bounded by

$$h_1(x) \le h_1^*(x) \le h_1(x) - \log E(q^{a(Y)}).$$

Eq. (9.4) provides the probability mass function of X^* as

$$\begin{aligned} f^*(x) &= S(x)S_G(x) - S(x+1)S_G(x+1) \\ &= S_G(x)(S(x) - S(x+1)) + S(x+1)[S_G(x) - S_G(x+1)] \\ &= S_G(x)f(x) + S(x+1)f_G(x). \end{aligned}$$

The usual hazard rate of X^* is

$$h^*(x) = \frac{f^*(x)}{S^*(x)} = \frac{f(x)}{S(x)} + \frac{S(x+1)}{S(x)} \frac{f_G(x)}{S_G(x)}$$
$$= h(x) + (1 - h(x))h_G(x).$$

The bounds for $h^*(x)$ are calculated using

$$h_G(x) = 1 - \frac{\sum_{y=0}^{\infty} q^{(x+1)a(y)} f(y)}{\sum_{y=0}^{\infty} q^{xa(y)} f(y)}$$
$$= 1 - \frac{E[q^{(x+1)a(Y)}]}{Eq^{xa(Y)}]}.$$

Since $h_G(x)$ is decreasing,

$$h_G(x) \leq h_G(0) = 1 - E[q^{a(Y)}].$$

Thus,

$$h(x) \leq h^*(x) \leq (1 - h(x))(1 - Eq^{a(Y)}) + h(x).$$

In practice, the distribution of Y is generally unknown and an appropriate choice needs to be made.

Example 9.1. Take the simple case when $a(Y) = Y$ where Y is distributed as Poisson with

$$f_Y(y) = e^{-\lambda} \frac{\lambda^y}{y!}.$$

Then,

$$S_G(x) = \sum_{y=0}^{\infty} q^{xy} e^{-\lambda} \frac{\lambda^y}{y!}$$
$$= e^{\lambda}(q^x - 1),$$

the survival function of a discretized negative Gompertz law. The corresponding hazard rate and the alternative hazard rate are, respectively,

$$h_G(x) = 1 - e^{-\lambda q^x (1-q)}$$

and

$$h_{1,G}(x) = \lambda q^x (1 - q).$$

Thus,

$$h_1^*(x) = \lambda q^x(1-q) + h_1(x)$$

and the original model is

$$h_1(x|Y=y) = y + h_1(x).$$

If, in addition, the distribution of X is geometric with parameter S, then

$$h_1^*(x) = \lambda q^x(1-q) - \log S.$$

In many cases, the ageing properties of X and X^* will greatly simplify the task of choosing appropriate models.

Theorem 9.1. *If X^* is IHR, then X is IHR, and when X is DFR, X^* is also DHR. Further, if X is bathtub shaped (BT), either X^* is DHR or BT, and if X is UBT, X^* is either IHR or UBT. Further, the change point when X is BT (UBT) is smaller (larger) than the change point in case when X^* is BT (UBT).*

Proof. Since $h_{1,G}(x)$ is decreasing in nature, the first part of the theorem follows from (9.6). To prove the second part, we write from (9.6) the following:

$$h_1^*(x+1) - h_1^*(x_1) = h_1(x+1) - h_1(x) + h_{1,G}(x+1) - h_{1,G}(x). \quad (9.7)$$

First, we note that $h_{1,G}(x)$ is decreasing for all x. When X is BT, there exists an $x_0 > 0$ such that $h_1(x)$ is decreasing in $x = 0, 1, \ldots, x_0 - 1$. Thus, $h_1^*(x)$ is decreasing for $x = 0, 1, \ldots, x_0 - 1$ and may increase in (x_0, ∞) since $h_1(x)$ is increasing in (x_0, ∞) because of BT shape or decrease in $[0, \infty]$ because of $h_G(x) > h(x)$ in (x_0, ∞). Thus, X^* is either DHR or BT. Similarly, if X is UBT, X^* is either IHR or UBT. The last part is obvious. □

Theorem 9.2. *If X^* is IHRA-1, then X is IHRA-1, and when X is DHRA-1, so is X^*.*

Remark 9.1. The proof is similar to that of the first part of Theorem 9.1. Further, Theorem 9.2 holds even if IHRA-1 is replaced by IHRA.

Theorem 9.3. *If X^* is NBUHR-1 (NBUHRA-1), then X is also NBUHR-1 (NBUHRA-1).*

Proof. Using (9.6) and the definition of NBUHR-1,

$$h_1^*(x) - h_1^*(0) = h_1(x) - h_1(0) + h_{1,G}(x) - h_{1,G}(0).$$

Since G is DHR, it is also NBUHR-1 and hence $h_{1,G}(x) - h_{1,G}(0) \leq 0$. Thus, $h_1(x) \geq h_1(0)$ or X is NBUHR-1. The proof for NBUHR-1 case is similar. □

Remark 9.2. X^* NBUHR (NBUHRA) $\Rightarrow X$ is NBUHR (NBUHR).

Theorem 9.4. *If X^* is NBU, X is NBU, and X^* is NWU if X is NWU.*

Proof. We have

$$\begin{aligned} S^*(x+t) - S^*(x)S^*(t) &= S(x+t)S_G(x+t) - S(x)S(t)S_G(x)S_G(t) \\ &= [S(x+t) - S(x)S(t)][S_G(x+t)] \\ &\quad + [S_G(x+t) - S_G(x)S_G(t)]S(x)S(t). \end{aligned}$$

Since G is DHR, it is NWU so that the second term is non-negative. The theorem now follows. □

We can employ stochastic orders for comparing two additive hazards models representing random variables (X_1, X_1^*, G_1) and (X_2, X_2^*, G_2), and these results are as follows:

(1) When $Y_1 =_{st} Y_2$, the model differences are solely due to X_1 and X_2. Then

$$X_1^* \leq_{hr} X_2^* \Leftrightarrow X_1 \leq_{hr} X_2$$

and

$$X_1^* \leq_{st} X_2^* \Leftrightarrow X_1 \leq_{st} X_2;$$

(2) When $X_1 =_{st} X_2$,

$$X_1^* \leq_{st} X_2^* \Leftrightarrow G_1 \leq_{st} G_2$$

and

$$X_1^* \leq_{hr} X_2^* \Leftrightarrow G_1 \leq_{hr} G_2;$$

(3)

$$\begin{gathered} Y_2^* \leq_{pgf} Y_1^* \Leftrightarrow G_1 \leq_{st} G_2, \\ Y_1 \leq_{pgf} Y_2, X_1 \leq_{st} X_2 \Rightarrow X_1^* \leq X_2^*, \\ X_1 \leq_{hr} X_2 \Rightarrow X_1^* \leq_{hr} X_2^*. \end{gathered}$$

9.2.3 Proportional Reversed Hazards Model

The structure of the proportional reversed hazard model (PRHM) is similar to that of the PH model, with the difference that the hazard rate is replaced by the reversed hazard rate. Suppose X and Y are two random variables with reversed hazard rate functions $\lambda_X(x)$ and $\lambda_Y(x)$. We say that X and Y satisfy the PRHM, with proportionality constant $C > 0$, if $\lambda_Y(t) = C\lambda_X(x)$. It is easy to see that PRHM is equivalent to

$$G(x) = [F(x)]^C, x = 0, 1, 2, \ldots,$$

where $G(\cdot)$ and $F(\cdot)$ are the distribution function of X and Y. Gupta and Gupta (2007) have pointed out that the PRHM is applicable in the case of spontaneous carcinogenics which allow for a pattern of term or growth kinetics. The properties and applications of PRHM, when the distribution function is continuous, has been discussed in Gupta et al. (1998), Di Crescenzo (2000), Gupta and Wu (2001), and Gupta and Gupta (2007).

In the case of discrete survival times, some basic results are given by Dewan and Sudheesh (2009). They considered the alternative reversed hazard rate in (2.59) given by

$$\lambda_1(x) = \log \frac{F(x)}{F(x-1)},$$

and postulate the PRHM as

$$G(x) = [F(x)]^\theta, \tag{9.8}$$

where, as before, $G(\cdot)$ and $F(\cdot)$ are the distribution functions of discrete lifetimes X and Y. This is equivalent to

$$\lambda_{1,Y}(x) = \theta\lambda_{1,X}(x).$$

The probability mass function of Y now becomes

$$g(x) = [F(x)]^\theta - [F(x-1)]^\theta = [F(x)]^\theta(1 - e^{-\theta\lambda_1(x)}).$$

Eq. (9.8) has a useful interpretation in terms of a parallel system. Since a parallel system will function whenever one of its components is working, the lifetime of the system is $Z = \max(X_1, \ldots, X_n)$, where $(X_1, \ldots, X_n)$ are the lifetimes of the components. When the X_i's are independent and identically distributed, the distribution of Z is F^n. Thus

$$F_Z(x) = [F(x)]^n$$

which conforms to the structure in (9.8). Dewan and Sudheesh (2009) have shown in this connection that

$$X \text{ is IRHR (DRHR)} \Leftrightarrow Y \text{ is IRHR (DRHR)}$$
$$X \text{ is IHR (DHR) and } \theta > 1 (\theta < 1) \Rightarrow Y \text{ is IHR (DHR)}$$

$$X \text{ is NBU (NWU) and } \theta > 1 (\theta < 1) \Rightarrow Y \text{ is NBU (NWU)}$$
$$X \text{ is IRMRL } \Leftrightarrow Y \text{ is IRMRL.}$$

If N has Sibuya distribution with

$$f(n) = (-1)^{n+1} \binom{\theta}{x}, \quad 0 < \theta < 1, n = 1, 2,$$

$Y = \min(X_1, \ldots, X_N)$ and N is independent of the X's, then Y admits PRHM. Some examples showing the limitations of PHM and the utility of PRHM in such cases can be seen in O'Neill (1991).

9.2.4 Proportional Mean Residual Life Model

Zahedi (1991) presented the *proportional mean residual life model*, (PMRLM) by replacing the hazard rate functions in PHM by the mean residual life function. Let X_0 be a random variable with mean residual life function $m_0(x)$ and survival function $S_C(x)$. Let $X_1, \ldots, X_n$ be independent random variables representing lifetimes of n individuals and $Z_{i1}, \ldots, Z_{ip}$ be a set of covariates with $\mathbf{Z}_i = (Z_{i1}, \ldots, Z_{ip})$. Then, the PMRLM requires that

$$m_i(x|\mathbf{Z}_i) = C(\mathbf{Z}_i) m_0(x), \tag{9.9}$$

where m_i is the mean residual life of X_i and $C_i = C(\beta_1 Z_{i_1} + \ldots + \beta_p Z_{i_p}) = C(\mathbf{Z}_i)$. One needs restrictions on the right side in order that m_i is a mean residual life. The model in (9.9) was introduced on the ground that the mean residual life function has more intuitive appeal than the hazard rate function, and that it does not require the existence of the density function and the conditions on the survival function can be expressed in terms of the mean. Gupta and Kirmani (1998) have studied some properties of the implications of PMRLM. For the analysis of the model in (9.9), we write it briefly as

$$m(x) = c m_0(x), \quad c > 0, \tag{9.10}$$

where c is independent of x and $m_0(x)$ is the baseline mean residual life function.

The relationship in (9.10) is used as the definition of PMRLM in the discrete case also where x takes values in N.

Example 9.2. Consider Waring baseline distribution with

$$f_0(x) = \frac{(a-b)(b)_x}{(a)_{x+1}}, \quad x = 0, 1, 2, \ldots, a > b > 0, \tag{9.11}$$

which has mean residual life function

$$m_0(x) = \frac{a+x}{a-b-1}.$$

Then,

$$cm_0(x) = \frac{c(a+x)}{a-b-1},$$

and the corresponding distribution is again Waring with

$$f_1(x) = \frac{(a_1-b_1)(b_1)_x}{(a_1)_{x+1}} \quad \text{where } a_1 = a,\ b_1 = b+(c-1)(a-1), \tag{9.12}$$

satisfying the PMRLM.

Remark 9.3. Generally, the PMRLM does not coincide with the PHM. In the above example, the hazard rate of (9.11) is

$$h_{1,0}(x) = -\log \frac{b+x}{a+x}$$

and that of (9.12) is

$$h_1(x) = -\log \frac{a-1-\frac{(a-b-1)}{c}+x}{a+x}.$$

We see that $h_1(x) \neq kh_{1,0}(x)$, for a constant k.

Some characteristics of the PMRLM are as follows:

(1) When the PMRLM holds, $X \leq_{mrl} (\geq_{mrl}) X_0 \Leftrightarrow X \leq_{hr} (\geq_{hr}) X_0$, where X and X_0 are, respectively, random variables with mean residual lives $m(x)$ and $m_0(x)$.

Proof. Using the identity in (2.21), we have

$$h(x) = \frac{1+m(x)-m(x-1)}{m(x)} = \frac{1+cm_0(x)-cm(x-1)}{cm_0(x)}$$

under the PMRLM. We can write

$$h(x) = h_0(x) + \frac{1-c}{m_0(x)}. \tag{9.13}$$

When $c \leq 1$,

$$m(x) \leq m(x_0) \Leftrightarrow h(x) \leq h_0(x),$$

and when $c \geq 1$,

$$m(x) \leq m(x_0) \Leftrightarrow h(x) \geq h_0(x).$$

The result stated above now follows; □

(2) The conditions for X and X_0 to have the same type of ageing are:

(i) If $c > 1(< 1), m_0(x)$ is convex and X_0 is IHR, IHRA, NBU (DHR, DHRA, NWU), then X is IHR, IHRA, NBU (DHR, DHRA, NBU);

(ii) If $c > 1(< 1), m_0(x)$ is concave and x_0 is DHR, DHRA, NWU (IHR, IHRA, NWU), then X is DHR, DHRA, NWU (IHR, IHRA, NBU).

Proof. We can write (9.13) as

$$\begin{aligned} h(x) &= h_0(x)\left[1 + \frac{1-c}{h_0(x)m_0(x)}\right] \\ &= h_0(x)\left[1 + \frac{1-c}{1 + m_0(x) - m_0(x-1)}\right]. \end{aligned}$$

If $c > 1(< 1)$ and $h_0(x)$ is increasing (decreasing), we can see that the second term on the right hand side is also increasing (decreasing) whenever $m(x)$ is convex (concave). This proves that X is IHR. The other cases and (ii) are proved using similar arguments; □

(3) The PMRLM can be written as

$$h_{E,X}(x) = kh_{E,X_0}(x), \quad k = +\frac{1}{c}, \tag{9.14}$$

where h_E is the hazard rate of the equilibrium distribution of the concerned random variable. Thus, PMRLM is a proportional hazard rate model in terms of the conventional definition of the discrete hazard rate. Several results are possible by virtue of the representation in (9.14) and properties of equilibrium distributions.

9.2.5 Proportional Odds Model

The notion of proportional odds model was introduced and studied by Bennett (1983), Kirmani and Gupta (2001), and Rossini and Tsiatis (1996). They argued that such a model is motivated by the following facts:

(1) In the PHM, $h_i(x)/h_0(x)$ is constant in x, thus ruling out situations in which the ratio of hazard rates converges to unity as $x \to \infty$;

(2) The assumption of constant hazard rates ratio becomes unreasonable when initial effects such as differences in the stage of disease or in treatment disappear with time;

(3) Demonstration of the effectiveness of a cure on mortality of a disease group compared to that of a disease-free control group requires that the hazard ratio converge to unity.

The proportional odds model takes the form

$$w_i(x) = w_0(x) \exp\left[\sum_{j=1}^{p} \beta_j Z_{ji}\right]$$

where $w(x)$ is the odds ratio of survival defined in (2.85) of the ith individual and $w_0(x)$ is the corresponding baseline odds ratio.

In this section, we look at the role and implications of the proportional odds model (POM) when survival times are discrete. To understand the structural properties of the POM, we write it as

$$w(x) = c w_0(x), \qquad c > 0, \tag{9.15}$$

where $w(x)$ and $w_0(x)$, as before, are the odds of survival of two lifetimes X and X_0 with survival functions $S(x)$ and $S_0(x)$, respectively. Then, from (2.85), we have

$$w(x) = \frac{S(x+1)}{1 - S(x+1)} \text{ and } w_0(x) = \frac{S_0(x+1)}{1 - S_0(x+1)}, \tag{9.16}$$

so that from (9.15) and (9.16),

$$S(x+1) = \frac{c S_0(x+1)}{1 - (1-c) S_0(x+1)}, \qquad x = -1, 0, 1, \ldots, \tag{9.17}$$

and

$$\begin{aligned} S_0(x+1) - S(x+1) &= S_0(x+1)\left[1 - \frac{c}{1 - (1-c) S_0(x+1)}\right] \\ &= \frac{S_0(x+1)(1 - S_0(x+1))}{1 - (1-c) S_0(x+1)} \\ &= \frac{(1-c)}{c}(1 - S_0(x+1)) S(x+1). \end{aligned} \tag{9.18}$$

From the above basic identities, we have the following implications of POM:

(1) If $0 < c \le 1$, $S_0(x) - S(x) \le \frac{1-c}{4c}$, and if $c > 1$, $S(x) - S_0(x) \le \frac{c-1}{4}$.

Proof. When $0 < c \leq 1$,

$$S_0(x) - S(x) = \frac{1-c}{c}(1 - S_0(x))S(x)$$
$$\leq \frac{1-c}{c}(1 - S_0(x))S_0(x), \text{ since } w(x) \leq w_o(x)$$
$$\leq \frac{1-c}{4c}.$$

When $c > 1$, consider $w_0 = \frac{1}{c}w$ and proceed similarly. □

Remark 9.4. The quantity $\frac{1-c}{4c}$ is the upper bound of the increase in the risk of dying by age x when POM is assumed with $c < 1$.

Remark 9.5. Eq. (9.17) gives the identity connecting the survival function under POM and the baseline distribution. It also enables the construction of a new distribution, which is more flexible than the original one in view of the new parameter c added. See Marshall and Olkin (1997) for the additional features of the new distribution in the continuous case. Eq. (9.17) is the discrete analogue of their result.

As an example of working with (9.17), we assume that X_0 is geometric. The resultant new model is specified by the survival function

$$S(x) = \frac{cq^x}{1-(1-c)q^x},$$

with hazard rate

$$h(x) = \frac{1-q}{1-(1-c)q^{x+1}}$$

which is more general than that of the geometric, which is constant;

(2) $\frac{F(x)}{F_0(x)}$ is decreasing when $0 < c \leq 1$, and increasing when $c > 1$;

(3) Using the identity in (2.91) between $w(x)$ and $h(x)$, we have

$$h(x) = h_0(x)\left(\frac{1+w_0(x)}{1+cw_0(x)}\right). \tag{9.19}$$

Hence, X does not admit a PH model, and it depends on $h_0(x)$ and a function that depends on the age. Also

$$w(x) > w_0(x) \Leftrightarrow h(x) < h(x_0), \quad c > 1,$$

and

$$w(x) \leq w_0(x) \Leftrightarrow h(x) \geq h(x_0), \quad c \leq 1;$$

(4) Since the reversed hazard rate

$$\lambda(x) = \frac{w(x-1) - w(x)}{1 + w(x)},$$

we can write

$$\lambda(x) = \lambda_0(x)\frac{c(1 + w_0(x))}{1 + cw_0(x)}$$

connecting the reversed hazard rates of X and X_0. Also,

$$\lambda_0(x) < \lambda(x) \Leftrightarrow w(x) > w_0(x), \quad c > 1,$$
$$\lambda_0(x) \geq \lambda(x) \Leftrightarrow w(x) \leq w_0(x), \quad c \leq 1.$$

Again, POM is not equivalent to a proportional reversed hazards model in terms of X and X_0;

(5) If $c > (< 1)$, the ratio of the hazard functions $\frac{h(x)}{h_0(x)}$ is increasing (decreasing) and converges to unity as $x \to \infty$, and $\frac{\lambda(x)}{\lambda_0(x)}$ tends to c as $x \to \infty$;

(6) Suppose POM holds with $c > 1$. If X_0 is IHR (IHRA, NBU), then X is IHR (IHRA, NBU). The proof follows from (9.19) and the fact that $\frac{1+w_0(x)}{1+cw_0(x)}$ is increasing, since $w(x)$ is a decreasing function. Similarly, if $0 < c < 1$, then

$$X_0 \text{ is DHR(DHRA, NWU)} \Rightarrow X \text{is DHR(DHRA, NWU)}.$$

Two other models with similar structure are the accelerated failure time model

$$X = \Psi(\mathbf{z})X_0 \tag{9.20}$$

and the competing risk model

$$h(x) = \sum_{i=1}^{n} h_i(x). \tag{9.21}$$

Of these, (9.20) is the simplest in which lifetime is stochastically increased or decreased according to whether $\Psi < 1$ or $\Psi > 1$. The competing risk model is based on (i) the life of individual is subject to several competing risks and (ii) life ends as it is exposed to one of the risks. For a detailed discussion of these models in continuous time, we refer to Newby (1988) and David and Moeschberger (1978).

9.3 SOCIAL SCIENCES

In this section, we deal with the impact of reliability concepts in various branches of social science. Unlike in survival analysis, the lifetime is not the random variable under consideration in these areas, but non-negative random variables where the notions of reliability are logical and interpretable. Most of the applications in economics is in the context of income analysis. In the early literature on income distributions, several discrete distributions were proposed. Among them are the zeta distribution with

$$f(x) = P(X = x) = [\zeta(\theta + 1)]^{-1} x^{-(\theta+1)}, \quad x = 1, 2, \ldots,$$

and the Zipf models with survival functions

$$S_1(x) = 1 + \left(\frac{x-a}{\sigma}\right)^{-\alpha}, \quad x = a, a+1, \ldots,$$

$$S_2(x) = \left[1 + \left(\frac{x-a}{\sigma}\right)^{\frac{1}{\beta}}\right]^{-1}, \quad x = a, a+1, \ldots,$$

and

$$S_3(x) = \left[1 + \left(\frac{x-a}{\sigma}\right)^{\frac{1}{\beta}}\right]^{-\alpha}, \quad x = a, a+1, \ldots .$$

The Simon distribution is specified by

$$f(x) = \frac{\theta^{-x}\Gamma(\alpha + x + 1)\Gamma(\beta + 1)}{{}_2F_1(\alpha + 1, 1, \frac{\beta}{\theta} + 1, \frac{\beta}{\theta} + 1, \theta^{-1})\Gamma(\frac{\beta}{\theta} + x + 1)\Gamma(\alpha + 1)},$$
$$\alpha > -1,\ \beta > -\theta,\ \theta > 1,\ x = 0, 1, 2, \ldots$$

As special cases, we have the Waring and Yule distributions mentioned in Chapter 3. Recently, modelling incomes is performed with the aid of continuous distributions like various Pareto forms, log-normal, beta, gamma, Weibull, Singh-Maddala, Dagum, Pareto-lognormal, etc., without much reference to discrete law; see Kleiber and Kotz (2003) for details. Quantile functions can also be employed to describe the patterns in income variation as is done in Tarsitano (2004) and Haritha et al. (2008). Accordingly, the reliability concepts associated with income analysis is confined to the continuous case, although many of them are valid for the discrete counterparts as well. Singh and Maddala (1976) obtained their family of income distributions from increasing hazard rate satisfying

$$h(x) = \alpha x^{\beta-1}(1 - F(x))^{\theta}, \quad \alpha, \beta, \theta > 0,\ x > 0,$$

while Dagum (1977) derived the income model based on the reversed hazard rate property

$$\lambda(x) = x^{-1}\beta\theta(1-(F(x))^{\frac{1}{\beta}}), \quad \beta > 0,\ \beta\theta > 1,\ x > 0,$$

both as continuous distributions, by solving for $F(x)$ from the above equations.

Curves that measure the inequality of incomes have been a topic of considerable interest ever since the work of Lorenz (1905). It is also used in the case of other economic variables distributed over a population that reflects the disparity in ownership of wealth among its members. The Lorenz curve is defined implicitly by

$$p = F(x) \text{ and } L(p) = F(x)\frac{E(X|X \leq x)}{\mu}, \ \mu = F(x). \tag{9.22}$$

From (9.22),

$$x = F^{-1}(p) = \begin{cases} \inf(x|F(x) \geq p), & 0 < p \leq 1, \\ \inf(x|F(x) > 0), & p = 0. \end{cases} \tag{9.23}$$

The inverse function is defined for discrete and continuous distributions. Various relationships between the Lorenz curve and reliability concepts are obvious from (9.22) such as

$$L(F(x)) = 1 - \frac{S(x)(m(x)+x)}{\mu} \text{ (Chandra and Singpurwalla, 1981)}$$

and

$$L(F(x)) = \frac{F(x)(x - r(x))}{\mu}.$$

A popular measure of inequality derived from the Lorenz curve is the Gini index given by

$$G = 1 - 2E[F_1(x)],$$

where $F_1(x) = \frac{F(x)E(X|X\leq x)}{E(X)}$ is called the first moment distribution corresponding to $F(x)$. Another index for measuring the inequality is the Pietra index

$$P = \frac{E|X-\mu|}{2\mu},$$

which is the maximum vertical deviation between $L(p)$ and $L(p) = p$. The properties of the Lorenz curve and Gini index, in terms of reliability concepts were studied by Pham and Turkkan (1994).

Another curve that is used instead of the Lorenz curve is the Bonferroni curve defined in the orthogonal plane $[F(x), B(F(x))]$, with

$$B(F(x)) = \frac{E(X|X \leq x)}{\mu}$$

where $F(x)$ is absolutely continuous with $E(X) < \infty$ and $F(x)$ is continuous and differentiable at least twice. By suitably modifying the conditions with discrete analogues, we can use it for discrete observations. It is easy to see that

$$B(F(x)) = [F(x)]^{-1} L(F(x)),$$
$$B(F(x)) = \frac{x - r(x)}{\mu}$$

and

$$r(x) = \mu B'(F(x))F(x).$$

When the Bonferroni curve lies completely above the unit square, then X is NBUE. For other properties and a defaulted description, see Giorgi and Crescenzi (2001a, 2001b). Pundir et al. (2005) have further shown that

$$B(F(x)) = \frac{1}{F(x)} - \frac{1}{\mu}\frac{S(x)}{F(x)}(m(x) + x).$$

Zenga (2007) proposed a curve to assess inequalities in income as

$$A(x) = 1 - \frac{E(X|X \leq x)}{E(X|X > x)}, \tag{9.24}$$

and pointed out that $A(x)$ satisfies several desirable properties for a discrete data set. The relationship $A(x)$ has with various reliability functions can be easily written down. Nair et al. (2012a) studied some properties of the curve $A(x)$ of which those related to reliability concepts are given below:

$$\begin{aligned} A(x) &= 1 - \frac{x + r(x)}{m(x) + x} \\ &= \frac{m(x) + x - \mu}{F(x)(m(x) + x)} \\ &= \frac{\mu - x + r(x)}{\mu - F(x)(x - r(x))}. \end{aligned}$$

The three different curves mentioned above are interrelated and each can be used to compare income distributions vis-a-vis the extent of inequality among them. All those aspects are studied in Pundir et al. (2005) and Nair et al. (2012a).

Another important class of inequality measures are called measures of poverty. These are based on a poverty line and concentrate on proportion of individuals below that line. Similarly, we can talk about affluence indices by considering the proportion of individuals in the population above an affluence line. Sen (1976), Takayama (1979) and Sen (1986, 1988) have dealt with such indices and their properties. Assuming the poverty line as $X = x$, the proportion of poor people is $F(x)$. The income gap ratio of the poor is defined as

$$\alpha(x) = 1 - \frac{E(X|X \le x)}{x},$$

which is the ratio of the deviation of the poverty line from the average income of the poor to the poverty line. It is closely related to the reversed mean residual life through the identity

$$N(x) = x(1 - \alpha(x)). \tag{9.25}$$

It has been shown by Nair et al. (2008) that when X is continuous, $\alpha(x)$ determines the distribution of incomes, and that $\alpha(x)$ is a constant characterizes the power distribution. Sen's (1976) poverty index is

$$p(x) = F(x)[\alpha(x) + (1 - \alpha(x))G(x)],$$

where $G(x)$ is the truncated Gini index

$$G(x) = 1 - E\left(X\left(1 - \frac{F(x)}{F(x)}\right)\middle| X \le x\right).$$

An interesting conclusion is that the Sen index is proportional to $F(x)$ for all $x > 0$ if and only if X has power distribution. In a similar manner, we can deal with affluence, wherein the income gap ratio is

$$\alpha^*(x) = 1 - \frac{x^*}{E(X|X > x^*)},$$

where $X = x^*$ is taken as an affluence line with people above x^* being assumed to be weakly or affluent and the average income of the affluent is

$$m(x^*) = E(X|X > x^*).$$

Elteto and Frigyes (1968) provided another set of measures of inequality. They considered

$$\mu = E(x), r = E(\mu - X|X \le \mu) \text{ and } m = E[X - \mu|X > \mu]$$

with interpretations that μ is the average income, r is the average income of the poorer than average individuals, and m is the average income of those richer than those with average income. The inequality measures proposed are

$$U = \frac{\mu}{r}, V = \frac{m}{r} \text{ and } W = \frac{m}{\mu},$$

whose connections with reliability functions are obvious.

Same kind of relationships are manifested in the distribution of wealth. Let X be the amount of wealth such as land holding with distribution function $F(x)$. Appropriate reliability notions were used by Bhattacharjee (1988) in his attempt to model the degree of inequality present in the pattern of ownership of wealth. The residual holding $(X - x|X > x)$ in excess of x has mean

$$m(x) = E(X - x|X > x).$$

Correspond it to the mean residual life of X. With the above interpretation of $m(x)$, the event of wealth possessed by the rich can be described by the behaviour of $m(x)$ for large values of x. Now, X is DHR if the proportion of progressively rich with residual holding in excess of any given amount increases with the level of affluence. Similarly, IMRL and NWUE can be interpreted. In view of these considerations, Bhattacharjee and Krishnaji (1985) proposed the following indices of inequality:

$$I_1 = \lim_{x\to\infty} \frac{m(x)}{\mu} \text{ and } I_2 = \lim_{x\to\infty} \frac{\ln(1 - F(x))}{\ln(1 - F_1(x))}$$

with $F_1(x) = \frac{1}{\mu}\int_0^x t\,dF(t)$. Under some general conditions, they showed that NWUE $\Rightarrow I_1 > 1$ if X is DHR, and $I_2 = 1$ if and only if the residual holding scaled by its mean converges to the exponential, and when $1 < I_2 < \infty$, $P(\frac{X}{x}|X > x)$ converges to the Pareto law $x^{-\alpha}$ with $\alpha = I_2/(I_2 - 1)$.

Belzunce et al. (1998) have more results in the same direction that enables classification of income distributions. They defined a non-negative random variable to be IPFR (increasing proportional failure rate) if $(\frac{X}{x_0}|X > x_0) >_{st} (\frac{X}{x_0^1}|X > x_0^1)$ for every $x_0 < x_0^1$. The IPRF variable is such that the probability of an income greater than x proportional to the affluence line x_0 is decreasing in x_0. Alternatively, X is IPRF if and only if $xh(x)$ is increasing in x. Similarly, X is said to be decreasing mean left proportional residual income (DML PRI) if $E(\frac{X}{x_0}|X > x_0)$ is decreasing in x_0. Also

(1) IPFR $\Rightarrow$ DMLPRI,

(2) If X is IPFR or DMLPRI, then $1 \le I_2 < \infty$.

Similarly, the right truncated residual income is $(\frac{X}{x_0^*}|X \le x_0^*)$. When X is an absolutely continuous non-negative random variable, it is DRPFR (decreasing right proportional failure rate) if $x\lambda(x)$ is decreasing in x. Defining decreasing mean right proportional income (DMRPRI) as saying that

$$E\left(\frac{X}{x^*}|X \le x_0^*\right)$$

is decreasing, it is shown that

$$X \text{ is DRPFR } \Rightarrow X \text{ is DMRPRI} .$$

In all these cases, the analogy with IHR (IRHR) and DMRL (DRMRL) classes are obvious.

An important problem in business activities is concerned with the pricing of a new durable product. Normally, the price as well as the sales of the product decline over time, irrespective of the discounts declared and the length of the finite planning horizon. Models are built up to relate the price and demand in which the hazard rate function plays a principal role. Bass (1987) proposed a model in which sales can increase to a peak and then decline. A model with a closed-form solution to this problem has been suggested by Sethi and Bass (2003). Suppose a customer buying the product at time x given that he has not bought it until now depends on the product price $p(x)$, the demand rate $D(p(x))$ and $F(x)$, the probability that a customer bought it by time x. Then the model prescribed is

$$h(x) = D(p(x)),$$

where $D(t) = a - bt$.

In political science, two important topics in which coherent structures find application are the famous *impossibility theorem* of Arrow (1951) and voting games. Of these, Arrow's theorem can be stated in verbal form in the following manner. When voters have three or more distinct options, no rank over voting system can convert the ranked preferences of individuals into a community wide ranking. For the validity of the theorem, the preference criteria of the voting shall be such that

(1) if every voter prefers alternative X over Y, then the group prefers X over Y;
(2) if every voter preference between X and Y, is unaltered the group preference should also be unaltered;
(3) no single voter is allowed the power to always determine group preferences, regardless of the other individual preferences.

Although the classical proof of Arrow's theorem is based on properties of complete pre-orders, Pechlivanides (1975) have given an alternative proof employing coherent-structure arguments in reliability theory. An exhaustive discussion of Arrow's theorem and a mathematical proof in terms of coherent structures is given in Bhattacharjee (1988).

The model for the voting games commences with a committee C of individuals and a coalition S, which is a subset of C. Each individual is allowed to vote yes or no on a resolution and a winning coalition is a coalition whose yes (no) votes ensures the committee passes (fails) a resolution. Let A be a set of winning coalitions and $g(\cdot)$ be a binary function such that

$$\begin{aligned} g(S) &= 1 \text{ if } S \in A, \\ &= 0 \text{ if } S \notin A. \end{aligned}$$

A voting game G is an ordered pair (N, A) such that

$$g(\phi) = 0, G(S) = 1 \text{ and } g(\cdot) \text{ is non-decreasing.}$$

The analogy between voting games is born from the fact that an individual, committee, outcome function and voting functions become synonymous with a component, system, structure function and reliability function in reliability structures. Thus, a voting game is mathematically equivalent to a coherent structure. An excellent discussion on this subject including the solution and various other matters, that brings about the role of reliability in voting games can be seen from Bhattacharjee (1988).

9.4 RISK ANALYSIS

Generally, risk refers to the underlying uncertainty of a given course of action, but is defined differently in different circumstances in which it is confronted. In finance, risk analysis refers to the uncertainty of forecasted cash flows, variations of portfolios and stock returns. Several measures of risk have been proposed in literature that focus on extreme and high consequence events. Financial risk measures are often interpreted as the amount of money to be held in reserve for a risky investment. A commonly used risk measure is the *value at risk* defined as

$$\text{VaR}_X(p) = F^{-1}(p),$$

where $F^{-1}(p) = \inf\{x|F(x) \geq p\}, 0 < p < 1$, is the quantile function of the random variable X. In the discrete case, one takes $F^{-1}(p)$ as the smaller integer value for which $F(x) \geq p$. We interpret VaR_p as the maximum risk, for a given time horizon, experienced by 100 $p\%$ of cases. It is also the minimum amount of capital that would be added to a risky instant so that the probability of loss does not exceed p. Further refinements to *VaR* gives certain other measures that have close proximity to reliability functions. One of them is the conditional tail expectations or the tail value at risk (TVaR) given by

$$\text{TVaR}_X(p) = E[X|X > F^{-1}(p)]$$

which, in reliability theory, is the conditional mean life function after time $F^{-1}(p)$. Another measure used in insurance and reinsurance is the stop-loss transform

$$\pi(x) = E[(X - x)_+]$$

which we encountered as the upper partial mean in the reliability context. If X denotes the risk for an insurance company, it bears the whole risk as long as $X < x$ and pass on to a re-insurance company the amount $X - x$ whenever $X > x$. In this case, $\pi(x)$ amounts for the expected cost for the reinsurance company. A new measure called the expected proportional short fall is defined by

$$EPS_X(p) = E\left[\left(\frac{X - F^{-1}(p)}{F^{-1}(p)}\right)_+\right],$$

where $(t)_+ = t$ if $t \geq 0$ and $(t)_+ = 0$ if $t < 0$ see Denuit et al. (2005) for the above measures and their relationships. We can see the close relationship between EPS(p) and the income gap ratio for the affluent. Apart from measuring risk, all the above measures are used to compare random risks for decision making, by invoking definitions of stochastic orders based on them. The use of coherent structures from reliability theory are also used in constructing models of risk; see Aven (2012). Discrete time risk models are discussed in Cai (2001) wherein premiums, claims and surplus are recorded only at times $n = 0, 1, 2, \ldots$. The probability that a company gets ruined, that is, the initial reserve plus accumulated premiums minus accumulated claims become negative, should be less than a tolerated percentage. A finite ruin probability is denoted by

$$\psi(u) = P(T < \infty) = P(\cup_{k=1}^{\infty}(Y_k < 0)),$$

where Y_n is an insurer's surplus at time n with $Y_0 = y$ and $T = \inf\{n : Y_n < 0\}$. Assume that Δ_n is the force of interest from time $n-1$ to n, which is a sequence of independent and identically distributed random variables and $Z = e^{\Delta_n}$. Denote by Y_n the total claim from time $(n-1)$ to n and by X_n the total premium over the same period. The sequence $< X_n >$ and $< Y_n >$ are independent and identically distributed as X and Y. Then, Cai (2001) has shown that the finite time ruin probability is governed by

$$\psi^*(u) \leq \inf_{y \geq 0}\left[\frac{S_2(y)}{S_1(x+y)}\right]^{-1}, \quad x \geq 0,$$

where S_1 is NWU and S_2 is NBU satisfying

$$E\left[\frac{S_2(\alpha X)}{S_1(\alpha Y Z^{-1})}\right] \leq 1, \quad \text{for } 0 < \alpha \leq 1.$$

Singpurwalla (2006) has discussed a parallel connection between reliability concepts with present value of a risk-free bond and the interest rates. A risk-free coupon assures the holder \$1 after time T from the time of purchase of the bond. If the bond is purchased at time t and it is held till time $t+T$, the purchase price $P(t,T)$ is known as the present value of the bond at time t. We have $P(t,0)=1$ and $P(t,T)$ decreases in T. The present value $P(t,T)$ depends on the interest rate $r(s)$ which changes with time. Then, Singpurwalla (2006) has argued that $P(t,T)$ has the same behaviour as that of a survival function and that of $r(s)$ as the case of a hazard rate. His arguments can be translated into discrete time as well. He has also developed the notions of IIR (increasing interest rate), IAIR (increasing average increasing rate), NWO (new worse than old) which are similar to the IHR, IHRA and NBU concepts in reliability.

We have provided only a few selected instances of the interface between reliability and risk. For a detailed study of these and other prospective connections between the two disciplines, we refer to Straub (1970), Denuit et al. (2005), and Zio (2007). Compound distributions and equilibrium compound distributions have an important role in representing stop-loss premiums of interest in connection with insurance claim modelling; see Willmot et al. (2005).

9.5 INFORMATION THEORY

In the conventional approach, the life distribution is characterized by reliability functions such as hazard rate, mean residual life, etc., and the ageing properties of a device are described in terms of the behaviour of these functions. Ebrahimi (1996) introduced an alternative methodology by characterizing the lifetime distribution in terms of Shannon's measure of uncertainty and that classes of life distributions, which are different from these based on reliability functions, can be obtained in such an approach. This work has stimulated a lot of research during the past decades, wherein various measures of uncertainty based on the residual life distributions have been proposed. In this section, we review some of the important results that concern discrete models.

Let X be a discrete lifetime, with survival function $S(x)$ and probability mass function $f(x)$. Then, Shannon's entropy of X is defined as

$$I(f) = -\sum_{x=0}^{\infty} f(x)\log f(x) = -E(\log f(X)).$$

The survival function and probability mass function of the residual life $(X-t \mid X>t)$ are

$$S_t(x) = \frac{S(x+t+1)}{S(t+1)} \text{ and } f_t(x) = \frac{f(x+t+1)}{S(t+1)}.$$

Hence,

$$I(f_t) = -\sum_{x=0}^{\infty} \frac{f(x+t+1)}{S(t+1)} \log \frac{f(x+t+1)}{S(t+1} \tag{9.26}$$

$$= -\frac{1}{S(t+1)} \sum_{x=t+1}^{\infty} f(x) \log \frac{f(x)}{S(t+1)}.$$

From (9.26), Rajesh and Nair (1998) found:

(i)

$$I(f_t) = -\log h(t+1) + \sum_{x=t+1}^{\infty} \frac{S(x+1)}{S(t+1)} \log \frac{h(x)}{h(x+1)(1-h(x))};$$

(ii)

$$I(f_{t+1}) = \frac{1}{1-h(t+1)} \{I(f_t) + h(t+1) \log h(t+1) + (1-h(t+1)) \log(1-h(t+1))\};$$

(iii) $I(f_t) = c$ for all t, where c is a constant if and only if X is geometric Belzunce et al. (2004) showed that if X has support $\{x_j : x_j < x_{j+1}\}$ and an increasing residual entropy $I(f_t)$, then the distribution of X is uniquely determined by $I(f_t)$.

There are some other results originating from the above definitions. From (9.26),

$$I(f_t) = -E[\log f(X)|X > t] + \log S(t+1)$$

$$I(f_{t+1}) - I(f_t) = (1-h(t+1)) \log(1-h(t+1)) + h(t+1)[I(f_{t+1}) + \log h(t+1)].$$

Ebrahimi (1996) argued that highly uncertain components or systems are inherently not reliable and therefore a better component is one which has lesser uncertainty in its residual lifetime. In this sense, life distributions can be classified according to the behaviour of $I(f_t)$. Thus, X has decreasing (increasing) uncertainty of residual by DURL (IURL) if $I(f_t)$ is decreasing (increasing) in $t \geq 0$. Notice that X is both DURL and IURL if X is geometric and that if X is IFR (DFR), then it is DURL (IURL).

In the same manner as entropy of residual life is of interest in reliability analysis, uncertainty in past (spent) life of a device is also important. Proceeding as in Di Crescenzo and Longobardi (2002) in the continuous case, we can define the entropy of past life $[t - X|X \leq t]$ as

$$I^*(f_t) = \sum_{x=0}^{t} \frac{f(x)}{F(t)} \log \frac{f(x)}{F(t)}. \tag{9.27}$$

Relationships between I^* and the reversed hazard rate $\lambda(t)$ can be derived as above.

The preceding results were extended to a more general information measure defined as

$$I_r(f_t) = \frac{1}{r} \sum_{x=t+1}^{\infty} \frac{f(x)}{S(t)} \left(1 - \frac{f^r(x)}{S^r(t)}\right),$$

or equivalently

$$1 - r\, I_r(f_t) = \frac{1}{S(t)^{r+1}} \sum_{x=t+1}^{\infty} [f(x)]^{r+1}.$$

If X has support of the form $(x_j : x_j < x_{j+1})$ and $I_r(f_t)$ is increasing in t then $I_r(f_t)$ uniquely determines the distribution of X (Belzunce et al., 2004).

Nanda and Paul (2006a) considered residual entropy of the first and second kinds of order r for a discrete random variable taking values $x_1, x_2, \ldots, x_n$. There are, respectively,

$$I_1(f_t) = \frac{1}{r-1}\left[1 - \sum_{x=t}^{n} \left(\frac{f(x)}{S(t)}\right)^r\right], \quad S(x) = \sum_{t=x}^{n} f(t),$$

and

$$I_2(f_t) = \frac{1}{r-1} \log \left(\sum_{x=t}^{n} \left(\frac{f(x)}{S(t)}\right)^r\right).$$

As $r \to 1$ in I_1 or I_2 we get the residual entropy $I(f_t)$, in which case n may be infinite. They established the following:

(a) The discrete uniform distribution, with support $(1, 2, \ldots, n)$, is characterized by decreasing first kind residual entropy

$$I_1(f_t) = \frac{1}{r-1}[1 - (n-t-1)^{1-r}], t = 1, 2, \ldots, n;$$

(b) The above uniform distribution is characterized by

$$I_2(f_t) = \log(n - t + 1), t = 1, 2, \ldots, n.$$

The entropy of past life in (9.27) has also been generalized on similar lines by Nanda and Paul (2006b) as

$$I_1^*(f_t) = \frac{1}{r-1}\left[1 - \sum_{x=0}^{t}\left(\frac{f(x)}{F(t)}\right)^r\right]$$

and

$$I_2^*(f_t) = \frac{1}{r-1}\log\left[\sum_{x=0}^{t}\left(\frac{f(x)}{F(t)}\right)^r\right].$$

If X has support $(0,1,2,\ldots,n)$, the uniform distribution is characterized through the properties

$$\textbf{(a)} \qquad I_1^*(f_t) = \frac{1}{r-1}[(1-(t+1)^{1-r}]$$

and

$$\textbf{(b)} \qquad I_2^*(f_t) = \log(t+1).$$

Following the idea of the residual entropy, Rao et al. (2004) introduced the concept of cumulative residual entropy, whose discrete version was studied by Bharatpour and Bami (2012). The definition of cumulative entropy is

$$\xi(X) = -\sum_{x=1}^{b} S(x)\log S(x), \qquad \text{where} \qquad b \le \infty.$$

They have shown that:

(i) If $Y = a + bx,\ a \ge 0,\ b > 0,$ then $\xi(Y) = b\xi(X)$;

(ii) $\xi(X) \ge Ce^{\frac{I(f_t)}{1-f(0)}}$, where

$$C = \exp\left[\frac{1}{1-f(0)}(f(0)\log f(0)) + (1-f(0))\log(1-f(0)) + \sum_{x=1}^{b} f(x)\log S(x)|\log S(x)|\right];$$

(iii) $E(|X-Y|) \le 2\xi(X)$;

(iv) $X \le_{st} Y \Rightarrow \xi(X) - \xi(Y) \le E(X)\log\frac{E(Y)}{E(X)}$. In particular,

$$X \le_{st} Y \Rightarrow \xi(X) \le \xi(Y) \text{ if } E(X) = E(Y);$$

(v) $\xi(X) = \sum_{x=1}^{b}\sum_{y=1}^{x} h(y)S(x)$;

(vi) $\xi(X) \geq E(m(X)), \quad m(x) = E(X - x | X \geq x)$;
(vii) If X is NWUE, $E(X) \leq E(m(X)) \leq \xi(X)$.

An extensive literature is available on different types of measures of uncertainty and their dynamic versions in the continuous case for which discrete analogues are yet to be found. The works of Kumar and Taneja (2011), Navarro et al. (2011), Abbasnejad (2011), Park et al. (2012), and Chemany and Bharatpour (2014) and their references give a good account of the recent developments on these topics.

9.6 MATHEMATICS AND STATISTICS

In this section, we highlight the impact of the ideas developed in reliability in some other disciplines. Mathematics and statistics richly benefited by various tools used in the course of solving problems in reliability analysis such as total positivity, Polya frequency functions, association of random variables, renewal theory, majorization, Schur functions, and so on. Boland and Proschan (1988) have discussed these aspects in detail. Generally, a function $K(x, y)$ of two real variables, ranging over a linearly ordered sets X and Y, is totally positive of order $r(TP_r)$ if

$$\begin{vmatrix} K(x_1, y_1) & K(x_1, y_2) & \ldots & K(x_1, y_m) \\ K(x_2, y_1) & K(x_2, y_2) & \ldots & K(x_2, y_m) \\ \vdots & \vdots & & \vdots \\ K(x_m, y_1) & K(x_m, y_2) & \ldots & K(x_m, y_m) \end{vmatrix} \geq 0$$

for all $1 \leq m \leq r$, $x_1 < x_2 < \ldots < x_m$, $y_1 < y_2 < \ldots < y_m$, $(x_i, y_i) \in (X, Y)$. The sets X and Y can be intervals or non-negative integers. If $K(x, y) = f(x - y)$, where f is a probability density (mass) function, we say that $f(\cdot)$ is Polya density (mass) function of order $r(PF_r)$. The use of TP_2 and PF_2 functions in reliability have been demonstrated in the earlier chapters. More applications in other areas like mathematical biology, approximation theory, and operator theory can be seen in Gasca and Micchelli (1996), Gori and Pitolli (2002) and Marshall et al. (2011).

Some results on majorization, Schur-convexity and Schur-constancy were discussed in connection with the Bayesian concepts of ageing in Chapters 7 and 8. Subsequently, their applications have spread over many areas in mathematics like combinational analysis, geometry, matrix theory and numerical analysis; for details, see Marshall et al. (2011). Association of random variables has triggered many interesting contributions in statistical theory. In its strongest version, $\mathbf{X} = (X_1, \ldots, X_p)$ is said to be associated if $cov(g_1(\mathbf{X}), g_2(\mathbf{X})) \geq 0$ for all non-decreasing functions g_1 and g_2 for which $E[g_1(\mathbf{X})g_2(\mathbf{X})] < \infty$. Some general results using associated random variables are as follows:

(1) If $X_1, \ldots, X_p$ are associated and $y_i = g_i(\mathbf{X})$ is non-decreasing for every $i = 1, 2, \ldots, r$, then

$$P(\mathbf{Y} \le \mathbf{y}) \ge \prod_{i=1}^{r} P(Y_i \le y_i)$$

and

$$P(\mathbf{Y}_i > \mathbf{y}) \ge \prod_{i=1}^{r} P(Y_i > y_i)$$

for all $\mathbf{y}$ in $\mathbb{R}^r$;

(2) If the X_i's in the above result are independent and $Y_i = \sum_{j=1}^{i} X_j$, then

$$P(\mathbf{Y} \le \mathbf{y}) \ge \prod_{i=1}^{p} P(Y_i \le y_i), \quad \mathbf{Y} = (Y_1, \ldots, Y_p),$$

and also

$$P[X_{(i_1)} > x_{i_1}, \ldots, X_{(i_k)} > x_{i_k}] \ge \prod_{j=1}^{k} P[X_{(i_j)} > x_{i_j}]$$

for every $1 \le i_1 < \ldots < i_k \le n$ and $x_{i_1} < \ldots < x_{i_k}$, where $X_{(i)}$ is the ith order statistic in a random sample $X_1, \ldots, X_n$.

The above result is used in the analysis of variance to test whether two hypotheses have the same error variance for each test. Another application is related to statistical mechanics. We say that $f(\mathbf{x})$ satisfies the Fortuim et al. (1971) condition if

$$f(\mathbf{x} \vee \mathbf{y}) f(\mathbf{x} \wedge \mathbf{y}) \ge f(\mathbf{x}) f(\mathbf{y})$$

for all $\mathbf{x}, \mathbf{y}$ in $\mathbb{R}^n$. If x, y are in the lattice N, μ is a non-negative function on N, and f and g are increasing functions,

$$\left(\sum f(x)g(x)\mu(x)\right)\left(\sum \mu(x)\right) \ge \left(\sum f(x)\mu(x)\right)\left(\sum g(x)\mu(x)\right).$$

Thus, the result says that if f and g are increasing or both decreasing, then the association is positive, and they are negatively associated otherwise. This result is fundamental in statistical mechanics, probabilistic combinatorics and random graphs wherein many increasing events are interpreted to be positively correlated. Total positivity is the strongest in a large hierarchy of association concepts and therefore it serves as a sufficient condition to establish positive association in terms of other weaker concepts which are sometimes difficult to prove.

Renewal theory plays an important role in various aspects of reliability theory. In order that a device or system is able to perform its intended function without disruption due to failure, several types of maintenance strategies are spelt out in reliability engineering. Two such strategies are age replacement policy and block replacement policy. If a unit is replaced upon failure or at age X, a specified value, it is an age replacement policy. In block replacement policy, a unit is replaced upon failure and at times $x, 2x, 3x, \ldots$. Then $N(x)$, the number of failures in $[0, x]$, with no planned failures is a renewal process and $N_1(x, X)$, the number of in service failures under the age replacement policy is also a renewal process. If $\bar{F}(t)$ is the survival function of T_i, the length of time between the $(i-1)$th and ith failures in the renewal process $N(t)$ and the distribution of T_i, the corresponding time between the $(i-1)$th and ith failure in the process $N_1(t, T)$ is given by the survival function

$$S_T(t) = [\bar{F}_{(T)}]^n \bar{F}(t - nT), \quad nT \leq t \leq (n+1)T, \; n = 0, 1, 2, \ldots .$$

The effectiveness of the age replacement policy is evaluated by comparing the reliability properties of $\bar{F}(t)$ and $S(t)$. The mean of the above distribution (called mean time to failure, MTTF) is given by

$$M(T) = \frac{1}{F(T)} \int_0^T \bar{F}(t)dt.$$

It is obvious that the nature of $M(T)$ is determined by the ageing patterns in F. Accordingly, ageing classes that contribute to decreasing MTTF have been investigated by Barlow and Proschan (1965) and Klefsjo (1982). Preservation of monotone MTTF under various reliability operations were studied in Knopik (2005) Asha and Nair (2010) have discussions. The conditions on MTTF that generate various ageing properties of F like IHR, IHRA, DMRL and NBUE, along with properties of stochastic orders that composes MTTF's of two distributions Marshall and Proschan (1972) determined the largest classes of life distributions for which age and block replacement policies diminish stochastically, or in expected value, the number of failures in service. The bounds on the probability that the total amount of insurance claims does not exceed the amount of premiums collected during a fixed time period was investigated by Straub (1970) if F is NBU or IHR. The residual life $u(t)$ at time $N(t)$ has distribution function (in the discrete case) as

$$A_t(u) = F(t+u) - \sum_{x=1}^{t-1} 1 - F(t+u-x)]m(x),$$

where $m(x) = \sum_{n=1}^{\infty} f_n(x)$, called renewal density, and $f_n(x)$ is the n-fold convolution with itself. The probability mass function of $U(t)$ becomes

$$a_t(u) = f(t+u) - \sum_{x=1}^{t-1} f(t+u-x)m(x)$$

when the inter-renewal time is geometric $a_t(u)$ is also geometric and is independent of t. Asymptotically,

$$\lim_{t\to\infty} A_t(u) = \frac{1}{\mu}\sum_{u=0}^{t-1}[1 - F(u)],$$

and the mean and variance are

$$\mu(u) = \frac{\sigma^2 + \mu^2}{2\mu} - \frac{1}{2}$$

and

$$v(u) = \frac{E(T^3)}{3\mu} - \left(\frac{\sigma^2 + \mu^2}{2\mu}\right)^2 - \frac{1}{12} \quad \text{(Heyman and Sobel, 1982).}$$

One can easily recognize that the asymptotic distribution is the univariate equilibrium distribution that appeared previously. Its multivariate extensions were also discussed in Chapters 7 and 8. The asymptotic joint distribution of age and residual life also have several implications in reliability studies, as can be seen from the discussions of Nair and Sankaran (2010b, 2014b) in the continuous case. Apart from reliability, equilibrium models have relevance in formulation of maintenance policies, income analysis (Kleiber and Kotz, 2003), and insurance (Willmot and Cai, 2001; Lin and Willmot, 1999, 2000). Many of the findings mentioned above do not have their discrete analogues due to slow progress in paving theoretical foundations in discrete-time reliability modelling. Much work still needs to be done in this regard!

References

Aarset, M.V., 1987. How to identify a bathtub hazard rate. IEEE Transactions on Reliability 36, 106–108.

Abbasnejad, M., 2011. Some characterization results based on dynamic survival and failure entropies. Communications of the Korean Statistical Society 18, 787–798.

Abouammoh, A.M., 1990. Partial orderings of discrete life distributions based on ageing. Unpublished manuscript.

Abouammoh, A.M., Hindi, M.I., Ahmed, A.N., 1988. Shock models with NBUFR and NBAFR survivals. Trabajosde Estadistica 3, 97–113.

Afendras, G., Balakrishnan, N., Papadatos, N., 2017. Orthogonal polynomials in the cumulative Ord family and its application to variance bounds. Statistics. In press.

A-Hameed, M.S., Proschan, F., 1973. Non-stationary shock models. Stochastic Processes and their Applications 1, 383–404.

A-Hameed, M.S., Proschan, F., 1975. Shock models with underlying birth process. Journal of Applied Probability 12, 18–28.

Ahmed, A.N., 1991. Characterization of beta, binomial and Poisson distributions. IEEE Transactions on Reliability 40, 290–293.

Akaike, H., 1974. A new look at statistical model identification. IEEE Transactions on Automation and Control 19, 716–723.

Aksenov, S.V., Savageau, M.A., 2005. Some properties of the Lerch family of discrete distributions. arXiv preprint. arXiv:math/0504485.

Alikhan, M.S., Kalique, A., Abouammoh, A.M., 1989. On estimating parameters of a discrete Weibull distribution. IEEE Transactions on Reliability 38, 348–350.

Allison, P.D., 1982. Discrete time methods for analysis of event histories. Sociological Methodology 13, 61–98.

Almalki, S.J., Nadarajah, S., 2014. A new discrete modified Weibull distribution. IEEE Transactions on Reliability 63, 68–80.

Alzaatreh, A., Lee, C., Famoye, F., 2012. On the discrete analogues of continuous distributions. Statistical Methodology 9, 589–603.

Alzaid, A.A., 1990. A weak quadrant dependence concept with applications. Communication in Statistics, Stochastic Models 6, 353–363.

Al-Zahrani, B., Dar, J.G., Daghistini, A., 2013. A note on closure properties of a class of discrete lifetime distributions. International Journal of Statistics and Probability 2, 42–46.

Anderson, J.E., Louis, T.A., Holm, N.V., Harvald, B., 1992. Time-dependent association measures for bivariate survival distributions. Journal of the American Statistical Association 87, 641–650.

Aranda-Ordaz, F.J., 1983. An extension of proportional hazard models for grouped data. Biometrics 39, 109–117.

Arrow, K.J., 1951. Social Choice and Individual Values. John Wiley & Sons, New York.

Asha, G., 1995. Some Bivariate Lifetime Models in Discrete Time. Ph.D. thesis. Cochin University of Science and Technology, Cochin, India.

Asha, G., Nair, N.U., 1997. Bivariate failure rates in discrete time. In: Abraham, B. (Ed.), Quality Improvement Through Statistical Methods. Birkhauser, Boston, pp. 339–350.

Asha, G., Nair, N.U., 2010. Reliability properties of mean time to failure in age replacement models. International Journal of Reliability, Quality and Safety Engineering 17, 15–26.

Asha, G., Nair, N.U., Sankaran, P.G., 2004. Discrete models characterized by bivariate conditional mean residual life. Calcutta Statistical Association Bulletin 55, 219–230.

Asha, G., Sankaran, P.G., Nair, N.U., 2003. Probability models with constant total failure rate. Communications in Statistics, Theory and Methods 32, 1089–1099.

Aven, T., 2012. Foundations of Risk Analysis. John Wiley & Sons, Hoboken, New Jersey.

Aven, T., Gaarder, S., 1987. Optimal replacement shock model. Journal of Applied Probability 24, 281–287.

Azadian, N., Amini, M., Bozorgnia, A., 2009. Some concepts of negative dependence for bivariate distributions and their relationships. Journal of Mathematical Extension 4, 43–59.

Bain, L.J., 1974. Analysis for the linear failure-rate life-testing distribution. Technometrics 16, 551–559.

Bain, L.J., 1978. Statistical Analysis of Reliability and Life-Testing Models. Marcel Dekker, New York.

Bairamov, I., Kotz, S., Kozubowski, T., 2003. A new measure of linear local dependence. Statistics 37, 243–258.

Balakrishnan, N., 1985. Order statistics from the half logistic distribution. Journal of Statistical Computation and Simulation 20, 287–309.

Balakrishnan, N., Cramer, E., 2008. Progressive censoring from heterogeneous distributions with applications to inference. Annals of the Institute of Statistical Mathematics 60, 151–171.

Balakrishnan, N., Kundu, D., Ng, H.K.T., Kannan, N., 2007. Point and interval estimation for a simple step-stress model with type-II censoring. Journal of Quality Technology 39, 35–47.

Balakrishnan, N., Pal, S., 2013. Lognormal lifetimes and likelihood-based inference for flexible cure rate models based on COM-Poisson family. Computational Statistics and Data Analysis 67, 41–67.

Balakrishnan, N., Pal, S., 2016. Expectation maximization-based likelihood inference for flexible cure rate models with Weibull lifetimes. Statistical Methods in Medical Research 25, 1535–1563.

Balakrishnan, N., Xie, Q., 2007. Exact inference for a simple step-stress model with type I hybrid censored data from the exponential population. Journal of Statistical Planning and Inference 137, 3268–3290.

Balakrishnan, N., Xie, Q., Kundu, D., 2009. Exact inference for a simple step-stress model from the exponential distribution under time constraint. Annals of the Institute of Statistical Mathematics 61, 251–274.

Balakrishnan, N., Ye, Y., Al-Jarallah, R., 2017. Likelihood inference for Type-I bivariate Polya-Aeppli distribution. Communications in Statistics – Simulation and Computation 46, 6436–6453.

Barber, J.S., Murphy, S.A., Axinn, W.G., Maples, J., 2000. Discrete-time multilevel hazard analysis. Sociological Methodology 30, 201–235.

Barlow, R.E., Mendel, M.B., 1992. De Finetti-type representations for life distributions. Journal of the American Statistical Association 87, 1116–1122.

Barlow, R.E., Mendel, M.B., 1993. Similarity as a probabilistic characteristic of aging. In: Barlow, R.E., Clarotti, C.A., Spizzichino, F. (Eds.), Reliability and Decision Making. Chapman and Hall, New York.

Barlow, R.E., Proschan, F., 1965. Mathematical Theory of Reliability. John Wiley & Sons, New York.

Barlow, R.E., Spizzichino, F., 1993. Schur-concave survival functions and survival analysis. Journal of Computational and Applied Mathematics 46, 437–441.

Barnett, V., 1979. Some outlier tests for multivariate samples. South African Statistical Journal 13, 29–52.

Bass, F.M., 1987. A new product growth model for consumer risk. Management Science 33, 451–466.

Bassan, B., Kochar, S.C., Spizzichino, F., 2002. Some bivariate notions of IFR and DMRL and related properties. Journal of Applied Probability 39, 533–544.

Bassan, B., Spizzichino, F., 1999. Stochastic comparisons for residual lifetimes and Bayesian notions of multivariate ageing. Advances in Applied Probability 31, 1078–1094.

Bassan, B., Spizzichino, F., 2005. Relations among univariate ageing, bivariate ageing and dependence for exchangeable lifetimes. Journal of Multivariate Analysis 93, 313–339.

Basu, A.P., Dhar, S.K., 1995. Bivariate geometric distribution. Journal of Applied Statistical Science 2, 33–34.

Basu, A.P., Ebrahimi, N., 1986. NHNBUE and HNWUE distributions: a survey. In: Basu, A.P. (Ed.), Reliability and Quality Control. North-Holland, Amsterdam, pp. 33–45.

Basu, A.P., Ebrahimi, N., Klefsjö, B., 1983. Multivariate harmonic new better than used in expectation distributions. Scandinavian Journal of Statistics 10, 19–25.

Bebbington, M., Lai, C.D., Wellington, M., Zitikis, R., 2012. The discrete additive Weibull distribution: a bathtub-shaped hazard for discontinuous failure data. Reliability Engineering and System Safety 106, 37–44.

Belzunce, F., Candel, J., Ruiz, J.M., 1998. Ordering and asymptotic properties of residual income distributions. Sankhyā, Series B 60, 331–348.

Belzunce, F., Navarro, J., Ruiz, J.M., Aguila, Y.D., 2004. Some results on residual entropy function. Metrika 59, 147–161.

Bennett, S., 1983. Analysis of survival data by proportional odds model. Statistics in Medicine 2, 273–277.

Berenhaut, K.S., Lund, R., 2002. Renewal convergence rates for DHR and NWU lifetimes. Probability in the Engineering and Informational Sciences 16, 67–84.

Berg, S., 1974. Factorial series distributions with applications to capture-recapture problems. Scandinavian Journal of Statistics 1, 145–152.

Bharatpour, S., Bami, Z., 2012. On discrete cumulative entropy. Journal of the Iranian Statistical Society 11, 203–215.

Bhattacharjee, M.C., 1988. Reliability ideas and applications in economics and social sciences. In: Krishnaiah, P.R., Rao, C.R. (Eds.), Handbook of Statistics, vol. 7. North-Holland, Amsterdam, pp. 175–213.

Bhattacharjee, M.C., 2002. Discrete convex ordered lifetimes: characterization, equivalence and applications. Sankhyā, Series B 65, 292–306.

Bhattacharjee, M.C., Krishnaji, N., 1985. DFR and other heavy tail properties in modelling the distribution of land in some alternative measures of inequality. In: Ghose, J.K. (Ed.), Statistics: Applications and New Directions. Indian Statistical Institute, Calcutta, India, pp. 100–115.

Bhattacharya, S.K., 1966. Confluent hypergeometric distributions of discrete and continuous type with applications to accident proneness. Calcutta Statistical Association Bulletin 60, 1060–1066.

Bilal, S., Hassan, A., 2006. On some models leading to quasi-negative binomial distribution. Journal of Korean Society for Industrial and Applied Mathematics 11, 15–29.

Bjerve, S., Doksum, K., 1993. Correlation curves: measures of association as functions of covariate values. The Annals of Statistics 21, 890–902.

Blischke, W.R., Murthy, D.N.P., 2000. Reliability: Modelling, Prediction and Optimization. John Wiley & Sons, New York.

Block, H.W., Savits, T.H., 1976. The IFRA closure problem. Annals of Probability 4, 1030–1032.

Block, H.W., Savits, T.H., Singh, H., 1998. The reversed hazard rate function. Probability in the Engineering and Informational Sciences 12, 69–90.

Blumenthal, S., 1967. Proportional sampling in life length studies. Technometrics 9, 205–218.

Boland, P.J., Proschan, F., 1988. The impact of reliability theory on some branches of mathematics and statistics. In: Krishnaiah, P.R., Rao, C.R. (Eds.), Handbook of Statistics, vol. 7. North-Holland, pp. 157–174.

Bouezmarni, T., Mesfioui, M., Tajor, A., 2009. On concordance measures for discrete data and dependence properties of Poisson model. Journal of Probability and Statistics 2009, 895742.

Bracquemond, C., Gaudoin, O., 2003. A survey on discrete lifetime distributions. International Journal of Reliability Quality and Safety Engineering 10, 69–98.

Bracquemond, C., Gaudoin, O., Roy, D., Xie, M., 2001. On some notions of discrete ageing. In: Irony, T., Xie, M. (Eds.), Hayakawa System and Bayesian Reliability, Essays in Honour of Professor Richard E Barlow on His 70th Birthday. In: Series on Quality, Reliability and Engineering Statistics, vol. 35. World Scientific Press, Singapore.

Brindley, E.C., Thompson, W.A., 1972. Dependence and ageing aspects of multivariate survival. Journal of the American Statistical Association 67, 822–830.

Brown, M., Rao, C.N., 1980. On the First Passage Time Distribution for a Class of Markov Chains. Technical Report No. 79. The Florida State University.

Bryson, M.C., Siddiqui, M.M., 1969. Some criteria for aging. Journal of the American Statistical Association 64, 1472–1483.

Buchanan, W.B., Singpurwalla, N.D., 1977. Some stochastic characterizations of multivariate survival. In: Tsokos, C.P., Shimi, I.N. (Eds.), The Theory and Applications of Reliability. Academic Press, New York, pp. 329–348.

Buiculescu, M., Petrehus, V., Zbaganu, G., 2011. Iterated integrated tail with periodic hazard rates. Math Reports 13, 127–139.

Cai, J., 2001. Discrete Time Risk Models Under Stochastic Forces of Interest. Ph.D. thesis. University of Melbourne.

Caramellino, L., Spizzichino, F., 1994. Dependence and aging properties of lifetimes with Schur-constant survival functions. Probability in the Engineering and Informational Sciences 8, 103–111.

Caramellino, L., Spizzichino, F., 1996. WBF property and stochastic monotonicity of the Markov process associated to Schur-constant survival functions. Journal of Multivariate Analysis 56, 153–163.

Castaner, A., Claramunt, M.M., Lefevre, C., Loisel, S., 2014. Discrete Schur-Constant Models. Report.

Castaner, A., Claramunt, M.M., Lefevre, C., Loisel, S., 2015. Discrete Schur-constant models. Journal of Multivariate Analysis 140, 343–362.

Castillo, J., Pérez-Casany, M., 1998. Weighted Poisson distributions for overdispersion and underdispersion situations. Annals of the Institute of Statistical Mathematics 50, 567–585.

Castillo, J., Pérez-Casany, M., 2005. Overdispersed and underdispersed Poisson generalizations. Journal of Statistical Planning and Inference 134, 486–500.

Castillo, X., Sieworek, D.P., 1981. Workload performance and reliability of digital computing systems. Proceedings of the FTCS 11, 84–89.

Cha, J.H., Finkelstein, M., 2012. On generalized shock models for deteriorating systems. Applied Stochastic Models in Business and Industry 29, 496–508.

Chakraborty, S., Chakravarty, D., 2012. Discrete gamma distribution: properties and parameter estimation. Communications in Statistics, Theory and Methods 41, 3301–3324.

Chandra, M., Singpurwalla, N.D., 1981. Relationships between some notions which are common to reliability theory and economics. Mathematics of Operations Research 6, 113–121.

Charalambides, C.A., 1986. Gould series distributions with applications to fluctuations of sums of random variables. Journal of Statistical Planning and Inference 14, 15–28.

Charalambides, C.A., 1990. Abel series distributions with applications to fluctuations of sample functions of stochastic processes. Communications in Statistics, Theory and Methods 19, 317–335.

Chatterjee, A., Mukherjee, S.P., 2001. Equilibrium distribution – its role in reliability. In: Balakrishnan, N., Rao, C.R. (Eds.), Advances in Reliability, Chapter 4. In: Handbook of Statistics, vol. 20. North-Holland, Amsterdam, The Netherlands, pp. 105–137.

Chemany, A., Bharatpour, S., 2014. A dynamic discrimination information based on cumulative residual entropy and its properties. Communications in Statistics, Theory and Methods 43, 1041–1049.

Chen, Y.Q., Rhode, C.A., Wang, M.C., 2002. Additive hazard models with latent treatment effectiveness. Biometrika 89, 417–931.

Cheng, S.W., Fu, J.C., Sinha, S.K., 1985. An empirical procedure for estimating parameters of mixed exponential life testing model. IEEE Transactions on Reliability 34, 60–64.

Chi, Y., Yang, J., Qi, Y., 2009. Decomposition of a Schur-constant model and its applications. Insurance: Mathematics and Economics 44, 398–408.

Chukova, S., Dimitrov, B., 1992. On distributions having the almost lack of memory property. Journal of Applied Probability 29, 691–698.

Chukova, S., Dimitrov, B., Garrido, J., 1993a. Renewal and non-homogeneous Poisson processes generated by distributions with periodic failure rates. Statistics and Probability Letters 17, 19–25.

Chukova, S., Dimitrov, B., Khalil, S., 1993b. Probability distributions similar to the exponential characterization. Canadian Journal of Statistics 21, 269–276.

Clayton, D.G., 1978. A model for association in bivariate life tables and its application in epidemiological studies of familial tendency in chronic disease incidence. Biometrika 65, 141–151.

Cohen, A., 1960. Estimation in the Poisson distribution when the sample values (c+1) are sometimes erroneously as c. Annals of the Institute of Statistical Mathematics 9, 181–193.

Colangelo, A., Scarsini, M., Shaked, M., 2005. Some notions of multivariate positive dependence. Insurance: Mathematics and Economics 37, 13–26.

Collett, D., 1994. Modelling Binary Data. Chapman and Hall, London.

Comptet, L., 1994. Advanced Combinatorics. D. Reidel, Dordrecht, The Netherlands.

Consul, P.C., 1989. Generalized Poisson Distributions: Applications and Properties. Marcel Dekker, New York.

Consul, P.C., 1990. Geeta distribution and its properties. Communications in Statistics, Theory and Methods 19, 3051–3068.

Consul, P.C., 1995. Some characterizations of the exponential class of distributions. IEEE Transactions on Reliability 44, 403–407.

Consul, P.C., Famoye, F., 2006. Lagrangian Probability Distributions. Birkhauser, Boston.

Consul, P.C., Jain, G.C., 1973a. A generalization of the Poisson distribution. Technometrics 15, 791–799.

Consul, P.C., Jain, G.C., 1973b. On some interesting properties of the generalized Poisson distribution. Biometrische Zeitschrift 15, 495–500.

Consul, P.C., Shenton, L.R., 1972. Use of Lagrangian expansion for generating generalized probability distribution. SIAM Journal of Applied Mathematics 23, 239–248.

Consul, P.C., Shenton, L.R., 1973. Some interesting properties of Lagrangian distributions. Communications in Statistics 2, 263–272.

Consul, P.C., Shenton, L.R., 1975. On the probabilistic structure and properties of discrete Lagrangian distributions. In: Patil, G.P., Kotz, S., Ord, J.K. (Eds.), Statistical Distributions in Scientific Work. D. Reidel Publishing Company, Boston, pp. 41–57.

Conway, R.W., Maxwell, W.L., 1962. A queueing model with state dependent service rates. Journal of Industrial Engineering 12, 132–136.

Cox, D.R., 1959. The analysis of exponentially distributed lifetimes with two types of failure. Journal of the Royal Statistical Society, Series B 21, 411–421.

Cox, D.R., 1972. Regression models and life tables. Journal of the Royal Statistical Society, Series B 34, 187–220.

Cox, D.R., Oakes, D., 1984. Analysis of Survival Data. Chapman and Hall, London.

Dagum, C., 1977. A new model for personal income distribution: specification and estimation. Economic Applique 33, 327–367.

David, H.A., Moeschberger, M.L., 1978. The Theory of Competing Risks. Griffin, London.

Davis, D.J., 1952. An analysis of some failure time data. Journal of the American Statistical Association 47, 113–150.

Deng, L., Chhikkara, R.S., 1990. On the characterization of the exponential distribution by independence of its integer and fractional parts. Statistica Neerlandica 44, 83–85.

Denuit, M., Dhaene, J., Goovaerts, M., Kass, R., 2005. Actuarial Theory for Dependent Risks: Measures, Orders and Models. John Wiley & Sons, Hoboken, New Jersey.

Deshpande, J.V., Singh, H., Bagai, I., Jain, K., 1990. Some partial orders describing positive ageing. Communications in Statistics, Stochastic Models 6, 471–481.

Dewan, I., Sudheesh, K.K., 2009. On Proportional (Reversed) Hazard Model for Discrete Data. Technical Report. Indian Statistical Institute, New Delhi, India.

Dhar, S.K., Balaji, S., 2006. On the characterization of a bivariate geometric distribution. Communications in Statistics, Theory and Methods 35, 759–765.

Di Crescenzo, A., 2000. Some results on the proportional reversed hazards model. Statistics and Probability Letters 50, 313–321.

Di Crescenzo, A., Longobardi, M., 2002. Entropy-based measure of uncertainty in past lifetime distributions. Journal of Applied Probability, 434–440.

Dimitrov, B., Chukova, S., Green Jr., D., 1997. Probability distributions in periodic random environment and their applications. SIAM Journal of Applied Mathematics 57, 501–517.

Dimitrov, B., Chukova, S., Khalil, Z., 1995. Definitions, characterization and structural properties of probability distributions similar to the exponential. Journal of Statistical Planning and Inference 43, 271–287.

Domma, F., 2011. Bivariate reversed hazard rate, notions and measures of dependence and their relationships. Communications in Statistics, Theory and Methods 40, 989–999.

Doray, L.G., Luong, A., 1995. Quadratic distance estimators of the zeta family. Insurance: Mathematics and Economics 16, 255–260.

Doray, L.G., Luong, A., 1997. Efficient estimators of the Good family. Communications in Statistics, Simulation and Computation 26, 1075–1088.

Ebrahimi, N., 1986. Classes of discrete decreasing and increasing mean residual life distributions. IEEE Transactions on Reliability 35, 403–405.

Ebrahimi, N., 1996. How to measure uncertainty in residual life distributions. Sankhyā, Series A 58, 48–56.

El-Neweihi, E., 1981. Stochastic ordering of a class of multivariate new better than used distributions. Communications in Statistics, Theory and Methods 10, 1655–1672.

Elabatal, I., Ahsanullah, M., 2012. On some properties of the discrete NBU class based on generating function order. Journal of Statistical Theory and Applications 11, 209–223.

Elffers, H., 1980. On interpreting product moment correlation coefficient. Statistica Neerlandica 34, 3–11.

Elteto, O., Frigyes, E., 1968. New income inequality measures as efficient tools for causal analysis and planning. Econometrica 36, 383–396.

Erdelyi, A., Magnus, W., Oberhettinger, F., Tricomi, F.G., 1953. Higher Transcendental Functions, vol. I. McGraw-Hill, New York.

Eryilmaz, S., 2012. Life behaviour of a system under discrete shock model. Discrete Dynamics in Nature and Society, 238641.

Eryilmaz, S., 2013. On the lifetime behaviour of a discrete time shock model. Journal of Computational and Applied Mathematics 237, 384–388.

Esary, J.D., Marshall, A.W., 1979. Multivariate distributions with increasing hazard average. Annals of Probability 7, 359–370.

Esary, J.D., Marshall, A.W., Proschan, F., 1973. Shock models and wear processes. Annals of Probability 1, 627–649.

Esary, J.D., Proschan, F., Walkup, D.W., 1967. Association of random variables with applications. Annals of Mathematical Statistics 38, 1466–1474.

Evans, D.L., Leemis, L.M., 2004. Algorithms for computing the distributions of sums of discrete random variables. Mathematics and Computer Modelling 40, 1429–1452.

Famoye, F., 1997. Generalized geometric distribution and some of its applications. Journal of Mathematical Sciences 8, 1–13.

Fan, J., Ghurye, S.G., Levine, R.A., 2000. Multicomponent lifetime distributions in the presence of ageing. Journal of Applied Probability 37, 521–533.

Finkelstein, M.S., 2002. On the reversed hazard rate. Reliability Engineering and System Safety 78, 71–75.

Finkelstein, M.S., Zarudnij, V.I., 2001. A shock process with non-cumulative damage. Reliability Engineering and System Safety 71, 103–107.

Fisher, R.A., 1934. The effects of methods of ascertainment upon the estimation of frequencies. Annals of Eugenics 6, 13–25.

Fortuim, C.M., Kastelyn, P.W., Ginibre, J., 1971. Correlation inequalities in some partially sets. Communications in Mathematical Physics 22, 89–103.

Fox, W.R., Lasker, G.W., 1983. The distribution of surname frequencies. International Statistics Review 51, 81–87.

Gasca, M., Micchelli, C.A., 1996. Total Positivity and Its Applications. Kluwer Academic Publishers, Boston, MA, pp. 133–155.

Ghitany, M.E., 2004. The monotonicity of the reliability measures of the beta distribution. Applied Mathematics Letters 17, 1277–1283.

Ghosh, M., Ebrahimi, N., 1981. Multivariate NBU and NBUE distribution. Egyptian Statistical Journal 25, 36–55.

Giorgi, G.M., Crescenzi, M., 2001a. A proposal of poverty measure based on Bonferroni inequality index. Metron 59, 3–16.

Giorgi, G.M., Crescenzi, M., 2001b. A look at the Bonferroni inequality measure in a reliability framework. Statistica 61 (4), 571–583.

Glanzel, W., 1987. A characterization theorem based on truncated moments and its applications some distribution families. In: Mathematical Statistics and Probability. In: Statistical Inference and Methods, vol. B. D. Reidel, Dordrecht, The Netherlands, pp. 75–84.

Glanzel, W., 1991. Characterization through some conditional moments of Pearson-type distributions and discrete analogues. Sankhyā 53, 17–24.

Glanzel, W., Teles, A., Schubert, A., 1984. Characterization by truncated moments and its application to Pearson-type distributions. Zeitschrift für Wahrscheinlichkeitstheorie und Verwandte Gebiete 66, 173–183.

Glaser, R.E., 1980. Bathtub related failure rate characterizations. Journal of the American Statistical Association 75, 667–672.

Goliforushani, S., Asadi, M., 2008. On the discrete mean past time. Metrika 68, 209–217.

Goliforushani, S., Asadi, M., 2011. Stochastic ordering among inactivity times of coherent systems. Sankhyā, Series B 73, 241–262.

Goliforushani, S., Asadi, M., Balakrishnan, N., 2012. On the residual and inactivity times of the components of used coherent systems. Applied Probability 49, 385–404.

Gomez-Deniz, E., Calderin-Ojeda, E., 2011. The discrete Lindley distribution: properties and applications. Journal of Statistical Computation and Simulation 81, 1405–1416.

Good, I.J., 1953. The population frequencies of species and the estimation of population parameters. Biometrika 40, 237–264.

Good, I.J., 1957. Distribution of word frequencies. Nature 179, 595.

Gori, L., Pitolli, F., 2002. On some applications of a class of totally positive bases. International Conference on Wavelet Analysis and Applications 25, 109–118.

Guess, F.M., Park, D.H., 1988. Modelling discrete bathtub and upside-down bathtub mean residual life functions. IEEE Transactions on Reliability 37, 545–549.

Gumbel, E.J., 1960. Bivariate exponential distributions. Journal of the American Statistical Association 55, 698–707.

Gupta, P.L., Gupta, R.C., Ong, S., Srivastava, H.M., 2008. A class of Hurwitz-Lerch zeta distributions and their applications in reliability. Applied Mathematics and Computation 196, 521–531.

Gupta, P.L., Gupta, R.C., Tripathi, R.C., 1997. On the monotonic properties of discrete failure rates. Journal of Statistical Planning and Inference 65, 255–268.

Gupta, R.C., 1974. Modified power series distributions and some of its applications. Sankhyā, Series B 36, 288–298.

Gupta, R.C., 1975. Some characterizations of discrete distributions by properties of their moment distribution. Communications in Statistics, Theory and Methods 4, 761–785.

Gupta, R.C., 1979. On the characterization of survival distributions in reliability by properties of their renewal densities. Communications in Statistics, Theory and Methods 98, 685–697.

Gupta, R.C., 2003. On some association measures in bivariate distributions and their relationships. Journal of Statistical Planning and Inference 17, 83–98.

Gupta, R.C., Gupta, P.L., Gupta, R.D., 1998. Modelling failure time data with Lehmann alternative. Communications in Statistics, Theory and Methods 27 (4), 887–904.

Gupta, R.C., Gupta, R.D., 2007. Proportional reversed hazards model and its applications. Journal of Statistical Planning and Inference 137, 3525–3536.

Gupta, R.C., Kirmani, S.N.U.A., 1990. The role of weighted of weighted distributions in stochastic modeling. Communications in Statistics, Theory and Methods 19, 3147–3162.

Gupta, R.C., Kirmani, S.N.U.A., 1998. On the proportional mean remaining life and its implications. Statistics 32, 175–187.

Gupta, R.C., Wu, H., 2001. Analyzing survival data by proportional reversed hazard model. International Journal of Reliability and Applications 2 (1), 1–26.

Gupta, R.P., Nair, N.U., Asha, G., 2006. Characterizing discrete life distribution by relations between reversed failure rates and conditional expectations. Far East Journal of Theoretical Statistics 20, 113–122.

Gupta, R.P., Sankaran, P.G., 1998. Bivariate equilibrium distribution and its application to reliability. Communications in Statistics, Theory and Methods 27, 385–394.

Gut, A., 2001. Mixed shock models. Bernoulli 7, 541–555.

Gut, A., 2005. Probability: A Graduate Course. Springer-Verlag, New York.

Haight, F.A., 1961. A distribution analogous to the Borel-Tanner. Biometrika 48, 167–173.

Haight, F.A., 1966. Some statistical problems in connection with word association data. Journal of Mathematical Psychology 3, 217–233.

Haight, F.A., 1969. Two probability distributions connected with Zipf's rank-size conjecture. Zastosowania Matematyki 10, 225–228.

Haritha, N.H., Nair, N.U., Nair, K.R.M., 2008. Modelling incomes using generalized lambda distributions. Journal of Income Distributions 17, 37–51.

Harris, R.A., 1970. Multivariate definition for increasing hazard rate distribution. Annals of Mathematical Statistics 41, 713–717.

Hassan, A., Bilal, S., 2008. On some properties of quasi-negative binomial distribution and its applications. Journal of Modern Applied Mathematical Methods 7, 616–631.

Hawkes, A.G., 1972. A bivariate exponential distribution with applications to reliability. Journal of the Royal Statistical Society, Series B 34, 129–131.

Hesselager, O., 1994. A recursive procedure for calculation of some compound distributions. ASTIN Bulletin 24, 19–32.

Heyman, D., Sobel, M., 1982. Stochastic Models in Operations Research, vol. 1. McGraw-Hill, New York.

Hitha, N., Nair, N.U., 1989. Characterization of some discrete models by properties of residual life function. Calcutta Statistical Association Bulletin 38, 151–154.

Hjorth, U., 1980. A reliability distribution with increasing, decreasing, constant and bathtub shaped failure rates. Technometrics 22, 99–107.

Hoffmann-Jorgensen, J., 1984. Stochastic Processes on Polish Spaces. Report. Aarhus University, Denmark.

Hoffman-Jorgensen, J., 1994. Probability with a View Towards Statistics. Chapman and Hall, London.

Hu, T.H., Min, M., Nanda, A.K., 2003. Moment inequalities for discrete ageing families. Communications in Statistics, Theory and Methods 32, 61–90.

Hurvich, C.M., Tsai, C.L., 1989. Regression and time series model selection in small samples. Biometrika 76, 297–307.

Hussain, T., Ahmad, M., 2014. Discrete inverse Rayleigh distribution. Pakistan Journal of Statistics 30, 203–222.

Irwin, J.O., 1968. The generalized Waring distribution applied to accident theory. Journal of the Royal Statistical Society A 131, 205–225.

Irwin, J.O., 1975a. The generalized Waring distribution, Part I. Journal of the Royal Statistical Society A 138, 18–31.

Irwin, J.O., 1975b. The generalized Waring distribution, Part II. Journal of the Royal Statistical Society A 138, 204–227.

Irwin, J.O., 1975c. The generalized Waring distribution, Part III. Journal of the Royal Statistical Society A 138, 374–384.

Jain, G.C., Consul, P.C., 1971. A generalized negative binomial distribution. SIAM Journal of Applied Mathematics 21, 501–513.

Jain, K., Nanda, A.K., 1995. On multivariate weighted distributions. Communications in Statistics, Theory and Methods 24, 2517–2539.

Jazi, M.A., Lai, C.D., Alamatsaz, M.H., 2010. A discrete inverse Weibull distribution and estimation of its parameters. Statistical Methodology 7, 121–132.

Jensen, P., 1989. Component failures based on flaw distributions. In: Proceedings of the Reliability and Maintenance Symposium, pp. 91–95.

Jiang, R., 2010. Discrete competing risk model with application to modeling bus-motor failure data. Reliability Engineering and System Safety 95, 981–988.

Jiang, R., Ji, P., Xiao, X., 2003. Ageing property of univariate failure models. Reliability Engineering and System Safety 79, 113–116.

Joe, H., 1997. Multivariate Models and Dependence Concepts. Chapman and Hall, New York.

Johnson, N.L., Kotz, S., 1975. A vector valued multivariate hazard rate. Journal of Multivariate Analysis 5, 53–56.

Johnson, N.L., Kotz, S., Balakrishnan, N., 1997. Discrete Multivariate Distributions. John Wiley & Sons, New York.

Johnson, N.L., Kotz, S., Kemp, A.W., 1992. Univariate Discrete Distributions, second edition. John Wiley & Sons, New York.

Kalbfleisch, J.D., Prentice, R.L., 2002. The Statistical Analysis of Failure Time Data. John Wiley & Sons, New York.

Katz, L., 1965. Unified treatment of a broad class of discrete probability distributions. In: Patil, G.P. (Ed.), Classical and Contagious Discrete Distributions. Statistical Publishing Society, Calcutta, India, pp. 175–185.

Keilson, J., Sumita, U., 1982. Uniform stochastic ordering and related inequalities. Canadian Journal of Statistics 10, 181–198.

Kemp, A.W., 1968a. A wide class of discrete distributions and the associated differential equations. Sankhyā, Series A 30, 401–410.

Kemp, A.W., 1968b. Studies in Univariate Discrete Distribution Theory Based on the Generalized Hypergeometric Function and Associated Differential Equations. Ph.D. thesis. Queens University of Belfast, Northern Ireland.

Kemp, A.W., 2004. Classes of discrete life distribution. Communications in Statistics, Theory and Methods 33, 3069–3093.

Kemp, A.W., 2008. The discrete half-normal distribution. In: Arnold, B.C., Balakrishnan, N., Sarabia, J.M., Minguez, R. (Eds.), Advances in Mathematical and Statistical Modeling. Birkhäuser, pp. 353–360.

Kemp, A.W., 2010. Families of power series distributions with particular references to the Lerch family. Journal of Statistical Planning and Inference 140, 2255–2259.

Khalique, A., 1989. On discrete failure-time distributions. Reliability Engineering and System Safety 25, 99–107.

Khorashadizadeh, M., Roknabadi, A.H.R., Borzadaran, G.R., 2010. Variance residual life function in discrete random ageing. Metron LXVII, 67–75.

Khorashadizadeh, M., Roknabadi, A.H.R., Borzadaran, G.R.M., 2012. Characterizations of lifetime distributions based on doubly truncated mean residual life and mean past to failure. Communications in Statistics, Theory and Methods 41, 1105–1115.

Khorashadizadeh, M., Roknabadi, A.H.R., Borzadaran, G.R.M., 2013a. Doubly truncated (interval) cumulative residual and past entropy. Statistics & Probability Letters 83, 1464–1471.

Khorashadizadeh, M., Roknabadi, A.H.R., Borzadaran, G.R.M., 2013b. Characterization of life distributions using log-odds rate in discrete aging. Communications in Statistics, Theory and Methods 42, 76–87.

Kirmani, S.N.U.A., Gupta, R.C., 2001. On proportional odds model in survival analysis. Annals of the Institute of Statistical Mathematics 53, 203–216.

Klefsjo, B., 1980. On a Class of Bivariate Life Distributions. Technical report. University of Umea, Sweden.

Klefsjo, B., 1981. HNBUE survival under some shock models. Scandinavian Journal of Statistics 8, 39–47.

Klefsjo, B., 1982. On ageing properties and total time on test transforms. Scandinavian Journal of Statistics 9, 37–41.

Kleiber, C., Kotz, S., 2003. Statistical Size Distributions in Economics and Actuarial Sciences. John Wiley & Sons, New York.

Knopik, L., 2005. Some results on ageing class. Control and Cybernetics 34, 1175–1180.

Kokonendji, C.C., Mizère, D., Balakrishnan, N., 2008. Connection of the Poisson weight function to overdispersion and underdispersion. Journal of Statistical Planning and Inference 138, 1287–1296.

Kokonendji, C.C., Senga Kiessé, T., Balakrishnan, N., 2009. Semiparametric estimation for count data through weighted distributions. Journal of Statistical Planning and Inference 139, 3625–3638.

Kotz, S., Balakrishnan, N., Johnson, N.L., 2000. Continuous Multivariate Distributions, vol. 1, second edition. John Wiley & Sons, New York.

Kotz, S., Johnson, N.L., 1991. A note on renewal (partial sums) distribution for discrete random variables. Statistics and Probability Letters 12, 229–231.

Kosambi, D.D., 1949. Characteristic properties of series distributions. In: Proceedings of the National Institute of Science, vol. 15. India, pp. 109–113.

Krishna, H., Pundir, P.S., 2009. Discrete Burr and discrete Pareto distributions. Statistical Methodology 6, 177–188.

Kulasekera, K.B., Tonkyn, D.W., 1992. A new discrete distribution, with applications to survival, dispersal and dispersion. Communications in Statistics, Simulation and Computation 21, 499–518.

Kumar, V., Taneja, H.C., 2011. Some characterization results on generalized cumulative residual entropy measure. Statistics & Probability Letters 81, 1072–1077.

Kupka, J., Loo, S., 1989. The hazard and vitality measures of ageing. Journal of Applied Probability 26, 532–542.

Lai, C.D., 2013. Issues concerning construction of discrete lifetime models. Quality Technology and Quality Management 10, 251–262.

Lai, C.D., Wang, D.Q., 1995. A finite range discrete life distribution. International Journal of Reliability, Quality and Safety Engineering 2, 147–160.

Lai, C.D., Xie, M., 2006. Stochastic Aging and Dependence for Reliability. Springer, New York.

Lehmann, E.L., 1966. Some concepts of dependence. Annals of Mathematical Statistics 37, 1137–1153.

Lerch, M., 1887. Note sur la function. Acta Mathematica 11, 19–24.

Li, M.S., 2011. The equilibrium distribution of counting random variables. Open Journal of Discrete Mathematics 1, 127–135.

Li, J., Dhar, S.K., 2013. Modelling with bivariate geometric distributions. Communications in Statistics, Theory and Methods 42, 252–266.

Lin, G.D., Hu, C., 2001. The Riemann zeta distribution. Bernoulli 7, 817–822.

Lin, X., Willmot, G.E., 1999. Analysis of defective renewal equation arising in ruin theory. Insurance: Mathematics and Economics 25, 63–84.

Lin, X., Willmot, G.E., 2000. The moments of the time of ruin, the surplus before ruin and the deficit at ruin. Insurance: Mathematics and Economics 27, 19–44.

Longford, N.T., 1989. Classes of Multivariate Exponential and Multivariate Geometric Distributions Derived from Markov Process. Ph.D. thesis. ETS Research Report Series.

Lorenz, M.O., 1905. Methods of measuring concentration of wealth. Publication of American Statistical Association 9, 209–219.

Lukacs, E., Laha, R.G., 1964. Applications of Characteristic Functions. Griffin, New York.

Magnus, W., Oberhettinger, F., Soni, R.P., 1966. Formulas and Theorems for the Special Functions of Mathematical Physics. Springer-Verlag, New York.

Mandelbrot, B., 1959. A note on a class of skew distribution functions: analysis and critique of a paper by H.A. Simmon. Information and Control 2, 90–99.

Mandelbrot, B., 1966. Information theory and psycholinguistics: a theory of words frequencies. In: Lazafeld, P., Henry, N. (Eds.), Readings in Mathematical Social Science. MIT Press, Boston.

Marsaglia, G., Tubilla, A., 1975. A note on the "lack of memory" of the exponential distribution. The Annals of Probability 3, 353–354.

Marshall, A.W., 1975. Some comments on the hazard gradient. Stochastic Processes and their Applications 3, 293–300.

Marshall, A.W., Olkin, I., 1967. A multivariate exponential distribution. Journal of the American Statistical Association 62, 30–44.

Marshall, A.W., Olkin, I., 1997. A new method of adding a parameter to a family of distributions with application to exponential and Weibull families. Biometrika 84, 641–652.

Marshall, A.W., Olkin, I., Arnold, B.C., 2011. Inequalities: Theory of Majorization and Its Applications, second edition. Springer-Verlag, New York.

Marshall, A.W., Proschan, F., 1972. Classes of distribution applicable in replacement with renewal theory implications. In: Proceedings of the Sixth Berkeley Symposium on Mathematical Statistics and Probability. California Press, Berkeley, CA, pp. 395–415.

Marshall, A.W., Shaked, M.A., 1982. Class of multivariate new better than used distributions. Annals of Probability 10, 259–264.

Marshall, A.W., Shaked, M.A., 1986. Multivariate new better than used distributions. Mathematics of Operations Research 11, 110–116.

Mendenhall, W., Hader, R.J., 1958. Estimation of parameter of a mixed exponential failure time distribution from censored life test data. Biometrika 45, 504–508.

Mi, J., 1993. Discrete bathtub failure rate and upside-down bathtub mean residual life. Naval Research Logistics 40, 361–371.

Mi, J., 1996. Bathtub failure rates and upside-down bathtub mean residual life. IEEE Transaction on Reliability 44, 388–391.

Minkova, L.D., 2002. A generalization of the classical discrete distributions. Communications in Statistics, Theory and Methods 31, 871–888.

Minkova, L.D., Balakrishnan, N., 2013. Compound weighted Poisson distributions. Metrika 76, 543–558.

Minkova, L.D., Balakrishnan, N., 2014a. On a bivariate Pólya-Aeppli distribution. Communications in Statistics, Theory and Methods 43, 5026–5028.

Minkova, L.D., Balakrishnan, N., 2014b. Type II bivariate Pólya-Aeppli distribution. Statistics and Probability Letters 88, 40–49.

Mohanty, S.G., 1966. On a generalized two-coin tossing problem. Biometrical Journal 8, 266–272.

Mosteller, F., Tukey, J.W., 1977. Data Analysis and Regression: A Second Course in Statistics. Pearson.

Muth, E.J., 1977. Reliability models with positive memory derived from mean residual life function. In: Tsokos, C.P., Shimi, I.N. (Eds.), Theory and Application of Reliability. Academic Press, Boston, pp. 401–435.

Nagaraja, H.N., 1983. Two characterizations of a bivariate geometric distribution. Journal of the Indian Statistical Association 21, 27–30.

Nair, K.R.M., Nair, N.U., 1988a. On characterizing a bivariate geometric distribution. Journal of the Indian Statistical Association 26, 45–49.

Nair, K.R.M., Nair, N.U., 1988b. On characterizing the bivariate exponential and geometric distributions. Annals of the Institute of Statistical Mathematics 40, 267–271.

Nair, N.U., 1983a. A measure of memory for some discrete distributions. Journal of the Indian Statistical Association 21, 141–147.

Nair, N.U., 1983b. On a distribution of first conception delays in the presence of adolescent sterility. Demography India 12, 269–275.

Nair, N.U., 1992. A characterization of the Gumbel's bivariate exponential distribution. Journal of the Indian Statistical Association 30, 99–102.

Nair, N.U., Asha, G., 1994. Characteristic properties of a bivariate geometric law. Journal of the Indian Statistical Association 32, 111–116.

Nair, N.U., Asha, G., 1997. Some classes of multivariate life distributions in discrete time. Journal of Multivariate Analysis 62, 181–189.

Nair, N.U., Asha, G., 2004. Characterization using failure and reversed failure rates. Journal of Indian Society of Probability and Statistics 8, 45–56.

Nair, N.U., Geetha, K.G., Priya, P., 1999. Modeling lifelength data using mixture distributions. Journal of the Japan Statistical Society 29, 65–73.

Nair, N.U., Geetha, K.G., Priya, P., 2000. On partial moments of discrete distributions and their applications. Far East Journal of Theoretical Statistics 4, 153–164.

Nair, N.U., Hitha, N., 1989. Characterization of discrete models by distribution based on their partial sums. Statistics and Probability Letters 8, 335–337.

Nair, N.U., Nair, K.R.M., 1990. Characterizations of a bivariate geometric distribution. Statistica 50, 247–253.

Nair, N.U., Nair, K.R.M., Haridas, H.N., 2008. Some properties of income gap ratio and truncated Gini coefficient. Calcutta Statistical Association Bulletin 60, 245–254.

Nair, N.U., Preeth, M., 2008. Multivariate equilibrium distributions of order n. Statistics and Probability Letters 78, 3312–3320.

Nair, N.U., Sankaran, P.G., 1991. Characterization of the Pearson family of distributions. IEEE Transactions on Reliability 40, 75–77.

Nair, N.U., Sankaran, P.G., 2010a. A new measure of association for bivariate survival data. Journal of Statistical Planning and Inference 140, 2569–2581.

Nair, N.U., Sankaran, P.G., 2010b. Properties of a mean residual life arising from renewal theory. Naval Research Logistics 57, 373–379.

Nair, N.U., Sankaran, P.G., 2010c. Some results on an additive hazard model. Metrika 75, 389–402.

Nair, N.U., Sankaran, P.G., 2013. Characterizations of discrete distributions using reliability concepts in reversed time. Statistics and Probability Letters 83, 1939–1945.

Nair, N.U., Sankaran, P.G., 2014a. Characterizations and time-dependent association measures for bivariate Schur-constant distributions. Brazilian Journal of Probability and Statistics 28, 409–423.

Nair, N.U., Sankaran, P.G., 2014b. Modelling lifetimes with bivariate Schur-constant distributions from renewal theory. Metron 72, 331–349.

Nair, N.U., Sankaran, P.G., 2014c. A family of bivariate exponential distributions and their copulas. Sankhyā, Series B 76, 1–18.

Nair, N.U., Sankaran, P.G., 2015a. Multivariate discrete scalar hazard rate. Statistical Methodology 27, 39–50.

Nair, N.U., Sankaran, P.G., 2015b. Odds function and odds rates for discrete lifetime distributions. Communications in Statistics, Theory and Methods 44, 4185–4202.

Nair, N.U., Sankaran, P.G., Balakrishnan, N., 2013. Quantile-Based Reliability Analysis. Birkhauser, Boston.

Nair, N.U., Nair, K.R.M., Sreelakshmi, N., 2012a. Some properties of the new Zenga curve. Statistica & Applicazioni 10, 43–52.

Nair, N.U., Sankaran, P.G., Preeth, M., 2012b. Reliability aspects of discrete equilibrium distributions. Communications in Statistics, Theory and Methods 41, 500–515.

Nair, N.U., Sankaran, P.G., Ramesh, N.P., 2016. Determination of hazard rate shape for discrete lives. International Journal of Reliability, Quality and Safety Engineering 23 (04), 1650015.

Nair, N.U., Sankaran, P.G., Ramesh, N.P., 2017a. Some properties of discrete bathtub-shaped distributions. Communications in Statistics, Theory and Methods 46 (11), 5384–5393.

Nair, N.U., Sankaran, P.G., Ramesh, N.P., 2017b. Discrete distributions with bathtub-shaped hazard rates. South African Statistical Journal 51 (1), 81–102.

Nair, N.U., Sudheesh, K.K., 2008. Some results on lower variance bounds useful in reliability modelling and estimation. Annals of the Institute of Statistical Mathematics 60, 591–603.

Nair, V.K.R., Nair, N.U., 1991. On conditional lack of memory. In: Proceedings of the National Symposium on Distribution Theory. Trivandrum, India, pp. 25–30.

Nakagawa, T., Osaki, S., 1975. The discrete Weibull distributions. IEEE Transactions on Reliability 24, 300–301.

Nanda, A.K., Paul, P., 2006a. Some results on generalized past entropy. Journal of Statistical Planning and Inference 136, 3659–3674.

Nanda, A.K., Paul, P., 2006b. Some results on generalized residual entropy. Information Sciences 176, 27–47.

Nanda, A.K., Sengupta, D., 1999. Logconcave and concave distributions in reliability. Naval Research Logistics 46, 419–433.

Nanda, A.K., Sengupta, D., 2005. Discrete life distributions with decreasing reversed hazard. Sankhyā, Series A 67, 106–125.

Nandi, S.B., Das, K.K., 1994. A family of Abel series distributions. Sankhyā, Series B 56, 147–164.

Navarro, J., Balakrishnan, N., Samaniego, F.R., 2008. Mixture representations of residual lifetimes of used systems. Journal of Applied Probability 45, 1097–1112.

Navarro, J., del Aguila, Y., Ruiz, J.M., 2001. Characterizations through reliability measures from weighted distributions. Statistical Papers 42, 395–402.

Navarro, J., Hernandez, P.J., 2004. How to obtain bathtub shaped failure rate models from normal mixtures. Probability in the Engineering and Informational Sciences 18, 511–531.

Navarro, J., Hernandez, P.J., 2008. Mean residual life functions of finite mixtures, order statistics and coherent systems. Metrika 67, 277–298.

Navarro, J., Ruiz, J.M., Aguila, Y., 2006. Multivariate weighted distributions: a review and some extensions. Statistics 40, 51–64.

Navarro, J., Rychlik, T., 2007. Reliability and expectation bounds for coherent systems with exchangeable components. Journal of Multivariate Analysis 98, 102–113.

Navarro, J., Samaniego, F.J., Balakrishnan, N., 2013. Mixture representations for the joint distribution of lifetimes of two coherent systems with shared components. Advances in Applied Probability 45, 1011–1027.

Navarro, J., Sarabia, J.M., 2010. Alternative definitions of bivariate equilibrium distributions. Journal of Statistical Planning and Inference 140, 2046–2056.

Navarro, J., Shaked, M., 2006. Hazard rates of order statistics and systems. Journal of Applied Probability 43, 391–408.

Navarro, J., Sunoj, S.M., Linu, M.N., 2011. Characterizations of bivariate models using dynamic Kullback-Leibler discrimination measures. Statistics and Probability Letters 81, 1594–1598.

Nekoukhou, V., Alamatsaz, M.H., Bidram, H., 2012. A discrete analog of the generalized exponential distribution. Communications in Statistics – Theory and Methods 41, 2000–2013.

Nekoukhou, V., Alamatsaz, M.H., Bidram, H., 2013. Discrete generalized exponential distribution of a second type. Statistics 47, 876–887.

Nelsen, R.B., 2005. Some properties of Schur-constant survival models and their copulas. Brazilian Journal of Probability and Statistics 19, 179–190.

Nelson, W., 1982. Applied Life Data Analysis. John Wiley & Sons, New York.

Neslehova, J., 2007. On rank correlation measures for non-continuous data. Journal of Multivariate Analysis 98, 544–567.

Newby, M., 1988. Accelerated failure time data for reliability data analysis. Reliability Engineering and System Safety 20, 187–197.

Noack, A., 1950. A class of random variables with discrete distributions. Annals of Mathematical Statistics 21, 137–142.

Nofal, Z.M., 2012. On a class of new better than used of life distributions. Applied Mathematical Sciences 6, 6809–6817.

Nooghabi, M.S., Roknabadi, A.H.R., Borzadaran, G.M., 2011. Discrete modified Weibull distribution. Metron LXIX, 207–222.

Noughabi, M.S., Roknabadi, A.H.R., Borzadaran, G.R.M., 2013. Some discrete lifetime distributions with bathtub-shaped hazard rate functions. Quality Engineering 25, 225–236.

O'Neill, T.J., 1991. The inverse proportional hazard model. Statistics and Probability Letters 12, 125–129.

Ord, J.K., 1967a. Graphical methods for a class of discrete distributions. Journal of the Royal Statistical Society, Series A 130, 232–238.

Ord, J.K., 1967b. On a system of discrete distributions. Biometrika 54, 649–656.

Ord, J.K., 1972. Families of Frequency Distributions. Griffin, London.

Osaki, S., Li, X., 1988. Characterizations of the gamma and negative binomial distributions. IEEE Transactions on Reliability 37, 379–382.

Padgett, W.J., Spurrier, J.D., 1985. On discrete failure models. IEEE Transactions on Reliability 34, 458–459.

Pakes, A.G., Navarro, J., Ruiz, J., del Aguila, Y., 2003. Characterizations using weighted distributions. Journal of Statistical Planning and Inference 116, 339–420.

Panaretos, J., 1989. On the evolution of surnames. International Statistical Review 51, 161–179.

Panjer, H.H., 1981. Recursive evaluation of a family of compound distributions. ASTIN Bulletin 12, 22–26.

Paranjpe, S.A., Rajarshi, M.B., 1986. Modelling non-parametric survivorship data with bathtub distributions. Ecology 67, 1693–1695.

Paranjpe, S.A., Rajarshi, M.B., Gore, A.P., 1985. On a model for failure rates. Biometrical Journal 27, 913–917.

Park, S., Rao, M., Shin, D.W., 2012. On cumulative Kullback-Leibler information. Statistics and Probability Letters 82, 2025–2032.

Parvardeh, A., Balakrishnan, N., 2013. Conditional residual lifetimes of coherent systems. Statistics and Probability Letters 83, 2664–2672.

Parvardeh, A., Balakrishnan, N., 2014. On the conditional residual life and inactivity time of coherent systems. Journal of Applied Probability 51, 990–998.

Parvardeh, A., Balakrishnan, N., 2015. On mixed δ-shock models. Statistics and Probability Letters 102, 51–60.

Patil, G.P., 1962. Certain properties of power series distributions. Annals of the Institute of Statistical Mathematics 14, 179–182.

Patil, G.P., 2002. Weighted distributions. In: El-Shaarawi, A.H., Peigorsch, W.W. (Eds.), Encyclopedia of Environmetrics, vol. 4. John Wiley & Sons, Chichester, pp. 2369–2377.

Patil, G.P., Rao, C.R., 1971. The weighted distributions: a survey and their applications. In: Krishnaiah, P.R. (Ed.), Applications of Statistics. North-Holland, Amsterdam, pp. 383–405.

Patil, G.P., Rao, C.R., 1978. Weighted distributions and size-biased sampling with applications to wildlife population and human families. Biometrics 34, 179–184.

Patil, G.P., Rao, C.R., Ratnaparkhi, M.V., 1986. On discrete weighted distributions and their use in model choice for observed data. Communications in Statistics, Theory and Methods 15, 907–918.

Paulson, A.S., Uppuluri, V.R.R., 1972. Characterization of the geometric distribution and a bivariate geometric distribution. Sankhyā, Series A 34, 88–91.

Pavlova, K.P., Cai, J., Willmot, G.E., 2006. The preservation of classes of discrete distributions under convolution and mixing. Insurance: Mathematics and Economics 38, 391–405.

Pechlivanides, P.M., 1975. Social Choice and Coherent Structures. Technical Report 75-14. Operations Research Center, University of California.

Pham, T.G., Turkkan, M., 1994. The Lorenz and the scaled total-time-on-test transform curves, a unified approach. IEEE Transaction on Reliability 43, 76–84.

Potter, U., Kopperschmidt, K., Nielsen, A.C., 2002. Covariate effect in periodic hazard rate model. In: Developments in Statistics. In: Metodolski zvezki, vol. 17, pp. 137–145.

Pril, N.D., 1984. Recursions for convolutions of arithmetic distributions. AUSTIN Bulletin 15, 135–139.

Priya, P., Sankaran, P.G., Nair, N.U., 2000. Factorial partial moments and their properties. Journal of the Indian Statistical Association 38, 45–53.

Proschan, F., 1975. Applications of majorization and Schur functions in reliability and life testing. In: Barlow, R.E., Fussel, A.B., Singpurwalla, N.D. (Eds.), Reliability and Fault Tree Analysis. Society for Industrial and Applied Mathematics, Philadelphia, pp. 237–258.

Pundir, S., Arora, S., Jain, K., 2005. Bonferroni curve and the related statistical inference. Statistics and Probability Letters 75, 140–150.

Qin, Y., Liu, K., Al-Jarallah, R., Balakrishnan, N., 2017. Inference for Type-II bivariate Polya-Aeppli distribution. Communications in Statistics – Simulation and Computation 46, 5981–5990.

Rajesh, G., Nair, K.R.M., 1998. Residual entropy function in discrete time. Far East Journal of Theoretical Statistics 2, 1–10.

Rao, C.R., 1965. On discrete weighted distributions arising out of methods of ascertainment. In: Classical and Contagious Distributions. Statistical Publishing Society and Pergamon Press, pp. 320–322.

Rao, M., Chen, Y., Vemuri, B.C., Wang, F., 2004. Cumulative residual entropy: a new measure of information. IEEE Transaction on Information Theory 50, 1220–1228.

Riordan, J., 1968. Combinatorial Identities. John Wiley & Sons, New York.

Rodrigues, F.W., Wechsler, S., 1993. A discrete Bayes explanation of a failure-rate paradox. IEEE Transactions on Reliability 42, 132–133.

Rodrigues, J., de Castro, M., Cancho, V.G., Balakrishnan, N., 2009. COM-Poisson cure rate survival models and an application to a cutaneous melanoma data. Journal of Statistical Planning and Inference 139, 3605–3611.

Roknabadi, A.H., Borzadaran, G.R.M., Khorashadizadeh, M., 2009. Some aspects of discrete hazard rate function in telescopic families. Economic Quality Control 24, 35–42.

Rossini, A.J., Tsiatis, A.A., 1994. A Semiparametric Proportional Odds Regression Model for the Analysis of Current Status Data. Technical Report. Department of Biostatistics, Harvard School of Public Health.

Rossini, A.J., Tsiatis, A.A., 1996. A semiparametric proportional odds regression model for the analysis of current status data. Journal of the American Statistical Association 91, 713–721.

Ruiz, J.M., Navarro, J., 1995. Characterization of discrete distributions using expected values. Statistical Papers 41, 423–435.

Salvia, A.A., Bollinger, R.C., 1982. On discrete hazard functions. IEEE Transactions on Reliability 31, 458–459.

Sandhya, E., Rajasekharan, K.E., 2012a. On discrete distributions with gaps having ALM property. Probability and Statistics Forum 5, 32–37.

Sandhya, E., Rajasekharan, K.E., 2012b. On the measure of memory of discrete distributions under change of origin and scale. International Journal of Mathematical Sciences and Engineering Applications 6, 293–301.

Sankaran, P.G., Gleeja, V.L., 2008. Association measure for bivariate lifetime data. Communications in Statistics, Theory and Methods 37, 3228–3249.

Sankaran, P.G., Nair, N.U., Ramesh, N.P., 2016a. Quantification of relative ageing in discrete time. Metron 74 (3), 339–355.

Sankaran, P.G., Ramesh, N.P., Nair, N.U., 2016b. Multivariate discrete reversed hazard rates. Communications in Statistics, Theory and Methods. http://dx.doi.org/10.1080/03610926.2016.1222430.

Sato, H., Ikota, M., Aritoshi, S., Masuda, H., 1999. A new defect distribution meteorology with a constant discrete exponential formula and its applications. IEEE Transactions on Semi Conductor Manufacturing 12, 409–418.

Scheaffer, R.L., 1972. Size-biased sampling. Technometrics 14, 635–644.

Schwarz, G.E., 1978. Estimating the dimension of a model. Annals of Statistics 6, 461–464.

Seal, H.A., 1947. A probability distribution of deaths at age x when policies are counted instead of lives. Scandinavisk Actuarietidsskrift 30, 18–43.

Sellers, K.F., Morris, D.S., Balakrishnan, N., 2016. Bivariate Conway-Maxwell-Poisson distribution: formulation, properties, and inference. Journal of Multivariate Analysis 150, 152–168.

Sen, A., 1976. Measurement of poverty: an axiomatic approach. Econometrica 44, 219–232.

Sen, A., Khattree, R., 1996. Length biased distributions, equilibrium distributions and characterization of probability laws. Journal of Applied Statistical Science 3, 239–252.

Sen, P.K., 1986. The Gini coefficient and poverty indexes: some reconciliations. Journal American Statistical Association 81, 1050–1057.

Sen, P.K., 1988. The harmonic Gini coefficient and affluence indexes. Mathematical Social Sciences 16, 65–76.

Sengupta, D., Chatterjee, A., Chakraborty, B., 1995. Reliability bounds and other inequalities for discrete life distributions. Microelectronics Reliability 35, 1473–1478.

Sengupta, D., Deshpande, J.V., 1994. Some results on relative ageing of two life distributions. Journal of Applied Probability 31, 991–1003.

Sethi, S.P., Bass, F.M., 2003. Optimal pricing in a hazard rate model of demand. Optimal Control, Applications and Methods 24, 183–196.

Shaked, M., Shanthikumar, J.G., 1991. Dynamic multivariate mean residual life functions. Journal of Applied Probability 28, 613–629.

Shaked, M., Shantikumar, G.J., Valdez-Toress, J.B., 1995. Discrete hazard rate functions. Computers and Operations Research 22, 391–402.

Shaked, M., Shanthikumar, J.G., 2007. Stochastic Orders. Springer, New York.

Shan, S., 2005. On the generalized Zipf distribution, part I. Information Processing and Management 41, 1369–1386.

Shmueli, G., Minka, T.P., Kadane, J.L., Borle, S., Boatwright, P., 2005. A useful distribution for fitting discrete data: revival of the Conway-Maxwell-Poisson distribution. Journal of the Royal Statistical Society, Series C 54, 127–142.

Sindhu, T.K., 2002. An Extended Pearson System Useful in Reliability Analysis. Ph.D. thesis. Cochin University of Science and Technology, Cochin, India.

Singer, J.D., Willet, J.B., 1993. It's about time, using discrete-time survival analysis to study duration and timing of events. Journal of Educational Statistics 18, 155–195.

Singh, S.K., Maddala, G.S., 1976. A function for the size distribution of incomes. Econometrica 44, 963–970.

Singpurwalla, N.D., 2006. Reliability and Risk, A Bayesian Perspective. John Wiley & Sons, Hoboken, New Jersey.

Skoulakis, G., 2000. A general shock model for a reliability system. Journal of Applied Probability 37, 925–935.

Smith, R.M., Bain, L.J., 1975. Analysis of linear failure rate life-testing distribution. Communications in Statistics 4, 469–481.

Spizzichino, F., 2001. Subjective Probability Models for Lifetimes. Chapman and Hall/CRC, Boca Raton, Florida.

Stein, W.E., Dattero, R., 1984. A new discrete Weibull distribution. IEEE Transactions on Reliability 33, 196–197.

Steutel, F.W., Thiemann, J.G.F., 1989. On the independence of integer and fractional parts. Statistica Neerlandica 43, 53–59.

Straub, E., 1970. Application of reliability theory to insurance. In: ASTIN, Colloquium. Renders, Denmark.

Su, J.C., Huang, W.J., 2000. Characterizations based on conditional expectations. Statistical Papers 41, 423–435.

Sudheesh, K.K., Nair, N.U., 2010. Characterization of discrete distribution by conditional variance. Metron LXVIII, 77–85.

Sun, K., Basu, A.P., 1995. A characterization of a bivariate geometric distribution. Statistics and Probability Letters 23, 307–311.

Sun, L., Park, D., Wang, M.C., 2006. The additive hazard model for recurrent gaps times. Statistica Sinica 16, 919–932.

Sunoj, S.M., Nair, N.U., 1998a. Characterization of some bivariate life distributions by weighted models. Journal for Applied Statistical Science 7, 239–248.

Sunoj, S.M., Nair, N.U., 1998b. Form-invariant bivariate weighted Pearson system. Far East Journal of Theoretical Statistics 2, 157–170.

Sunoj, S.M., Nair, N.U., 1999. Bivariate distributions with weighted marginals and reliability modelling. Metron LVII, 117–126.

Takayama, N., 1979. Poverty, income inequality and their measures: Professor Sen's axiomatic approach reconsidered. Econometrica 47, 747–760.

Tang, L.C., Lu, Y., Chew, E.P., 1999. Mean residual life of lifetime distributions. IEEE Transactions in Reliability 48, 73–78.

Tanguy, C., 2009. Mean time to failure for periodic failure rate. Proceedings of the SSARS 2, 361–365.

Tanner, J.C., 1953. A problem of interface between two queues. Biometrika 40, 58–69.

Tarsitano, A., 2004. Fitting generalized lambda distribution to income data. In: COMPSTAT 2004 Symposium. Springer, Berlin, pp. 1861–1867.

Vilaplana, J.P., 1987. The Hurwitz distribution. In: Puri, M.L., Vilaplana, J.P., Wertz, W. (Eds.), New Perspectives in Theoretical and Applied Statistics. John Wiley & Sons, New York.

Wang, F.K., 2000. A new model with bathtub-shaped hazard rate using an additive Burr XII distribution. Reliability Engineering and System Safety 70, 305–312.

Wang, Y., Hossain, A.M., Zimmer, W.J., 2003. Monotone log odds rate distributions in reliability analysis. Communications in Statistics, Theory and Methods 32 (11), 2227–2244.

Wang, Y., Hossain, A.M., Zimmer, W.J., 2008. Useful properties of the three parameter Burr XII distribution. In: Ahsanullah, M. (Ed.), Applied Statistics Research Progress. Nova Science Publishers, Hauppauge, NY, pp. 11–20.

Willet, J.B., Singer, T.D., 1993. Investigating onset, cessation, relapse and recovery: why should and how you can use discrete-time survival analysis to examine event occurrence. Journal of Consulting and Clinical Psychology 61, 952–965.

Willmot, G., Lin, X., 1994. Lundberg bounds for tails of compound distributions. Journal of Applied Probability 31, 743–756.

Willmot, G.E., Cai, J., 2001. Ageing and other distributional properties of discrete compound geometric distributions. Insurance: Mathematics and Economics 28, 361–379.

Willmot, G.E., Drekic, S., Cai, J., 2005. Equilibrium compound distributions and stop-loss moments. Scandinavian Actuarian Journal 1, 6–24.

Willmot, G.E., Panjer, H.H., 1987. Difference equation approaches in the evaluation of compound distributions. Insurance: Mathematics and Economics 6, 43–56.

Willmot, G.E., Cai, J., 2000. On classes of life distributions with unknown age. Probability for the Engineering and Informational Sciences 14, 473–484.

Witherell, C.E., 1994. Mechanical Failure Avoidance. McGraw-Hill, New York.

Wong, K.L., 1988. Off the bathtub on to the roller-coaster curve electronic equipment failure. In: Proceedings of the Reliability and Maintenance Symposium, pp. 356–363.

Wong, K.L., 1991. The physical basis for the roller-coaster hazard rate curve for electronics. Quality and Reliability Engineering International 7, 489–495.

Woodward, J.A., Palmer, C.G.S., 1997. On the exact convolution of discrete random variables. Applied Mathematics and Computation 83, 69–77.

Xekalaki, E., 1983. Hazard functions and life distributions in discrete time. Communications in Statistics, Theory and Methods 12, 2503–2509.

Xekalaki, E., 1984. The bivariate generalized Waring distribution and its application to accident theory. Journal of the Royal Statistical Society, Series A 147, 488–498.

Xekalaki, E., 1985. Some bivariate extension of the generalized Waring distribution. Studia Scientiarum Mathrmaticarum Hungarica 20, 173–181.

Xekalaki, E., 1986a. The bivariate Yule distribution and some of its properties. Statistics 17, 311–317.

Xekalaki, E., 1986b. The multivariate generalized Waring distribution. Communications in Statistics, Theory and Methods 15, 1047–1064.

Xie, M., Gaudion, O., Bracquemond, C., 2002a. Redefining failure rate functions for discrete distributions. International Journal for Reliability, Quality and Safety Engineering 9, 275–285.

Xie, M., Tang, Y., Goh, T.N., 2002b. A modified Weibull extension with bathtub-shaped failure rate function. Reliability Engineering and System Safety 76, 279–285.

Yeh, H., 2002. Six multivariate Zipf distribution and their properties. Statistics and Probability Letters 56, 131–141.

Zacks, S., 1984. Estimating the shift to wear-out of systems having exponential-Weibull life. Operations Research 32, 741–749.

Zahedi, H., 1985. Some new classes of survival distribution functions. Journal of Statistical Planning and Inference 11, 171–188.

Zahedi, H., 1991. Proportional mean remaining life model. Journal of Statistical Planning and Inference 29, 221–228.

Zenga, M., 2007. Inequality curve and inequality index based on the ratios between the lower and upper arithmetic means. Statistica and Applicazioni 5, 3–27.

Zhang, Z., 2010. Mixture representations of inactivity times of conditional coherent systems and their applications. Journal of Applied Probability 47, 876–885.

Zimmer, W.J., Wang, Y., Pathak, P.K., 1998. Log-odds rate and monotone log-odds rate distributions. Journal of Quality Technology 30, 376–385.

Zio, E., 2007. An Introduction to the Basis of Reliability and Risk Analysis. World Scientific Publishing, Singapore.

Zipf, G.K., 1949. Human Behaviour and the Principle of Least Effort. Addison-Wesley, Cambridge, MA.

Zornig, P., Altmann, G., 1995. Unified representation of Zipf distributions. Computational Statistics and Data Analysis 19, 461–473.

Index

C

D

E

F

M

O

P

Q

R

S

T

U

V

W

FSC
www.fsc.org
MIX
Papier | Fördert
gute Waldnutzung
FSC® C083411

Kolibri 360 GmbH
Ferdinand-Jühlke-Straße 7
99095 Erfurt, Deutschland
produktsicherheit@kolibri360.de